Measuremen
and Comput

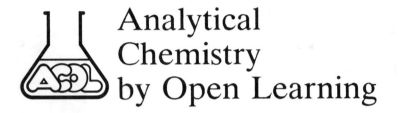

Analytical Chemistry by Open Learning

Project Director
BRIAN R CURRELL
Thames Polytechnic

Project Manager
JOHN W JAMES
Consultant

Project Advisors
ANTHONY D ASHMORE
Royal Society of Chemistry

DAVE W PARK
Consultant

Administrative Editor
NORMA CHADWICK
Thames Polytechnic

Editorial Board
NORMAN B CHAPMAN
Emeritus Professor,
University of Hull

BRIAN R CURRELL
Thames Polytechnic

ARTHUR M JAMES
Emeritus Professor,
University of London

DAVID KEALEY
Kingston Polytechnic

DAVID J MOWTHORPE
Sheffield City Polytechnic

ANTHONY C NORRIS
Polytechnic of the South Bank

F ELIZABETH PRICHARD
Royal Holloway and Bedford
New College

Titles in Series:

Measurement, Statistics and Computation

Analytical Chemistry by Open Learning

Authors:
DAVID McCORMICK
Manchester Polytechnic

ALAN ROACH
Paisley College of Technology

Editor:
NORMAN B. CHAPMAN

on behalf of ACOL

Published on behalf of ACOL, Thames Polytechnic, London
by
JOHN WILEY & SONS
Chichester · New York · Brisbane · Toronto · Singapore

Published by permission of the Controller of
Her Majesty's Stationery Office

Library of Congress Cataloging in Publication Data:

McCormick, David.
   Measurement, statistics and computation.
   (Analytical Chemistry by Open Learning)
   1. Chemistry, Analytic—Statistical methods—Programmed instruction.
2. Chemistry, Analytic—Measurement—Programmed instruction.
3. Chemistry, Analytic—Data processing—Programmed instruction.
4. Chemistry, Analytic—Programmed instruction.
Roach, Alan.   II. Chapman, N. B. (Norman Bellamy), 1916–   .   III. ACOL
(Project)   IV. Title.   V. Series: Analytical Chemistry by Open Learning
(Series)

QD75.4.S8M4   1987   543   87–10470

ISBN 0 471 91366 9
ISBN 0 471 91367 7 (pbk.)

British Library Cataloguing in Publication Data:

McCormick, David.
   Measurement, Statistics and computation.—(Analytical Chemistry).
   1. Chemistry, Analytic—Statistical methods
   I. Title   II. Roach, Alan   III. Chapman, Norman B.
   IV. ACOL   V. Series

519.5'024541   QD75.4.S8

ISBN 0 471 91366 9
ISBN 0 471 91367 7 Pbk.

Printed and bound in Great Britain

Analytical Chemistry

This series of texts is a result of an initiative by the Committee of Heads of Polytechnic Chemistry Departments in the United Kingdom. A project team based at Thames Polytechnic using funds available from the Manpower Services Commission 'Open Tech' Project has organised and managed the development of the material suitable for use by 'Distance Learners'. The contents of the various units have been identified, planned and written almost exclusively by groups of polytechnic staff, who are both expert in the subject area and are currently teaching in analytical chemistry.

The texts are for those interested in the basics of analytical chemistry and instrumental techniques who wish to study in a more flexible way than traditional institute attendance or to augment such attendance. A series of these units may be used by those undertaking courses leading to BTEC (levels IV and V), Royal Society of Chemistry (Certificates of Applied Chemistry) or other qualifications. The level is thus that of Senior Technician.

It is emphasised however that whilst the theoretical aspects of analytical chemistry can be studied in this way there is no substitute for the laboratory to learn the associated practical skills. In the U.K. there are nominated Polytechnics, Colleges and other Institutions who offer tutorial and practical support to achieve the practical objectives identified within each text. It is expected that many institutions worldwide will also provide such support.

The project will continue at Thames Polytechnic to support these 'Open Learning Texts', to continually refresh and update the material and to extend its coverage.

Further information about nominated support centres, the material or open learning techniques may be obtained from the project office at Thames Polytechnic, ACOL, Wellington St., Woolwich, London, SE18 6PF.

How to Use an Open Learning Text

Open learning texts are designed as a convenient and flexible way of studying for people who, for a variety of reasons cannot use conventional education courses. You will learn from this text the principles of one subject in Analytical Chemistry, but only by putting this knowledge into practice, under professional supervision, will you gain a full understanding of the analytical techniques described.

To achieve the full benefit from an open learning text you need to plan your place and time of study.

- Find the most suitable place to study where you can work without disturbance.

- If you have a tutor supervising your study discuss with him, or her, the date by which you should have completed this text.

- Some people study perfectly well in irregular bursts, however most students find that setting aside a certain number of hours each day is the most satisfactory method. It is for you to decide which pattern of study suits you best.

- If you decide to study for several hours at once, take short breaks of five or ten minutes every half hour or so. You will find that this method maintains a higher overall level of concentration.

Before you begin a detailed reading of the text, familiarise yourself with the general layout of the material. Have a look at the course contents list at the front of the book and flip through the pages to get a general impression of the way the subject is dealt with. You will find that there is space on the pages to make comments alongside the text as you study—your own notes for highlighting points that you feel are particularly important. Indicate in the margin the points you would like to discuss further with a tutor or fellow student. When you come to revise, these personal study notes will be very useful.

∏   When you find a paragraph in the text marked with a symbol such as is shown here, this is where you get involved. At this point you are directed to do things: draw graphs, answer questions, perform calculations, etc. Do make an attempt at these activities. If necessary cover the succeeding response with a piece of paper until you are ready to read on. This is an opportunity for you to learn by participating in the subject and although the text continues by discussing your response, there is no better way to learn than by working things out for yourself.

We have introduced self assessment questions (SAQ) at appropriate places in the text. These SAQs provide for you a way of finding out if you understand what you have just been studying. There is space on the page for your answer and for any comments you want to add after reading the author's response. You will find the author's response to each SAQ at the end of the text. Compare what you have written with the response provided and read the discussion and advice.

At intervals in the text you will find a Summary and List of Objectives. The Summary will emphasise the important points covered by the material you have just read and the Objectives will give you a checklist of tasks you should then be able to achieve.

You can revise the Unit, perhaps for a formal examination, by re-reading the Summary and the Objectives, and by working through some of the SAQs. This should quickly alert you to areas of the text that need further study.

At the end of the book you will find for reference lists of commonly used scientific symbols and values, units of measurement and also a periodic table.

Contents

Study Guide

Consider the scope of the activities in which analytical chemists the world around are engaged. The starting-point is usually that someone – we shall call him the client – is confronted with a situation. It may be that the fish in a river are dying, that a chemical product (perhaps a drug) is no longer being manufactured to its usual specification, or that there is a need to discover whether or not a fire was the result of arson. The client decides, sooner rather than later, we hope, that the situation would benefit from the involvement of an analytical chemist. In discussion, analytical problems emerge, and in the light of all the circumstances of the case, procedures and methods suggest themselves for the solution of the problem.

The relationship between the client and the chemist is clear. The client needs to make rational decisions. To do this, trustworthy information is needed. Analytical chemists are not the only people involved in providing help to such clients. However, when the chemical aspects of a situation have been identified, the information given by the analytical chemist is of paramount importance. In choosing the procedures and methods for the analysis, the chemist will have taken into account matters such as the required sensitivity and accuracy of the analysis. The chemist must also be concerned with questions such as whether the analysis is being carried out on samples which are representative of the whole. It may be that meaningful information can be provided only if several samples are analysed. It may be good enough to carry out one determination on each sample; alternatively, the client's requirement with regard to the uncertainty in a measurement may be such that several determinations on each sample are needed. In making decisions about such matters, the chemist relies heavily on that branch of mathematics known as statistics. The major purpose of this Unit is to introduce you to the application of statistical ideas in the context of analytical chemistry.

There are, however, *three* words in out Unit title – Measurement, Statistics, and Computation. It is not our aim to introduce you to a formal and rigorous exposition of statistical theory in the abstract. Above all, analytical chemists are practical people. The great weight of our expertise and experience is directed to the competent handling of our sample, and then to subjecting it to a carefully selected procedure for making *measurements*. Reaching that stage is the hard job, but it is also important to ensure that we can extract the maximum benefit from our labours. This Unit is thus fundamentally about drawing quantitative conclusions from experimental measurements, assessing the value of our results, and suggesting at times additional work which might usefully be undertaken.

The methods of *statistics* give us the tools we need for this task. In introducing you to these methods our aims will be threefold; that you should:

(*a*) be able to apply a given test competently;

(*b*) understand clearly the issues addressed by a test;

(*c*) develop, through practice, a 'feeling' for the significance of the conclusions you reach.

We shall not be concerned with formal derivations, and we shall keep mathematical manipulations to a minimum. Nor do we intend to give an exhaustive survey of methods. The Unit concentrates on helping you to obtain a good practical grasp of a limited, but useful, range of methods.

The third word in our title is *computation*. The microcomputer revolution is completely transforming our field of work, as it is many others. Analytical instruments are more and more commonly interfaced with computers. Quality control measurements are becoming increasingly computer-automated. The quantity of analytical data which we can expect to have to handle is increasing substantially. Also, you will find, statistical manipulation can involve a lot of rather tedious arithmetic, as well as a certain amount of intensive activity scouring through tables. Much checking and re-checking is required, to make sure that no slips in arithmetic have occurred. This work lends itself admirably to a computer approach. Once our data is inside our computer, we can, with relative ease, perform a variety of different tests, investigate the effects of rejecting suspect measurements, or of including others, etc. So to study this Unit you need access to a computer. We make no assumptions about *which* computer, as there are very many different models of microcomputer, any of which will suffice. If you have your own machine at home you have a head start. Otherwise you should arrange to obtain access to a machine at work or at college. (An interactive terminal linked to a mainframe computer, will, of course, also fill the bill).

If you have had no previous experience of computing, the last paragraph may have come as a shock to you. But it needn't. There are no two ways about it, computers are becoming more and more significant to our work. This Unit gives you an opportunity to make a beginning. Our approach is to use the computer to extend rather effortlessly your experience in applying the methods of statistics to the analysis of the results of measurement. At the same time, we use our need to learn about statistics as a vehicle to help you to become familiar with using a computer. It may be that you already have had considerable experience with computers. Equally, it may be that you have had none. To cope with both possibilities, our practical introduction to using a machine is given in an Appendix. The time which you will need to devote to studying that depends on your previous experience.

Now just what previous knowledge do we expect of you, if you are to undertake our foray into statistics and computation? The answer is – very little! If you have a basis in mathematics up to GCE O-level or equivalent, that will be quite adequate. The processes we have to use are essentially arithmetical. The only other material which you may need at hand for occasional reference is the user's guide of the micro-computer you use. There are many available text books covering the area of interest to us, but the time to refer you to those is when you have completed the Unit.

This does not necessarily mean that you will find the work easy. Understanding what exactly it is that you are doing in statistics requires clear and careful thinking. Great pains must be taken to develop a thorough insight into a number of concepts which on first acquaintance may be found quite confusing. For this reason we shall start slowly by discussing relatively familiar terms such as *accuracy, precision* and *errors*; after this, we shall introduce statistical ideas such as *sample, population, confidence interval*. Later in the Unit we shall discuss matters such as *correlation* and *regression*. By then, we believe, you will be quite competent in the use of statistics in analytical chemistry.

If you find the going easy to begin with, please don't try to cut corners. Attempt all the exercises: that is most important. The danger is that you may *think* you understand concepts which you may have only grasped superficially. Working through examples, and checking your responses, will help to ensure that this does not happen. Statistics can be misused, and they often are, hence the famous maxim about the three categories of lies – 'lies, damned lies and statistics'. If you understand statistical methods properly, and if you apply them properly, they will never deceive you. On the contrary, statistical methods provide the *only* basis for making a sure-footed evaluation of a set of measurements. As for the computing, you really should find that you can take it in your stride. If you are a beginner you must be prepared to work through the Appendix rather thoroughly. You should then find it quite straightforward to tackle the computation in the main body of the Unit. You will quite soon appreciate what a useful tool the computer can be!

Because of the nature of the ground we are covering, the general style of this Unit is a little different from most of the others in the Analytical Chemistry series. You will find relatively fewer 'normal' SAQs, but that does not mean that you will be left as a passive reader. On the contrary, much of the time the argument is developed by discussing specific examples. If you want to make a success of your study it is important that *you* work through the details, and, *inter alia*, check our arithmetic. This work supplements the tasks explicitly set for you in the text, in SAQ's, and in computer exercises.

So to business! First tackle our Appendix, which is reprinted from the *Microprocessor Applications* volume in this series. When you have dealt with that you will be ready to start on the main body of the Unit. It is important to study that in the order in which it is presented. As you start, we wish you good luck!

Bibliography

There are very many books which deal with topics which are of interest in this Unit, and these approach their subject matter from a great diversity of points of view. Other sources are, we feel, best consulted after you have worked your way through to the end of the present Unit. At this point, in Section 8.7, we are in a position to comment rather more fully about the available literature. You will find that we there highlight the following representative texts.

1. J. C. Miller and J. N. Miller, *Statistics for Analytical Chemistry*, Ellis Horwood, 1984.

2. C. Chatfield, *Statistics for Technology*, Chapman and Hall, 1983.

3. D. Cooke, A. H. Craven and G. M. Clarke, *Basic Statistical Computing*, Edward Arnold, 1982.

4. I. M. Kolthoff and P. J. Elving, *Treatise on Analytical Chemistry*, Part 1, Volume 1, John Wiley, 1975.

5. O. L. Davis and P. L. Goldsmith, *Statistical Methods in Research and Production*, Longman, 1984.

6. R. E. Walpole and R. H. Myers, *Probability and Statistics for Engineers and Scientists*, Collier McMillan, 1985.

7. E. L. Grant and R. S. Leavenworth, *Statistical Quality Control*, McGraw-Hill, 1980.

8. C. Liteanu and I. Rica, *Statistical Theory and Methodology of Trace Analysis*, Ellis Horwood, 1980.

9. M. R. Spiegel and R. W. Boxer, Schaums's Outline Series: *Theory and Problems of Statistics*, McGraw-Hill, 1972.

Acknowledgements

Figures 6.1a, 6.b, 6.6c and 6.6d are taken from the *British Regional Heart Survey*, as published in the *British Medical Journal*, May 1980.

1. Accuracy and Precision

The symbol for the Society of Analytical Chemistry (SAC) contains two words – 'Accuracy and Precision'. To those not engaged in scientific or technological measurement it might be thought that the SAC has made use of two words where one would have been sufficient. To the layman the words accuracy and precision are interchangeable. To the analytical chemist, however, the two words have quite different meanings. It is the purpose of this Part of the course both to define and to give quantitative meaning to these terms. We shall also investigate and classify various categories of error which occur in experimental measurements. Let us make a start.

1.1. REPLICATE MEASUREMENTS AND THE ARITHMETIC MEAN

All chemists are aware that when we carry out determinations we often do more than one measurement, ie we do *replicate measurements*. Practically every student of chemistry first experiences this in the context of titrimetric analysis. With this in mind, we shall use analytical results obtained by the titrimetric method as the basis for the ensuing discussion.

We start with the averaging of a series of replicate measurements. The most commonly used averaging procedure is to calculate the *arithmetic mean*. This is done by adding together individual values, $x_1, x_2, \ldots, x_n$ for a series of measurements and dividing this sum by the number of measurements, n.

In mathematical notation we give the arithmetic mean $\bar{x}$, in equation 1.1.

$$\bar{x} = \frac{x_1 + x_2 + x_3 + .. + x_n}{n} = \frac{\sum\limits_{i=1}^{i=n} x_i}{n} \qquad (1.1)$$

Let us put this relationship to use. Suppose that we had to determine the alkali concentration of a solution and that in order to do so we carried out five replicate titrations with standard acid. Let us assume that the values obtained were: 20.02 cm^3, 20.00 cm^3, 20.02 cm^3, 20.04 cm^3, 20.02 cm^3. Now, almost without thinking, we would use these values to obtain the arithmetic mean of the titres (from now on called the mean titre), ie

$$\bar{x} = \frac{\sum\limits_{i=1}^{i=n} x_i}{n} = \frac{100.10}{5} = 20.02 \text{ cm}^3$$

This value would then be used in our calculations of the alkali concentration. For example, if the acid-base reaction had a 1:1 stoichiometry, we would use

$$M_{alkali} = \frac{M_{acid} V_{acid}}{V_{alkali}}$$

The question now arises. Why did we use the mean value in our calculation? The simple answer is that we believe *the mean value to be the best estimate of the experimental titre*. This is not the same as saying that the mean value is closest to the true value, ie that it gives the best result. It may well have been that the true alkali concentration would have been obtained from a titre of 20.10 cm^3. This being so, we would have obtained a result closer to the truth by using the highest experimental titre (20.04 cm^3), rather than the mean titre, (20.02 cm^3). However, we were in no position to pick and choose from our titres: we were dealing with an unknown. All that we had to work on was our experimental results. Since there was variation in these, we chose to work with the mean because we believed it reflected better the results of our measurements. Without wishing to labour the point, we repeat that the mean is used because it is taken to be the best estimate of our experimental measurements.

Of course many questions still remain, the most fundamental of which is probably the relationship between the mean experimental value and the true value of an unknown. The term which describes this relationship is *accuracy*, to which we now turn our attention.

1.2. ACCURACY

Accuracy refers to *how near a measurement is to the true, accepted, or known (correct) value.* It thus embodies the very purpose of quantitative analytical chemistry, ie, to obtain the trustworthy results which make rational decisions possible. An analysis is not an end to itself: it is a prelude to an interpretation which is followed by a decision on the most appropriate course of action. Both the interpretation and what follows depend upon having the correct data, and upon having measurements which are accurate. On some occasions the accuracy of a measurement can be a matter of life or death: this is a situation which clinical chemists, amongst others, meet daily.

1.2.1. Absolute Error

Let us assume that the correct value of some determinand is T and that on measurement we obtain a value x. How can we indicate the accuracy of our measurement? Recalling that accuracy refers to the closeness of a result to the correct value, an obvious way would be to take the difference between these values ie $x - T$. This is known as the *absolute error*.

$$Absolute\ error\ =\ E_{abs}\ =\ (x\ -\ T)$$

If the result was obtained by replicate measurements, we would use the mean $\bar{x}$, and our absolute error would be given by $E_{abs} = (\bar{x} - T)$. *This is the absolute error of the mean.* (Do not call it the mean absolute error!)

Let us explore the extent to which E_{abs} is useful in aiding our understanding of the term *accuracy*. Suppose that three chemists analyse a salt whose known calcium content is 31.29%, and obtain calcium values of 31.25%, 31.31% and 31.33% respectively. We process their results in the table below.

| | Percentage of Calcium | | |
Chemist	Observed Value	Correct Value	Absolute Error
1	31.25	31.29	−0.04
2	31.31	31.29	+0.02
3	31.33	31.29	+0.04

∏   Check the arithmetic. Do you agree with the *signs* quoted? What is their significance?

The definition of E_{abs} is such that a positive value indicates an overestimate of T and a negative value an underestimate. From the calculated absolute errors, and accepting a direct relationship between accuracy and absolute error, we are able to make the following comments.

(*a*) The second chemist has the most accurate result; it is the result with the smallest absolute error.

(*b*) The first and the third chemist have results which are equally accurate; both have results with the same magnitude of absolute error, even though they are in opposite directions.

In the light of these comments, is it safe to say that we can tell how accurate a result is by observing the magnitude of the absolute error? Before answering this question, let us consider the absolute error associated with the determination of the calcium content of a different material by a fourth chemist. This time let us assume that the known value for the percentage of calcium is 0.04 (ie, 4 parts in 10 000) and that an analysis yields the result 0.02: the absolute error is thus $0.02 - 0.04 = -0.02$. By chance this is the same absolute error as that which was obtained by the second chemist in his determination of the calcium content of the salt. Is the accuracy of the two measurements the same? A moment's reflection on these two different analyses, and their associated absolute errors, should convince you that absolute error is an inadequate tool for measuring accuracy.

In the first example, the absolute error of 0.02 was associated with a determination which gave a result quite close to the accepted value; 31.31 compared with 31.29. In the second example, the absolute error of 0.02 is associated with a result, 0.02, which is only one half of the known value, 0.04. Clearly we need some other measure of accuracy.

1.2.2. The Relative Error

The *relative error* or as some prefer to call it, the *relative absolute error* is defined as the difference between the observed value and the true value divided by the true value, ie

$$E_{rel} = \frac{E_{abs}}{T} = \frac{x - T}{T} \qquad (1.2)$$

The relative error is therefore expressed here as a fraction of the true value.

Since most chemists prefer to think in terms of percentages rather than fractions we now introduce the *percentage relative error*, viz

$$\% \ E_{rel} = \frac{E_{abs} \times 100}{T} = \frac{(x - T) \times 100}{T} \qquad (1.3)$$

We now apply this relationship to the examples given above illustrating the inadequacy of comparing accuracies by comparison of the absolute errors.

Statistic	Chemist 2	Chemist 3	Chemist 4
Observed value, x	31.31	31.33	0.02
Correct value, T	31.29	31.29	0.04
Absolute error, $(x - T)$	+0.02	+0.04	-0.02
% Relative error, $\dfrac{100 \ (x - T)}{T}$	$\dfrac{0.02 \times 100}{31.29}$ $= 0.06$	$\dfrac{0.04 \times 100}{31.29}$ $= 0.13$	$\dfrac{-0.02 \times 100}{0.04}$ $= -50$

We now see the worth of the percentage relative error. We have been able to give a quantitative measure of the concept of accuracy and to use it to compare the accuracies of different measurements. A moment's thought should serve to convince you that we could use the % E_{rel} to compare the accuracies obtained by different methods of measurement even if the results were reported in different units. For example, compare the accuracies obtained in the following experiment. A chemist is asked to determine the calcium concentration of a solution by a gravimetric method and by a titrimetric method. The solution is known to contain 0.3206 g calcium dm^{-3}: in molarity units, this is 0.0080 M. The result of the gravimetric determination was a mean concentration of 0.3236 g dm^{-3} of calcium. The titrimetric method yielded a mean calcium concentration of 0.0081 M. A comparison of the accuracies of the two methods appears below.

	Gravimetric Method	Titrimetric Method
Known value	0.3206 g dm^{-3}	0.0080 M
Experimental value	0.3236 g dm^{-3}	0.0081 M
% E_{rel}	$\dfrac{100\,(0.3236 \; - \; 0.3206)}{0.3206}$ $= 0.94$	$\dfrac{100\,(0.0081 \; - \; 0.0080)}{0.00800}$ $= 1.25$

There is not much in it, but it appears that the gravimetric determination was the more accurate. Did you think before we did the calculation that the more accurate method was likely to be the titrimetric method? If you did, it was probably because you simply took a quick look at the numbers and decided that the titrimetric method, which had numbers which agreed up to the fourth place of decimals, looked the more accurate. Watch out! The titrimetric method has values given to two significant figures; and these disagree in the second significant figure. The gravimetric values are to four significant figures with disagreement at the third figure. If you are worried about significant figures, don't be! We will deal with them later. The important message is to do your sums!

1.3. REPLICATE MEASUREMENTS AND PRECISION

In introducing the arithmetic mean we gave the results of five replicate titrations: they are given below along with the mean value.

Measurement	Titre/cm^3
1	20.02
2	20.00
3	20.02
4	20.04
5	20.02

Mean titre $= 20.02$ cm^3

As can be seen from a quick examination of these values, they are all very close to one another. The closeness of the results is what one would expect from an analytical chemist working as carefully as possible, with a well established titrimetric method. *The term that is used when describing the closeness of replicate measurement is precision.* From a casual observation of the titre values above, one could say that the chemist who performed the titrations was a precise worker or alternatively that his results were of high precision. Our task now is to discuss ways in which we can give quantitative meaning to the idea of precision.

1.3.1. Mean Deviation

One way of describing the precision of a set of measurement is to calculate how near, on average, each of the measurements is to their mean, ie to calculate the *mean deviation of the measurements from their mean*. For simplicity, this is called the *mean deviation*.

First, let us express the deviation of an individual measurement from the mean by $d_i = x_i - \bar{x}$.

We will now sum these deviations, ie

$$\sum_{i=1}^{i=n} d_i = \sum_{i=1}^{i=n} (x_i - \bar{x}) \tag{1.4}$$

∏   Do it for the 5 replicate titration values above! What answer
       do you get? Does this surprise you?

A little thought here brings to light a problem. Since some of the
individual measurements are less than the mean and some greater,
we have some deviations with a negative sign and some with a positive sign. The result of this is that the summation above will give
zero. The mean is designed to this end! In order to overcome this
we sum the deviations without regard to sign, ie we use the absolute
value (modulus) for the deviations.

$$\sum_{i=1}^{i=n} |d_i| = \sum_{i=1}^{i=n} |x_i - \bar{x}| \qquad (1.5)$$

If we divide this sum by the number of deviations, n, we have the
mean deviation, $\bar{d}$.

$$\bar{d} = \frac{\sum_{i=1}^{i=n} |d_i|}{n} = \frac{\sum_{i=1}^{i=n} |x_i - \bar{x}|}{n} \qquad (1.6)$$

Let us immediately use this relationship to process the titration data
presented above. This is done in the table below:

Measurement	x_i	$x_i - \bar{x}$	$|d_i|$
1	20.02	20.02 − 20.02	0.00
2	20.00	20.00 − 20.02	0.02
3	20.02	20.02 − 20.02	0.00
4	20.04	20.04 − 20.02	0.02
5	20.02	20.02 − 20.02	0.00

$$\bar{x} = \frac{80.10}{5} = 20.02 \qquad\qquad \bar{d} = \frac{0.04}{5} = 0.008$$

We are now in a position to consider quantitatively the precision
of the replicate measurements. We can say that the precision, as
measured by the mean deviation, is 0.008 cm^3.

This means that if we wished to report our results, we could say that the mean of the five titres was 20.02 cm^3 and that the mean deviation from the mean was 0.008 cm^3. Note: The units for the mean deviation are the same as those for the mean. It is often considered preferable to report measurement precision as a mean deviation relative to the mean. We therefore have the relative mean deviation, given by RMD = $\bar{d}/\bar{x}$, or the percentage relative mean deviation, given by: % RMD = 100 $\bar{d}/\bar{x}$.

In the example given above we then have:

$$\text{RMD} = \bar{d}/\bar{x} = \frac{0.008}{22.02} = 3.6 \times 10^{-4}$$

$$\% \text{ RMD} = \text{RMD} \times 100 = 3.6 \times 10^{-2} \approx 0.04\%.$$

Reporting measurements and the associated precision we can say that the mean of five titrations was 22.02 cm^3, with a relative mean deviation of 0.04%. As stated earlier this is very good precision indeed. It is probably as good as we could ever expect from such a series of measurements. The perfectionists amongst us might ask why it is that all the titrations were not of exactly the same value. The only answer that will be given at present is that such an occurrence would be very rare. Those of you experienced in replicate measurements should try to recall how often it has happened to you. We shall return to this issue, when we come to discuss random errors.

1.3.2. Standard Deviation

Having mastered the mean deviation and the relative mean deviation as measures of precision, we can introduce a much more widely used measure, the standard deviation. This is determined from equation 1.7.

$$s = \sqrt{\frac{\sum_{i=1}^{i=n}(x_i - \bar{x})^2}{n-1}} = \sqrt{\frac{\sum_{i=1}^{i=n}d_i^2}{n-1}} \qquad (1.7)$$

Before utilising this relationship, let us give it a moment's thought. You will see that the expression in the bracket contains the mean, $\bar{x}$, and that the mean is substracted from each individual result, x_i, to give the difference, d_i. Recall that we did the same manipulation in our calculation of the mean deviation. These differences are now squared. This ensures that when we come to the summation, the result will be a positive number. Recall that we did something similar in the mean deviation calculation when we took the absolute value of the differences, before summation. The next step is the division by $(n\text{-}1)$. (You might ask here, why not divide by n? We will delay answering this question for the time being!) The final step in the calculation of the standard deviation is to take the square root of the quotient. Keeping in mind that we have previously squared the differences, you will see that the operation of taking the square root is likely to give an answer, which is similar in magnitude to that of the mean deviation, and which like the mean deviation will have the same units as the mean. However, the result here is called the *standard deviation*. As you will soon discover, the standard deviation is an important entity in statistics: its widespread use is related to its statistical significance. Let us now calculate the standard deviation of the five titres which have previously attracted our attention. The titre values are processed in the table below.

Measurement	x_i	$d_i = (x_i - \bar{x})$	d_i^2
1	20.02	0.00	0.0000
2	20.00	-0.02	0.0004
3	20.02	0.00	0.0000
4	20.04	$+0.02$	0.0004
5	20.02	0.00	0.0000

$$\bar{x} = 20.02 \qquad\qquad \sum_{i=1}^{i=5} d_i^2 = 8 \times 10^4$$

$$s = \sqrt{\frac{8 \times 10^{-4}}{(5-1)}} = 0.014$$

The precision as measured by the standard deviation is 0.014 cm^3. The result could be reported as a mean titre of 20.02 cm^3 with a standard deviation of 0.014 cm^3.

We can again use a relative measure for our quantitative expression of precision. The relative standard deviation, RSD, is given by: RSD $= s/\bar{x}$. The percentage relative standard deviation is given by %RSD $= 100\ s/\bar{x}$. In our example, the relative standard deviation is

$$\frac{0.014}{20.02} = 7.0 \times 10^{-4}.$$

The percentage relative standard deviation is 7.0×10^{-2}, ie 0.070%. These small values for the RSD and the % RSD reflect the excellent measurement precision evident in the five titres. We shall return to a consideration of precision later. For the moment, this is a suitable occasion for you to assess your progress.

SAQ 1.3a In the determination of lead in a sample of water by atomic absorption spectrophotometry, four absorbance readings were taken: their values were 0.207, 0.210, 0.208, and 0.211.

Calculate:

(i)   the mean absorbance value,

(ii)   the mean deviation and the percentage relative mean deviation,

(iii) the standard deviation and the percentage relative standard deviation.

SAQ 1.3a

SAQ 1.3b   Use a computer program to repeat the calculations of SAQ 1.3a above. You should already
have created a suitable program when studying
Section 2.4 of the Appendix. It requires only
very minor alteration to produce all required
quantities.

1.4. DOING THE CHORES BY COMPUTER

You will probably have found it harder work to solve the lead analysis problem of SAQ 1.3 by using the computer than when you did it by hand. True; but if we now asked you to investigate a fresh set of results the computer method would win – you would merely have to retype the DATA then RUN it. Also, having shown that the computer gets the right answer once, we can rely on it to do so again, provided only that we insert the new data correctly. Hand calculations will be error-prone on all occasions and at all stages. If we have doubts about what data we actually did use, with the computer we can simply LIST the DATA lines to check. If necessary, we could easily edit any particular datum and then reRUN in a flash. Again, the SAQ problem required us to handle only four values; if we had 20 readings it would take much more work to do a hand calculation, but it would be very easy to do it by using the program. Another advantage of the computer approach is that we can readily check what happens if we change our data, perhaps rejecting a suspect reading or, alternatively, including the results of additional replicate measurements.

We are dealing at this stage with a rather simple calculation. Things will become more involved as we go on. In other words the advantages of a computer-centred approach will grow. What we shall aim to do as the unit progresses is steadily to build together a *package* of programs, which will perform for us most of the tasks in which we shall be interested. Developing these programs will make it much easier to investigate a wide range of examples, and it will remove a heavy burden of arithmetic, thus freeing us to concentrate on the significance of our answers rather than on the mechanics of obtaining them. It will make it easier for you to apply statistics to measurements you obtain in your normal work. Now, almost all good things can be over-indulged in. It is important that you do get *some* experience in actually doing the arithmetic manually, so that you will be sure to get a clear grasp of what is involved. That is why we set you the hand calculation SAQ 1.3a before the computer version SAQ 1.3b.

This is a suitable point to begin the development of our statistical package. Our first component will simply be a general program to calculate the mean, $\bar{x}$, and standard deviation, s, of a set of measurements. Fine you may say, we already have that as a result of tackling SAQ 1.3b. True, but we would like you to enter on your computer the following alternative program to cope with the same task. The reason for doing this is that we shall gradually add to this new program, transforming it, in the long run, into a quite powerful and versatile tool. The program you should type is given below.

```
  80 DIM X(30)
 100 READ N
 110 A=0
 120 FOR I=1 TO N
 130    READ X(I)
 140    A=A+X(I)
 150    NEXT I
 160 M=A/N
 170 A=0
 180 FOR I=1 TO N
 190    A=A+(X(I)-M)↑2
 200    NEXT I
 210 S=SQR(A/(N-1))
 220 R=100*S/M
1640 REM Show sample mean and st dev
1650 PRINT "No of values  = ";N
1660 PRINT "Sample mean = ";INT(M*1000+.5)/1000
1670 PRINT "Est. st. dev.  = ";INT(S*1000+.5)/1000
1680 PRINT "% Rel.st.dev. = ";INT(R*1000+.5)/1000
4000 DATA 4
4010 DATA 0.207,0.210,0.208,0.211
9999 END
```

The differences from the earlier program are relatively minor but nonetheless will prove helpful in the long run. The most fundamental change is that the measurements X are now read into an array. Notice that line 80 sets the dimensions of the array. As the program stands it will handle only up to 30 replicate values. This will suffice for our purposes and in any case presents no real problem; if we wish to find s and %RSD for a set of 48 values we would simply change line 80 to read, say, DIM X(50). Other changes to the original program are very minor. The line numbers have been chosen to leave room for future additions. Our sum is accumulated into variable A, leaving symbol S to represent only the calculated standard deviation. We have dropped all reference to the mean deviation and the %RMD. Finally, we have complicated the PRINT statements, by using the INT function to control the number of decimals printed (Appendix 2.2.6).

RUN the program and check that the results agree with those in SAQ 1.3a. If you have difficulties check over your typing very carefully. If the fault still eludes you, consult the checklist 'Eliminating Gremlins' (Section 1.9.1).

Once the program is giving satisfactory answers *SAVE* it. We shall refer to this program as "STAT1" whenever we ask you to retrieve it in future. You can of course save it under a different name if you choose. We shall be using the program again in section 1.5 below.

When we ran the program it produced the following output:

> No of values   = 4
> Sample mean = 0.209
> Est. st. dev.   = 2E-3
> % Rel.St.dev. = 0.874

Notice that the INT statements have led to the output being rounded to three places of decimals (not necessarily to three significant figures). Try editing line 1670 replacing 1000 by 10000 each time it occurs. What effects does this have on the results when you run the program? (You should find the standard deviation written out as 0.0018 or 1.8E-3.) It will often be useful for you to change INT functions in this way. We suggest you experiment with making further changes (eg try using 100, or 1000000, try dropping the $+.5$) until you are sure you fully understand how INT operates.

Before we move on we should briefly comment on what we are about as we begin to build our set of programs. It is almost certain that there is a standard 'statistics package' (or several such) available for your particular computer. You may even have had experience in using such a package. No matter, we are not trying to compete with your software supplier. We wish you to use the programs we give you as you work through this unit. You will be asked to incorporate each segment as we give it, and we want you to ring the changes as you try out a variety of exercises. That way you will pick up more about statistics, and you should also become more adept with your computer.

On the other hand, the programs we give have been tailored to be simple in structure and to involve relatively little typing at the keyboard. The result is that they are neither as efficient nor as near to being foolproof as would be typical of a commercial package. Once you have mastered this unit, and if you find yourself a heavy user of statistical methods, we would recommend that you then learn to use such a package. Stick with us meantime, however!

1.5. COMPARING ACCURACY AND PRECISION

Let us assume that it was decided to assess the ability of four chemists to carry out a titration. The procedure called for the chemists to transfer an aliquot of a given solution (by using a given transfer-pipette) to a conical flask, to add a stated indicator, and to titrate (by using a given burette) with a given titrant. The concentration of both the titrant and the titrand were known to the assessors and should have resulted in a titre of 23.04 cm^3. The glassware used by each of the chemists was made to the same specifications. The results obtained by each of the chemists are presented in Fig. 1.5a.

Chemist 1		Chemist 2	
Measurement	Titre/cm^3	Measurement	Titre/cm^3
1	23.09	1	22.81
2	23.04	2	22.96
3	23.07	3	23.18
4	23.08	4	23.13
Chemist 3		Chemist 4	
Measurement	Titre/cm^3	Measurement	Titre/cm^3
1	22.82	1	22.74
2	22.84	2	22.61
3	22.87	3	22.95
4	22.83	4	22.82

Fig. 1.5a. *The replicate titration results of four different chemists. (The known true titre is 23.04 cm^3)*

∏ Use your program "STAT1" for each set of data. Calculate (*i*) the mean, (*ii*) the % E_{rel}, (*iii*) the standard deviation, (*iv*), the % RSD.

Note (*a*) you will have to add one or two lines to generate % E_{rel},
(*b*) you will require four separate runs, one for each chemist.

Your results should agree with those in Fig. 1.5b below.

	Chemist 1		Chemist 2
$\bar{x}$	$= 23.07$ cm^3	$\bar{x}$	$= 23.02$ cm^3
% E_{rel}	$= 0.13\%$	% E_{rel}	$= -0.09\%$
s	$= 0.022$ cm^3	s	$= 0.169$ cm^3
% RSD	$= 0.094\% \approx 0.1\%$	% RSD	$\approx 0.73\%$
	Chemist 3		Chemist 4
$\bar{x}$	$= 22.84$ cm^3	$\bar{x}$	$= 22.78$
% E_{rel}	$= -0.87\%$	% E_{rel}	$= -1.13\%$
s	$= 0.022$ cm^3	s	$= 0.143$ cm^3
% RSD	$= 0.095\% \approx 0.1\%$	% RSD	$\approx 0.63\%$

Fig. 1.5b. *Analysis of the titration values of the four chemists of Fig. 1.5a*

Comment: In our own program we dealt with % E_{rel} by adding the following lines.

```
230 READ T
240 E1=100*(M-T)/T
250 PRINT "% Rel. error = ";INT(E1*1000+.5)/1000
4020 DATA 23.04
```

T represents the true result and E1 is our symbol for % E_{rel}. *T* is read from the new DATA item at the end of the program. Of course, we also adjusted the data values on line 4010 to correspond to those given in Fig. 1.5a for each chemist in turn. By running the program for each set of data you should obtain the results listed in Fig. 1.5b.

The inclusion of % E_{rel} could, if one wished, be handled by adding only a single line to "STAT1", eg that below.

230 PRINT "% Rel. error = ";100*(M-23.04)/23.04

∏   A discussion of the results obtained by our chemists is in order. Before continuing with it, answer the following questions.

Which chemist has results that can be described as:

(*i*)   accurate but imprecise?

(*ii*)   precise but inaccurate?

(*iii*)   accurate and precise?

(*iv*)   inaccurate and imprecise?

Your answers should have been 2, 3, 1 and 4 respectively. The first chemist's results are both accurate (% E_{rel} = 0.13) and precise (% RSD = 0.094). The second chemist also has a small relative error (−0.09%) and insofar as this is a measure of accuracy we might be tempted to say that this is worthy of praise! However, before we start handing out compliments, let us reflect on the precision of his results. Here we see that compared with the results of the first chemist the replicate titres are quite imprecise (% RSD = 0.73). You might ask whether this is important. He did after all end up with a mean titre which was even closer to the truth than that of the first chemist. This being so, does precision (or lack of it) matter? Let us be unequivocal in our answer. It matters!

Let us, in a manner of speaking, accompany the second chemist as he carried out his work. Had he carried out only the first titration, his result (22.81 cm³) would have been over 0.2 cm³ in error. He was, however, under instructions to carry out four replicate measurements. His second titration differed from the first by 0.15 cm³. Even schoolchildren can do better than this with titrations which involve well-behaved reactions. Nevertheless, had he decided to stop here, the mean titre would still have given a result which was 0.16

cm^3 in error. The third titration gave a result, 23.18 cm^3, which differed from the first by nearly 0.4 cm^3. This sort of occurence would stop any conscientious chemist dead in his tracks! At the very least, he would start to have doubts about the worth of any of the titre values. The fourth titration yields a result which, although close to that of the third, is still markedly different from those of the first two. Taking the mean of the four titre values happens to give a result which is quite accurate. However, it is an accuracy that is almost an accident. Had only the first two or the last two titrations been carried out the resulting accuracy would have been considerably less. We can hardly praise the work of someone whose accuracy is accidental. *On a consistent basis, accuracy without precision is impossible.* We know from the work of the first chemist what sort of precision is possible with the titrimetric method being utilised. Our second chemist is not working at this level of precision. This time his accuracy is good. On some future occasion it is highly likely it will not be so. Rather than praise, our second chemist is deserving of some enforced education!

Let us now turn our attention to the work of the third chemist. His precision (% RSD = 0.095) is as good as that of the first chemist. His accuracy (% E_{rel} = −0.87), however, is not nearly as good. What can we say about his work? Had we observed him carrying out the determinations it is likely that we would have noticed the sort of careful worker who can usually be counted on to obtain precise results. Unless we were observing very closely, it is unlikely that we would have noticed the cause of his obtaining a titre consistently about 0.2 cm^3 lower than that which was expected. What has happened to our third chemist is what happens to all chemists at one time or another; *he has obtained high precision without accuracy.* This is always a possibility. We shall need to spend time discussing the matter. Before we do, let us pass some comments on the fourth chemist. He obtained replicate titres which have relatively poor precision. (% RSD = 0.63) and were also relatively inaccurate (% E_{rel} = −1.13). His lack of precision compared with chemists one and three certainly suggests the need for some re-education. The fact that all his results were, to a greater or less extent, on the low side suggests that like the third chemists there is something in his method, beside his lack of precision, which is causing this negative bias.

Now attempt the following two SAQs. They will illustrate that your computer program can make light work of dealing with larger sets of results. They will also highlight the need for continual thoughtfulness, both about whether it is appropriate at all to treat a given set of data in a particular way and about the significance of the results obtained. Once you have done these we shall put aside the computer for a while, and take a closer look at errors and their origins.

SAQ 1.5a   It is ordinarily a good thing to have as many replicate measurements as possible when attempting a determination. Four chemists have between them provided a total of 16 measurements for a titration. Their values are listed in Fig. 1.5a, and the 'true' result is known to be 23.04 cm^3. Use your program "STAT1" to calculate the mean, the standard deviation and the %RSD for the 16 values taken as a single set. Comment on the value of doing this.

SAQ 1.5b   The following titration values were all obtained by a single analyst, applying a standard technique to separate aliquot samples of a homogeneous starting material. The procedure involves a number of pretreatment stages before the final titration, and it is these which are the main source of the spread in the results.

Titre/cm^3: 27.86, 27.68, 28.02, 28.05, 27.60, 27.55, 27.03, 27.76, 27.73, 27.75, 27.77, 27.53.

Twelve values are reported in all. Use program "STAT1" to compute the mean, $\bar{x}$, the standard deviation, s, and the %RSD for the whole set. Compare the results with what would have been obtained if the investigation had been stopped (*i*) after the first *two* measurements, (*ii*) after the first *four*, (*iii*) after the first *six*.

We set SAQs for a purpose. Here we wanted you to have a little more practice in using program "STAT1". We shall not use the computer again till we are well into Part 2 of the Unit. More important, though, we want to get you used to the rather capricious way in which nature can treat us, even when we perform very careful and well-designed experiments. Later we shall want to determine by just how much we can expect our results to become more reliable if we carry out a given number of careful repeat measurements. First, however, we must return to the wider scene and discuss in more depth the various kinds of error which we are likely to meet.

1.6. ERRORS

Analytical measurement is concerned with obtaining trustworthy results which make rational decision-making possible. Any measurement on an unknown is an estimate of the truth. When we communicate the results of our measurements, ie our estimate of the truth, we must accompany it with an estimate of the possible errors associated with it. Without such an estimate of error, our result is of relatively little value in aiding rational decision-making. A simple example will probably convince you of the validity of this statement. Suppose that we were required to assess the purity of a material. Let us say that after analysis we reported that the purity was 100%. Any chemist who took our report seriously would be making a blunder. Why? Because he has no idea of the errors associated with our measurement. If we had reported that the purity probably lay in the range 99.98–100.00%, then it would be safe for almost all purposes to assume that the material was pure. Had we reported that the purity probably lay in the range 95–100%, then it would be anything but safe to assume that the material was pure. It could be as low as 95% purity. Of course, it all depends on the use to which a result is put. It might be that a reported purity of within 5% is good enough, ie if the purity of the material is greater than about 95%, the material is acceptable. This simply underlines the point being made. Results must be communicated with some estimate of the possible error. Most professional analysts realise this. They are also aware, from bitter experience, and their clients often do not appreciate this point. It is the task of the client to state what uncertainty is tolerable in a determination. The duty of the analyst is, after deciding whether he can attain to this objective, to respond with an estimate of the possible error accompanying the result obtained. We must now turn our attention to the errors present in measurements.

1.6.1. Random Errors

We mentioned earlier that in a series of replicate measurements it would be very rare for all the results to have exactly the same value. A chemist working as carefully as possible, under as near identical conditions as imaginable, will still obtain results which differ amongst themselves, and which are scattered about the mean of the results. Some of the results will be larger than the mean, some smaller; some results will be very close to the mean, others relatively further away from it. There is no way that we can predict a result exactly or, equivalently, predict the sign and magnitude of its difference from the mean result. The sign and magnitude of differences occur in a random fashion. It is for this reason that we say that the *scatter* or *variability* of the results is attributable to *random errors*. Random errors are sometimes alternatively called *indeterminate errors*. The term *indeterminate* is appropriate in the sense that we cannot predict exactly a result from an examination of other results. The random error of an individual measurement is indeterminate. The scatter of results about the mean can be quantified by measures such as the relative mean deviation or the standard deviation. *Thus, measures of precision are also measures of random error.*

Recall the results of the first and third chemists in Section 1.5 where accuracies and precisions were compared. Both had standard deviations of about 0.02 cm^3 (% RSD $\approx$ 0.1). We said of these chemists that their precision – closeness of replicate measurements – was good; that they had obtained the sort of precision that could be expected of experienced chemists working carefully. We can now say that the random errors, as measured by the standard deviations, associated with the series of operations carried out by these two chemists are about 0.2 cm^3 (% RSD $\approx$ 0.1). We must now try to identify the possible sources of this random error.

1.6.2. Sources of Random Error

What are the sources of random error? Let us try to identify some of them and attempt where possible to estimate the uncertainty that each might bring into a measurement. Our chemists carried out a classical procedure which utilised a pipette and a burette. the pipette is used first: it is overfilled then drained to the mark. We have to estimate when the meniscus of the solution and the mark on the pipette coincide. How good are we at this task? The answer is – quite good but never consistently perfect! We always believe we are at the mark, but sometimes we are actually a little higher, sometimes a little lower, and on occasions, right on. However, since we never really know whether we are right on the mark or not, there is an uncertainty associated with our delivered volume. For a 25 cm^3

pipette this uncertainty is probably in the range ± 0.005–0.008 cm^3. The uncertainty in delivered volume is a source of random error associated with the use of a pipette. As a percentage of the delivered volume, this uncertainty might be small (eg $\pm 100 \times 0.008/25.00 = \pm 0.03\%$), but it is important to acknowledge that it exists.

No matter how experienced we are in pipetting we shall never be able to eliminate the error in it, nor to know in a particular case its magnitude or direction. There is in our work a *personal uncertainty*. This type of uncertainty is also present in the use of the burette. Here, *both at the start and the finish of the titration*, we are required, in the light of the position of the meniscus, to read the burette. (The burette – let us assume that it is the familiar 50 cm^3 type – is calibrated in 0.1 cm^3 divisions.) We can probably do this with an uncertainty somewhere in the range ± 0.01–0.02 cm^3. This uncertainty is always present. We cannot remove it. It contributes to the scatter of results which are obtained when replicate determinations are carried out, ie it is a source of random error. In a titre of say 25.00 cm^3, the reading uncertainty might be as high as ± 0.04 cm^3 (± 0.02 cm^3 for the initial reading and ± 0.02 cm^3 for the final reading). As a percentage of the titre this is $\pm(100 \times 0.04/25.00) = \pm 0.16\%$. This again is a small percentage, but it gives some idea of why the results of replicate titrations show scatter.

There are other sources of random error present in a titration besides the uncertainties mentioned above. For example, during the course of a series of replicate measurements the temperature might fluctuate slightly. This would cause slight variations in the volumes associated with the use of the pipette and burette. This leads to random errors which are attributable to an *instrumental uncertainty*. Let us be clear about this point. Each time we read an instrument we are subject to personal uncertainties, which are a source of random error. In addition to this, we are now saying that variability in measurements can occur because of random deviations in the instrument. The effects of such variations on titrations might generally be so small as to be ignored.* However, in some analytical

---

* The coefficient of cubical expansion of water is approximately 0.025% per degree change in temperature; that of volumetric glassware about a tenth of this. Therefore, fluctuations in the temperature of ± 1 °C will affect the volume measurement of, say, 10.00 cm^3 of solution by approximately ± 0.002 cm^3.

---

procedures, instrumental uncertainty is a major source of random error and it is important to be aware of its presence.

Another source of random error in titrations is associated with the use of the pipette. Experienced chemists touch the pipette against the side of the conical flask and permit it to drain for a set time. However the way in which they do this and the time which is allowed for drainage may vary, even if only slightly. This variation can cause a variation in the volume delivered and thus is a source of random error. Such sources of random error are known as *methodological uncertainties*, or simply *method uncertainties*. Method uncertainties affect every procedure or manipulation involved in analysis. We shall return to them shortly.

From the above it is seen that uncertainties – personal, instrumental and methodological - are the sources of the random scatter of results obtained when replicate determinations are carried out. This scatter can be quantified be measures of precision, such as the standard deviation or the relative standard deviation. Our first and third chemists obtained standard deviations of approximately ± 0.02 cm^3 (% RSD ≈ 0.1). Does this seem appropriate for the tasks they performed? Let us do some estimating. First, let us assume that the only source of random error is associated with the uncertainty in reading the burette. For our purpose, this is not such a bad assumption. It means that we have assumed that our chemists' method and instrument uncertainties are negligible: this is reasonable for careful chemists carrying out their determination under identical conditions. It also means that we are neglecting the uncertainty in using the pipette: this seems reasonable since the uncertainty in using a pipette is only about half that associated with a burette reading, and the burette has to be read twice. Now the reading uncertainty for a burette was estimated to be in the range of ± 0.01–0.02 cm^3, for a single reading, and double that when we take into account the initial and the final reading; ie uncertainty in the range ± 0.02–0.04 cm^3. Simply by inspection one can see that this is of the same order as the standard deviations that were obtained. This is not surprising: uncertainty is a source of random error; the standard deviation is a measure of the effect of random error.

It is about time that we attempted an exercise. We will concentrate on the contribution to random error made by reading uncertainties. However, it is important not to forget that other types of uncertainties can be equally or even more important contributors to random error.

∏   Estimate the % RSD associated with each of the following:

(*i*)   a titre of volume 20.00 cm^3, where the reading uncertainty of the burette is approximately ± 0.02 cm^3;

(*ii*)   a titre of volume 2.00 cm^3 by using the same burette as in (*i*);

(*iii*) a weighing of 1.0000 g, where the reading uncertainty is ± 0.0002 g;

(*iv*)   a weighing of 0.0010 g made by using the same balance as in (*iii*).

(*i*)   The percentage uncertainty is obtained by adding the two reading uncertainties (initial and final reading) dividing by the volume and multiplying by 100. It is therefore

$$\frac{\pm 0.04 \text{ cm}^3}{20.00 \text{ cm}^3} \times 100\% = \pm 0.2\%$$

This serves as an approximation to the % RSD.

There are several reasons why you might not have got this answer. The first is 'our friend' the pocket calculator! The second is that you might have forgotton to double the reading uncertainty. Related to this is a third reason: that you are not happy with doubling the reading uncertainty. If you fall into this category, please be patient.

(*ii*)   The same sort of calculation is in order here, ie % uncertainty

$$= \frac{\pm 0.04 \text{ cm}^3}{2.00 \text{ cm}^3} \times 100 = 2\%$$

Our estimate for the % RSD is therefore of this order, ie 2%. The reasons for an incorrect response are probably as in (*i*). Comparing the answers obtained for (*i*) and (ii) brings to light a very important point. It is that although the reading uncertainty remains a constant for a given burette, the random error, as expressed by the relative standard deviation, gets larger the smaller the titre. It is for this reason that analytical chemists always tend to avoid working in such a way that there is a small titre. What is true for burettes is of course true for other instruments. A smaller burette with an associated smaller reading uncertainty can be used if titrant volumes are unavoidably small. For example a 10 cm^3 burette with a reading uncertainty of ± 0.01 cm^3 would, for a titre of 2.00 cm^3, give a % uncertainty of

$$\frac{\pm 0.02 \ cm^3}{2.00 \ cm^3} \times 100 = 1\%.$$

(*iii*) The reading uncertainty is ± 0.0004 g (don't forget that an initial and a final reading of the balance are required). The % reading uncertainty is

$$\frac{\pm 0.0004 \ g}{1.0000 \ g} \times 100 = 0.04\%.$$

This is the order of random error which we can expect when using the balance for weighings of about 1 g.

(*iv*) The % reading uncertainty is

$$\frac{\pm 0.0004 \ g}{0.0010 \ g} \times 100 = 40\%!$$

The analytical balance is one of the post precise instruments in any laboratory. However, like all instruments, the more you push it to its limits, the less its precision.

In all the examples given above, we have doubled the reading uncertainty to take into account both the initial and the final reading. Actually the situation is not quite as bad as that.

By doubling the reading uncertainty we have been rather pes-
simistic. We have assumed that the worst is always going to happen;
that, for example, an underestimate of 0.0002 g in an initial reading
of the balance will combine with an overestimate of 0.0002 g in the
final reading to give a combined error of $+0.0004$ g. However, if
both readings were underestimated by 0.0002 g, the combined effect of the final result (ie on the difference between the readings)
would give an accumulated error of zero. Other combinations are
also possible. With this in mind, we now state without proof a relationship which comes from statistics. It is that the *most probable*
uncertainty – as measured by the overall standard deviation – is obtained by taking the square root of the sum of the squares of the
individual standard deviations.

$$s_T = \sqrt{s_1^2 + s_2^2 + s_3^2 \ldots + s_n^2} \qquad (1.8)$$

$$= \sum_{i=1}^{i=n} s_i^2$$

By substituting reading uncertainties for standard deviations in the
above question we obtain the following results.

(*i*) $\qquad s_T = \sqrt{(0.02)^2 + (0.02)^2} = 0.028 \text{ cm}^3$

$\qquad \% \text{ RSD} = \dfrac{0.028 \text{ cm}^3}{20.00 \text{ cm}^3} \times 100 = 0.14\%$

(*ii*) $\qquad s_T = 0.028 \text{ cm}^3$

$\qquad \% \text{ RSD} = \dfrac{0.028 \text{ cm}^3}{2.00 \text{ cm}^3} \times 100 = 1.4\%$

(*iii*) $\qquad s_T = \sqrt{(0.0002)^2 + (0.0002)^2}$

$\qquad = 0.00028 \text{ g}$

$\qquad \% \text{ RSD} = \dfrac{0.00028 \text{ g}}{1.0000 \text{ g}} \times 100 = 0.028\%$

(*iv*) $s_T = 0.00028$ g

$$\% \text{ RSD} = \frac{0.00028 \text{ g}}{0.0010 \text{ g}} \times 100 = 28\%$$

In the light of these new calculations, it is hoped that if you were unhappy with the simple doubling of reading uncertainty you are more content now that we have stated the 'proper' way to make these estimates!

It was always clear that the worst need not happen. Successive random errors will not always act in the same direction, sometimes they will tend to cancel one another. Nevertheless, the overall probable standard deviation is increased when two or more individual random errors each contribute to it. Where the overall error results from summing (or subtracting) a number of contributing random errors, then the effect on the standard deviation is given by Eq. 1.8.

Eq. 1.8 perhaps takes a little swallowing. It boldly states that the overall standard deviation, s_T, is given by the square root of the sum of the squares of each individual contributing standard deviation. Now it is not our purpose to go into the awkward business of giving formal statistical proofs. On the whole, we shall be prepared to accept on trust the assurances of statisticians that this or that result is valid. It does concern us, however, to try to persuade ourselves, where possible, that a result at least seems reasonable. Eq. 1.8 might not seem so at first sight. It helps a little, though, if we square both sides:

$$s_T^2 = s_1^2 + s_2^2 + s_3^2 + \ldots + s_n^2 \tag{1.9}$$

The individual standard deviations are not additive, but their squares are! The square of the standard deviation is an important quantity in its own right; it is known as the *variance*. The expression for calculating the variance is given simply by squaring each side of Eq. 1.7.

$$\text{Variance} = s^2 = \frac{\sum\limits_{i=1}^{i=n} d_i^2}{n - 1} \tag{1.10}$$

The variance is effectively the mean-square deviation of the measurements. When a result arises from the sum or difference of subsidiary readings or operations then the overall variance is given by the sum of the variances from each contribution. For our titre value, potential contributions to the overall variance (and hence to the overall standard deviation) include filling and emptying a pipette, reading a burette scale (twice) and closing the burette tap exactly at the end-point. We delayed introducing the term variance because definitions were coming quite thick and fast earlier on. The term will, however, reappear later in the Unit. It features prominently in statistics texts.

1.6.3. Random Error and Careful Work

In our discussions of random error we have used the results obtained by the first and the third chemist; let us now turn our attention to the second and the fourth chemist. As has been discussed previously these workers obtained precisions, as measured by the standard deviation or percentage relative standard deviation, which were not nearly as good as those obtained by the other chemists, see below.

Chemist	% RSD
1	0.094
2	0.73
3	0.095
4	0.63

What further comments can we make about the relative performance of these chemists? Insofar as measures of precision are measures of random error we can say that random errors associated with the work of chemist 2 and chemist 4 are much higher than they need be. The other two chemists have demonstrated that the random errors associated with the titration can be as low as $\approx 0.1\%$. Why have the other two chemists obtained random errors 6 or 7 times larger than this? The simple answer is carelessness. We can assume that all four chemists worked under identical conditions and with identical equipment. This means that we could expect instrument uncertainties to be approximately the same for all four workers. This suggests that personal and method uncertainties are the sources of the large differences in random error which were obtained. If chemist 2 and chemist 4 took a rather casual approach to reading their pipettes and burettes, to transferring aliquots of titrand, to operating their burettes and in deciding the exact point where their indicator changed colour, then it is highly likely that the precision they obtained would be relatively poor. As had been stated previously chemist 2 and chemist 4 are in need of some enforced education! Careful work results in better precision: careful work minimises random error.

1.6.4. Precision, Reproducibility, and Repeatability

All chemists, at one time or another, have used the terms *reproducible* and *reproducibility* when talking about the closeness to one another of the results of replicate determinations. When it is said that results are 'highly reproducible' it means that the results are very close to one another and would therefore give relatively small values for measures of precision. We have now learned that small values for measures of precision imply that the random errors in a procedure are small. In replicate titrations we have seen that a careful worker can obtain results which are very close to one another, ie, results which are highly reproducible. We have explained this by examining some of the sources of random error in a titration. We have seen that a careful worker minimises personal and method uncertainties and that the instrument uncertainties associated with the procedure are small. Let us now ask and answer an interesting question.

What would happen to the precision of a series of replicate titrations if they took place over perhaps a day or two rather than an hour or two, as is usual. It is highly probable that all readers will say that the precision would get worse; that is, the results would be more scattered. During the course of a day or two, it is likely that the laboratory conditions would change significantly; for example it may be cold in the morning, and possibly very hot in the afternoon. Furthermore, it is probable that different volumetric glassware would be used; over an extended period, someone always borrows someone else's burette and pipette! Other things might happen, eg a different bottle of indicator solution might have to be used. It is not surprising therefore that readers instinctively and correctly feel that the precision will be worse. It is for this reason that two terms are now used when talking about precision. The first is *repeatability*. This refers to the situation where replicate determinations are done *within-run*. The chemist who carried out his replicate titrations one after another is asked how '*repeatable*' his results were. The other term is the familiar *reproducibility*. This term is used when discussing what is known as *between-run precision*, ie in our example, the case where different times and different equipment were involved. With these definitions in mind, one can always expect a measure of repeatability for a given method to have a lower value than an equivalent measure of reproducibility. Having said all of this, do not be surprised if you find chemists still using the word 'reproducibility' when actually talking about 'repeatability'. Old habits die hard! Let us leave random errors for the moment and turn our attention to systematic errors.

1.7.   SYSTEMATIC ERRORS

Again, recall the comparisons we made of the accuracy and precision obtained by the four chemists carrying out a simple titration (Fig. 1.5b). Focus your attention on the work of the first and the third chemist. They obtained the same measurement precision, which demonstrated rather good repeatability. However, the first chemist had an accuracy ($\% E_{rel} = 0.13\%$) which was significantly better than that of the third chemist ($\% E_{rel} = 0.87\%$). The third chemist obtained consistently a titre lower than expected. His precision suggests a careful worker; his accuracy suggests that something is not right with his work, that somewhere there is a source of error which is resulting in low titres. Let us discuss some possibilities.

The glassware may be incorrectly calibrated. This would lead to a pipette, burette, or volumetric flask giving a volume different from that indicated by its graduation. It is rare nowadays to have volumetric glassware which has been incorrectly marked. However, glassware is calibrated at a standard temperature, usually 20 °C. At any other temperature the volume will not be that indicated by the graduations. If we carried out a series of titrations at a fixed temperature which was different from the calibration temperature then titres would be either all too low or all too high. We leave it to you as an exercise in logic to deduce whether, for example, high temperature will result in a low titre value or a high one. (Don't forget that you have to take into account all the glassware and both solutions!) The important thing here is that a departure from the calibration temperature will lead to what is known as a *systematic error*. With systematic errors the results are shifted in one direction or another, ie they are all either lower or higher than they should be: systematic errors are unidirectional. The systematic error just discussed can be classified as an *instrumental error*. As with random errors, systematic errors can be categorised as being instrumental, methodological or personal.

The use of the pipette introduces a possible *methodological error*. If the pipette is not given enough time to drain fully but is drained to the same extent each time, then we shall have a systematic error in one procedure – the aliquots on which we carry out our titrations will always be smaller than they should be. An extreme example of this would be if a pipette was of the 'blow-out' variety and was treated as if it was of the drainage type (the opposite is also a possibility). Some people are not charitable enough to classify this sort of error as being a systematic error – they call it a blunder! Systematic errors of the *methodological type* also arise from the use of indicators. For example, if we titrate a 25 cm^3 aliquot of approximately 0.1 M-ethanoic acid (acetic acid) with approximately 0.1 M-sodium hydroxide, the equivalence-point is located quite accurately with phenol red: the end-point is observed about 0.02 cm^3 before the equivalence-point (a systematic error of magnitude $E_{rel} = -(0.02/25.00) \times 100 = -0.08\%$; note the unidirectional nature of the error). If bromocresol purple is used to monitor the equivalence-point, the end-point is observed about 0.8 cm^3 before the equivalence-point (% $E_{rel} = -3.2\%$). Clearly using the second indicator would be wrong. However, even with the 'good' indicator a systematic error is evident. Obviously, some indicators have a colour change which is only witnessed after the equivalence-point has been passed. Therefore indicators can cause either a negative or a positive bias in the result.

Personal errors associated with the use of indicators are also possible. For example, many people are either colour blind or have poor colour discrimination. Such people carrying out a titration might experience large systematic error: either a positive or a negative bias is possible in their work. Poor colour discrimination is more common than many people believe; perhaps 10% of the population should not be employed in any task which calls for good colour discrimination. There are tests designed to identify such people. Have you had one? There are many possibilities for systematic personal error. In discussing random errors we mentioned the personal error which arises from reading uncertainty. Reading a scale or graduation can also cause a systematic error. Some people perhaps, because of shortcomings in their education, always read the level of liquid in a pipette or burette either incorrectly high or incorrectly low (this is in addition to the reading uncertainty). Nearly all of us, according to those who study such things, seem to possess what is known as number bias - an example is the common preference for even numbers rather than odd. (You can argue with yourself as to whether this results in a random or a systematic error.) The important thing is that as with random errors there are instrumental, methodological and personal sources of systematic error. Any or all the categories of systematic error might operate so as to give a chemist consistently low (or consistently high) titre values. Random errors can be minimised by careful work. What can we do about systematic errors?

1.7.1. The Treatment of Systematic Errors

As has been mentioned previously, the presence of random errors can be seen in the scatter of results obtained in replicate measurements. Careful work can minimise such errors. However, the amount of random error in an individual measurement cannot be ascertained. Such error fluctuates in a random fashion; it is indeterminate. Careful work is also necessary when dealing with systematic errors. However, no matter how experienced and careful the worker, systematic errors, which can sometimes be quite large, still occur. Much of the excitement (and frustration) of analytical chemistry is associated with the detection and elimination of systematic error. The fact that systematic errors can, in principle, be detected, measured, and eliminated, has led to their being called *determinate* errors. Let us attempt some generalisations about the types of systematic error and point out, in a general way, how they are treated.

1.7.2. Instrumental Errors

These, as their name implies, are associated with the instruments
employed in a determination. It would be a mistake, however, to
interpret this as meaning that the instrument is to blame. Instrumental errors are a result of our misplaced faith in the accuracy
of instruments. Volumetric glassware can, as has been discussed,
deliver or contain volumes different from those indicated by the
graduations. A major cause of this is that we may operate at temperature other than the calibration temperature. If we operate at
such temperatures we should calibrate our glassware appropriately.
If we do not, we can expect a systematic error. When was the last
time you calibrated some glassware?

Our belief in the truth of what instruments tell us is, perhaps, an
occupational hazard. Nevertheless, all of us know, from bitter experience, that all instruments, at one time or another, show systematic
errors. Caution should be our watchword: regular checking and if
necessary recalibration should become second nature.

1.7.3. Personal Errors

These could be termed errors due to human bias. When an analyst
makes a measurement, he is often required to make a judgement.
His ability to do so is improved if he is aware that he could be biased in his judgement. A person with poor colour perception who
is aware of that fact can often easily eliminate the bias which results
from that disability when he carries out titrations. A second indicator whose colour change can be accurately observed by him might
be available. Alternatively, the titration might be monitored poten-
tiometrically. A human bias that arises out of a physical disability is
easy to appreciate. Other types of bias, such as prejudice in favour
of even numbers, might not be. With all types of human bias, the
first step on the path to eliminating it is to be aware of its existence.
What is true of life in general is true of analytical chemistry!

1.7.4. Methodological Errors

These are due to inherent imperfections in the analytical method or to shortcomings in our application of the method. This type of systematic error causes us the most problems and therefore offers us the greatest challenges. Let us consider a simple example: the familiar acid-base titration. This method capitalises on the fact that a marked change in the pH of a titrated solution takes place in the region of the equivalence-point. We monitor the change with an indicator which changes colour at a pH very close to that of the equivalence-point, ie we try to get the end-point as close to the equivalence-point as possible. Even with a 'good' indicator, a small but significant volume of titrant may be consumed in excess of that required simply to react with the species being determined. The aqueous solution may have a small additional buffer capacity and a finite quantity of titrant is needed to react with the indicator itself. We thus have a further systematic error. Well, you are probably saying this presents no problem, a blank correction is in order. You are right! We have learned how to eliminate the bias caused by having an imperfect indicator. However, not all systematic errors due to method are so easy to detect and eliminate.

Let us continue by looking further at the titrimetric method. Often a titrand solution contains constituents other than that to be determined, which consume the titrant and hence cause a systematic error. Easy, you say. Perhaps. We first investigate a wide range of species for their possible interference effects. For those that do interfere, we then devise procedures for their removal or sequestration. Analytical chemists have been doing this sort of thing for years and are very good at it. It sounds easy, but of course it isn't. Furthermore, we cannot check every conceivable species for its interference effects. Moreover, it is not only the titrant which can be interfered with; the indicator too is subject to interference effects. In the light of this we must accept that when we come to an unknown there is always an element of doubt about whether or not we have a potential source of methodological bias. This is not to say that we lack procedures that help us to minimise our doubts. A quick look at two such procedures is in order.

(*i*) *The use of standard materials*

Here a material of known composition which is similar to that of the sample, both in the proportion of desired constituent and of other constituents (the matrix) is analysed. The lack of any systematic error is taken as evidence that systematic error will be absent when we analyse an unknown.

(*ii*) *Comparison of methods*

It is unlikely that two independent methods, which are different in all their procedures, would lead to the same methodological systematic errors. Consequently, if the results of a new method are compared with those of an established method, on which bias is thought not to exist, then agreement between the two methods may be taken as an indication that the new method lacks bias. (Keep this approach in mind when you come to later material on statistical significance.)

1.7.5. Blunders

It is not unknown for analytical chemists to make mistakes which could best be described as silly! Examples of such mistakes are: recording a number incorrectly, using a wrong scale (eg reading absorbance on the 0 to 2 scale rather than the 0 to 1 scale), and getting 'sums' wrong! All these obviously lead to errors. We classify such errors as *blunders*. In much of what follows we pretend that they do not exist! However, in practice they are, as everyone knows, always a possibility.

Let us have another SAQ. This one is of a descriptive nature. It is a bit different, but have a go.

SAQ 1.7a	Although the gravimetric method of analysis was crucial to the development of chemistry, it is, even in the hands of experienced analysts, often subject to large errors. $\longrightarrow$

SAQ 1.7a (cont.)

Gravimetric analysis often involves steps similar to the following.

(*i*) A weighed amount of a sample containing the determinand is dissolved in a suitable solvent and the conditions of the solution (eg pH) are appropriately adjusted.

(*ii*) A weighed amount of the precipitant is dissolved in a suitable solvent to give a solution of known concentration.

(*iii*) The solution containing the precipitant is added slowly and with stirring to the determinand solution, which is often at an elevated temperature. This is continued until the precipitant is in excess. The solution is allowed to cool.

(*iv*) The precipitate formed in (*iii*) is filtered off and washed with a suitable solvent.

(*v*) The washed precipitate is heated or dried to constant weight and by using the appropriate stoichiometric relationships the percentage of determinand in the sample is calculated.

Try to identify which of the steps are likely to introduce significant errors (you can assume that blunders are absent) and suggest the category of error which is likely to be present.

SAQ 1.7a

1.8. THE STRUCTURE OF ERROR

A result can be subject to *random errors, systematic errors, and blunders*. The absolute error can be expressed as a sum of contributions from these sources. For a single measurement,

$$E_{abs} = x - T = e_r + e_s + e_b \qquad (1.9)$$

whilst following replicate measurements

$$E_{abs} = \bar{x} - T = e_r + e_s + e_b \qquad (1.10)$$

where E_{abs} = the absolute error,
       $\bar{x}$ = the mean of replicate determinations,
       T = the true value,
       e_r = the random error,
       e_s = the systematic error,
       e_b = error due to blunders.

Let us assume that blunders are absent and therefore write $\bar{x} - T = e_r + e_s$. Let us try to visualise the significance of this equation with the aid of some diagrams.

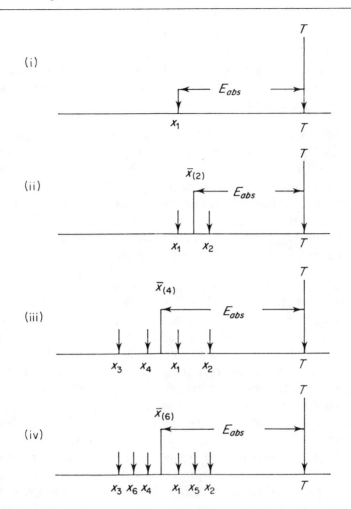

Fig. 1.8a (*i*)–(*iv*).   *Successive measurements x_i, in a case where the true value is T*

In Fig. 1.8a results are recorded on the x axis and compared with the correct value T. In Fig. 1.8a(i) a single result, x_1, is seen to be different from T. A second determination gives the result, x_2; the mean of the two results is represented by $\bar{x}_{(2)}$ Fig. 1.8a(ii). We see that $\bar{x}_{(2)}$ and T do not coincide, ie that error is present. You might think here that the mean is closer to the truth than was the first result and that this is a good reason for doing two measurements. Stop having such thoughts! The reason for taking the mean of replicates is that it represents the best estimate of our experimental result. If you still believe that one gets closer to the correct value by doing replicates, than examine Fig. 1.8a(*iii*). Here we have carried out two more determinations, and represented the mean of all four measurements by $\bar{x}_{(4)}$. Note that $\bar{x}_{(4)}$ *is further* from the correct value than was $\bar{x}_{(2)}$. Are you convinced now?

In Fig. 1.8a(*iv*) the results of six determinations are recorded and the mean is represented by $\bar{x}_{(6)}$ (Yes, we presented a nice neat picture and yes, life is rarely like this!) What is the significance of Fig. 1.8a(*iv*)? Here, we are attempting to show a situation where increasing the number of measurements does not seem appreciably to change the value of the mean (we have actually arranged things to show no change in the value of the mean in going from four measurements to six). Now let us start again!

If we obtained in practice a situation such as that in Fig. 1.8a(*i*) we should be aware of error but have no knowledge whether it was mainly random, mainly systematic, or a mixture of both. By doing repeated measurements, we start to realise that we have a bias (a systematic error) in our measurements. This is not the only information we gain. We also start to get some idea, from the scatter of the results, of the random error present: we shall return to this shortly. Furthermore, by the time we have made, say, six measurements we start to appreciate the extent of the bias in our system: a good estimate of it would be $\bar{x}_{(6)} - T$. All this needs further elaboration. If we make repeated measurements on an unknown and see that the mean value of our measurements changes little with more measurements, than the difference between the mean and the correct value must be a measure of the bias in our measurements. Consider a situation where the scatter of results is the same as in Fig. 1.8a(*iv*) but where bias is absent: this situation is represented in Fig. 1.8b.

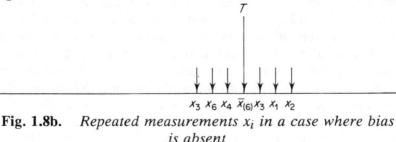

Fig. 1.8b.   *Repeated measurements x_i in a case where bias is absent*

Comparing Fig. 1.8b with Fig. 1.8a(iv) should convince you that $\bar{x}_{(6)} - T$ gives a good estimate of the systematic error present. But, you might say, we were told that $\bar{x} - T$ is a measure of both systematic and random error. So you were and so it is! It is simply that by the time we reach the situation in Fig. 1.8a(iv) the average random error is approaching zero.

Let us continue the argument. Suppose that another measurement were made. What value would you expect it to have? Let us accept that the bias is given by $\bar{x}_{(6)} - T$ and represent it in Fig. 1.8c(i). Now again we ask the question, what value would you expect a new measurement to have? An answer would be, a value close to that of the known correct value minus the bias. Good, but not good enough. How close to the value mentioned is close? Well, you might reply, we cannot say for sure because we are now dealing with random error. You are right, we cannot say for sure. We can, however, suggest from the information available in Fig. 1.8a(iv) that the next measurement is likely to be the range spanned by the earlier six results, as indicated in Fig. 1.8c(ii). Can we do any better than this? Well, we could start to be like gamblers and play the odds. This might lead us to suggest that the next measurement will quite probably be in the rather narrower range indicated in Fig. 1.8c(iii). Here, we have calculated the mean deviation from the mean in Fig. 1.8a(iv) and drawn a range lying within this amount on either side of the value which indicates the extent of bias. When we discussed mean deviations we mentioned that the standard deviation was considered a better measure of precision. (We never said why then, nor shall we now!) In the light of this we have drawn a probability range for the next measurement based on the standard deviation – see Fig. 1.8c(iv). You will notice that this is a little broader than the range based on the mean deviation. By now you are probably tired of estimating where the next measurement will fall. Alright see Fig. 1.8c(v). Well, we were close! Never mind, the important thing is that on this occasion we can see clearly how our error $x - T$ was made up of a bias component and a component due to random error.

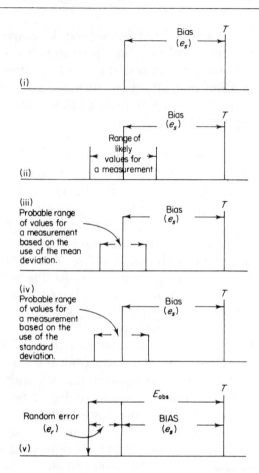

Fig. 1.8c.(*i*)–(*v*). *Predicting and analysing the result of a 7th measurement, following upon the 6 results recorded in Fig. 1.8a*

Fig. 1.8a(*ii*) has been drawn again as Fig. 1.8d so as to indicate the systematic and random components present in the error, E_{abs} = $\bar{x}$ − T.

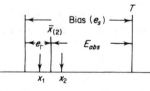

Fig. 1.8d. *Analysis of the first two readings from Fig. 1.8a, where the random error partially compensates for the bias*

We have been discussing a situation where the systematic component of the error is greater than that of random component. Obviously, the opposite may be true, or the systematic error may even be zero as in Fig. 1.8b. Such situations can be examined quite nicely by simple statistical arguments. You will have to wait a little before we do so.

Have a quick look at Fig. 1.8e and then answer the SAQ which follows.

Random errors	Systematic errors
Also known as indeterminate errors.	Also known as determinate errors.
Can never be eliminated but can be minimised by careful work.	Can, in principle, be located and eliminated.
Recognised by the variability in replicate determinations, ie, by the scatter of results about their mean.	Recognised by the lack of agreement between the mean of a series of replicate determinations and the correct value. This is true provided the random error is not so large as to obscure the bias.
Quantified by measures of precision such as the standard deviation or the relative standard deviation.	Quantified by measures such as the absolute error or the relative error in the mean. This is true provided that enough measurements have been so as to have a true estimate of the bias.
Sources include personal instrumental, and methodological uncertainties.	Sources include personal, instrumental, and methodological bias.

Fig. 1.8e. *A summary of random and systematic errors*

SAQ 1.8a   Do you consider the following statement to be
                 (*i*) true, (*ii*) sometimes true (*iii*) false.

                 If systematic error is removed, then the results
                 will be accurate.

1.9.   SUMMARY

Half the battle in mastering statistics involves a very careful use of
language. We have introduced a rather large number of terms in
this first Part of our Unit. It is a quite useful review of our work so
far simply to list the most important of them; our cast list in order
of appearance, as it were. Before going on, you should confirm that
you have a clear understanding of the meaning of each of the terms
below.

accuracy	precision
replicate measurements	arithmetic mean
absolute error	relative error
mean deviation	relative mean deviation
standard deviation	relative standard deviation
random error	indeterminate error
systematic error	determinate error
personal error	methodological error
instrumental error	variance
reproducibility	repeatability
blunder	bias

We have done the groundwork for the deeper and more quantitative discussion which is to follow. We have begun to make use of the computer as a tool to help us with our labours. Most important, we hope that we have made it clear that in this field, when you make your measurements, and when you scrutinise them, you must at all times keep your critical faculties alive. Sources of error are many and varied: with care some may be minimised and others may be identified and compensated. Chance plays a role, and may mislead us if we are not careful.

Section 1.9.1 below is provided for reference in case you get 'stuck' when you are computing. Once you build up a little experience (of getting stuck!) you will find you become quite adept at resolving such difficulties.

Objectives

As a result of completing Part 1 you should feel able to:

- give clear definitions of the terms listed in Section 1.9, and distinguish without difficulty between the different concepts they involve;

- justify the central importance of the experimental mean and of the standard deviation when assessing the results of analytical measurements;

- use your computer program "STAT1" to derive values for the mean, the standard deviation and the percent relative standard deviation for a sample set of replicate measurements;

- describe in some detail the general structure of errors; list distinct types of source which contribute to random errors and to systematic errors; comment on general strategies through which the influence of errors might be minimised; and apply such an error analysis in particular practical situations.

1.9.1. Eliminating 'Gremlins' from your Programs

Everyone makes mistakes from time to time when computing. They
may result from mistyping or from misconstrued logic. They show
up when your program breaks down or when it produces manifestly silly answers. The following checklist should help you to track
down and eradicate such 'gremlins' when they strike you, as they
inevitably will.

(*a*) If your computer program breaks down during a run, most machines will tell you at which line number the breakdown occurred. This information, together with any comment the machine may make (eg 'divide by zero at line 145') can provide an
important clue to the source of the trouble. Quite often, however, the actual error may have occurred at an earlier point,
but was not such as to cause breakdown until the later line was
reached.

(*b*) Remember that any variable used in a calculation (ie one which
appears to the right of an equals sign) ought to have been assigned a value before the calculation was attempted.

(*c*) Beware lest you confuse 1(unity) with I, or 0(zero) with O.

(*d*) Try running an offending program with very simple data, so that
you can cross-check its results.

(*e*) Add extra lines to print intermediate results so that you can pin
down at what stage things begin to go awry.

(*f*) Remember always that it is *you* who is misbehaving. The computer is a mere automaton which follows slavishly, but with re-
markable reliability, precisely the instructions you have given
it. This is not quite the same as carrying out the instructions
which, perhaps you had *meant* to give it!

(*g*) If, when pursuing tactic (*e*), you find that values are being presented on the screen faster than you can read them, then add lines to halt the program temporarily at suitable points. Use GET$ or its equivalent (see Appendix 2.2.4). Alternatively, you can add a line such as INPUT A which will cause the computer to wait for data till you press the 'return' key.

2. Probability and the Distribution of Error

Introduction

Analytical chemists, like other scientists and technologists, work enveloped in imperfections and uncertainties. Measurements are never perfect and the information gained from one or more experiments is never complete.

Consequently, the conclusions drawn from experiments and the actions which necessarily follow always carry with them an element of doubt. In contending with this doubt, the skilled analyst brings into play his judgement – an invaluable asset. In relatively simple situations the analyst knows his judgements to be correct: decisions are made and no sleep is lost! Unfortunately, not all situations are simple and some decisions can make the difference between profit and loss, or even between life and death. Here, as the experienced analytical chemist would be the first to acknowledge, any aid to making judgements is welcome. Statistics is such an aid. Amongst other things, it enables us to maximise the information obtainable from experimental results and to minimise the chances of drawing the wrong conclusions from them.

One must point out immediately that statistics does not make decisions for us. We are always in charge! Statistics acts as an unprejudiced adviser. We can accept or reject the advice it gives us: the choice is ours. However, it's nice to know that when we have to make difficult decisions, statistics is there to help us.

In formal education, courses in statistics often follow a detailed examination of the fundamentals of probability. We, too, will start with a discussion of the concept of probability. However, our treatment will be at an elementary level and will move into statistics as soon as possible. Let us get underway.

2.1. PROBABILITY

Let us start by deciding upon a scale for probabilities. The most convenient and perhaps the most obvious scale is 0 to 1. A probability of 0 means that an event will never occur. A probability of 1 means that an event is certain to occur. Benjamin Franklin once said that 'In this world nothing is certain but death and taxes.' We will dodge the controversial issue of taxes and stick to death! The probability of death is undoubtedly 1 ($P = 1$); the probability of never dying is, unfortunately, 0 ($P = 0$). The probabilities of events that are in the range between never ($P = 0$) and always ($P = 1$) are expressed as fractions or as decimals. For example, if a perfectly balanced coin is tossed, the probability of its landing face upwards is 1/2. Notice how easy it was to agree on this probability. We know that there are only two outcomes when a coin is tossed; either a head or a tail. (We ignore the possibility of its standing on its edge.) We know that neither of the outcomes is preferred over the other, so the probability of either outcome must be the same, ie the probability of a head equals the probability of a tail. Finally, even though we might not have realised we were doing it, we decided that in order to give magnitude to the probability of a head (or a tail) we would divide this one outcome by the number of possible outcomes, ie we divided 1 by 2. Yes, perhaps we have laboured the point! It is easy. Let us keep it so for a while.

∏ What is the probability of throwing a three with an unweighted die?

The number of possible outcomes is six; the die has six faces, numbered 1 to 6. When a die is thrown all outcomes are equally probable and the probability of any one of them is 1/6. Therefore we can write the answer to our question as P (of a three) = 1/6.

∏　　　What is the probability of drawing a jack of diamonds from a fresh unmarked pack of cards?

There are 52 cards in a pack and therefore 52 possible outcomes, all of which have equal probability. The probability of any one outcome is 1/52. Therefore P (jack of diamonds) = 1/52.

So far, so good. Now let us formalise something which is almost intuitive. It is that the sum of individual probabilities of *all* possible outcomes must equal 1. We can write this as:

$$P_1 + P_2 + P_3 \ldots + P_n = \sum_{i=1}^{i=n} P_i = 1 \qquad (2.1)$$

Some examples should convince you of the truth of this equation.

The probability of obtaining a head *or* a tail when a coin is tossed is:

$$P_{head} + P_{tail} = 1/2 + 1/2 = 1$$

(Note that even if the coin was not perfectly balanced the probability of obtaining a head or a tail would still be 1.)

The probability of obtaining a 1 or a 2 or a 3 or a 4 or a 5 or a 6 when a fair die is thrown is:

$$P_{one} + P_{two} + P_{three} + P_{four} + P_{five} + P_{six} =$$
$$1/6 + 1/6 + 1/6 + 1/6 + 1/6 + 1/6 = 1$$

We may not be interested in the probability of all possible outcomes; (after all, this always equals 1), but rather in the probability of a few of the outcomes.

For example, what is the probability of obtaining either a 3 or a

4 when a die is thrown. Well, if the probability of throwing a 3 is 1/6 and the probability of throwing a 4 is also 1/6, it seems to make sense to sum the individual probabilities. That is, the probability of throwing a 3 or a 4 is:

$$P_{three} + P_{four} = 1/6 + 1/6 = 2/6 = 1/3$$

2.1.1. The Addition Law

The above example and those where all possible outcomes were considered demonstrate the use of the *addition law. This law states that if the probabilities of n mutually exclusive events $E_1, E_2, E_3 \ldots E_n$ occurring are $P_1, P_2, P_3 \ldots P_n$, then there is a probability of $P = P_1, + P_2 + P_3 + \ldots P_n$ that any one of these events will occur.* The term *mutually exclusive* refers to events that cannot occur simultaneously – the reason being that the occurrence of any one of them precludes the occurrence of the others. Let us put this law to use whilst keeping our eye on the term 'mutually exclusive events'. When we toss a coin, we are aware that the outcome is either a head or a tail: both cannot occur simultaneously, they are mutually exclusive. The same was true of the outcome of a roll of the die, the different outcomes, which we can call events, are mutually exclusive. If we draw one card from a pack, each possible outcome or event is mutually exclusive.

In answer to the question what is the probability of obtaining a 3 or a 4 when a die is rolled, we summed the probabilities of each outcome, ie 1/6 + 1/6 + 2/6. We were justified in doing this because the outcomes were mutually exclusive.

Consider now a slightly more difficult question. It is what is the probability of obtaining one head when two coins are tossed? First we will list the possible outcomes. These are: HH, TT, HT and TH. The probability of each of these four outcomes is 1/4. Each outcome is mutually exclusive; we cannot simultaneously obtain say a TT and a TH. The event in which we are interested – obtaining one head – can be arrived at by two outcomes, ie by HT and by TH. The probability of this event is therefore $P_{HT} + P_{TH} = 1/4 + 1/4 = 1/2$.

2.1.2.   The Multiplication Law

Let us take a rest from the addition law and turn our attention to another law which features in probability. It is the multiplication law. Let us consider an example of its utilisation before stating it. First a question – what is the probability of obtaining two heads in two tosses of a coin? Almost intuitively, we write down that the probability of $1/2 \times 1/2 = 1/4$. What have we done? We have simply multiplied the probability of obtaining a head in the first throw by the probability of obtaining a head in the second throw. Here, we have, perhaps without knowing it, applied the *Multiplication Law*. *This law states that if the probabilities of n independent events are given by* $P_1, P_2, P_3, \ldots P_n$, *then the probability that all the events will occur is given by* $P = P_1 \times P_2 \times P_3 \times \ldots \times P_n$.

Events are said to be independent if the occurrence of one does not affect the occurrence of another. When we toss a coin for the second time it (the coin) is oblivious of the result of our first toss (!): the events are independent. The probability of obtaining two heads in two tosses of a coin is $1/2 \times 1/2 = 1/4$. Three obviously independent events would be to toss a head, roll a six and draw a ten of clubs! What is the probability of these three events occurring together?

$$P = P_{(head)} \times P_{(six)} \times P_{(10clubs)} = 1/2 \times 1/6 \times 1/52 = 1/624$$

Well, we have done enough for you to understand some simple probability ideas. Assess your progress by attempting the following questions.

SAQ 2.1a	(*i*) What is the probability of throwing a 4 with a fair die?
	(*ii*) What is the probability of throwing either a 4 or a 5 with a fair die?
	(*iii*) A die is rolled and a 4 is obtained. What is the probability of throwing a 5 on the next roll of the die? $\longrightarrow$

SAQ 2.1a
(cont.)

(*iv*) What is the probability of throwing a 4 with one die and a 5 with a second die?

(*v*) A coin is tossed three times in a row. What is the probability of obtaining:

(*a*) 3 heads,
(*b*) 2 heads and a tail,
(*c*) 2 of a kind?

2.1.3. The Probability Tree

The tossing of a coin is a good example of situations where there are equal probabilities for each of two mutually exclusive events. Here are two others.

A bucket contains an equal number of red and white balls. The first ball to be picked out will be red(white) – the probability = 1/2.

The next child to be born anywhere on earth will be a girl(boy) – the probability = 1/2.

(If you are not happy with this example, be patient.)

In working with such situations, we often use what is known as a probability tree. One has been constructed in Fig. 2.1a. below. It refers to the tossing of a coin, but it could be used in any situation where two events of equal probability are mutually exclusive.

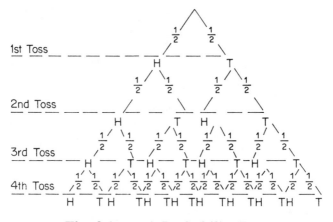

Fig. 2.1a. *A Probability Tree*

As can be seen from the 'tree', it is possible to calculate the probability of any outcome by simply following the appropriate branch and multiplying probabilities as we go along. For example, the probability of tossing three heads in a row is 1/2 × 1/2 × 1/2 = 1/8. Another example, the probability of tossing a head, then a tail, then a tail and finally a head is $1/2 \times 1/2 \times 1/2 \times 1/2 = 1/16$. (Yes. things do get complicated near the bottom of the tree. Why do you think that a tree for five tosses of the coin was not constructed?)

It is useful to analyse the 'tree' a little further. In Figs. 2.1b, 2.1c and 2.1d., we have constructed trees for the first, second and third tosses respectively. For each toss the probability of a given outcome, the probability of each event (arrived at by summing the probabilities of the relevant outcomes), and the ratios of the probabilities of the events are given. Work through each of the 'trees' and then read the comments which follow.

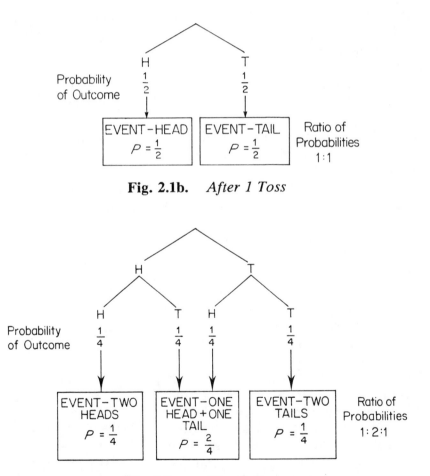

Fig. 2.1b. *After 1 Toss*

Fig. 2.1c. *After 2 Tosses*

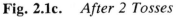

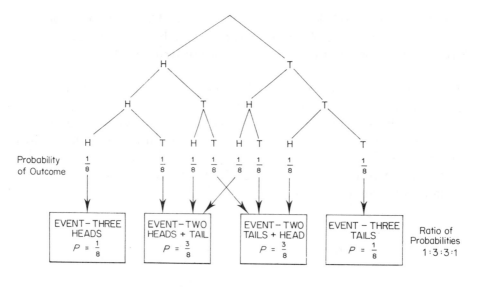

Fig. 2.1d.   *After 3 Tosses*

The 'tree' representing the first toss (Fig. 2.1b) is straightforward. There are two outcomes, each representing an event – a head or a tail. Both events have the same probability therefore the ratio of the probabilities is 1 : 1. With the second toss, we see that there are four outcomes – two heads, a head and a tail, a tail and a head, and two tails. The probability of the event two heads (or two tails) is 1/4; the event one head and one tail can be reached by two outcomes, hence its probability 2/4. The ratio of probabilities is therefore 1 : 2 : 1. Similar reasoning shows that after the third toss the ratio of the probabilities of the events three heads, two heads and one tail, two tails and one head, and three tails is 1 : 3 : 3 : 1. Thus it can be seen that the 'probability tree' gives us not only the probability of a given outcome, but also the ratio of probabilities for the events.

Obviously the more tosses of a coin that we have to contend with the more complicated the 'tree' and the more likely we are to make a mistake in utilising it. Try, when you have an hour or two to spare, to construct a probability tree which can handle ten tosses of a coin. Fortunately for us there is some relatively simple mathematics which when mastered, will make such problems easy to handle. Let us turn our attention to the binomial distribution.

2.1.4. The Binomial Distribution

The binomial distribution formula, theorem, or expansion, are names given to the algebraic formula in Eq. 2.2.

$$(p + q)^n = p^n + np^{n-1}q + \frac{n(n-1)}{2!} p^{n-2}q^2$$

$$+ \frac{n(n-1)(n-2)}{3!} p^{n-3}q^3 \ldots + npq^{n-1} + q^n \quad (2.2)$$

[p and q are real numbers, and n is a positive integer, $n! = n(n-1)(n-2) \ldots 1$].

The distribution has a rather interesting history; names such as Isaac Newton and Blaise Pascal are associated with it. Some believe that the distribution was discovered in the study of games of chance. Even if this is not true, the distribution has found its place in such studies for a long time. One of the easiest games of chance to study (as has probably become apparent!) is the tossing of coins. Let us return to this game.

We start our discussion in a general way. We let p be the probability that an event will occur, and $q = (1 - p)$ the probability that an event will not occur: the overall probability is $p + q = 1$. In the special case where the probability of an event occurring equals the probability of an event not occurring we have $p = q = 1/2$. This is the situation which obtains when we toss a fair coin; $P_{\text{head}} = P_{\text{tail}} = 1/2$. We will write this as $(p + q) = 1$.

$$1/2 \quad + \quad 1/2 \quad = \quad 1$$

probability + probability = overall
of a head   of a tail   probability

When we toss a fair coin twice, we have seen (Fig. 2.1b) that the probabilities of the events 2 heads, 1 head + 1 tail, and 2 tails, are 1/4, 1/2 and 1/4 respectively. Let us see if these probabilities are obtainable from the expansion of $(p + q)^2$.

$$(p + q)^2 = p^2 + 2pq + q^2$$

$$= (1/2)^2 + 2(1/2).(1/2) + (1/2)^2$$

$$= 1/4 + 2(1/4) + 1/4 = \qquad 1$$

or $\qquad P_{2H} + P_{H+T} + P_{2H} =$ overall probability

Success(!), the probabilities are extractable from the expansion. What is good for two tosses should be good for three!

$$(p + q)^3 = p^3 + 3p^2q + 3pq^2 + q^3$$

$$= (1/2)^3 + 3(1/2)^2.(1/2) + 3(1/2).(1/2)^2 + (1/2)^3$$

ie $\qquad = 1/8 + 3(1/8) + 3(1/8) + 1/8 = \qquad 1$

$$P_{3H} + P_{2H+T} + P_{H+2T} + P_{3T} = \quad \text{overall} \\ \text{probability}$$

Now compare these probabilities with those obtained by utilising the probability tree, Fig. 2.1c. Again success!

Would you now like to write out the probabilities for the possible events when a coin is tossed ten times? You wouldn't! Why so? It is an easy question to answer. You would have no trouble at all in writing down the values of p and q, ie p^{10}, p^9q^1, p^8q^2, etc. However, it is a bit troublesome to work out the coefficients for each of the terms.

Let us now introduce you to Pascal's triangle (see Fig. 2.1e.)

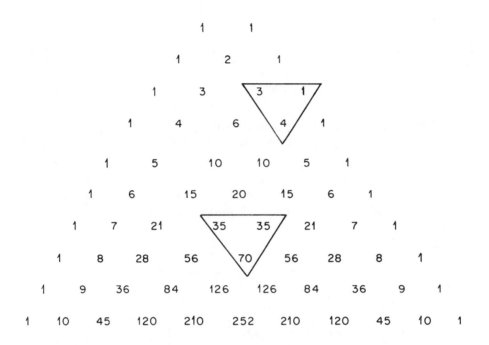

Fig. 2.1e. *Pascal's triangle and the binomial distribution*

From an examination of the triangle, it is seen that each number in a row is made up by summing the two numbers above it, eg 3 + 1 = 4 or 35 + 35 = 70. What is the significance of the numbers in a row? For a start, examine the numbers in the second row: the numbers are 1, 2 and 1. Note that these numbers are the coefficients for the expansion $(p + q)^2$. Look at the third row: the numbers are 1, 3, 3, 1. These numbers are the coefficients for the expansion of $(p + q)^3$. Need we say more? If you want the coefficients for the expansion of $(p + q)^{10}$ simply write down the numbers in the tenth row of the triangle, ie 1, 10, 45, 120, 210, 252, 210, 120, 45, 10, 1.

We can now write out the binomial expansion of $(p + q)^{10}$ with relative ease. It is

$$1p^{10} + 10p^9q^1 + 45p^8q^2 + 120p^7q^3 + 210p^6q^4 +$$

$$252p^5q^5 + 210p^4q^6 + 120p^3q^7 + 45p^2q^8 + 10pq^9 + 1q^{10}.$$

Easy! Now for the tossing of a fair coin $p = q = 1/2$. Therefore, we have $p^{10} = p^9q^1 = p^8q^2 = \ldots = (1/2)^{10}$

ie in each case the part of the term which follows the coefficient is $(1/2)^{10}$ which equals 1/1024. We can now write down the probabilities of all of the events. They are given in Fig. 2.1f.

Event	Probability
10 heads	1/1024
9 heads + 1 tail	10/1024
8 heads + 2 tails	45/1024
7 heads + 3 tails	120/1024
6 heads + 4 tails	210/1024
5 heads + 5 tails	252/1024
4 heads + 6 tails	210/1024
3 heads + 7 tails	120/1024
2 heads + 8 tails	45/1024
1 head + 9 tails	10/1024
10 tails	1/1024

Fig. 2.1f. *Probabilities of possible events when a coin is tossed ten times in a row (or when ten separate coins are tossed)*

Whilst we are busy congratulating ourselves on mastering Pascal's triangle and the binomial distribution when applied to the tossing of a fair coin, let us make a few more points. The number 1024 can be obtained by adding all the coefficients together: it represents all possible outcomes. The number of possible outcomes which lead to a given event is simply obtained from the coefficient of a term. For example, there are 45 ways in which we can obtain 2 heads + 8 tails.

The ratio of the coefficients gives us the ratio of the probabilities of the possible events. With all this in mind, try the following exercise.

∏ What are the events and their probabilities when a fair coin is tossed 6 times in a row?

From the sixth row of the Pascal's Triangle we obtain the numbers 1, 6, 15, 20, 15, 6, 1. When added together these numbers give 64. This information is now used to construct a probability table.

Event	Outcomes	Probability of Event	Ratio of Probabilities
6 heads	1	1/64	1
5 heads + 1 tail	6	6/64	6
4 heads + 2 tails	15	15/64	15
3 heads + 3 tails	20	20/64	20
2 heads + 4 tails	15	15/64	15
1 head + 5 tails	6	6/64	6
6 tails	1	1/64	1

2.1.5. Probability and Relative Frequency

So far in our discussions about probability we have hidden from view something of great importance. Let us make amends! Say that we were given a coin and asked to test whether or not it is a 'fair' one, ie that it is not weighted so as to give, say, more heads than tails. How would we go about carrying out this test? The simple answer is that we would start tossing the coin and recording the results. Pick up a coin and try it! We did and after 10 tosses had obtained 4 heads and 6 tails. Was the coin fair? Yes, you are right, we have not got enough experimental evidence to make such a judgement; more tosses are required. How many tosses would you consider to be adequate in order to make the judgement in question? One hundred?

One thousand? Let us say that having carried out a '1,000 tosses-experiment', the results obtained were 510 heads and 490 tails. Is the coin fair? Some might say yes and give as their reason the fact that the relative frequencies of heads 510/1000 and tails 490/1000 are close to the probabilities expected for a fair coin, ie $P_{(head)} = P_{(tail)} = 1/2$. Others might argue that a larger number of tosses – 10,000 or perhaps 1,000,000(!) – is required. Note that the argument here is about what constitutes a large number. Regardless of one's opinion on this matter, there appears to be an acceptance of the idea that the relative frequencies obtained after a large number of tosses, or after what is known as 'a long run', will give good approximations to the true probabilities. This amounts to a relative frequency definition of probability.

The way of testing whether or not a coin is fair is to toss it a large number of times. If the resulting relative frequencies are very close to 1/2, then the probabilities are very close to 1/2 and this is what is meant by a fair coin. If on the other hand the relative frequencies show a significant departure from equality, eg relative frequency of heads $= 3/4$ and relative frequency of tails $= 1/4$, then the assigned probabilities would be $P_{(head)} = 3/4$ and $P_{(tail)} = 1/4$ and it would be silly to believe that the coin was fair.

In the same way as we have defined probabilities in terms of long-run relative frequencies, we can identify the relative frequency of an event, as being that which would occur in the long run given the probability of the event. For example, we have seen that when a coin is tossed ten times the probability of obtaining 5 heads and 5 tails is 252/1024. This means that if we tossed a coin ten times, then repeated the experiment again and again until we had a very large number of sets of ten tosses, the relative frequency of the event, 5 heads and 5 tails, would be 252/1024. Of course, if the long-run relative frequency (which from now on we will simply call the relative frequency) of an event can be obtained from the probability of the event, then the ratio of the relative frequencies of several events can be obtained from the ratio of their probabilities. For example, the probabilities of the events $(7H + 3T)$ and $(6H + 4T)$ when a fair coin is tossed ten times, are 120/1024 and 210/1024 respectively. Since the ratio of the probabilities is $120:210$, the ratio of the relative frequencies of $(7H + 3T)$ and $(6H + 4T)$ is $120:210$.

2.2. PROBABILITY AND RANDOM ERROR

2.2.1. The Tossing of Coins and Random Error

By now you are probably asking why we seem to be obsessed with the tossing of coins and for that matter with the binomial distribution. Well, for a start the binomial distribution has helped us to calculate the probabilities of the possible events when coins are tossed. Yes, this still requires an explanation of our coin-tossing obsession. Be patient! We have been discussing a rather special case of the expansion: it is where $p = q = 1/2$. This is the situation which exists when a fair coin is tossed. It is also the situation which obtains when random error is analysed. Think about it! For simplicity assume a measurement system where systematic error is absent. Random error is, of course, always present. It causes results to be either greater than or less than the true value. (Keep in mind that systematic error is absent.) What is the probability of having a random error which causes a result to be larger than the true value? Yes, it is exactly the same as the probability of having a random error that leads to a result which is smaller than the true value. This is the nature of random error. It can result in either a positive or negative deviation and both types of deviation are equally probable,

ie $p_{\text{positive deviation}} = p_{\text{negative deviation}} = 1/2.$

Now do you see it? The seeming obsession with the tossing of coins is because the case of either a head or a tail is analogous to the case of either a positive deviation or a negative deviation. We shall drive the point home in the ensuing argument.

2.2.2. Probability, Relative Frequency and Random Error

Let us assume a somewhat idealised measurement procedure. It is one in which two uncertainties, $\pm U_1$ and $\pm U_2$, contribute to the random error. (Keep in mind the significance of the 'plus or minus' sign — it means that the uncertainties have an equal probability of being either positive or negative.) We further assume that the two uncertainties are equal in magnitude*, ie $U_1 = U_2 = U$.

We have two uncertainties which are equal in magnitude $U_1 = U_2 = U$. The ways in which these uncertainties could combine, the resulting values of the random error (the outcomes), the number of ways in which a value can be realised (the number of different outcomes which lead to an event), the probability (and the relative frequency) of the events and the probability (and relative frequency) ratios are given in Fig. 2.2a. below.

Combinations of Uncertainties	Value of Random Error	Number of Ways	Probability or Relative Frequency	Probability or Relative Frequency Ratio
$+U_1 + U_2$	$+2U$	1	1/4	1
$+U_1 - U_2$	0			
		2	2/4	2
$-U_1 + U_2$	0			
$-U_1 - U_2$	$-2U$	1	1/4	1

Fig. 2.2a.   *The combination of two uncertainties of magnitude, U*

* If you are a little upset by this model for random error, you are in good company! In the light of all that was said about the origin of random error in 1.6.2., the thought that only a few uncertainties contribute to the random error and that these uncertainties are fixed in magnitude is a lot to swallow. Why then are we introducing this model? The answer is simple. We wish to make a connection between the tossing of coins and random error and to derive a distribution curve for the values of random error which results from the model. This distribution curve helps us visualise the real world of random error. If this does not satisfy you, then be a little patient: the 'truth' is revealed in 2.2.4.

From the table it is seen that the random error in our measurement system could have values $+2U$, 0, or $-2U$ and that the relative frequency of these values are respectively 1/4. 2/4 and 1/4. The probability ratios, like ratios in general, are useful in visualising a situation. Here they also bring to our attention the fact that combining the two uncertainties has similarity to the tossing-of-a-coin-twice experiment. Take a quick look at either the appropriate probability tree or Pascal's triangle and note the probability ratio for the resulting events when a coin is tossed twice; they are $1 : 2 : 1$. We shall make use of our experience with the tossing of coins as we proceed. However, let us now continue with the matter to hand – the random error which results from two uncertainties. The key information in Fig. 2.2a. can be represented graphically. One representation is in the form of a *bar chart*, Fig. 2.2b.

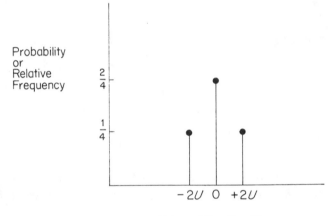

Fig. 2.2b. *Bar chart of the random error resulting from two uncertainties of U*

An alternative representation of this information is in the form of a *histogram*, Fig. 2.2c.

Note that in the histogram, the rectangles have been constructed so that their bases are of equal width and are centred at the appropriate value of the random error. The heights of the rectangles are

determined by the probability of relative frequency of a given value
of random error. This type of histogram is known appropriately as a
probability or relative frequency histogram. It contain a rather important idea, – the use of areas to represent probabilities. A little
explanation is now in order. Let us arbitrarily assign the bases of
each of the rectangles unit width. If this is done, then the rectangle
associated with random error $2U$ has area equal to $1/4 \times 1 = 1/4$.
This value is the probability of a random error of $2U$ occurring; it
is also the relative frequency of this random error. Similarly, the
areas of the other rectangles, $1/2$ and $1/4$ are the probabilities of the
occurrence of random errors 0 and $-2U$ respectively.

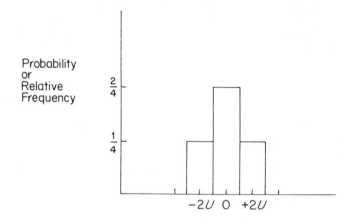

Fig. 2.2c.   *A probability histogram of the random error resulting
from the combination of two uncertainties of magnitude U*

The sum of the areas of all the rectangles in Fig. 2.2c. is $1/4 + 1/2
+ 1/4 = 1$. This represents the probability of obtaining a random
error of either $+2U$, 0 or $-2U$. We must, of course, always obtain
one or other of these values for the random error. Remember, that
always indicates that we have a probability of 1.

We will now set ourselves the task of constructing a probability histogram for the combination of four uncertainties, $\pm U_1$, $\pm U_2$, $\pm U_3$,
and $\pm U_4$, which are equal in magnitude. The relevant information
is presented first in the form of a table, Fig. 2.2d.

Combination of Uncertainties	Value of Random Error	Number of Ways	Probability or Relative Frequency	Probability or Relative Frequency Ratio
$U_1 + U_2 + U_3 + U_4$	$+4U$	1	$\dfrac{1}{16}$	$\dfrac{1}{16}$
$-U_1 + U_2 + U_3 + U_4$				
$+U_1 - U_2 + U_3 + U_4$	$+2U$	4	$\dfrac{4}{16}$	$\dfrac{4}{16}$
$+U_1 + U_2 - U_3 + U_4$				
$+U_1 + U_2 + U_3 - U_4$				
$-U_1 - U_2 + U_3 + U_4$				
$-U_1 + U_2 - U_3 + U_4$				
$-U_1 + U_2 + U_3 - U_4$	0	6	$\dfrac{6}{16}$	$\dfrac{6}{16}$
$+U_1 - U_2 - U_3 + U_4$				
$+U_1 - U_2 + U_3 - U_4$				
$+U_1 + U_2 - U_3 - U_4$				
$+U_1 - U_2 - U_3 - U_4$				
$-U_1 + U_2 - U_3 - U_4$	$-2U$	4	$\dfrac{4}{16}$	$\dfrac{4}{16}$
$-U_1 - U_2 + U_3 - U_4$				
$-U_1 - U_2 - U_3 + U_4$				
$-U_1 - U_2 - U_3 - U_4$	$-4U$	1	$\dfrac{1}{16}$	$\dfrac{1}{16}$

Fig. 2.2d. *Information for the probability histogram in Fig. 2.2e*

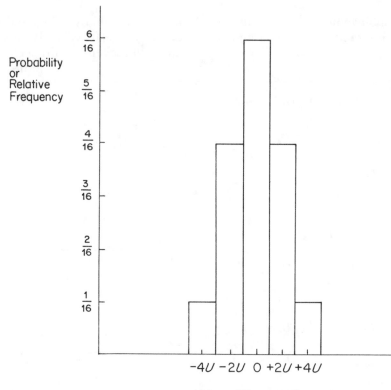

Fig. 2.2e.   *A probability histogram of the random error resulting from the combination of four uncertainties of magnitude U*

The corresponding histogram is given in Fig. 2.2e. Note that the possible values for the random error are 0, $\pm 2U$ and $\pm 4U$. Can you discover why we cannot have random error $\pm 1U$ or $\pm 3U$? Hint: look at the way the uncertainties have to be combined. Have you got it? Good!

Note also that we could have used a Pascal's triangle to obtain the information presented in Fig. 2.2d. The fourth row of the triangle gives us the number of ways (outcomes) in which the values (events) can be realised, ie 1, 4, 6, 4 and 1 for the values $4U$, $2U$, 0, $-2U$

and $-4U$ respectively. The total number of ways is obtained by summing the outcomes, ie $1 + 4 + 6 + 4 + 1 = 16$. Therefore the probability of the values $4U, 2U, 0, -2U$ and $-4U$ are 1/16, 4/16, 6/16, 4/16 and 1/16 respectively.

We now use this probability histogram as the basis for an exercise.

∏   What is the probability of a random error of:

    (i)   either $-4U$ or $-2U$,

    (ii)   either $-2U$ or 0 or $+2U$.

(i)   1/16 + 4/16 = 5/16. We have simply added up the areas of the appropriate blocks in the histogram of Fig. 2.2e.

(ii)   4/16 + 6/16 + 4/16 = 14/16.

We now construct a probability histogram for the combination of ten uncertainties which can all take on values of $\pm U$. We are doing so only because it might be useful for further discussion! Reading off the tenth row of the Pascal's triangle we obtain the numbers 1, 10, 45, 120, 210, 252, 120, 45, 10 and 1.

These numbers give the number of outcomes which lead to the random error values $10U$, $8U$, $6U$, $4U$, $2U$, 0, $-2U$, $-4U$, $-6U$, $-8U$ and $-10U$ respectively. The total number of outcomes is 1024. Therefore the probabilities of the random error values as listed are: 1/1024, 10/1024, 45/1024, 120/1024, 210/1024, 252/1024, 210/1024, 120/1024, 45/1024, 10/1024, and 1/1024, respectively!

The resulting probability histogram appears in Fig. 2.2f.

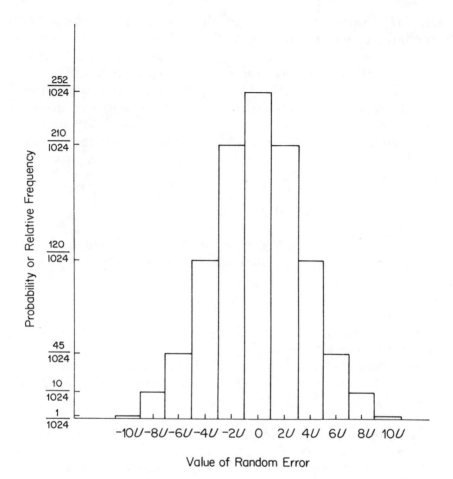

Fig. 2.2f. *Probability histogram for the random error resulting from the combination of ten uncertainties of magnitude U*

Let us comment on the histograms we have constructed. First, note that they are symmetrical about the position of random error equal to zero. Relatedly, the probabilities of a given magnitude of random error are, irrespective of sign, equal. Next, note that the random error value of zero has the greatest probability.*

Finally, as we increase the number of uncertainties contributing to the random error, the resulting histogram becomes increasingly bell-shaped. To help us to visualise this, the probability histogram for ten uncertainties has been redrawn in Fig. 2.2g with the midpoints of the top of each rectangle joined together in a continuous curve. Of course, we have no right to draw a continuous curve for what is

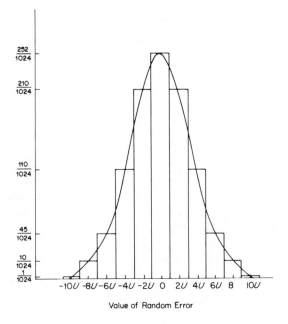

Value of Random Error

Fig. 2.2g. *The probability histogram shown in Fig. 2.2f. with the mid-points of the rectangles joined together*

---

* We deliberately chose to work with an even number of uncertainties. This ensured that the value of random error having the greatest probability was zero. Had we chosen to have an odd number of uncertainties, the symmetry around the central position of the histogram would still be observed, however there would not have been any random error value of zero. For example, with three uncertainties the values of the possible random errors are $-3U$, $-U$, $+U$ and $+3U$ with probabilities of 1/8, 3/8, 3/8 and 1/8 respectively.

a discontinuous distribution. However, we have done it to make a rather important point, this is that the greater the number of uncertainties contributing to the random error, the more the histogram takes on the appearance of a bell-shaped curve. This would not surprise any mathematician. *They know that when* n *in the binomial distribution becomes very large* (n → ∞) *then a continuous curve, which is known as the normal or gaussian distribution results.*

In the next section we shall use the computer to allow you to look at the evolution of the histogram for the binomial distribution towards this distinctive bell-shape. We shall, however, shirk the job of proving formally that it becomes exactly equivalent mathematically to the normal distribution. We just ask you to accept that it does so on the basis of the diagrams you see. It is a matter of some significance that we have now made a first mention of the normal distribution: our concern with it is about to become almost overwhelming!

2.2.3. Shape of the Binomial Distribution – a Computer Exercise

We have drawn the binomial distribution for one or two examples. The computer program below will generate histograms for combinations of any number of our model uncertainties (or, equivalently, for any number of tosses of a fair coin). You should therefore be able to see how the shape passes smoothly over to the characteristic bell-shape of the normal distribution as the number of uncertainties is increased. The program is quite short and should not take you long to type and run.

The program is as follows:

```
10  REM Scaled binomial distribution
20  DIM C(200)
30  PRINT"No of trials (max=200, 0 to stop)";
40  INPUT M
50  IF M>200 THEN 30
60  IF M<1 THEN 250
70  J=INT(M/2)
80  C(J)=30
90  C(M-J)=30
```

```
100  IF M=1 THEN 150
110  FOR I=J-1 TO 0 STEP-1
120      C(I)=C(I+1)*(I+1)/(M-I)
130      C(M-I)=C(I)
140      NEXT I
150  FOR I=0 TO M
160      C(I)=INT(C(I)+.5)
170      PRINT TAB(1);2*I-M;"U";TAB(7);
180      IF C(I)=0 THEN 220
190      FOR K=1 TO C(I)
200          PRINT "*";
210          NEXT K
220      PRINT
230      NEXT I
240  GOTO 30
250  END
```

Type it in and RUN it. In response to the question 'No of trials?' enter 10. Check if the resulting histogram agrees with Fig. 2.2f turned on its side. If not, list your program and examine it carefully for typing mistakes, then rerun. The histogram is scaled so that the central maximum is 30 asterisks in height. (We have assumed that your computer screen shows at least 37 characters on a line. If your screen is narrower or much broader than this, then you can replace the number 30 on line 80 and line 90 by the number of characters in a full screen line less, say, 7.) The heights of all other entries on the histogram are plotted in due proportion to the central peak, rounded to the nearest whole number of asterisks. For 10 trials the output is:

```
-10U
 -8U  *
 -6U  *****
 -4U  *************
 -2U  ************************
  0U  ******************************
  2U  ************************
  4U  *************
  6U  *****
  8U  *
 10U
```

You can run the program for any number uncertainties up to 200. (You could change this limit to, say, 500 simply by substituting that value in place of 200 on lines 20, 30 and 50.) Run the program for progressively larger sets: 1, 2, 3 ... etc. Watch the progressive development of the shape of the distribution towards the bell-shape of the normal curve. Note that the program is not limited to an even number of uncertainties.

If you run the program for more than about 20 trials, the histogram will eventually become too long for a single screen, and the earlier part of it will scroll up rather quickly out of view. Most computers have mechanisms for holding up the scrolling process. For example, on the BBC micro, holding down simultaneously the shift and the control key acts as a brake on the screen display. You might find it useful to look up how your own machine does this, before running the program for a set of, say, 100 trials. Another way of delaying the scrolling is to add a line such as the following.

225 IF (I-INT(I/10)*10)=0 THEN A$ =GET$

This would stop the output after every tenth line. Pressing the space bar would reveal the next ten lines etc. (see Appendix 2.2.4).

How can we check that the program is not cheating? Unless you know a little about binomial coefficients it is not obvious how the histogram heights have been calculated. You can check the validity of the program by making the following additions.

```
 25  DIM D(200)
145  C0=C(0)
155      D(I)=C(I)/C0
232  FOR I=0 TO M
234      PRINT D(I)
236      NEXT I
```

The new lines rescale the coefficients so that the first value is 1. Rerun the program for 10 trials. After the histogram has been shown the numbers 1, 10, 45, 120, 252, 210, 45, 10, 1 should be listed. Check that these agree with Fig. 2.2f and with the tenth row of

Pascal's triangle (Fig. 2.1e). Rerun the modified program for any other chosen number of trials. Compare the reported coefficients with the corresponding row of Pascal's triangle. Use the program for 20 trials, for 50 trials, etc. Notice how large the Pascal coefficients become in the centre of the distribution. Each coefficient gives the total number of ways in which positive and negative uncertainties can be combined to achieve the corresponding outcome.

The binomial distribution has played a useful role in helping us to establish the idea of a distribution and of its relevance to the study of errors of measurement. To do this we had to restrict ourselves to a simplified model of random errors, one where we had in each case a fixed and immutable uncertainty of $\pm U$. We are now ready to abandon our simple model and to face the ordinary situation where not only the sign, but also the magnitude of the random error will vary from one measurement to the next. In taking this step we also abandon the binomial distribution. We have seen how, as the number of trials becomes large, the binomial distribution approaches the normal distribution in form. As we open out our discussion it is the normal distribution which comes to dominate our attention.

2.2.4. Random Error and the Normal Distribution Curve

In the preceding section, we have assumed that a relatively small number of fixed uncertainties contribute to the random error in a measurement. The consequence of this assumption was that random error takes on a binomial distribution; only certain values of random error are possible and the probabilities of these values are given by the binomial expansion. Now it is a matter of experience that if we make a large number of replicate measurements, the results are only discontinuous because of the limitations of the measurement device. For example, two titres might be 20.00 cm^3 and 20.01 cm^3. This does not mean that there are no titres possible between these two results. It is simply that our instrument is incapable of being so discriminating. In principle all values between 20.00 cm^3 and 20.01 cm^3 are possible. With this in mind, let us try a new mathematical model for random error. Let us assume that in a measurement there is an infinitely large number of very small uncertainties of varied magnitude; some people call them elementary errors. Each of these elementary errors contributes, with an equal probability of being either positive or negative, to the random error. What is the consequence of this model of random error? You have guessed it! *Random error is normally distributed*. The probability or frequency distribution for random error takes on the appearance of the curve shown in Fig. 2.2h.

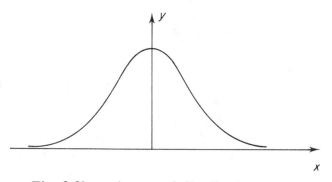

Fig. 2.2h.   *A normal distribution curve*

We cannot prove that random error is distributed in this fashion − we need infinite time to make an infinite number of measurements! Nevertheless, it is generally accepted that a normal distribution is the best model for random error. We shall proceed on this basis. What follows is an introduction to the normal curve and some of its features. Once we have a feeling for the normal curve we shall start applying it to random error and then to analytical chemistry in general.

2.3.   THE NORMAL DISTRIBUTION

There is a mathematical function which is defined by

$$y \;=\; \mathrm{f}(x) \;=\; \frac{1}{\sigma\sqrt{2\pi}}\, \exp\!\left\{ \frac{-(x \,-\, \mu)^2}{2\sigma^2} \right\} \qquad (2.3)$$

(Don't get upset by this equation! Hardly any analytical chemists remember it!) When this function is plotted the curve obtained is as in Fig. 2.3a.

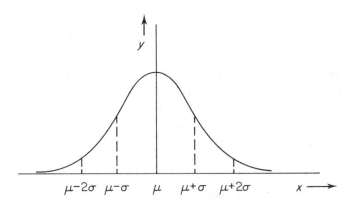

Fig. 2.3a.   *The normal curve represented by Eq. (2.3)*

As can be seen from the plot, the curve is symmetrical about the position $x = \mu$; ie μ is the mean value of x. At $x = \mu$, y is a maximum. At a value of x equal to $(\mu - \sigma)$ and to $(\mu + \sigma)$ the curve has what is known as a point of inflexion; it is here that the curve changes from being concave downwards to concave upwards. The value σ is the *standard deviation*. We can therefore say that the curve has a point of inflexion at plus and at minus one standard deviation from the mean. Interestingly, if tangents to the curve are drawn at the points of inflexion, they cut the x-axis at the point $x = (\mu - 2\sigma)$ and at $x = (\mu + 2\sigma)$. The curve itself meets the x-axis at plus and at minus infinity. In other words the x-axis is its asymptote. This means that when we plot it (unless we have a very large piece of paper!) the curve should not be shown cutting the x-axis.

In order to find the area under the curve, ie the area bounded by the curve and the x-axis, we need to carry out the following integration.

$$\text{Area} = \int_{-\infty}^{\infty} f(x)\,dx = \frac{1}{\sigma\sqrt{2\pi}} \int_{-\infty}^{\infty} \exp\left\{\frac{-(x - \mu)^2}{2\sigma^2}\right\} dx \quad (2.4)$$

Are you getting a little worried? Don't be! We do not need to be skilled with the integral calculus. All that is required is the result of the integration and the result is one that you will never forget; the area is unity. Let us repeat ourselves. The area under a normal curve is 1.0. Keep this in mind as we discuss a few normal curves.

2.3.1. Some Normal Curves

A normal curve is characterised by its values of μ and σ, ie by the mean and its standard deviation. In Fig. 2.3b we see two normal curves which have the same standard deviation but different means.

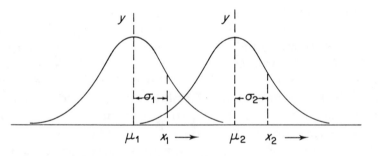

Fig. 2.3b.   *Normal curves with $\mu_1 < \mu_2$ and $\sigma_1 = \sigma_2$*

The two curves have symmetry axes at different positions, but otherwise are identical in form. The areas under the curves are, of course, equal, ie $\text{area}_1 = \text{area}_2 = 1.0$.

In Fig. 2.3c we see two normal curves which have the same mean but different standard deviations.

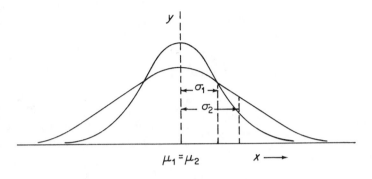

Fig. 2.3c. *Normal curves with $\mu_1 = \mu_2$ and $\sigma_1 < \sigma_2$*

Here we see that the curves have the same symmetry axis but that the curve with the larger standard deviation has a lower maximum and is more spread out. Can you think why it is that the curve with the larger standard deviation has a lower maximum as well as being more spread out? Good! It has to do with the area under a normal curve being equal to one. The curve with the larger standard deviation must be more spread out. Since the area under both curves is the same, ie equal to 1, it follows that the curve with the larger standard deviation must have less height at its maximum point. In Fig. 2.3d we see two normal curves with different means and different standard deviations.

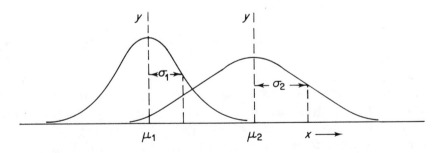

Fig. 2.3d. *Normal curves with $\mu_1 < \mu_2$ and $\sigma_1 < \sigma_2$*

Here the two curves have symmetry axes at different positions and have different widths, but the area under each curve is 1.0.

2.3.2. Areas Under a Normal Curve

As you will soon discover, it is most important to be able to determine the area bounded by a normal curve, the x-axis and given values of x. Consider for example, in Fig. 2.3e below, the area of the shaded region.

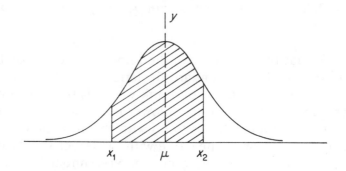

Fig. 2.3e.   *Area under a normal curve between $x = x_1$ and $x = x_2$*

The equation for determining this area is

$$\text{Area} = \frac{1}{\sigma\sqrt{2\pi}} \int_{x_1}^{x_2} exp\left\{\frac{-(x-\mu)^2}{2\sigma^2}\right\} dx \qquad (2.5)$$

Unfortunately, it is far more difficult to carry out the integration than to write the integral down! To make things even worse, having determined the area between x_1 and x_2 for a normal curve of given μ and σ, we have to face the fact that there is an infinite number of normal curves, each characterised by their own values of μ and σ and that we need to be able to determine areas under portions of the normal curve for all of them. Once again, do not be worried! Others have faced this problem and have come up with procedures for determining areas, which are easy to carry out. The first thing that is required is to transform the normal curve in question into what is known as a *standardised normal curve*. What is done is to transform each value of x to a 'z-value' by carrying out the operation in (2.6).

$$z = \frac{x - \mu}{\sigma} \qquad (2.6)$$

This change in the representation of the normal curve is shown in Fig. 2.3f.

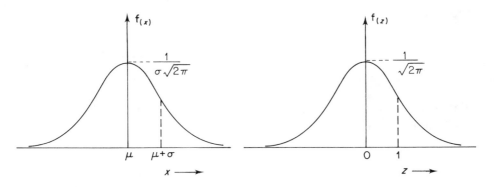

Fig. 2.3f. *(i) The normal curve. (ii) The standardised normal curve*

Note that the standardised normal curve

$$f(z) = \frac{1}{\sqrt{2\pi}} \exp(z^2/2)$$

has mean equal to zero and standard deviation equal to unity. The transformation of $z = (x - \mu)/\sigma$ means that z is dimensionless, it measures deviation from the mean in multiples of the standard deviation. The position $x = (\mu + \sigma)$ in the normal curve becomes $z = (\mu + \sigma - \mu)/\sigma = 1$ in the standardised normal curve.

The total area under the standardised normal curve is, as might be expected, one. Therefore, in summary, *the standardised normal curve has mean equal to zero, a standard deviation of unity and a total area equal to 1.0*. All normal curves can be transformed into standardised normal curves and all standardised normal curves are, as the name implies, the same. Now if someone was to work out the areas under portions of the standardised normal curve and present the results in the form of a table, then the rest of us would be saved the trouble of carrying out the appropriate integration. This has been done. Fig. 2.3g presents the results of carrying out the integration

$$\frac{1}{\sqrt{2\pi}} \int_z^\infty \exp(-z^2/2)\, dz$$

This gives us the area in the 'tail' of standardised normal curves, ie the shaded area as illustrated at the top of Fig. 2.3g.

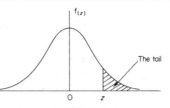

z	.00	.01	.02	.03	.04	.05	.06	.07	.08	.09
0.0	.5000	.4960	.4920	.4880	.4840	.4801	.4761	.4721	.4681	.4641
0.1	.4602	.4562	.4522	.4483	.4443	.4404	.4364	.4325	.4286	.4247
0.2	.4207	.4168	.4129	.4090	.4052	.4013	.3974	.3936	.3897	.3859
0.3	.3821	.3783	.3745	.3707	.3669	.3632	.3594	.3557	.3520	.3483
0.4	.3446	.3409	.3372	.3336	.3300	.3264	.3228	.3192	.3156	.3121
0.5	.3085	.3050	.3015	.2981	.2946	.2912	.2877	.2843	.2810	.2776
0.6	.2743	.2709	.2676	.2643	.2611	.2578	.2546	.2514	.2483	.2451
0.7	.2420	.2389	.2358	.2327	.2296	.2266	.2236	.2206	.2177	.2148
0.8	.2119	.2090	.2061	.2033	.2005	.1977	.1949	.1922	.1894	.1867
0.9	.1841	.1814	.1788	.1762	.1736	.1711	.1685	.1660	.1635	.1611
1.0	.1587	.1562	.1539	.1515	.1492	.1469	.1446	.1423	.1401	.1379
1.1	.1357	.1335	.1314	.1292	.1271	.1251	.1230	.1210	.1190	.1170
1.2	.1151	.1131	.1112	.1093	.1075	.1056	.1038	.1020	.1003	.0985
1.3	.0968	.0951	.0934	.0918	.0901	.0885	.0869	.0853	.0838	.0823
1.4	.0808	.0793	.0778	.0764	.0749	.0735	.0721	.0708	.0694	.0681
1.5	.0668	.0655	.0643	.0630	.0618	.0606	.0594	.0582	.0571	.0559
1.6	.0548	.0537	.0526	.0516	.0505	.0495	.0485	.0475	.0465	.0455
1.7	.0446	.0436	.0427	.0418	.0409	.0401	.0392	.0384	.0375	.0367
1.8	.0359	.0351	.0344	.0336	.0329	.0322	.0314	.0307	.0301	.0294
1.9	.0287	.0281	.0274	.0268	.0262	.0256	.0250	.0244	.0239	.0233
2.0	.02275	.02222	.02169	.02118	.02068	.02018	.01970	.01923	.01876	.01831
2.1	.01786	.01743	.01700	.01659	.01618	.01578	.01539	.01500	.01463	.01426
2.2	.01390	.01355	.01321	.01287	.01255	.01222	.01191	.01160	.01130	.01101
2.3	.01072	.01044	.01017	.00990	.00964	.00939	.00914	.00889	.00866	.00842
2.4	.00820	.00798	.00776	.00755	.00734	.00714	.00695	.00676	.00657	.00639
2.5	.00621	.00604	.00587	.00570	.00554	.00539	.00523	.00508	.00494	.00480
2.6	.00466	.00453	.00440	.00427	.00415	.00402	.00391	.00379	.00368	.00357
2.7	.00347	.00336	.00326	.00317	.00307	.00298	.00289	.00280	.00272	.00264
2.8	.00256	.00248	.00240	.00233	.00226	.00219	.00212	.00205	.00199	.00193
2.9	.00187	.00181	.00175	.00169	.00164	.00159	.00154	.00149	.00144	.00139
3.0	.00135									
3.1	.00097									
3.2	.00069									
3.3	.00048									
3.4	.00034									
3.5	.00023									
3.6	.00016									
3.7	.00011									
3.8	.00007									
3.9	.00005									
4.0	.00003									

Fig. 2.3g. *Area in the tail of a standardised normal distribution*

Let us get some experience in using the table.

∏   What is the area of the tail of a standardised normal curve
        when $z = 1.00$?

Look down the left-hand column until you come to $z = 1.0$. Immediately to the right of this number you will see the number 0.1587;
this is the area of the tail when $z = 1.00$. (If we had wanted the
area of the tail associated with $z = 1.01$ we would have read off the
number 0.1562. Check it!)

With this result in mind, examine the standardised normal curve
below.

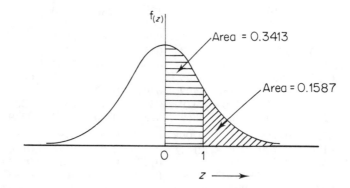

Fig. 2.3h.   *Some areas associated with a standardised
normal curve*

First, note the area in the tail, it is 0.1587. Next, since the area under
half the curve is 0.5000, the area bounded by the curve, the z axis
and $z = 0$ and $z = 1$ is $0.5000 - 0.1587 = 0.3413$. Got it? Yes,
using the table makes the calculation of areas under a standardised
normal curve relatively easy. Let us carry on a little.

∏   What is the area under the normal curve between $z = -1$
        and $z = 1$?

We are attempting to calculate the shaded area indicated in Fig. 2.3i below.

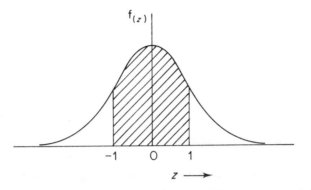

Fig. 2.3i. *The area under the standardised normal curve between $z = -1$ and $z = +1$*

Keeping in mind the symmetry of a normal curve about its mean, it is easy to calculate this area; it is $0.3413 + 0.3413 = 0.6826$. This, in other words, means that 68.26% of the total area under a standardised normal curve lies between $z = -1$ and $z = +1$.

It is time for a SAQ.

SAQ 2.3a
What percentage of the total area under a standardised normal curve lies between $z = -2$ and $z = +2$?

SAQ 2.3a

Be sure to read the response to SAQ 2.3a before continuing.

We are now relatively proficient at obtaining the area in the tail of a standardised normal curve from the table and at using this area to calculate other areas. Now let us return to the reason why we need a standardised normal curve. Let us say that we have a normal curve whose mean, μ, is 175 and whose standard deviation, σ, is 25. We have sketched the normal curve below in Fig. 2.3k.

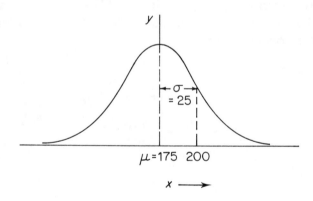

Fig. 2.3k.   *The normal curve with $\mu = 175$ and $\sigma = 25$*

Let us now ask a question about this normal curve. What is the area under the curve from $x = 175$ to $x = 200$?

We will start by transforming the normal curve into a standardised normal curve. When x is 175 in the normal curve, z in the standardised normal curve is $(x - \mu)/\sigma = (175 - 175)/25 = 0.$

When x is 200, z is $\dfrac{x - \mu}{\sigma} = \dfrac{200 - 175}{25} = 1$

The area bounded by $z = 0$ and $z = 1$ has previously been calculated to be 0.3413. This means that in the normal curve under examination, the area bounded by $x = 175$ and $x = 200$ is also 0.3413. Would you like to state the area under the normal curve bounded by $x = 150$ and $x = 200$? Yes, it is easy! The answer is 0.6826.

∏ A normal curve has $\mu = 100$ and $\sigma = 15$. What is the area under the curve bounded by $x = 75$ and $x = 140$?

First we will represent the problem in a sketch.

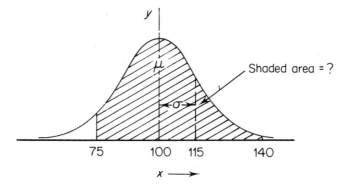

Fig. 2.31. *The area between $x = 75$ and $x = 140$ for a normal curve with $\mu = 100$ and $\mu = 15$*

We will now transform this normal curve into a standardised normal curve.

When $x = 75$, $z = \dfrac{x - \mu}{\sigma} = \dfrac{75 - 100}{15} = \dfrac{-25}{15} = -1.67$

When $x = 100$, $z = \dfrac{x - \mu}{\sigma} = \dfrac{100 - 100}{15} = 0$

When $x = 140$, $z = \dfrac{x - \mu}{\sigma} = \dfrac{140 - 100}{15} = 2.67$

This means that we need to calculate the shaded area under the standardised normal curve below.

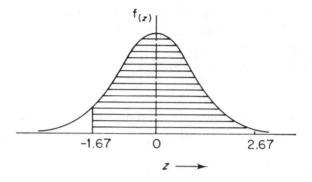

Fig. 2.3m.   *The area between z = -1.67 and z = $+2.67$ for a standardised normal curve*

Let us start by calculating the area in the tail when $z = 2.67$. From the tables the area in the tail for $z = 2.67$ is 0.0038. This means that the area from $z = 0$ to $z = 2.67$ is $0.5000 - 0.0038 = 0.4962$.

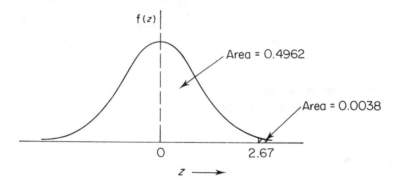

Fig. 2.3n.   *The area between z = 0 and z = $+2.67$*

Now we will calculate the area in the tail when $z = 1.67$. What, you might be asking, has happened to the minus sign? Well, keep in mind the symmetrical nature of the normal curve. If we know the area in tail when $z = +1.67$, then we immediately know the area in the tail (the shaded area indicated below) for $z = -1.67$.

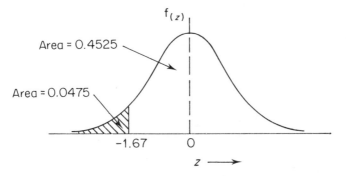

Fig. 2.3o. *The area between z = 1.67 and z = 0*

From the table, for $z = 1.67$ the area in the tail is 0.0475. This means that the area bounded by $z = -1.67$ and $z = 0$ is 0.5000 − 0.0475. Our problem calls for us to calculate the area under the curve from $z = -1.67$ to $z = +2.67$. This is obtained by adding the area bounded by $z = 1.67$ and $z = 0$ to the area bounded by $z = 0$ and $z = 2.67$. This is 0.4525 + 0.4962 = 0.9487, ie 94.87%.

We can now state that for the normal curve with $\mu = 100$, $\sigma = 15$, the area bounded by $x = 75$ and $x = 140$ is 0.9487 or 94.87% of the total area. This problem is similar in nature to many encountered in statistics in general and in analytical chemistry in particular. You are strongly advised to go through it again. If you have followed it without much difficulty, then you have mastered one of the most difficult parts of introductory statistics. Congratulations!

An SAQ is in order.

SAQ 2.3b

A normal curve has $\mu = 800$ and $\sigma = 10$. What is the area under the curve bounded by $x = 805$ and $x = 830$?

SAQ 2.3b

2.4. THE NORMAL CURVE AND RANDOM ERROR

Now is a good time to take stock of our progress. We have famil-
iarised ourselves with the normal curve and know that it is characterised by its mean, μ and its standard deviation, σ. The total area
under a normal curve is 1.0. We have learned how to calculate the
area bounded by the curve, the x axis and two given values of x.
The procedure involves first transforming our normal curve into a
standardised normal curve by use of the operation $z = (x - \mu)/\sigma$.
The z values associated with the given x values are calculated and
the table is then used to obtain the 'area in the tail' associated with a
z value as a fraction of the whole area. This permits the calculation
of the required area. A question now arises. Why have we taken so
much trouble to learn how to calculate areas under portions of a
normal curve? In order to answer this question, let us go back to
where we first introduced the normal curve; it was when we stated
our belief that random error is normally distributed. What does this
mean? It means that if, in some determination we carried out a very
large number of replicate measurements and presented the results in
the form of a frequency distribution histogram, the histogram would
match the characteristic bell-shape of the normal curve. Now, since
the scatter of results about their mean is attributable to random
error, we can say that random error is normally distributed. The
frequency distribution curve for measurements with random error,

like any other normal curve, has a mean, μ and a standard deviation, σ. It also, like any normal curve, has some other interesting characteristics which are based on areas under the curve. In Fig. 2.4a we see a frequency distribution curve for measurements with random error.

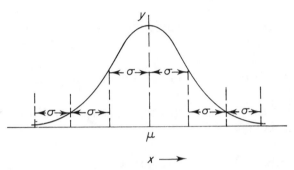

Fig. 2.4a. *A frequency distribution curve for measurements subject to random error*

Transforming this normal distribution curve into a standardised curve we obtain Fig. 2.4b.

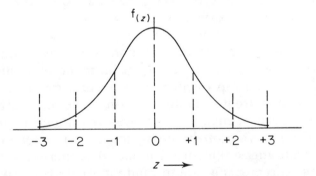

Fig. 2.4b. *The standardised form of the curve in Fig. 2.4a*

We can now use the z values to look up areas in the tail of the curve and hence calculate other areas. Let us do so and make some comments which, as you will soon discover, are of great significance.

When $z = +1$, the area in the tail is 0.1587. The area bounded by $z = -1$ and $z = +1$ is 0.6826. Returning to our original distribution curve for a very large number of measurements, we see that the area under the curve in the range $\mu \pm \sigma$ is 0.6826, ie 68.26%.

Now for the comments!

The fact that the area under the curve in the range $\mu \pm \sigma$ is 0.6826 means that:

(*a*) 68.26% of all measurements lie in the range $\mu \pm \sigma$,

(*b*) the probability of a single measurement falling in the range $\mu \pm \sigma$ is 68.26%.

At last a number of topics previously introduced are coming together! We introduced probability concepts and probability histograms and showed how areas could be used to represent probabilities. We talked about 'long-run' relative frequencies and their equivalence to probabilities. Our histograms then started to be called either probability or relative frequency histograms. We used the binomial distribution as a model for random error and plotted histograms for random error. As with other probability histograms, the total area of the random-error histograms was equal to 1 – recall that the probability of all possible events is 1.

We then stated our belief that random error was normally distributed: this meant that a very large number of replicate measurements could be represented by a normal curve. The normal curve represents a frequency distribution curve for a large number of replicate measurements. The total area under a normal curve is 1. The normal curve is a probability distribution curve for a large number of replicate measurements. The fact that there is variability in replicate measurements is due to random error. The probability or frequency distribution curve for measurements can also be thought of as a 'random-error curve'. Our ability to calculate areas under the normal curve enables us to make statements about measurements which differ in value because of random error. These statements can be either about long-run relative frequencies, eg 68.26% of all measurements lie in the range $\mu \pm \sigma$, or about probabilities, eg the probability of a single measurement lying in the range $\mu \pm \sigma$ is 68.26%.

Let us work out some more relative frequencies and probabilities. When $z = 2$, the area in the tail of the standardised normal curve is 0.0228. Therefore the area under the curve bounded by $z = 0$ and $z = 2$ is 0.4772. The area bounded by $z = -2$ and $z = +2$ is therefore 0.9544. The significance of this for our measurements is that 95.44% of all measurements lie in the range $\mu \pm 2\sigma$. The probability of a single measurement lying in the range $\mu \pm 2\sigma$ is 95.44%.

Similar reasoning shows that 99.74% of all measurements lie in the range $\mu \pm 3\sigma$. The probability of a single measurement lying in the range $\mu \pm 3\sigma$ is 99.74%.

If we know the probability of a single measurement lying within a given range, we obviously know the probability of a single measurement lying outside of the given range. For example, there is a 95.44% probability that a single measurement is in the range $\mu \pm 2\sigma$: the probability of a measurement falling outside of this range is (100.00 − 95.44)% = 4.56% (or as a decimal 0.0456). A related question is what is the probability of a single measurement being greater than $\mu + 2\sigma$? A moment's thought should convince you that the probability is 0.0456/2 = 0.0228 (or as a percentage, 2.28%). The probability of a single measurement being less than $\mu - 2\sigma$ is also 0.0228. A further moment's thought should convince you that this figure (0.0228) is simply the area in a tail when $z = \pm 2.00$. Keep this in mind as we tackle the following problem.

∏   What range, symmetrical about the mean, embraces 95% of all measurements for a measurement procedure with mean μ and standard deviation σ?

We will express the range in the usual fashion, ie as the mean plus or minus a certain number of standard deviations. How do we get the appropriate number of standard deviations? The answer is – by knowing the appropriate z value: for 95.44% of all measurements z was 2 and the range is $\mu \pm 2\sigma$. We now want the range which embraces 95% of all measurements. This means that we want to know the z value which leaves 5% of all measurements in *both* tails. In one tail we will have 2.5% of all measurements; as a decimal this is 0.025. If we consult the table for the z value which gives an area in the tail equal to 0.025, we find that the answer is $z = 1.96$. This immediately gives us the range (symmetrical about the mean) which embraces 95% of all measurements; it is $\mu \pm 1.96\sigma$. Whilst we are at it, let us not miss the opportunity to make some equivalent comments about the measurements. There is a 95% probability that a single measurement lies in the range $\mu \pm 1.96\sigma$ and a 5% probability that a single measurement falls outside of the range. There is a 2.5% probability (0.025 as a decimal) that a single measurement is greater than $\mu + 1.96\sigma$ and an equal probability that a single measurement is less than $\mu - 1.96\sigma$. Repetition is a favourite method of teachers!

2.4.1. Further Comments on the Normal Distribution Model

So far, in our application of the normal curve to measurements and random error, we have dodged a few points. We will deal with them now. The first is that we have not drawn attention to the fact that the normal distribution model for random error suggests that some measurements could have an associated infinite random error! Obviously, this is a limitation in the model! This does not mean that we cease to believe in its almost general validity. The model suggests that only 0.27% of measurements fall outside the range $\mu \pm 3\sigma$ (0.006% outside the range $\mu \pm 4\sigma$). We can accept this sort of statement without accepting that some measurements will have infinitely high random error. As ever with models there, are obvious limitations.

The second point is that with a normal distribution model of random error we are unable to assign a probability to a given value of a measurement occurring. We use areas to estimate the probability of a single measurement lying in a range, eg in the range $\mu - \sigma$ to $\mu + \sigma$. However, we cannot estimate the probability of a single measurement having the value of say, μ. Why not? Because we need an area to estimate a probability and the infinitesimally narrow line representing the position $x = \mu$ has no width!

We could benefit from an example.

∏   A direct reading pH-meter under rigorously controlled con-
     ditions, was used to make a very large number of pH mea-
     surements on a solution (Don't ask why!). Computer analysis
     of the measurements showed that the mean pH was 5.000 and
     the standard deviation was 0.005 pH unit.

     (*i*) What is the probability of a single measurement giving
            a pH value greater than 5.010?

     (*ii*) Give a range, symmetrical about the mean, within which
            95% of all measurements lie.

(*i*) We will represent the problem with a sketch.

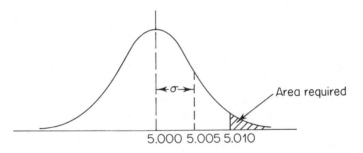

Fig. 2.4c.   *Distribution followed by a very large number of replicate
pH measurements, showing the tail above pH = 5.010*

When $x = 5.010$, $z = (5.010–5.000)/0.0052 = 2$. The area in a
tail for $z = 2$ is 0.0228 or 2.28%. There is a 2.28% probability
that a single measurement will have a pH greater than 5.010.

(*ii*) The area in two tails is to be 5% of the whole. (The area between the tails is 95% of the whole.) The area of one tail is
2.5% of the whole (or as a decimal 0.025 – see Fig. 2.4d). The
z value for an area of 0.025 is 1.96. The required range is thus
$\mu \pm 1.96\sigma$, ie the pH range is 5.000 $\pm$ 1.96 (0.005), or from
4.990 to 5.010.

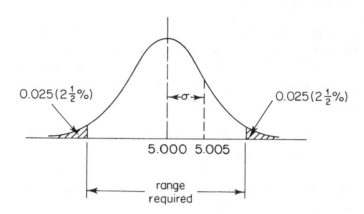

Fig. 2.4d.   *Finding the range in which 95% of measurements lie*

2.5.   FINDING THE AREAS BY COMPUTER

Instead of continually consulting the table in Fig. 2.3g to find values for the area of the tail of the standardised normal curve, we can use a computer program to do the job for us. We shall be meeting a number of other tables as we go on and it will be useful to build into our computer package the facility to automate the kind of calculations we have been looking at. This marks the second stage of building our main statistics package. In Part 1 of this Unit we constructed a program "STAT1", which obtained the mean and standard deviation of a given set of measurements. The calculations we wish to do here are, for the moment, quite independent of the previous work. Nevertheless, we ask you to type in the new program as an addition to "STAT1". The reasons for working this way will become clear later when we develop the package further.

Load program "STAT1" into your computer, then type in the following additional lines:

```
  10  READ D1,D2,D3,D4,D5,D6
  20  DATA 4.98673470E-2,2.11410061E-2,3.2776263E-3,
                        3.80036E-5,4.88906E-5,5.3830E-6
 510  PRINT"Give z ";
 520  INPUT Z
 530  GOSUB 1000
 540  PRINT"Prob of EXCEEDING this value is ";
                        INT(P*100000 + .5)/1000;"%"
 550  GOTO 9999
1000  REM One-tail normal: z - ->prob
1010  Z9 = ABS(Z)
1020  P  = 1 + Z9*(D1 + Z9*(D2 + Z9*(D3 + Z9*(D4 + Z9*
                        (D5 + Z9*D6)))))
1030  P = .5/P↑16
1040  P = SGN(Z)*(P-.5) + .5
1050  RETURN
```

Note: for clarity's sake, we have broken the long lines 20, 540 and 1020 into two. They should, however, be typed into your computer as single continuous lines, with no intervening spaces or carriage returns.

You will need to take special care to type lines 20 and 1020 correctly. This program replaces the table in Fig. 2.3g by a complicated formula involving the parameters D1 to D6. The values of these parameters are read from line 20 and the formula itself is applied on lines 1010 to 1040. Line 1020 could have been written without brackets as follows.

```
1020  P = 1 + D1*Z9 + D2*Z9↑2 + D3*Z9↑3 + D4*Z9
                        ↑4 + D5*Z9↑5 + D6*Z9↑6
```

The form given earlier is more efficient for the computer to apply.

RUN the program. When it asks "Give *z*", enter the value 1. The computer should respond:

Prob for EXCEEDING this value is 15.866%.

If it does not give this result check for mistyping, especially of lines 20 and 1020. If you still have difficulties consult Section 1.9.1.

Let us briefly describe what the program has done. It has accepted your input z-value through line 520. The instruction GOSUB 1000 on line 530 acts in the same way as would GOTO 1000, but with one notable difference. The program jumps to line 1000 and the following few lines compute the tail area P corresponding to the z-value of your input. When the RETURN statement is met, however, on line 1050, control is returned to line 540 (the line after the original GOSUB), and the answer is printed out. We shall make use of GO-SUB again in future additions. The GOTO 9999 instruction on line 550 is a temporary measure to cause the program to stop without completing also the original task achieved by "STAT1". (The output from "STAT1" all comes from later line numbers which are now skipped past.) In due course we shall build a 'menu' section into the program to allow you to specify which calculation you want to do in any given RUN. That's for later, though.

Now let's get back to considering the result you have got. You gave $z = 1$ and were told that the probability of EXCEEDING that value is 15.866%. Compare this with Fig. 2.3g. For $z = 1$ the table quotes a probability of 0.1587, or 15.87%. This is in full agreement with the computer result, though rounded to one significant figure fewer.

Try to catch the program out (or the table?). Re-run the program, giving various z-values and check the program's answer against the table. Search for discrepancies. We failed to find any. You now have an alternative to looking up the table. In fact the program is more powerful, in that you can generate values which are not tabulated.

∏   (i)   Find the area of the tail for $z = 2/3$, as exactly as you can,

       (ii)   Fill in the missing row for $z = 3.01, 3.02 \ldots 3.09$,

       (iii) Examine the probability of making very large random errors. Obtain the values for $z = 2, 3, 4, 5, 6$.

(i)   For $z = 0.6666667$, the program gives a probability of 25.249%. This is in pretty well exact agreement with the answer obtained by interpolating in the table. The table gives 25.46% for $z = 0.66$ and 25.14% for $z = 0.67$, a difference of 0.32%.

To interpolate we subtract two thirds of this difference from the first value, giving $25.46 - 0.21 = 25.25\%$.

(*ii*) For $z = 3.05$, for instance, we get a probability of 0.114%. This is a little more accurate than the value obtainable from the table by interpolation. From the table, one would estimate, half way between the values quoted for $z = 3.0$ and $z = 3.1$. That gives a value of 0.116%.

(*iii*) The PRINT statement on line 540 truncates the probability to three places of decimals (as a percentage). Values quoted for $z = 5$ and $z = 6$ are reported as zero. If, however, we add a temporary line

545 PRINT P

we get the value untruncated. In this way we got the following results.

z	2	3	4	5	6
probability	2.3E-2	1.3E-3	3.2E-5	3.0E-7	1.3E-9

Remove line 545 again once you have completed this exercise.

∏ Suppose you can carry out a given titration with a standard deviation of 0.03 cm^3. Suppose you carry out 100 replicate titrations every day, 5 days a week, 50 weeks a year over a career spanning 40 years (always with the same equipment and with the same stock solutions at the same temperature). How many times are you likely to make a random error as big as 0.18 cm^3?

Remember we are talking of random errors – we rule out the possibility of ever making a blunder! We are looking for an error in which $z = 0.18/0.03 = 6$, or more. Now the program tells us that the area of the tail of the normal distribution beyond $z = 6$ is 1.3E-9. We must double this to take account also of the tail below $z = -6$; ie we must count underestimates as well as overestimates. The upshot is that our normal distribution model gives a total probability of 2.6E-9, ie we expect the event to occur on average a little more frequently than twice in every thousand million measurements, or about once in every 400 million repeats.

Our life's work will involve us in $100 \times 5 \times 50 \times 40$ titrations, a total of 1 million in all. The odds are, therefore, that we will *never* even once make a random error of 6 standard deviations or more! If we recruit 400 colleagues and we all spend our working lives repeating this titration, then the normal distribution model predicts that on average we would expect that the event would occur just once amongst all our accumulated work.

In Section 2.4.1 above we described it as a weakness of the normal distribution model that it predicated that there was a finite probability of making an unreasonably large random error. Finite this probability may be, but we hope that by now you are convinced that for practical purposes it is negligible!

So far we have used the program simply to do the job of looking up the table. We have to give it a value of z each time, ie we have to utilise the standardised normal distribution. Clearly we could make the program do this work also.

Consider for instance the data we had in SAQ 1.5b, where we gave 12 replicate measurements from an analysis as follows.

Final titre/cm^3: 27.86, 27.68, 28.02, 28.05, 27.60, 27.55, 27.03, 27.76, 27.73, 27.75, 27.77, 27.53.

We commented that these readings were representative of a series where the 'true' experimental result is 27.66 cm^3 and the technique as applied had a standard deviation of 0.32 cm^3. Let us examine how representative the reported data are of such an experiment.

Type the following new lines into your program.

```
 50  READ M9,S9
390  PRINT"- - - - - - - - - - - - - - - - - - - - - - - - - - - - - - - - - - - - - - -"
400  PRINT"1=normal tail   2=z tail   3= stop";
410  INPUT I
420  IF I>1 THEN 500
```

```
 430  PRINT"Mean = ";M9;"   st.dev. = ";S9
 440  PRINT"Give value to assess";
 450  INPUT V
 460  Z=(V-M9)/S9
 470  PRINT"Gives z = ";INT(Z*1000+.5)/1000
 480  GOTO 530
 500  IF I>2 THEN 9999
 550  GOTO 390
3900  DATA 27.66,0.32
```

The 'true' experimental mean and standard deviation are given by M9 and S9 respectively. These are read on line 50 from the new DATA line 3900. We now have introduced the beginnings of a 'menu'. Lines 390 to 410 allow you to choose between two calculations, the one we are proposing to tackle now, labelled 'normal tail', and what we were doing above, '*z* tail'. Notice that line 550 has been changed. When a calculation is completed control is now sent back to the menu.

When the program is RUN, if we enter 1 for the choice of option, the values of the 'true' mean and standard deviation are written out (line 430) and you are asked to give a value to be 'assessed'. Whatever value you type in is INPUT via line 450. The program then calculates and prints out the area in the upper tail corresponding to the value specified.

Test your program by running it, with a value of 27.66. As this co-incides with the mean, the probability calculated for the tail should be 50.000%. If you do not get this answer you will need to check your altered program to find what you have mistyped. Once you are happy the program is working properly, save it. We shall call it "STAT2". We shall be using this program again later, when we shall add still further to it.

∏   Use the modified program to find the probabilities of making measurements:

    (*i*)   above 28.00 cm^3,

    (*ii*) below 27.00 cm^3,

(*iii*) in the range 27.50 - 27.80 cm^3,

Comment in the light of the actual set of measurements reported above.

The answers are very easily obtained!

For (*i*) you should find there is a probability of 14.4% of obtaining a value over 28.00 cm^3. That is about a 1 in 7 chance. In fact there were 2 such values in the series of 12 reported, an incidence of 1 in 6. Pretty close! However, before getting carried away by the success of this look what would have happened if we had taken only the first four values. We would still have 2 values over 28.00, ie 2 out of 4, or 50%. This seems rather out of line, but it does happen. We must always remember that the probabilities we calculate give us the relative frequency of an occurrence *in the long run*. The smaller the number of readings we take, the more likely is it to be unrepresentative.

For (*ii*) you should find there is a probability of 98.04% *of exceeding* 27.00 cm^3. Therefore the chance of getting a value below 27.00 cm^3 is (100 − 98.04) = 1.96%. This is about one chance in 51. Not unexpectedly there was no such value in our set of 12 readings (but do notice that there was a near miss, at 27.03 cm^3).

For (*iii*) you should find that the probability of obtaining a result above 27.50 cm^3 is 69.15%, whilst that of obtaining a value above 27.80 cm^3 is 33.09%. Hence the probability of obtaining a value between 27.50 and 27.80 cm^3 is (69.15 − 33.09) = 36.06%. This is equivalent to a frequency of occurrence of about 4 times in 11. In the set of 12 measurements reported above, 8 out of 12 actually lie in this range. As we keep saying, such things can happen. It can be shown that obtaining so many of our set of 12 readings in this range is something which should only happen to us 3 times in a hundred. Well, this is one such time. It is all a question of how extreme an occurrence is. It can be shown that the chance of getting as many as 10 out of 12 in this range is only one in a thousand. That has not happened to us, and it rarely will!

We could test other predictions of the normal distribution curve for our experiment against the 12 results actually reported. Usually we find the results fit quite closely with what is typically expected. If we dream up enough different tests, however, we should expect to find a few respects in which our particular set of results is unusual. Test (*iii*) above is a case in point.

For reference, we give below a full listing of "STAT2". You could use this to cross-check your own version if you corrupt it in any way.

```
 10 READ D1,D2,D3,D4,D5,D6
 20 DATA 4.98673470E-2,2.11410061E-2,3.2776263E-3,
                 3.80036E-5,4.88906E-5,5.3830E-6
 50 READ M9,S9
 80 DIM X(30)
100 READ N
110 A=O
120 FOR I=1 TO N
130    READ X (I)
140    A=A+X(I)
150    NEXT I
160 M=A/N
170 A=O
180 FOR I=1 TO N
190    A=A+(X(I)-M)↑2
200    NEXT I
210 S=SQR(A/(N-1))
220 R=100*S/M
390 PRINT"- - - - - - - - - - -  - - - - - - - - - - - - - - - - - - - - - - -"
400 PRINT"1=normal tail   2= z tail   3= stop";
410 INPUT I
420 IF I>1 THEN 500
430 PRINT"Mean = ";M9;"  st. dev. = ";S9
440 PRINT"Give value to assess";
450 INPUT V
460 Z=(V-M9)/S9
470 PRINT"Gives z = ";INT(Z*1000+.5)/1000
480 GOTO 530
500 IF I>2 THEN 9999
```

```
510  PRINT"Give z ";
520  INPUT Z
530  GOSUB 1000
540  PRINT"Prob of EXCEEDING this value is ";
                              INT(P*100000+.5)/1000;"%"
550  GOTO 390
1000 REM One-tail normal: z --> prob
1010 Z9=ABS(Z)
1020 P =1+Z9*(D1+Z9*(D2+Z9*(D3+Z9*(D4+Z9*
                              (D5+Z9*D6)))))
1030 P=.5/P↑16
1040 P=SGN(Z)*(P-.5)+.5
1050 RETURN
1640 REM Show sample mean and st dev
1650 PRINT "No of values = ";N
1660 PRINT "Sample mean = ";INT(M*1000+.5)/1000
1670 PRINT "Est. st. dev.= ";INT(S*1000+.5)/1000
1680 PRINT "%Rel.st.dev. = ";INT(R*1000+.5)/1000
3900 DATA 27.66,0.32
4000 DATA 12
4010 DATA 27.86,27.68,28.02,28.05,27.60,27.55,27.03,27.76,
                              27.73,27.75,27.77,27.53
9999 END
```

2.6.   A TASTE OF THINGS TO COME!

If the normal distribution curve was a model for random error only, it would still be worthwhile learning something about it. However, it is believed that the normal curve describes the distribution of many other variables. Two widely differing examples can be given. It is believed that IQ (ie how well people do on IQ tests!) is normally distributed. There is evidence that the weights of bags of crisps (made by the same manufacturer) are normally distributed. You might ask, 'What have these examples got to do with analytical chemistry?' Before we start to hint at the relevance of these examples to analysis, let us make one thing clear. Not all distributions are normal. Take a quick look at any introductory Physical Chemistry text-book! You will see that the famous Maxwell–Boltzmann distribution of molecular energies is not a normal distribution. It is skewed to the side of lower energies. Check it! Now to return to our examples. Let us consider the example of bags of crisps! Don't laugh! You might be amazed at the amount of high technology that goes into the bagging of crisps. The name of the game is to get the right weight of crisps into the bag, ie if the bag states the contents to be 30.0 g then there must be as close to 30.0 g in the bag as is possible. Of course, it could be arranged to put about 35 g of crisps into each bag. This would guarantee the 30 g, but it would be a waste of money for the manufacturer. Think of how many bags of crisps you consume in a year! Modern bagging technology – computer controlled – can put 30.0 g in each bag with a standard deviation of 0.25 g. With this in mind, let us state a problem.

∏ A manufacturer claims that the mean weight of his bags of crisps is 30.0 g with a standard deviation of 0.25 g. What is the probability that a single bag of crisps contains less than 29.50 g?

First, we will assume that the weights are normally distributed. We will then calculate a z value associated with the weight 29.50 g.

$$z = \frac{x - \mu}{\sigma} = \frac{29.50 - 30.00}{0.25} = -2.0$$

Immediately, we know the area in the tail. It is 0.02275. This is the probability that a bag of crisps has a weight less than 29.50 g. In percentage terms it is 2.275%. Did you follow this? If so, let us continue. What is the probability of taking, at random, two bags of crisps from the manufacturers warehouse and finding that both of them weighed less than 29.50 g?

Recalling the tossing of coins, we know that the multiplication law is in order. The probability is $0.02275 \times 0.02275 = 0.00052$ or 0.052%. Now let's say that you were to carry out this experiment and found that indeed each of the two bags of crisps did weigh less than 29.50 g. What comment would you make? The manufacturer was cheating? The expensive technology wasn't up to specification? Your balance wasn't properly calibrated? Why make any comment at all? The answer to this is simple. You have used the normal distribution curve to show that the probability of having two bags, chosen at random, each contain less than 29.50 g is very small. Hence the event is most unlikely. It could happen – but! Have you followed this? We have used statistics in a situation that is quite common in industry. A sample is taken and as a result of measurements on the sample decisions have to be made. Are you starting to get a feel for statistics? The problem just worked gives a taste (!) of things to come. From now on it should be downhill all the way! Well, nearly downhill all the way.

Objectives

As a result of completing Part 2 you should feel able to:

● apply confidently an understanding of the connection between the probability of any particular occurrence and the relative frequency with which it is expected to occur 'in the long run';

● apply, as appropriate, the addition and multiplication laws of probability;

● be happy with the idea that uncertainties in experimental measurement processes lead to a *distribution* of error in the results obtained;

● sketch roughly the shapes of the binomial distribution and the normal distribution curves and recognise the relevance of the normal distribution to the random errors typical in measurements in analytical chemistry;

● recognise that the probability of obtaining a result in a certain range is given by a corresponding area under a normal distribution curve;

● recognise the usefulness of the standardised normal distribution curve, and be able to derive the value of the parameter z of the standardised normal distribution which corresponds to value x in another normal distribution of known mean and standard deviation;

● make use of the table of z (Fig. 2.3g), giving one-tail areas under the standardised normal distribution;

● for a measurement process where the true mean and standard deviation are known, use either the table of z or your version of program "STAT2" to deduce the probability that the result of a measurement would lie in any given interval.

3. Samples, Estimation and Hypothesis Testing

Introduction

It is difficult to overestimate the contribution made by analytical chemistry to the development of science and technology, to our understanding of the natural and man-made world, and to the health and well-being of modern societies. The ability of the 'lab boys' to discover the constituents present in a sample (qualitative analysis) and to determine how much of each constituent is present (quantitative analysis) is generally accepted. It is part of popular culture. Almost everyone believes that Sherlock Holmes, with his retorts, condensers, and Bunsen burners, will ultimately find the material clue. Furthermore, when Mr. Spock activates some futuristic sensors to ascertain whether the atmosphere of a distant planet is safe for Earthlings (and Vulcans), there are few who cry out 'Space traveller, beware!'.

The general public may be cynical about the activities of, for example, politicians, doctors and lawyers: they seem however, to have faith in analytical chemists! We analytical chemists are, of course, cynical about ourselves. We know that all measurements are subject to error, that the information gained from analytical determinations is never complete, and consequently that any action taken in the light of such determinations must always carry with it some doubt. An important factor in this thinking is the knowledge that we rarely analyse the whole of a material under investigation: we analyse a sample of the whole. Often this sample is a very small fraction of the material under investigation. A lake is to be investigated for its sulphate content, a few litres are collected for analysis; a large quantity of a high-performance alloy arrives at a jet-engine factory, a few hundred grams are sent to the quality-control laboratory so that the amount of each of several deleterious impurities may be determined. We are all familiar with the way in which we generalise about the nature of a material in the light of our analysis of a sample of it. We also know that such generalisations may have major implications. This Part of the Unit will familiarise you with the statistical ideas which support the process of generalising from the analysis of samples. It will also introduce you to statistical procedures which assist in the making of decisions in the light of analytical determinations. This material is at the very heart of the use of statistics in analytical chemistry. You will, on a few occasions, have to struggle a little to visualise what is going on. The struggle is worthwhile. When you have mastered this Part of the Unit, you will rarely be put off by statistical arguments again.

3.1. POPULATION, RANDOM SAMPLES AND SAMPLE SIZE

We have, over the years, become quite familiar with opinion polls. It is worthwhile reviewing the purpose of such polls and the vocabulary they use. They start with the purpose of *estimating* how a well-defined part of society feels about a certain issue. 'Which political party should govern the country?' and 'Should smoking be banned in public buildings?' are typical of the questions asked in such polls. The questions asked help to define the part of society to be polled. For example, it makes sense when trying to estimate the voting preferences of a society to poll only potential voters. Once the appropriate part of society is defined, it is given a name: it is called the *population*. *A population comprises the set of all possible observations of the type with which we are concerned*. For the voting-preference poll, the population is the opinions of all the potential voters. Since the number of such people is large (in the UK over 30 million) it is impractical to ask everyone their opinion. Because of this, the pollsters are forced to *sample* the population. *A sample is a subset of a population.* Instead of asking the opinions of millions of people, the pollsters ask only a few thousand people their opinions. They do so in the belief that the relative frequencies of the various opinions which occur in the sample will be a good estimate of opinion in the population. The number of people actually polled is called the *sample size. The population comprises the set of all possible observations; the sample is a subset of the population and has a size which is determined by the number of observations made.* Immediately, we see a problem. If the results of observations made on a sample are to be used to generalise about the population as a whole, then we must be terribly careful that the sampling procedure is fair, ie that it does not cause bias in the estimate. The best way to avoid such bias is to take what is known as a *random sample.* What is meant by the term random sample? *It is a sample taken in such a way that every observation in the population has an equal probability of being included.* It is relatively easy to define a random sample; it is much more difficult to arrive at one in practice. Some people, including a few chemists, believe that it is a sample taken in a haphazard fashion. It isn't! Such a sampling procedure has many hazards! It can result in a sample which is not representative of the whole population. Rather than taking a sample in a haphazard fashion, what is required is a well-designed sampling programme. Such programmes have been designed for sampling the voting preference of a population: not at all an easy task. They have also been designed, for example, for sampling lakes, batch and continuous processes in chemical industry, biological tissues. For those of you engaged in sampling, this is familiar territory. This is an area crucial to the work of an analyst, so much so that another volume in the Analytical Chemistry series, *Samples and Standards*, is devoted entirely to it. We shall now move on.

A *random sample* is chosen from a *population* in order to *estimate* some *property of the population.* Inherent in this statement is a belief that the value(s) of measurements made on a sample are a reflection of the value(s) which would occur if the same measurements were made on the whole population. In a voting preference poll, the population has the same number as the total number of voters. However, the voters are not, in a statistical sense, the population. The pollsters are interested in who the voters would vote for. The population is therefore identified as the totality *of opinions.* The sample is a small portion of the population and the number of people's opinions actually recorded is the *sample size*. If we were engaged in estimating the weights of bags of crisps, we should first have to define the population. Is this the weight of each for all the bags produced on a given day? in a given week? or held in the manufacturers warehouse? With this done, we should next need a well-designed sampling programme, which of course, would depend upon how we have defined the population. One element of this programme would be the decision as to the number of bags to be weighed, ie the sample size. Now let us consider estimating the sodium hydroxide content of a large quantity of caustic soda solution. It is easy to define the population: it is all the possible measurements we could carry out on the caustic soda solution. Again, our population is a totality – in this example of measurements, in the first example of opinions.

We next need a sampling programme. Let us assume that one exists and that it calls for us to collect, in a well-defined manner, 100 cm^3 of the solution. What is the sample size? No, it is not the 100 cm^3. The sample size has nothing to do with the physical dimensions of the sample. Sample size refers to the number of measurements or observations made. If we titrate two aliquots of the sample of caustic soda solution, then the sample size is 2; if we titrate three aliquots of the sample, then the sample size is 3. This is something that is a little confusing to those starting a study of statistics. However, once explained, the concept of sample size is easy to understand. Let us try to confuse you!

∏ A portion of 250 cm^3 of wine is taken from a large vat. The solution is analysed by glc for its ethanol content. The procedure involves injecting 0.5 μl of the solution onto a polar column, evaluating the area of the ethanol peak on the output chart, then using an area *versus* percentage-ethanol calibration curve. The determination is carried out in triplicate. What is the sample size?

Yes, it is 3! If you thought it was either 250 cm^3 or 0.5 μl, then re-read the last few paragraphs very carefully indeed.

Now, in all of the above there were questions that should have been asked. Let us ask a few important ones. First, accepting the concept of sampling a population, carrying out measurements on the sample of some property and using the results of these measurements to estimate the value of this property in the population as a whole, begs the question: how good is the estimate? Secondly, a sample appears to be characterised by its size: how important is the sample size? Does it affect the validity of an estimate? If it does, how do we go about deciding on the size of a sample? We will now concentrate on providing answers to these questions.

3.1.1.   The Central-limit Theorem

There is a theorem whose consequences are really quite profound. It is known as the central-limit theorem. It states that 'if random samples, of size n, are taken from a population with mean μ and standard deviation σ, the distribution of the sample means (the $\bar{x}$' s) will be approximately normal with mean μ and standard deviation $\sigma/\sqrt{n}$.'

The approximation improves as n increases. It is worthwhile going through this theorem in a mechanistic way. We assume that a population exists and that it has mean μ and standard deviation σ. Note that we have *not* said that the population must be normally distributed: populations have means and standard deviations whatever the form of their distribution. We now take a random sample from the population, carry out n measurements on it, and record the mean ($\bar{x}$) of these measurements. We will keep on taking random samples from population, carrying out n measurements, and recording the mean of the measurements. Ultimately, we shall have recorded enough of these means to consider plotting them in the form of a frequency distribution histogram. When we do this, it is found that if the population is normally distributed, then the distribution of sample means is indeed normal with mean μ and standard deviation $\sigma/\sqrt{n}$. Further, it is found that even if the population is not normal, the distribution of means is close to being normal with mean μ and standard deviation $\sigma/\sqrt{n}$. This closeness to normality of the distribution of sample means is an amazing result. Its significance will become apparent later. Let us represent what has been said above in graphical form.

In Fig. 3.1a we show a population which is normally distributed with mean μ and standard deviation σ.

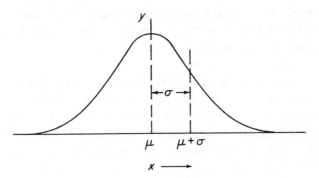

Fig. 3.1a. *A normally distributed population*

In this distribution, all the possible measurement values appear on the x axis; the measurement values are the random variable.

In Fig. 3.1b we show the distribution curve for the means of random samples drawn from the above population.

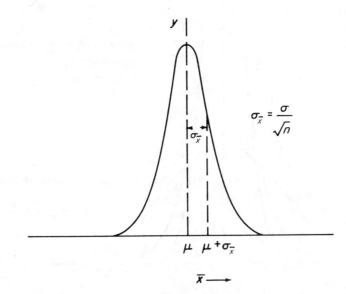

Fig. 3.1b. *The distribution of the means of random samples of size n drawn from a population of mean μ and standard deviation σ*

For this distribution, the random variable is the value of the mean, $\bar{x}$: this random variable is plotted on the $\bar{x}$-axis. The standard deviation of this normally distributed variable is given by $\sigma/\sqrt{n}$. The relationship $\sigma_{\bar{x}} = \sigma/\sqrt{n}$ results in the distribution curve for sample means always being thinner than the corresponding curve for the population, except when $n = 1$.

In Fig. 3.1c you can see a normally distributed population and, plotted in the same graph, the distribution of sample means for samples of different sizes. Notice how the distribution of sample means curves become thinner as n increase. This is because $\sigma_{\bar{x}} = \sigma/\sqrt{n}$, whilst the area under the normal curves must always equal one.

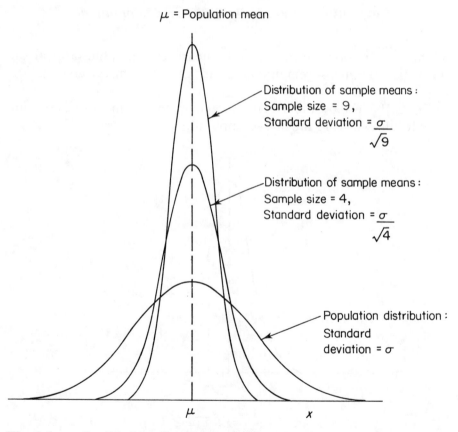

Fig. 3.1c. *Population distribution and distribution of sample means for sample sizes 4 and 9*

Let us apply these ideas to some examples.

∏ The weights of bags of crisps are normally distributed with a mean of 25.0 g and standard deviation 0.25 g. What is the mean and the standard deviation of the distribution of the mean weights of random samples of four bags of crisps?

The mean of the distribution of means is 25.0 g the standard deviation is $\sigma/\sqrt{n} = 0.25/\sqrt{4} = 0.125$ g.

∏ A very large number of pH measurements was carried out on a solution: the mean pH was found to be 6.800 and the standard deviation 0.006 pH unit. (The variation is due to random error, which is normally distributed). Describe the distribution of means curve for samples of size 9, and for samples of size 36.

For samples of size 9, we obtain a normal distribution with mean 6.800 pH units and standard deviation equal to $\sigma/\sqrt{n} = 0.006/\sqrt{9} = 0.002$ pH unit.

For samples of size 16, we have a normal distribution with mean equal to 6.800 pH units and standard deviation equal to $\sigma/\sqrt{n} = 0.006/\sqrt{36} = 0.001$ pH unit.

Note that when the sample size increases, the value for $\sigma_{\bar{x}}$ decreases. We shall refer to the implications of this later.

3.1.2. A Demonstration

The central-limit theorem asserts that sample means will fit approximately to a normal distribution even when the population being sampled is itself not normally distributed. This is a quite surprising result and in this section we ask you to demonstrate its validity by using your computer.

Computers can generate *random numbers*. In particular, in BASIC, the function RND(1) will, on each occasion it is used, generate a random number between 0.0 and 1.0. Different values are obtained each time, and any value between 0 and 1 is equally likely to turn up.

Try it. Type in the following 3-line program.

```
10  FOR I=1 TO 20
20  PRINT RND(1)
30  NEXT I
```

RUN the program. RUN it again. Each time you will get a list of 20 numbers, a different list each time, numbers jumbled in an arbitrary sequence, all lying between 0 and 1.

Let us regard RND(1) as a computer experiment, one in which a measurement is made. The expected distribution of values of RND(1) is shown in Fig. 3.1d(i). The distribution is rectangular, there is no central peaking. The value of the probability density is 1.0 everywhere between RND(1) = 0 and RND(1) = 1; hence, as required, the area under the graph is unity.

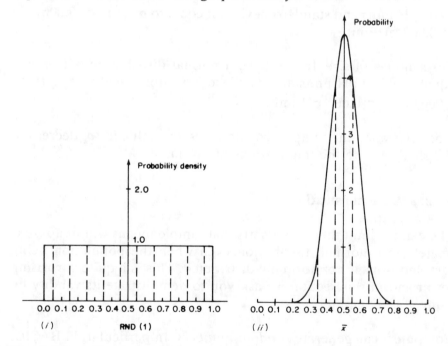

Fig. 3.1d. *(i) Distribution of RND(1); (ii) Distribution of $\bar{x}$, the mean of 12 measurements of RND(1)*

Let us check that this is indeed the distribution produced. Let us round off each value obtained for RND(1) to one decimal place. We can than count the frequency with which each of the possible values 0.0, 0.1, 0.2, etc arises. We could simply run our 3-line program over and over again, and count up how many of the values produced lie in each range and use this information to plot a histogram. It is easier, however, to make the computer do the hard labour. The following program will do it.

```
 10 DIM N(10)
 20 PRINT"    band       total"
 60 FOR I=1 TO 1000
 70     X=RND(1)
 80     J=INT(X*10+.5)
 90     N(J)=N(J)+1
100     NEXT I
110 FOR I=0 TO 10
120     PRINT I/10,N(I)
130     NEXT I
140 END
```

The loop between line 60 and line 100 is gone through 1000 times. On each occasion RND(1) is measured and rounded to one significant figure. The array N is used to count the number of occurrences of each of the possible results. Thus N(0) records the number of values which are rounded to 0.0, N(1) the number rounded to 0.1, etc. Look carefully at how INT works on line 80. A RND(1) value of 0.463, for instance, is rounded to 0.5; on line 80 it would lead to J = INT(4.63+.5) = INT(5.13) = 5. Hence, as required, on line 90 array element N(5) is increased by 1[*]. The second loop in the program prints out the total number of values found in each band.

---

[*] Note that for the program as given assume that when the RUN instruction is given, all values in array N are set initially to zero. Nearly all machines do this. If, however, you have problems with this program, type in the following extra lines.

```
30 FOR I=0 TO 10
40 N(I)=0
50 NEXT I
```

RUN the program. (It may take a few seconds before it prints its results.) We got the following results.

Band	Total
0	49
0.1	92
0.2	110
0.3	109
0.4	103
0.5	96
0.6	91
0.7	102
0.8	97
0.9	96
1.0	55

Now what should we expect from Fig. 3.1d(i)? An equal number in each band? Not quite that, because the end bands are only half as broad as the others (the broken vertical lines in Fig. 3.1d(i) separate the bands we have selected). The nine middle bands should crop up equally often, but we should expect only half as many values falling in each end band. In short the prediction is that from our 1000 measurements, 100 should fall in each of the nine middle bands, and 50 in the bands at each end.

Well, our results are not far from this. Are yours? Re-run the program. The actual values change from run to run. You should be used to that by now; samples vary as always. But the middle values should be mainly in the range 90–110, only rarely should any fail outside the range 80–120. The end values will almost always lie between 40 and 60. A particular entry which is larger than expected in one run will, on some other occasions, turn out smaller than expected. All in all, the results should seem generally consistent with Fig. 3.1d(i).

So RND(1) is not normally distributed. Now what about *means* of samples of RND(1)? From the symmetry of Fig. 3.1d(i), the mean value, μ, of RND(1) is clearly 0.5. It can also be proved that the standard deviation is given by $\sigma = 1/\sqrt{12}$*. If we take the mean $\bar{x}$ of a sample of n values of RND(1), the central-limit theorem asserts that $\bar{x}$ will be approximately normally distributed, about $\mu = 0.5$, with a standard deviation $\sigma_{\bar{x}} = \sigma/\sqrt{n}$. For convenience we shall take $n = 12$, so we shall expect $\sigma_{\bar{x}} = (1/\sqrt{12})/\sqrt{12} = 1/12$.

Add the following lines to your program.

```
62  X=0
64  FOR K=1 TO 12
66      X=X+RND(1)
68      NEXT K
70  X=X/12
```

Now, when you RUN the programme on each of the 1000 passes through the main loop starting on line 60, variable X will be computed as the mean of 12 readings of RND(1). RUN the program. (It is now doing quite a lot of work, so you will probably notice a significant delay before the results are printed. On a BBC Model-B micro this program takes about 45 seconds to run.)

The results of this change in your program should be dramatic. The distribution should show a very marked central peaking, matching Fig. 3.1d(ii) in contrast to Fig. 3.1d(i). The normal distribution expected, with $\mu = 0.5$ and $\sigma_{\bar{x}} = 1/12$ predicts approximately the following results.

---

* By adding two further lines to your program, you could check the assertion that the standard deviation of RND(1) is $1/\sqrt{12}$. Suitable lines are given below.

```
75  S=S+(X-.5)↑2
135  PRINT"st.dev. = ";SQR(S/1000)
```

The former line adds up the sum of the squares of the deviations from the mean μ and the latter then calculates and prints the standard deviation of the 1000 values sampled. Now $1/\sqrt{12} = 0.2887$. If you RUN the modified program it will usually return a value within 2–3% of this. Delete the extra lines before moving on.

Band	Total
0	0
0.1	0
0.2	1
0.3	35
0.4	238
0.5	452
0.6	238
0.7	35
0.8	1
0.9	0
1	0

Your values should follow this pattern. RUN the program again to see that the same basic picture re-emerges, and that the fluctuations sample by sample are relatively minor. Only once in about 40000 runs should you ever expect to find a non-zero total for any of the bands 0.0, 0.1, 0.9 or 1.0.

SAQ 3.1a

This is a revision exercise on deriving probabilities from a normal distribution curve. The means, $\bar{x}$, for samples of 12 values for RND(1) are expected to fit a normal distribution with mean $\mu = 0.5$ and with standard deviation $\sigma_{\bar{x}} = 1/12$. Find the probability that a given value of $\bar{x}$ lies in each of the following ranges: 0.00–0.05, 0.05–0.15, 0.15–0.25, 0.25–0.35, 0.35–0.45, 0.45–0.55, 0.55–0.65, 0.65–0.75, 0.75–0.85, 0.85–0.95 and 0.95–1.00.

This question could be done by hand, by using the table in Fig. 2.3g. It is much easier, however, to use program STAT2 which you developed in Part 2. (The data on line 3900 of that program will need to be adjusted.)

SAQ 3.1a

3.1.3. The Distribution of Sample Means: Areas, Frequencies and Probabilities

From our study of the properties of normal curves we have found it possible to make the following types of statement for normally distributed variables.

(*a*) Since the area under the curve in the range $\mu \pm 1.96\,\sigma$ is 95% of the whole area, 95% of all measurements must lie in the range $\mu \pm 1.96\,\sigma$. (Recall when $z = 1.96$ the area in a tail is 0.025: two tails have area $= 0.05$.)

(*b*) Since the area under the curve in the range $\mu \pm 1.96\,\sigma$ is 95% of the whole area, there is a 95% probability that a single measurement will fall within this range.

Just to refresh your memory, note that in (*a*) we have used areas and relative frequencies interchangeably, and that in (*b*) we have used areas and probabilities interchangeably.

Now what was true for individual measurement will be true for sample means. In Fig. 3.1e we see a curve for the distribution of sample means.

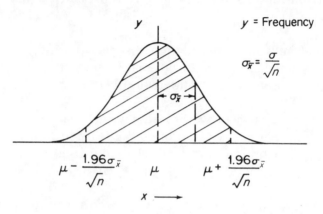

Fig. 3.1e. *A frequency distribution curve for sample means*

Let us attempt some exercises.

∏   What percentage of all sample means of size n lie in the range $\mu \pm 1.96\, \sigma/\sqrt{n}$?

We are dealing with a normally distributed variable – the sample means. The means are distributed with mean μ and standard deviation $\sigma_{\bar{x}} = \sigma/\sqrt{n}$.

Since the area under the curve in the range $\mu \pm 1.96\, \sigma_{\bar{x}}$ or $\mu \pm 1.96\, \sigma\sqrt{n}$ is 95% of the total area, 95% of the sample means of size n must lie in the range $\mu \pm 1.96\, \sigma/\sqrt{n}$.

∏   What is the probability of a sample mean of size n lying in the range $\mu \pm 1.96\, \sigma/\sqrt{n}$.

Since the area under the curve in the range $\mu \pm 1.96\, \sigma/\sqrt{n}$ is 95% of the total area, there is a 95% probability that a sample mean of size n will fall within this range.

Do you see what we have done? Yes, we have simply said that what was appropriate for the normal distribution (μ, σ) should be appropriate for the normal distribution $(\mu, \sigma/\sqrt{n})$. Incidentally, we have just introduced a notation for normal curves. It is $N(\mu, \sigma)$, where μ is the population mean and σ is the population standard deviation. For a distribution of sample means we represent the normal curve by $N(\mu, \sigma/\sqrt{n})$.

The parallel between a population distribution and a distribution of sample means has enabled us to make rapid progress. We will go back a little, so that we can understand better what we have done.

In Part 2 of the Unit we dealt with what we now call populations. The random variable was the value of x. We learned how to transform any normal distribution curve into a standardised normal distribution by using the operation $z = (x - \mu)/\sigma$.

This enabled us to calculate the z values corresponding to given x values. This in turn permitted us to use the table of areas in the tail associated with a z value, or to make use of the corresponding computer program. From tail areas we found it possible to obtain any area in which we had an interest. Areas, frequencies and probabilities were then used interchangeably in order to answer various types of questions. We are now dealing with a distribution of sample means. This distribution is normal with mean μ and standard deviation $\sigma/\sqrt{n}$. The random variable here is $\bar{x}$. In order to make use of the table or of our program we need to transform the normal distribution into a standardised one. The transformation is brought about by the operation

$$z = \frac{\bar{x} - \mu}{\sigma_{\bar{x}}} = \frac{\bar{x} - \mu}{\sigma/\sqrt{n}}$$

We can use this to calculate a z-value associated with a given $\bar{x}$-value. For example, the z-value associated with an $\bar{x}$ value of $\mu + 1.96\,\sigma/\sqrt{n}$ is

$$\frac{(\mu + 1.96\,\sigma/\sqrt{n} - \mu)}{\sigma/\sqrt{n}} = \frac{1.96\,\sigma/\sqrt{n}}{\sigma/\sqrt{n}} = 1.96$$

If we have the z value, we can obtain the area in the tail. For $z = 1.96$, it is 0.025. Check it! Keeping in mind the symmetry of a normal distribution curve, we can deduce that the area under a curve in the range $\mu \pm 1.96\, \sigma/\sqrt{n}$ is $1 - 2(0.025) = 0.95$, or 95%.

The consequences of this are that (*a*) 95% of all samples of size n have means which lie in the range $\mu \pm 1.96\, \sigma/\sqrt{n}$ and (*b*) there is a 95% probability that a sample of size n drawn from the population has a mean which will lie in the range $\mu \pm 1.96\, \sigma/\sqrt{n}$.

SAQ 3.1b

It is found from long experience that a machine produces components whose lengths are normally distributed with mean 20.00 cm and standard deviation 0.15 cm. A random sample of 9 components is taken from the production line.

(*i*)   What is the probability that the mean of the sample will be greater than 20.05 cm?

(*ii*) What percentage of samples of size 9 will have a mean length less than 20.05 cm?

(*iii*) What percentage of samples of size 9 will have a mean length in the range 20.00 $\pm$ 0.10 cm?

3.2. ESTIMATING A POPULATION MEAN FROM A SAMPLE MEAN

This is a topic for which you have probably been waiting! It is at the heart of what analysts do every day of the working week. By the time you have finished the topic, you will understand much of the statistical basis of what you do in the lab! We will start by assuming that we have a normally distributed population, $N(\mu, \sigma)$. The distribution of sample means is also normal, $N(\mu, \sigma/\sqrt{n})$. The distribution is plotted in Fig. 3.2a.

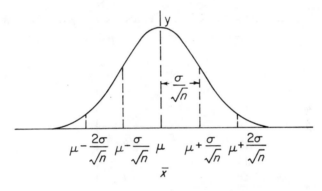

Fig. 3.2a. *The distribution of sample means*

We have indicated on the plot the positions of sample means which are one and two standard deviations from their mean – keep in mind that we are dealing with the standard deviation for the distribution of the sample means. Let us assume that we *do not* know the population mean, but that we *do* know the standard deviation. 'What are they talking about?' you might well ask. How can one know the population standard deviation without knowing the population mean? A good question! We shall not answer it for the moment! We shall, however, say that it is possible. Patience!

Suppose we now take a sample from the population, carry out n measurements, and calculate the mean of these. What will be the value of this mean? Let us answer the question in probability terms. There is a 68.26% probability that it falls in the range $\mu \pm \sigma/\sqrt{n}$ and a 95.44% probability that it falls in the range $\mu \pm 2\sigma/\sqrt{n}$.

A probability of 95.44% seems like good odds: let us assume that the sample mean just measured was in this range. In Fig. 3.2b we show the distribution of sample means curve and indicate the range $\mu \pm 2\sigma/\sqrt{n}$ by shading.

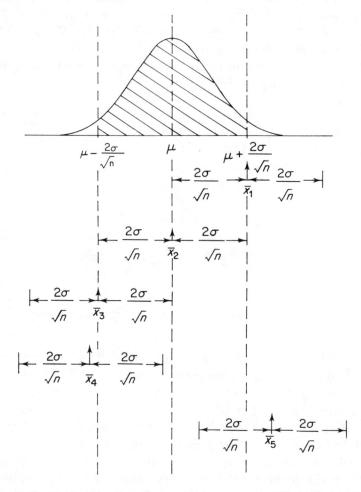

Fig. 3.2b.   *Estimating a population mean (95.44% level of confidence)*

We also indicate, below the curve, three examples of sample means which lie in the range (though the first and third are out the extreme edges of it). Below these are two examples of sample means which lie outside of the stated range. We would expect to obtain sample means outside of the range 4.56% of the time. You will see that we have drawn a range about each sample mean of width $\pm 2\sigma/\sqrt{n}$. Why? Because we are going to *estimate our population mean* as being in the range $\bar{x} \pm 2\sigma/\sqrt{n}$. Why? Patience!

You will see that for the first three examples of sample means our estimate that the population mean lies in the range $\bar{x} \pm 2\sigma/\sqrt{n}$ is correct. Since these examples of sample means are of the type which will occur 95.44% of the time, it is evident that when we estimate that a population mean is in the range $\bar{x} \pm 2\sigma/\sqrt{n}$, we are right 95.44% of the time. Being right so often can make one arrogant! We must not forget, as is evident from the last two examples of sample means, that we shall be wrong 4.56% of the time. What is the practical significance of the above? It is that if we know the standard deviation of a population, and then determine the mean of a sample of size n, we can estimate the population mean as lying in the range $\bar{x} \pm 2\sigma/\sqrt{n}$ and be right 95.44% of the time. We can write the equation for the estimate as: $\mu = \bar{x} \pm 2\sigma/\sqrt{n}$. If we report our result as $\bar{x} \pm 2\sigma/\sqrt{n}$, we have a 95.44% *level of confidence* that we are right. The range $\bar{x} \pm 2\sigma/\sqrt{n}$ is known as the 95.44% *confidence interval*.

You might ask here 'What is so special about a 95.44% level of confidence and a 95.44% confidence interval? The answer is of course, nothing! We decided to look at a range, symmetrical about the population mean, within ± 2 standard deviations. The instant we decided upon this, everything else followed. The range $\mu \pm 2\sigma/\sqrt{n}$ embraces 95.44% of the area under the curve and hence 95.44% of the sample means. It followed that there was a 95.44% probability that the population mean was in the range $\bar{x} \pm 2\sigma/\sqrt{n}$.

What is the confidence interval associated with a 95% confidence level? The answer is $\bar{x} \pm 1.96\,\sigma/\sqrt{n}$. Let us go through the procedure which gives us this result. We first need to know the area in two tails of a normal curve which leaves 95% of the area under the curve between the tails. That is easy enough; each tail must have 2.5% of the total area or as a decimal 0.025. We next ask, what z value gives such an area in a tail? It is 1.96. We then substitute this value into the general equation for a confidence interval:

$$\mu = \bar{x} \pm z\sigma/\sqrt{n} \qquad\qquad (3.1)$$

We therefore have

$$\mu = \bar{x} \pm 1.96\,\sigma/\sqrt{n}$$

This is a 95% confidence interval. We are 95% confident that the population mean lies in the range $\bar{x} \pm 1.96\,\sigma/\sqrt{n}$. Examine the table in Fig. 3.2c and check that we have deduced correctly the confidence intervals associated with given confidence levels.

Confidence Level	Area in one tail	z Value	Confidence interval
90%	5% or 0.05	1.645	$\bar{x} \pm 1.645\,\sigma/\sqrt{n}$
95%	2.5% or 0.025	1.960	$\bar{x} \pm 1.960\,\sigma/\sqrt{n}$
99%	0.5% or 0.005	2.576	$\bar{x} \pm 2.576\,\sigma/\sqrt{n}$

Fig. 3.2c. *Confidence intervals deduced for particular confidence levels*

Examination of the table emphasises an important feature of confidence levels and confidence intervals. It is that the wider the confidence interval the greater the confidence we can have in our estimate. For example, we are 90% confident that the population mean lies in the range $\bar{x} \pm 1.645\,\sigma/\sqrt{n}$, but 95% confident that it lies in the range $\bar{x} \pm 1.96\,\sigma/\sqrt{n}$. This seems to make sense. If we widen the range, we should be more confident. It is analogous to the stance we would take on watching a professional darts player. We might not be too confident that he would hit the 'bull'; we are quite confident that he will hit the board!

Π A manufacturer of crisp-bagging machinery guarantees that the standard deviation associated with the weight of crisps placed in bags is ±0.20 g. A random sample of 4 bags is taken from a production line and the mean weight is found to be

24.85 g. Estimate the weight of crisps in the bags produced during that production run, at the 95% confidence level.

First, the information we have been given is: $\sigma = 0.20$ g, $n = 4$ and $\bar{x}$ = 24.85 g: the confidence level is 95%, hence z = 1.96. Substituting these values into the equation $\mu = \bar{x} \pm 2\sigma/\sqrt{n}$ we obtain:

$$\mu = 24.85 \pm \frac{1.96(0.20)}{\sqrt{4}}$$

$$= 24.85 \pm 1.96(0.10)$$

$$= 24.85 \pm 0.196 = 24.85 \pm 0.20 \text{ g}$$

This means that we are 95% confident that the population mean is in the range 24.85 ± 0.20 g. Easy isn't it! Well perhaps for you, but what about the works manager who asked for the estimate? He is aware that the company sells the crisps with 25 g stamped on the bag. The result from the test, ie 24.85 ± 0.20 g, is compatible with this figure. However, he is still a little worried. Why might he be worried? Well, suppose he had asked for the test result to be reported at the 90% confidence level. The range quoted would then have been 24.85 ± 0.16 g. (Check this, it is simply a matter of repeating the above calculation, but using the value 1.645 from Fig. 3.2e in place of 1.96.) 25.0 g is still within this range, but it is getting rather close to the edge! Let us take stock. The mean resulting from the test was on the low side. There are, however, two possible reasons for this. It may simply be a chance result from our random sample or it may be the possibility we are worried about, that the true population mean μ has dropped below 25.0 g. The works manager does not want to shut down and adjust his bagging machinery unnecessarily. Sampling could well be the sole cause of the discrepancy. A result like this, due to sampling, would not be so rare as to have less than a 5% chance of occurring (we are within the 95% confidence range). However, 25.0 g is right on the edge of the 90% confidence range. We therefore have a discrepancy of a size which might be expected to occur only around one time in ten (ie with around 10% probability).

'Take another sample,' orders the works manager, 'and this time make it a large one!' This is duly done. A sample of 36 bags is taken; the mean weight is found to be 24.88 g. At the 95% confidence level, the population mean is estimated as

$$24.88 \pm \frac{1.96(0.20)}{\sqrt{36}} = 24.88 \pm 0.065 \text{ g}$$

Now do you see why he had reason to be worried? 25.0 g now lies clearly outside the 95% confidence range. (It also lies outside the 99% confidence range, as you can check if you replace the value 1.96 in the calculation by the value 2.576 from the bottom row of Fig. 3.2c.) This is just the sort of thing that could give the company a bad name! The machinery will need to be adjusted.

What lessons we have learned from this example! First, we can, for a given level of confidence, narrow down the confidence interval by taking a larger sample. This is a consequence of $\sqrt{n}$ in the term $\sigma/\sqrt{n}$. The sample size is something we can control. Bigger samples (shorthand for larger sample sizes) assist us by narrowing the confidence interval. Unfortunately, bigger samples take more time and hence cost more money. The choice is ours.

Secondly, when we estimate a population mean, we should never forget that to use only the result for the *sample mean* is to court trouble. The estimate of the *population mean* is $\bar{x} \pm z\sigma/\sqrt{n}$. The sample mean from the first test was 24.85 g and we were 95% confident that the population mean was in the range 24.85 $\pm$ 0.20 g, a range within which our product specification lies. When we took a second sample, its mean, 24.88 g, was closer to specification than that for the first sample. Had we stopped short, we might have been tempted to say that since the mean for the first sample seemed consistent with the specification, an even closer value for the second sample was confirmation that things were OK. Of course, we did not stop with $\bar{x}$, we knew that our estimate carries with it a calculable tolerance; that is the $\pm z\sigma/\sqrt{n}$ term. Its contribution to the outcome was conclusive!

The third lesson that we can learn from this example is that we have to decide at what level of confidence we should work. Changing the level of confidence required can change the conclusions reached as to whether things seem to be satisfactory or not. What confidence level should we choose in a given case? The answer is that circumstances will dictate which level of confidence is most appropriate. That is the only answer you are getting for the time being! Remember, patience! We shall certainly have more to say about making tests of this kind, and about drawing conclusions from them.

You will have noticed that for the crisp-bagging machinery we were able to give the population standard deviation for the weight because of the manufacturer's guarantee, ie that $\sigma = 0.20$ g. A large number of experiments must have been carried out in order to permit such a guarantee. Phrased another way, we know σ even though we do not know μ because of long experience with the machinery. This is similar to the example of the machine producing components whose lengths were normally distributed (SAQ 3.1b). The population mean was not known but long experience gave knowledge of the population standard deviation.

The same thing may occur in analytical chemistry. We might know a standard deviation associated with a given procedure, perhaps some titrimetric method, without having any knowledge of the population mean of the material to which the method is to be applied. How would we know σ? Again, by long experience gained in applying the method. In laboratories which routinely carry out standard analytical methods on materials which vary only in the relative concentrations of the components, it is not at all uncommon for the analyst to know the variability of the results, ie to know the population standard deviation. Values may be recorded in the laboratory's handbook of methods. Consult such a handbook, and it is likely that you will find examples where, for a given method applied to determine a given species present at a specified approximate concentration level, the precision will be quoted. This will be specified either in the form of a standard deviation, or as a relative standard deviation.

Thus it is indeed often possible to know a population standard deviation without knowing the population mean. In essence this is because often the standard deviation depends only on the method used. It may be very little influenced by the particular (unknown) material to which the method is applied in a given determination.

SAQ 3.2a

A manufacturer has developed a modification of the familiar combustion-analysis apparatus. It permits replicate determinations to be carried out quickly and it is claimed that it will determine the percentage of carbon in an organic compound with a standard deviation of 0.20%. A chemist decided to evaluate the apparatus. He asked a colleague to provide him with a 'pure' organic compound but not to inform him of its identity. The chemist subsequently carried out four determinations on the compound which yielded the following results: 68.83% C, 68.85% C, 68.83% and 68.82% C. Give an estimate, at the 95% confidence level, of the percentage of carbon in the compound.

Now we shall let you in on a secret, the compound whose analysis appeared in SAQ 3.2a was primary-standard grade benzoic acid, which should have a carbon content of 68.85%. Our chemist was quite peased when he was told that fact: the 95% confidence interval (68.83 ± 0.20%) embraced the expected value. He nontheless thought that it should be possible to narrow the 95% confidence interval to ±0.10% C. The 'signpost' to doing this lies in the term $z\sigma/\sqrt{n}$. To make this term equal to 0.10%, all that we need do is to increase the sample size, n. We proceed as follows.

$$z\sigma/\sqrt{n} \;=\; 0.10$$

$$\frac{1.96(0.20)}{\sqrt{n}} \;=\; 0.10$$

$$\sqrt{n} \;=\; 1.96(0.20)/0.10 \;=\; 3.92 \quad \therefore \; n \;=\; 15.37$$

Rounding up to the nearest whole number we obtain $n = 16$. Therefore if one wants to use the new apparatus to determine a sample mean and then to use this value to estimate the percentage of carbon in a compound, with a confidence interval of ±0.10% at the 95% confidence level, it is necessary to have a sample of size 16! Are you starting to feel that statistics might be useful to you? If your answer to this question is yes, then there is a strong likelihood that you will soon start to consult some of the textbooks on statistics which are recommended in the bibliography. When you do, you will encounter a notation for describing confidence levels and confidence intervals which you might as well learn now. In Fig. 3.2d you will see a standardised normal distribution curve with the area between the tails designated 1-α. The area in a tail is thus $\alpha/2$.

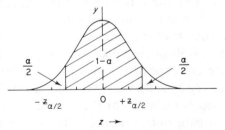

Fig. 3.2d. *A standardised normal distribution curve introducing parameter α*

The z value associated with an area in the tail of $\alpha/2$ is designated $z_{\alpha/2}$. Armed with this notation we can make the following comments:

(i) $(1-\alpha) \times 100\%$ of the means of random samples of size n lie in the range

$$\mu \pm z_{\alpha/2}(\sigma/\sqrt{n})$$

(ii) If $\bar{x}$ is the mean of a random sample of size n from a population with standard deviation σ, then there is a $(1-\alpha) \times 100\%$ level of confidence that the population mean lies in the range $\bar{x} \pm z_{\alpha/2}(\sigma/\sqrt{n})$. This range is known as the $(1-\alpha) \times 100\%$ confidence interval for the estimate of the population mean.

(iii) Rephrasing (ii), when we use $\bar{x}$ as an estimated value for the population mean we are $(1-\alpha) \times 100\ \%$ confident that the error in doing so will not exceed $\pm\ z_{\alpha/2}(\sigma/\sqrt{n})$. (Think about this.)

We shall make use of this notation from now.

3.2.1. The Effect of Systematic Error on Estimates

In this section of the unit we have been discussing the mean values for random samples. Such values have been stated to be normally distributed. They are random variables whose frequency distribution curve is normal. Random effects in the processes of bagging crisps or cutting components to length results in a variability in the measured property. The same is true of the measurements made in analytical chemistry; here the random effects are called random errors. In accepting that such random effects can result in the mean values for random samples being distributed normally, we have been able to quantify the concept of uncertainty. The significance of this in analytical chemistry is as follows. We take a random sample from what is known as the population or the whole, we make n measurements on it (n replicate determinations for those who are happier talking chemistry than statistics) and calculate the mean value of these measurements: the mean is taken because we consider it to be the best estimate of our experimental results. We could stop here, but we are too knowledgeable to do so. We know that n determinations on another random sample would probably have resulted in a slightly different value for the mean. The mean values for random samples are normally distributed. We therefore decide to incorporate our knowledge of the uncertainty present in determinations into statements we make about the relationship between the sample mean and the mean of the population as a whole. We say that we are $(1-\alpha) \times 100\%$ confident that the mean of the population is in the range $\bar{x} \pm z_{\alpha/2}(\sigma/\sqrt{n})$. We *do not* say that the population mean is $\bar{x}$: such a statement would be out of character for an analytical chemist! Now, *if systematic error is absent*, our estimate of the population mean is synonymous with an estimate of the truth. We could, in the absence of systematic error rightfully make statements such as 'we are 95% confident that the true carbon content of the material under investigation lies in the range 68.83 $\pm$ 0.20% (see SAQ 3.2a). Why is it that we do not make such statements. It surely must be because we are aware of the possibility of systematic error, of a bias in our measurement system. Consider the distribution curves in Fig. 3.2e.

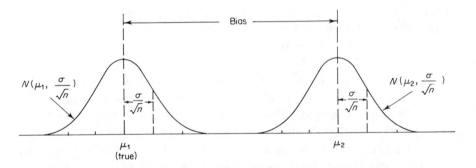

Fig. 3.2e. *The effect of bias on distribution of sample means curve*

The distribution of sample means curve on the left, $N(\mu_1, \sigma/\sqrt{n})$, shows how the mean values determined by a measurement system without systematic error would distribute themselves. The mean of the curve coincides with the true value, μ_1. The curve on the right, $N(\mu_2, \sigma/\sqrt{n})$, represents the distribution of the means if the measurement system had the same precision as the first measurement system but had a bias equal to $(\mu_2 - \mu_1)$. This reminds us of an important point, ie that a population refers to the totality of possible measurements. If we use measurement system 1, our means would be distributed as $N(\mu_1, \sigma/\sqrt{n})$. If we used measurement system 2 on exactly the same material our means would be distributed as $N(\mu_2, \sigma/\sqrt{n})$. For the first distribution, we can call the population mean the true population mean: the second distribution of means has of course a population mean μ_2; because of systematic error its population mean is *not* the true population mean.

Removing systematic error is a major preoccupation of analytical chemists. If systematic error is absent, then our estimates, based on

$$\mu = \bar{x} \pm z_{\alpha/2}(\sigma/\sqrt{n}),$$

will be unbiased and we can be $(1-\alpha) \times 100\%$ confident that they embrace the 'truth'. If systematic error is present, we can still make estimates, but they will be biased. How can we be certain that our estimates are not biased? The answer is that when dealing with unknowns we can never be absolutely certain that our measurements are unbiased. There are procedures which help us to search for the presence of systematic errors (see Section 1.7.4). These procedures make use of statistical approaches similar to those we have been using. We shall give examples of the use of statistics in the search for systematic error later in this Part of the Unit.

∏   After long experience with new combustion-analysis apparatus, a chemist is satisfied that the manufacturer's claim that 'it will determine the percentage of carbon in an organic compound with a standard deviation of 0.20%' is essentially true. (It should be understood that the manufacturer was not simply showing off about the precision of his apparatus. The phrase 'it will determine' has the implication that the determination will be accurate.) The chemist now wishes to use the apparatus for the determination of the carbon content of some novel inorganic complexes. He chooses a model compound, which is known to have a carbon content of 38.91%, and carries out 4 replicate determinations on it. The mean carbon content is found to be 38.55%. Comment on this result.

The information given is $\bar{x} = 38.55\%$, $\sigma = 0.20\%$ and $n = 4$. At the 95% confidence level $z = 1.96$. The estimate for the mean carbon content of the material is as follows:

$$\mu = x \pm z(\sigma/\sqrt{n})$$

$$= 38.55 \pm \frac{1.96(0.2)}{\sqrt{4}}$$

$$= 38.55 \pm 0.196 = 38.55 \pm 0.20$$

We are thus 95% confident that the mean carbon content is in the range $38.55 \pm 0.20\%$. Now the known carbon content is 38.91%. What comments can be made? In the first instance there are only two: either the material analysed was not pure, or the combustion method tends to give low results when used to determine the carbon content of certain inorganic complexes.

Where does the chemist go from here? More investigations of a chemical nature are obviously required. Statistics have been employed and have indicated that something is wrong. For the time being, it is 'back to the bench'!

3.3. ESTIMATING A POPULATION STANDARD DEVIATION FROM A SAMPLE STANDARD DEVIATION

It has probably not escaped your notice that we have made a great effort to convince you that it is possible to know a population standard deviation without knowing the population mean. Why did we make the effort? Because we felt that you, an experienced chemist, might need to be convinced! Why? Because we knew that you were as aware as we were that the usual situation in analysis is that we are ignorant not only of the population mean but also of the standard deviation! Unknowns are unknowns! Now, this might appear to be bad news. We have been making considerable capital out of the knowledge that we are $(1-\alpha) \times 100\%$ confident that the population mean lies in the range $\bar{x} \pm z_{\alpha/2}(\sigma/\sqrt{n})$ and we now admit that it is more common than not to be ignorant of σ. What can we do about our ignorance? Well, we know how to calculate a *sample standard deviation*. The equation is:

$$s = \sqrt{\frac{\sum(x_i - \bar{x})^2}{n - 1}} = \sqrt{\frac{\sum(d_i)^2}{n - 1}} \qquad (3.2)$$

(See Section 1.3.2). Perhaps we may use the sample standard deviation as an estimate of the population standard deviation, ie $s \approx \sigma$. This would lead to the following relationship:

$$z = \frac{\bar{x} - \mu}{\sigma/\sqrt{n}} \approx \frac{\bar{x} - \mu}{s/\sqrt{n}} \qquad (3.3)$$

This in turn would lead to a $(1-\alpha) \times 100\%$ confidence interval for the population mean being suggested as:

$$\mu \approx \bar{x} \pm z_{\alpha/2}(s/\sqrt{n})$$

In practice, this would mean that our procedure in estimating a population mean would be as follows. We would carry out n replicate measurements and calculate the mean and standard deviation of the sample. We would choose a level of confidence and then substitute the appropriate values into the above approximate equation. If we did this, how good would our estimate be? The answer is surely to be found in the question 'How good an estimate of σ is s?' Let us play with some numbers and see what happens. We start by inventing a normal population; its mean is 24.86 and its standard deviation 0.04. Let us assume that the population refers to the totality of titration values (in cm^3) which could be obtained by a measurement system without a systematic error. Nearly all (actually 99.73%) of the titre values would be in the range $\mu \pm 3\sigma$, ie 24.86 $\pm 3(0.04)cm^3$. Random error causes this spread of results and whilst we can invent a titration procedure without systematic error, it would be foolish to pretend that we can have a procedure without random error. Let us carry out two titrations (sample size 2) on the material. As experienced chemists we are not surprised when we obtain titres of 24.80 cm^3 and 24.82 cm^3. The mean titre is 24.81 cm^3 and the standard deviation is 0.014 cm^3. How good an estimate of σ is s? The answer is that on this occasion, it is a gross *underestimate* since $\sigma = 0.04$ cm^3. Nonetheless, if we were in a position where σ was unknown to us, then s would be the best indication available to us of the value of σ. Thus if we followed equation (3.3) above we would be led to suggest that at the 95% confidence level the population mean lies in the range below

$$24.81 \pm \frac{1.96(0.014)}{\sqrt{2}} = 24.81 \pm 0.02 \text{ cm}^3$$

Since here we do know the population mean we can see that this statement is plainly wrong! Yes, that is the point being made. Since s in this case is such a bad estimate of σ, we have a very poor estimate of μ. However, you might also be saying to yourself, 'they are cooking the books'. Yes, we admit it! However, the situation which has just been described does happen, and if it does we can be misled into making a poor estimate of the population mean. Another measurement possibility is: first titre $= 23.86 \text{ cm}^3$, second titre $=24.78 \text{ cm}^3$. (If you believe in random error, you have to believe that it could happen!) The consequences of these titre values are $\bar{x} = 24.82$, $s = 0.057$, and at the 95% confidence level, the population mean lies in the range:

$$24.82 \pm \frac{1.96(0.057)}{\sqrt{2}} = 24.82 \pm 0.08 \text{ cm}^3.$$

Yes, this time we got it right! If we take a couple more measurements, would we get it right again? We don't know! We obviously have a problem. We would like to make use of the sample standard deviation, but from the examples given, the sample standard deviation seems to vary considerably. This could result in some estimates being good and some poor.

Have we misled you with our examples? Could it be that matters are seldom as bad as we have portrayed them? Perhaps we have exhibited prejudice in our choice of values. Perhaps we need the assistance of an unbiased observer! We must convince you of the validity of the argument we are making and, even more important, we must give you a 'feel' for the way in which all experimental work may be buffeted by the vagaries of random error. One way to do this would be to send you to the laboratory to carry out a large number of sets of replicate titrations on the same unknown. We

can, however, give you the same experience with much less effort
by using a computer program. After all, we have already seen (in
Section 3.1.2 above) that we can use our computer to simulate the
process of making measurements which are subject to random error.
In ten minutes at your keyboard we can review rather effortlessly a
range of typical titration results that it might take a week to generate
in the lab.

3.3.1. Gaining Experience

The program below will allow you to simulate the results of sets of
replicate titrations which are subject to random error. It should only
take you a couple of minutes to type it into your machine.

```
10   READ M9,S9,N
20   PRINT
30   PRINT
40   PRINT"true mean = ";M9;"   true s.d. = ";S9
50   M1=0
60   S2=0
70   PRINT"- - - - - - - - - - - - - - - - - - - - - - - - - - - - - -"
80   PRINT ;N;" readings:";
90   FOR I=1 TO N
100      Z=-6
110      FOR J=1 TO 12
120         Z=Z+RND(1)
130      NEXT J
140      X=M9+Z*S9
150      M1=M1+X
160      S2=S2+X*X
170      PRINT" ";INT(X*100+.5)/100;
180      NEXT I
190  M1=M1/N
200  S1=SQR((S2-N*M1*M1)/(N-1))
210  PRINT
220  PRINT"- - - - - - - - - - - - - - - - - - - - - - - - - - - - - -"
230  PRINT"calc mean = ";INT(M1*100+.5)/100;
240  PRINT" calc s.d. = ";INT(S1*1000+.5)/1000
```

```
250  E = ABS(M1-M9)
260  PRINT"Error in mean = ";INT(E/S1*10 + .5)/10;"*calc s.d."
270  PRINT"                    ";INT(E/S9*10 + .5)/10;"*true s.d."
280  END
290  DATA 24.86,0.04,2
```

About the program. You can treat this program as a 'black box' if you like. If you wish to do so simply type it into your machine as given, skip the next three paragraphs, and move on directly to running simulated titration experiments.

As usual the program has been given in a form which economises on typing. There are, for instance, no REM statements. One or two niceties of detail which an ideal program would have included are omitted for simplicity's sake. The only awkward expressions involve the use of function INT to round numbers printed to the screen. The first line READs the values of M9, S9 and N from the very last DATA line. These variables represent, respectively, the 'true' (ie population) mean, the 'true' (population) standard deviation, and the proposed number of replicate titrations to be performed.

Look closely at lines 100-140. They do almost the same job as lines 62-70 of the program we ran in Section 3.1.2 above. There, by taking the *mean* of 12 values of RND(1) we showed we obtained a normally distributed variable, with a mean of 0.5 and standard deviation of 1/12, *viz*, $N(0.5, 1/12)$. By taking instead the *sum* of 12 values of RND(1) (ie if we do not divide by 12 to get the mean) we would get a normal variable with mean 6 and standard deviation 1, ie both 12 times bigger than before, giving $N(6,1)$. By subtracting 6 from the total (see line 100), this distribution is shifted to mean 0 without affecting the standard deviation, in other words we now have the standardised normal distribution of variable z, viz, $N(0,1)$. Line 140 then sets a value for x by using $x = \mu + z\sigma$, where x is a random measurement from the normal population for our titre values.

The main loop, from line 90 to line 180, samples N such values for x, and keeps a running record of the sum of these (M1) and of the sum of their squares (S2). Once all N values have been obtained the sample mean and standard deviation (S1) are calculated on lines

190 and 200. These are printed out, followed by the absolute error (E) as between the calculated mean and the true mean, expressed as a multiple of the calculated standard deviation and of the true (population) standard deviation*.

Look at the last line of the program. The values 24.86 and 0.04 are the population mean and standard deviation for the simulation to be run, and each time the program is RUN a random sample of 2 titration results will be obtained from the population, and the results analysed. By repeatedly typing RUN very many independent random samples can be investigated. You can change line 290 to set whatever values you like for the population mean and standard deviation, and for the number of replicates to be studied.

RUN the program once as it stands. When we did this we got the following output.

True mean	= 24.86	True s.d.	= 4E-2
2 readings: 24.88 24.87			
calc mean	= 24.88	calc s.d.	= 4E-3
Error in mean	=	4.1*calc. s.d.	
	=	0.4*true s.d.	

You will get a similar layout of results, but for a quite different random sample. The error in mean is the difference between the calculated mean and the 'true' population mean. In our example this is 0.02, and this difference is 4.1 times the standard deviation,

---

* Very occasionally the program may break down with an error at line 260 if the calculated standard deviation S1 turns out to be zero. This does no harm, but can happen with $N = 2$; it means that the RND(1) has come up with the same z-value twice in succession, in which rare case any error in the mean will be an infinite multiple of the calculated standard deviation.

s, calculated from the two measurements listed. Now keep in mind that we tentatively suggested that we might attempt to estimate μ as $\bar{x} \pm zs/\sqrt{n}$. For a 95% confidence interval $z = 1.96$. For a sample of size 2, $z/\sqrt{n} = 1.96/\sqrt{n} = 1.39$. Hence, if our suggested formula using z was valid, only once in twenty times would the error in the mean exceed 1.39 times the calculated standard deviation. Our example above would have to be a very rare event indeed; it would not be expected to occur even once in a million times! Notice, however, that our error is only 0.4 times the true standard deviation, well within the 95% confidence bracket.

The reason for this effect is very easy to pin down. Our example above is an instance where the calculated standard deviation is a serious underestimate of the true σ. It isn't always thus, it is very often an overestimate! Examine your own calculation. Is your estimate of s an overestimate or an underestimate, or does it give the correct value? RUN the program a number of times. Note the variability of the outcome, and you should notice in particular that the error in the mean is quite frequently considerably more than 1.39 times the calculated s-value.

We must remember that in a real laboratory situation we don't know the values of the 'true' mean and standard deviation. We have to rely on our (random) sample to derive an appropriate confidence interval for the true mean. To do this, as you will soon find out, we have to use, instead of z, a new table of multipliers. When we have only 2 replicates, the 95% confidence interval is $\bar{x} \pm (12.71/\sqrt{2})s$, ie $\bar{x} \pm 8.99s$. Thus it should happen only about 1 in 20 times that the 'Error in the mean' is more than 9 times the 'calc. s.d.'. On the other hand, in terms of the *true* standard deviation the 95% confidence range is $\bar{x} \pm (1.96/\sqrt{2})\sigma$, ie $\bar{x} \pm 1.39\sigma$.

Thus only about 1 time in 20 should 'the error in mean' be as much as 1.4 times the 'true s.d.'.

Try it and see. RUN the program 20 times and count how often, if ever, each of the respective limits of 9 times and 1.4 times is breached. On average you should find *one* occurrence of each. Chance as ever plays its part though, and a given selection of 20 replicates will quite frequently yield 0 or 2 occurrences of either event (or with increasing rarity 3 or even 4). You will see that in the great majority of runs the error in the mean is a great deal smaller than the 95% threshold.

It is worth carrying out many repetitions to study the variability of the outcome. If the calculated standard deviation is an overestimate, then the error in the mean will turn out to be a *smaller* multiple of 'calc. s.d.' than of 'true s.d.'. However, on those occasions where *s* is an underestimate of σ, it quite frequently happens that the error is a *much larger* multiple of 'calc s.d.' than of 'true s.d.'. It is in order to cope with this situation that our 95% limits have had to be so dramatically widened.

You might very probably tire of typing RUN all the time. You can save yourself that effort by typing additional lines:

```
 15  FOR K=1 TO 1000
272  INPUT A$
274  IF A$="S" THEN 280
276  NEXT K
```

If you RUN the program now, after viewing the results from one sample it will be sufficient to press the return key to run the next sample. You would type S (& "return") to stop the process. This is only one of many small changes you might consider making. For instance you might like to add lines that allowed a count to be made of the number of times the 95% limits are breached for each criterion.

With this program you can rapidly build up experience of the capricious way in which chance affects even the most careful experiments. We would like you now to sample the improvements brought about by increasing the number of replicates. This involves changing line 290, in particular trying larger values for *N* (the final number in the DATA line). We would like to leave you in the driving seat for this further work, though we suggest that you try two specific investigations:

(a) Try repetition for two or three higher values of N. New 95% threshold multipliers are given below:

No. of replicates, N	Multiplier
2	$12.71/\sqrt{2}$
3	$4.30/\sqrt{3}$
5	$2.78/\sqrt{5}$
10	$2.26/\sqrt{10}$
20	$2.09/\sqrt{20}$
60	$2.00/\sqrt{60}$

(We shall say more about the origin of these multipliers later.) Check how often they are breached.

(b) Try a series of runs in which you focus your attention purely on the calculated mean and standard deviation values. Try steadily increasing N. You should find that, as N increases, the calculated mean fairly rapidly settles down to the true value but that the standard deviation may still be a little wayward even after 30 or 40 replicates have been included.

If you wish to change N run by run this can conveniently be done by typing new program lines:

```
12 PRINT "How many replicates";
14 INPUT N
```

This will allow you to type a value of your choice for each RUN, and will override the value set in the DATA statement.

Do you remember the discussion in Section 2.3 before we introduced the program? We generated the data that we discussed there by running the program with the true mean set at 24.86 cm^3 and the true standard deviation at 0.040 cm^3. We set $N = 50$ but 'doctored' the program to print out the calculated mean and standard deviation at each step. Some of our results were as follows.

No. replicates complete	calc. mean	calc s.d.
2	24.81	0.014
5	24.83	0.019
10	24.85	0.025
20	24.86	0.030
30	24.86	0.037
50	24.86	0.038

Note the slowness of the convergence of the standard deviation. Of course it does not always happen like this, but you should be able to generate similar examples for yourself.

3.4.   THE *t*-DISTRIBUTION

We have been discussing the distribution of sample-means (sample size n) for a population with mean μ and standard deviation σ. We learned that the distribution of the sample means is normal with mean μ and standard deviation $\sigma/\sqrt{n}$. We used the operation $z = (\bar{x} - \mu)/(\sigma/\sqrt{n})$ to transform the distribution into a standardised normal distribution, $N(0,1)$. In carrying out this operation we switched from having to deal with the variable $\bar{x}$ to the variable z. The advantage of this was that we could make use of a table giving the 'area in a tail' for specified values of z. In consequence of this we were able to state that $(1-\alpha) \times 100\%$ of all sample-means fall in the range $\mu \pm z_{\alpha/2}\sigma/\sqrt{n}$. In absence of knowledge of the value of μ we could, on the basis of measurements of x, state that we were $(1-\alpha) \times 100\%$ confident that the population mean lay in the range $\mu = \bar{x} \pm z_{\alpha/2}\sigma/\sqrt{n}$.

We then moved on to a common situation, where we were ignorant not only of the value of μ but also of the population standard deviation σ. We are now forced to use s, the sample standard deviation, as an estimate of σ. When we do this, however, it has become clear that we can no longer simply use values for $z_{\alpha/2}$ to define the $(1-\alpha) \times 100\%$ confidence interval; the range within which μ can be bracketed has to be widened. We have seen why this is so. The sample standard deviation does not behave as we would wish it to!

It takes a rather large sample size before it can be relied on to settle down to a value close to the population standard deviation.

We need a new relationship to give confidence limits, to replace $\mu = \bar{x} \pm z_{\alpha/2}\sigma/\sqrt{n}$. Well, we cannot improve on $\bar{x}$, it is the best estimate for μ we can obtain from our measurements. In ignorance of the value of σ, there is no alternative but to use s in its place. The sample size n might have something to offer us. When n becomes large, s settles down and approximates to σ. Unfortunately, the practical consequences of always making n large are not too appealing. Most of us are used to doing 2, 3 or 4 replicate determinations. The thought of doing 30, 40 or 50 is frightening! Nonetheless, let us keep n in mind for later comments. But what we really need is a replacement for z. When we use s, the z variable is no longer applicable, it is in general too tolerant.

We have already hinted during the computer exercise that we do have available a statistical parameter to replace z. It is labelled t. The *t distribution* was introduced by W.S. Gosset and first reported in 1908. Gosset worked for an Irish brewery (you don't have to be a genius to guess which one!). The brewery had a policy which forbade the publication of research work by its employees. Because of this, Gosset published his paper under the pseudonym 'Student'. This led to the distribution becoming known as the Student *t*-distribution, or *t*-distribution for short. What does the distribution look like? The equation for it is given below.

$$y = \frac{y_0}{[1 + t^2/(n-1)]^{n/2}} \tag{3.4}$$

Remember when we said that one should not be worried by the equation for the normal distribution. 'Hardly any analytical chemists remember it!' Well, the same advice can be given for this equation. This time though, one can confidently state that many analytical chemists have never even seen the equation!.

In the equation y_0 is a constant whose value depends on n in such a way that the area under the curve is always one. That's a good start! Probability distribution functions should have the total area under the curve equal to 1.00. The n is any positive integer. When *n* is very large (> about 30) the curve looks almost identical to the standardised normal curve, $N(0,1)$. That too is a good thing; for large *n* the curve will approach the *z* distribution. The curve keeps its symmetry about its mean for all values of *n*. As *n* gets smaller, the curve becomes flatter (or if you wish thicker). Nevertheless, it maintains the property that the total area under the curve is unity, Fig. 3.4a.

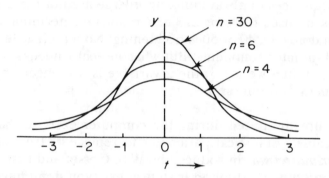

Fig. 3.4a. *The t-distribution for various values of n*

Now in a normal curve 95% of the *x*-values fall in the range $\mu \pm 1.96\ \sigma$. In a standardised normal distribution curve 95% of the values fall between $z = -1.96$ and $z = +1.96$. The reason for this, in both cases, is that 95% of the total area under the curve is found in the stated ranges. In a *t*-distribution, when *n* is *very* large, 95% of the total area lies between $t = -1.96$ and $t = +1.96$. Good! When *n* is 20 it turns out that 95% of the total area under the curve is found in the range $t = -2.09$ to $t = +2.09$. When *n* is 5 then 95% of the total area under the curve is found in the range $t = -2.78$ to $t = +2.78$. Three examples are enough! As you might have inferred from Fig. 3.4a, as *n* gets smaller the range of *t* (symmetrical about the mean) which embraces 95% of the area under the curve gets larger. We can find ranges for any value of n and any percentage of the total area by consulting the t-table, Fig. 3.4b. Before we get on to learning about the table, let us make an important comment – it looks as though the t-distribution is going to be of considerable use!

3.4.1. The Percentile Values for the t-distribution

Learning how to use the table for the t-distribution curve requires only a little more effort than was necessary in becoming familiar with the normal distribution table. Let us make a start. A table for the percentile values for the t-distribution is in Fig. 3.4b.

Examine it by first considering the column on the left-hand side of the table; it is headed ν. The symbol ν stands for *degrees of freedom*. We shall not concern ourselves with the meaning of degrees of freedom for the time being. All that need be said here is that until we tell you otherwise $\nu = n - 1$, where n is our number of replicate determinations*. Now if we state a value for n, we immediately have a value for ν, eg, if $n = 26$, $\nu = 25$. That is easy enough! Now consider the top row of the table; it has numbers ranging 0.20 to 0.001. These numbers represent areas in the tail of a t-distribution. We shall look at a number that all students of statistics remember; it is 0.025. Note that this number comes fourth as we read across the top row. Now look at the column under the number 0.025; it starts with the value 12.706 and ends with the value 1.960. *These values represent the values of t associated with an area of 0.025 in the tail of a t distribution for each value of ν*. These values are referred to as the *percentiles of t* corresponding to a tail area of 0.025, or 2.5%.

---

* Because, eventually, we shall have an occasion to tell you otherwise, and because ν is the parameter which determines the t-value, we ought, strictly, to have used ($\nu + 1$) in place of n in Eq. 3.4

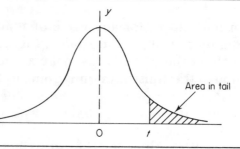

	\multicolumn{7}{c	}{Area in tail of *t*-distribution}
	0.20	0.10	0.05	0.025	0.01	0.005	0.001
ν							
1	1.376	3.078	6.314	12.706	31.821	63.657	318.310
2	1.061	1.886	2.920	4.303	6.965	9.925	22.327
3	0.978	1.638	2.353	3.182	4.541	5.841	10.215
4	0.941	1.533	2.132	2.776	3.747	4.604	7.173
5	0.920	1.476	2.015	2.571	3.365	4.032	5.893
6	0.906	1.440	1.943	2.447	3.143	3.707	5.208
7	0.896	1.415	1.895	2.365	2.998	3.499	4.785
8	0.889	1.397	1.860	2.306	2.896	3.355	4.501
9	0.883	1.383	1.833	2.262	2.821	3.250	4.297
10	0.879	1.372	1.812	2.228	2.764	3.169	4.144
11	0.876	1.363	1.796	2.201	2.718	3.106	4.025
12	0.873	1.356	1.782	2.179	2.681	3.055	3.930
13	0.870	1.350	1.771	2.160	2.650	3.012	3.852
14	0.868	1.345	1.761	2.145	2.624	2.977	3.787
15	0.866	1.341	1.753	2.131	2.602	2.947	3.733
16	0.865	1.337	1.746	2.120	2.583	2.921	3.686
17	0.863	1.333	1.740	2.110	2.567	2.898	3.646
18	0.862	1.330	1.734	2.101	2.552	2.878	3.610
19	0.861	1.328	1.729	2.093	2.539	2.861	3.579
20	0.860	1.325	1.725	2.086	2.528	2.845	3.552
21	0.859	1.323	1.721	2.080	2.518	2.831	3.527
22	0.858	1.321	1.717	2.074	2.508	2.819	3.505
23	0.858	1.319	1.714	2.069	2.500	2.807	3.485
24	0.857	1.318	1.711	2.064	2.492	2.797	3.467
25	0.856	1.316	1.708	2.060	2.485	2.787	3.450
26	0.856	1.315	1.706	2.056	2.479	2.779	3.435
27	0.855	1.314	1.703	2.052	2.473	2.771	3.421
28	0.855	1.313	1.701	2.048	2.467	2.763	3.408
29	0.854	1.311	1.699	2.045	2.462	2.756	3.396
30	0.854	1.310	1.697	2.042	2.457	2.750	3.385
40	0.851	1.303	1.684	2.021	2.423	2.704	3.307
60	0.848	1.296	1.671	2.000	2.390	2.660	3.232
120	0.845	1.289	1.658	1.980	2.358	2.617	3.160
∞	0.842	1.282	1.645	1.960	2.326	2.576	3.090

Fig. 3.4b.   *The value of t associated with a given area in the tail of a t-distribution and given degrees of freedom, ν*

∏   What is the value of t associated with an area in the tail of a t-distribution of 0.025 when $n = 26$.

When $n = 26$, $\nu = 25$. Reading across from $\nu = 25$ until we reach the column headed 0.025 we see that the t-value is 2.060. We shall now represent this information graphically (see Fig. 3.4c).

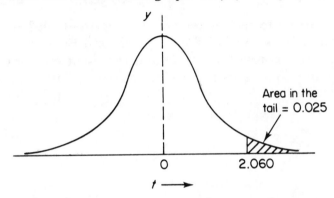

Fig. 3.4c.   *The t-distribution for n = 26 (ν = 25), illustrating the t-value for an area in the tail of 0.025*

∏   What is the t value corresponding to an area in the tail of 0.05 in a t distribution with $n = 13$?

When $n = 13$, $\nu = 12$. The t-value for an area in the tail of 0.05 and $\nu = 12$ is 1.782. This information is shown graphically in Fig. 3.4d.

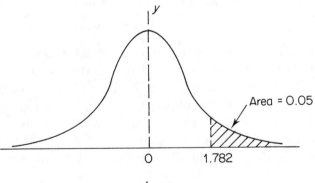

Fig. 3.4d.   *The t-distribution for n = 13, showing 0.05 of the total area in a tail*

∏   What is the area in the tail when $t = 2.776$ in a t-distribution
         with $n = 5$?

When $n = 5$, $\nu = 4$. Reading across the row for $\nu = 4$ until we
come to $t = 2.776$ we see that this t-value appears in the column
headed 0.025; this value is the required area in the tail.

Are you starting to feel better about using the table? If so, let us
start to move back towards familiar territory! First let us sketch a
t-distribution for $n = 5$ ($\nu = 4$) and represent on it the t-values
corresponding to an area in *each* tail of 0.025. This is done in Fig.
3.4e.

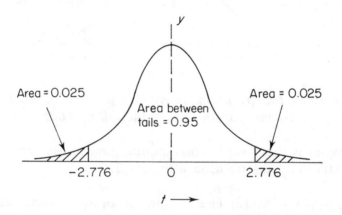

Fig. 3.4e.   *A t-distribution for $n = 5$ ($\nu = 4$) showing 0.025 of the
total area in each tail*

From Fig. 3.4e we can see that 0.95 or 95% of the total area under
the t-distribution curve for $n = 5$ lies in the range $t = -2.776$ to
$t = +2.776$. We have enough familiarity with this sort of statement
to realise that it is synonomous with saying that 95% of all t-values
lie in the range $t = -2.776$ to $t = +2.776$. Yes, we are using the
t-table in a way which is analogous to our use of the standardised
normal distribution. Let us do another example to drive the point
home!

∏   Give the range, symmetrical about $t = 0$ (the mean of the
      t-distribution) which embraces 95% of all t-values when n
      $= 31$.

For $n = 31$, $\nu = 30$. When the area between the tails is 95%, the
area in one tail is 2.5% or 0.025. The t value for $\nu = 30$ and area in
a tail equal to 0.025 is 2.042. The required range symmetrical about
$t = 0$ embracing 95% of all t values is thus from $t = -2.042$ to
$t = +2.042$.

If you have been able to follow this, then you are undoubtedly
ready to make the connection between the t-distribution and the
task which has been the concern of this section of the Unit, ie estimation.

3.4.2.   The t-distribution in Statistical Estimation

Let us start with a brief revision of our work with the normal distribution and estimation. Some variables, eg random error, are normally distributed, $N(\mu, \sigma)$. Normal distribution curves can be transformed into the standardised normal curve, $N(0,1)$, by use of the
operation $z = (x - \mu)/\sigma$. Another way of saying this is that the
variable z is normally distributed with mean 0 and standard deviation 1. The standardised normal curve is useful because the areas in
a tail associated with given values of z have been determined and
appear in a table. Since areas and probabilities or relative frequencies are related, we can make use of the table in answering such
questions as 'If measurements are normally distributed with mean
μ and standard deviation σ, what percentage of measurements lie
in the range $\mu \pm z\sigma$?'

The Central Limit Theorem informs us that for random samples of
size n drawn from a population $N(\mu, \sigma)$, the sample means are also
normally distributed, $N(\mu, \sigma/\sqrt{n})$. Furthermore, it tells us that this
is approximately true even if the parent population is not itself normally distributed. This information together with the standardising
transformation $z = (\bar{x} - \mu)/(\sigma/\sqrt{n})$ permitted us to estimate a
population mean from the measured value of a sample mean. Generally, we are $(1-\alpha) \times 100\%$ confident that a population mean lies
in the range $\bar{x} \pm z_{\alpha/2}\sigma/\sqrt{n}$.

We next acknowledge that we are often ignorant not only of μ but also of σ. This makes an estimate of μ based on $\bar{x}$ more awkward. We are forced to use s, the sample standard deviation, as an estimate of σ, the population standard deviation. We spent some time convincing ourselves that to do this may be problematical. The sample standard deviation may vary considerably from sample to sample: it can sometimes be much smaller than the population standard deviation and sometimes much larger. It does settle down to approximate to the population standard deviation as the sample size becomes large, but before that its behaviour is quite capricious. This is a pity! It results in our being unable to write out confidence intervals in the form $\mu = \bar{x} \pm z_{\alpha/2} \, s/\sqrt{n}$. Another way of saying this is that $(\bar{x} - \mu)/(s/\sqrt{n})$ is not a normally distributed variable.

This does not mean that sample means are not normally distributed. If we knew σ, we could write

$$z = \frac{\bar{x} - \mu}{\sigma/\sqrt{n}}$$

and be confident that z was a normal variable. What it does indicate is that s, except when n is large, is not a good estimate of σ. The operation $(x - \mu)/(s/\sqrt{n})$ depends not only on the value of $\bar{x}$ but also on the value of s, which can vary considerably except when n is large. Now there is a distribution curve which is similar to the normal distribution curve in that it is symmetrical about its mean and has area under the curve equal to one. The equation for this distribution curve, unlike that of the normal distribution curve, involves an integer ν. When ν is large, the curve is identical with a normal distribution curve. As ν becomes smaller, the shape of this distribution curve becomes flatter. This curve is known as the t-distribution. Here comes the statement you have been waiting for! *It is believed that $(\bar{x} - \mu)/(s/\sqrt{n})$ is a variable which follows a t distribution for $\nu = n - 1$, ie that $t = (\bar{x} - \mu)/(s/\sqrt{n})$.* There are many consequences of this. The one that concerns us here is the consequence for confidence intervals.

Let's try it! A population of measurements has a mean of 24.86 cm^3 and a standard deviation 0.04 cm^3. Unfortunately, a certain chemist is *not* aware of either of these values. He carries out two measurements and obtains the values 24.80 cm^3 and 24.82 cm^3. (You might think this problem is familiar: it is, see Section 3.3.) The mean value for this sample of size 2 is 24.81 cm^3: the sample standard deviation works out to be 0.014 cm^3. (Check it!) We shall write a 95% confidence interval for the population mean. It is obtained from Eq. 3.5:

$$\mu = \bar{x} \pm t\,s/\sqrt{n} \qquad (3.5)$$

How shall we get the appropriate value for t? For $n = 2$, $\nu = 1$. For $\nu = 1$ and area in a tail = 0.025, $t = 12.706$. Check this from Fig. 3.4b. The 95% confidence interval is thus

$$\mu = 24.81 \pm \frac{12.706(0.014) \text{ cm}^3}{\sqrt{2}}$$

$$= 24.81 \pm 0.13 \text{ cm}^3$$

Since we happen to know that the population is 24.86 cm^3, we can see that the estimate of it, a 95% confidence interval based on the relevant distribution, has actually embraced it. Success! How often do you think estimates based on $\bar{x} \pm 12.706\ s/\sqrt{2}$ would embrace the true value of the population mean? Yes you are right; in the long run 95% of such estimates would embrace the true population mean. (This of course assumes that systematic error is absent.)

Think back to the computer exercise in Section 3.3.1. There you tested precisely this point. You had the computer choose for you random samples, of size 2, drawn from a population of mean 24.86 cm^3 and standard deviation 0.04 cm^3. From these 2 values it calculated the mean $\bar{x}$ and the standard deviation s. It then printed out, on each occasion, by what multiple of s the calculated mean $\bar{x}$ differed from the true mean μ. If that multiple was smaller than $12.706/\sqrt{2}$, then the population mean did indeed fall within the range specified by our 95% interval based on $t = 12.706$. Remember you checked that this did work roughly 95% of the time. Of course, you were doing a series of random experiments, hence you were only able to verify that this was *roughly* so. Our belief in the precise t-value rests in the last analysis on our trust in the formal statistical conclusions of Gosset and others. The mathematical details need not concern us because we are in the business of *applying* the tools of statistics, not that of deriving them from first principles. Our faith in t is reinforced by the successful conclusions we are able to draw by using it.

'But,' you might be saying, 'the value of t, for $n = 2$ and an area in the tail of 0.025, is such a large number'. Yes, it is, that is just the point we have been making. We believe that the statistic $(\bar{x} - \mu)/(s/\sqrt{n})$ follows a t-distribution. When $n = 2$ the corresponding curve for t is very flat, and the values which correspond to an area of 0.95 between the tails are $t = \pm 12.706$.

Things do improve fairly rapidly if we increase n. Not only do we benefit from the divisor $\sqrt{n}$ in $\pm ts/\sqrt{n}$, but also the appropriate value of t itself decreases because $\nu(viz\ n - 1)$ is increased.

You have believed in z, believe also in t!

SAQ 3.4a	A barrel of wine was analysed for its ethanol content by gas chromatography. Four replicate measurements were made on a random sample of the wine. The results were: 14.16, 14.23, 14.20 and 14.29% of ethanol (v/v).

(*i*) Calculate the sample mean and the sample standard deviation.

(*ii*) Give a 90% and a 95% confidence interval for the ethanol content of the wine.

(*iii*) If it was known from past experience that the precision of the chromatographic method, as measured by the standard deviation, is 0.065% ethanol (v/v), how would this affect your reporting of the 95% confidence interval for the ethanol content?

SAQ 3.4a

3.4.3. Confidence Intervals: Some Loose Ends

Before leaving confidence levels and confidence intervals, there are three points which must be raised. First, you will recall that when we know the population standard deviation σ, we are $(1-\alpha) \times 100\%$ confident that the population mean lies in the range given below:

$$\mu = \bar{x} \pm z_{\alpha/2}\sigma/\sqrt{n}.$$

When σ is unknown, we shall for the sake of consistency, express our confidence intervals in a similar fashion, ie the $(1-\alpha) \times 100\%$ confidence interval is given by $\mu = \bar{x} \pm t_{\alpha/2}\sigma/\sqrt{n}$ where $t_{\alpha/2}$ is the value for $(n\text{-}1)$ degrees of freedom. Now, a slight problem arises here. The $t_{\alpha/2}$ refers to the t value associated with an area of $\alpha/2$ in one of the tails of the t-distribution. Unfortunately, many authors call the area in a single tail of a t-distribution α instead of $\alpha/2$. Confusion is therefore a distinct possibility! It need not be so. To be forewarned is to be forearmed! Practice in using t-tables will eventually overcome any initial confusion.

The second point is also a matter of notation. We have been making statements such as 'the value of t for $n = 4$ (ie $\nu = 3$) and an area in the tail of 0.025 is 3.182'. We can simplify this to $t_{0.025,3} = 3.182$. It is an easy notation to remember. Generally it is

$$t_{area\ in\ the\ tail,\ degrees\ of\ freedom}.$$

Finally, the extreme values or end values in a confidence interval are called the *confidence limits*. If a confidence interval is written as 14.22 ± 0.09% ethanol (v/v), then the confidence limits are 14.13% and 14.31% ethanol (v/v).

We shall reinforce this part of our studies with a simple exercise.

∏   In each of the following give a 90%, 95% and 99% confidence interval for the population mean. State the confidence limits.

(*i*) $\bar{x} = 2.75\%$, $\sigma = 0.060\%$ and $n = 4$.

(*ii*) $\bar{x} = 0.1040M$, $s = 0.0009M$ and $n = 9$.

(*iii*) $\bar{x} = 400\ \mu gl^{-1}$, $s = 20.5\ \mu gl^{-1}$ and $n = 30$.

(*i*)   The relationship $\mu = \bar{x} \pm z_{\alpha/2}\sigma/\sqrt{n}$ should be used:

  – for a 90% confidence interval, $z_{\alpha/2} = 1.645$

  – for a 95% confidence interval, $z_{\alpha/2} = 1.960$

  – for a 99% confidence interval, $z_{\alpha/2} = 2.576$

The confidence intervals are therefore:

$$90\%,\ \mu = 2.75 \pm \frac{1.645(0.060)}{\sqrt{4}} = 2.75 \pm 0.05\%$$

(Confidence limits 2.70 and 2.80%)

$$95\%,\ \mu = 2.75 \pm \frac{1.96(0.060)}{\sqrt{4}} = 2.75 \pm 0.06\%$$

(Confidence limits 2.69 and 2.81%)

$$99\%, \; \mu \; = \; 2.75 \pm \frac{2.576(0.060)}{\sqrt{4}} \; = \; 2.75 \pm 0.08\%$$

(Confidence limits 2.67 and 2.83%)

(*ii*) The relationship $\mu \; = \; \bar{x} \pm t_{\alpha/2} s / \sqrt{n}$ should be used.

- for a 90% confidence interval with 8 degrees of freedom we require $t_{0.05,8}$. This is 1.860.

- for a 95% confidence interval with 8 degrees of freedom we require $t_{0.025,8}$. This is 2.306.

- for a 99% confidence interval with 8 degrees of freedom we require $t_{0.005,8}$. This is 3.355.

The confidence intervals are therefore:

$$90\%, \; \mu \; = \; 0.1040 \pm 1.860 \, \frac{0.0009}{\sqrt{9}} \; = \; 0.1040 \pm 0.0006 M$$

$$95\%, \; \mu \; = \; 0.1040 \pm 2.306 \, \frac{0.0009}{\sqrt{9}} \; = \; 0.1040 \pm 0.0007 \; M$$

$$99\%, \; \mu \; = \; 0.1040 \pm 3.355 \, \frac{0.0009}{\sqrt{9}} \; = \; 0.1040 \pm 0.0010 \; M$$

(*iii*) For a 90% confidence interval with $\nu = 29$,

$$t_{0.05,29} \; = \; 1.699$$

Now $\mu \; = \; \bar{x} \pm ts / \sqrt{n}$

ie $\mu \; = \; 400 \pm \dfrac{1.699(20.5)}{\sqrt{30}}$

$$= \; 400 \pm 6.4 \; \mu \mathrm{gl}^{-1}$$

It is interesting to reflect upon the confidence interval which would have been obtained had we used the normal distribution relationship for the confidence interval. That is, if we had used $z_{0.05} = 1.645$ instead of $t = 1.699$.

The resulting confidence interval is given below.

$$400 \pm \frac{1.645(20.5)}{\sqrt{30}} = 400 \pm 6.2 \ \mu gl^{-1}$$

This is not too different from the confidence interval obtained by using the t-value. Why is this? Because as n gets bigger the t-distribution approximates to the z-distribution. When $n = 30$, the t distribution and the standardised normal distribution are quite similar. It is for this reason that many statisticians advise that the normal tables may be used when $\nu > 30$.

In this exercise, we shall continue to use the t-tables; you need the practice!

For a 95% confidence interval with $\nu = 29$,

$$t_{0.025,29} = 2.045.$$

Therefore the confidence interval is

$$\mu = \bar{x} \pm ts/\sqrt{n}$$

$$= 400 \pm \frac{2.045(20.5)}{\sqrt{30}}$$

$$= 400 \pm 7.7 \ \mu gl^{-1}$$

For a 99% confidence interval with $\nu = 29$,

$$t_{0.005,29} = 2.756$$

$$\mu = \bar{x} \pm ts/\sqrt{n}$$

$$= 400 \pm \frac{2.756(20.5)}{\sqrt{30}}$$

$$= 400 \pm 10.3 \ \mu gl^{-1}$$

3.4.4. Programming the *t*-calculation

In the last exercise we were dealing with a sample size of $n = 30$. In practice, we would first have to evaluate the sample mean $\bar{x}$ and the standard deviations before we could start looking up the *t*-value and calculating the confidence range for μ. It can all become rather hard work. It is time to develop our computer package a stage further.

Remember that in Part 2 we were able to replace the need to look up the *z*-table by a few lines of programming. We can do the same thing for *t*. First, however, we should emphasise the fundamental difference in presentation between the *z*- and the *t*-table. They work in opposite directions. In Fig. 2.3g you read off *the probability corresponding to a given value of z*, whereas in Fig. 3.4b you obtain *the t-value corresponding to a given probability*. The reason behind this change of approach is that there is a different *t*-distribution for each value of ν and so, in order to present the data for *t* on a single page it has been necessary to be highly selective in the number of values quoted. In fact only 7 values are given for each individual value of ν. In this context it has become generally accepted that the most useful thing to do is to quote *t*-values corresponding to certain selected probability levels (ie corresponding to certain set tail-areas). This is what has been done, and the fact that standard tables are conventionally set out in this fashion has influenced the way in which statistical tests are commonly applied. It would take some 30 pages of tables to present information on *t* in as full detail as that given earlier for *z*.

With a computer to hand we need not be so limited. The first task we shall set ourselves is to show that we can very easily derive the area (ie the probability) corresponding to any chosen *t*-value. Very few lines of programming are required to achieve this.

Perhaps we overstate our claims just a little. We shall have to pay a price for calculating *t* with such brevity of programming. The results produced will not be quite so accurate as they were for *z*. This is, however, a very minor quibble; the results will be fully adequate for ordinary situations. It would be perfectly feasible to construct our program so as to avoid even the smallest discrepancies, but that would have involved you in much more typing. So even if you are a perfectionist, you might be prepared to excuse us for the approach we are taking.

One final preliminary. Our new programming involves taking a natural logarithm. Some versions of BASIC obtain $\ln(x)$ as LOG(X), others as LN(X). Find out which applies on your own machine. This can be done simply by running the one-line program:

    10 PRINT LOG(10)

If the answer obtained is 1.0 then LOG gives logs to base 10 and LN will be the function we need below. If the answer given is 2.303 then LOG is the function we want, in which case you should type LOG where we use LN in the programming below.

Here goes then – load program "STAT2" into your computer. If you have any doubts about the state of your version of this program you can cross-check it against the listing given at the end of Section 2.5. The first thing we shall do is 'fix' the program so that it does our new calculation instead of its previous tasks. Type:

    400 GOTO 630

The effect of this line is to avoid offering the previous menu and to direct attention to a *t*-calculation which will be programmed starting on line 630. Later on we shall insert a new and fuller menu starting on line 400. In the meantime, however, let us set up our *t*-calculation by typing:

```
 630  PRINT"Give nu, t";
 640  INPUT K,T
 650  GOSUB 1200
 660  GOTO 540
1200  REM t            t --> prob
1210  IF ABS(T)<3 THEN 1230
1220  T=T*(1+.037*LN(ABS(T))↑2.5/(2*K-1))
1230  Z=SGN(T)*(8*K+1)*SQR(K*LN(1+T*T/K))/(8*K+3)
1240  GOSUB 1000
1250  RETURN
```

Lines 630 and 640 arrange for the input of ν and t values (inside the program ν is represented by symbol K). The GOSUB statement takes the action on to line 1200. The formulae on lines 1220 and 1230 derive, from the specified values of T and K, a value Z which lies approximately at the corresponding position in a standardised normal distribution. The existing subroutine from line 1000 is then used to calculate the upper tail probability associated with this z-value. We make no attempt to justify the formulae used. We simply leave you to judge the program by the results it produces. The RE-TURN on line 1250 transfers execution back to line 660 (the line following the earlier GOSUB). This in turn passes control to line 540, a line which was already present and which will print out the value of the probability.

RUN the program. For ν and t enter the values 1, 12.706. Our computer responded:

Probability of EXCEEDING this value is 2.505%

Does your value agree? If not, repeat the run (in case you inadvertently typed in the wrong data values). If you still disagree, look carefully for an error in the new lines you have typed, especially lines 1210–1230.

Why did we choose such a strange value of t for which to test your program? Look back again at the t-table (Fig. 3.4b), at the first row. Yes, for one degree of freedom 12.706 is the tabulated value for t corresponding to a probability of 2.500%! The program has reproduced this result to within 2 parts in 1000.

Test the program against other values in the t-table. For instance, try $\nu = 6$, $t = 1.440$ and see how close the calculated probability is to 0.10. You should find that you get quite good agreement with the table no matter which entry you choose to test. To reproduce the bottom row of the table simply choose a very large number of degrees of freedom, such as $\nu = 1000$.

You will notice that the program leaves you stuck in an infinite loop! After each calculation it asks you to give it the next one. To stop the program you will have to press the 'escape' key, or its equivalent on your computer. This rather frowned-upon feature will be removed later.

Now the program is much less limited than the table, as we can easily illustrate.

∏   (*i*)   Show that $t = 0$ corresponds to a probability of 50% whatever the value of ν is.

      (*ii*) For 10 degrees of freedom, what t-value corresponds to a probability of 0.0005?

      (*iii*) Find the probability associated with $t = 2.5$ for $\nu = 1$. Repeat for $\nu = 5, 10, 20, 30, 1000$. Comment.

      (*iv*) Find the probability corresponding to $t = -2.0$, for $\nu = 5$.

(*i*)   The fact that a probability of 50% is found for $t = 0$ reflects the fact that the t-distribution is symmetric about $t = 0$. We should remind ourselves that the probability we obtain on each occasion represents the chance of occurrence of a t-value *larger* than the value quoted. The fact that for $t = 0$ the probability is 0.5 means that, whatever the value of ν, there is always a 50% probability of a positive t and hence, also, a 50% chance of a negative t.

(*ii*) Here you would have to experiment with different t-values, keeping $\nu = 10$, to try to obtain from the program a probability as close as possible to 0.0005 (ie 0.05%). This should lead you to a t-value of approximately 4.58. The hit-and-miss process necessary to find the answer to a problem like this is something we shall improve upon below.

(*iii*) This problem asked us to find the probability of occurrence of a t-value greater than 2.5, for various values of ν. You should obtain:

ν	1	5	10	20	30	1000
Probability	12.5%	2.7%	1.6%	1.1%	0.91%	0.63%

This set of results illustrates the steady narrowing of the *t*-distribution towards the standardised normal curve as the number of degrees of freedom grows (see Fig. 3.4a). The area in the tail of the *t*-distribution above $t = 2.5$ falls by a factor of about 20 between $\nu = 1$ and $\nu \to \infty$. Notice that the tail area diminishes most rapidly at low values of ν, but note also that there is still a significant effect even after $\nu = 30$.

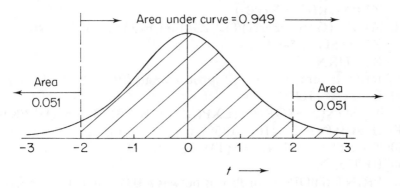

Fig. 3.4f. *The t-curve for $\nu = 5$ and the areas associated with $t = +2.0$ and $t = -2.0$.*

(*iv*) With a negative value of *t*, we obtain a probability greater than 0.50. In this case, for $t = -2.0$ and $\nu = 5$, the program gives a probability of 0.949. In other words 94.9% of the time *t* will exceed -2.0. The 'upper tail' of the distribution has grown into the major part of it. The chance of obtaining a *t*-value *more negative* than -2.0 will be 0.051 (ie $1.000 - 0.949$). This latter value is the area in the lower tail below $t = -2.0$ (see Fig. 3.4f). Again you could check the symmetry of the *t*-distribution by testing $t = +2.0$ for $\nu = 5$. You should find indeed that the resulting probability is 0.051.

So far so good, but we would like to be able to automate the calculation of confidence limits for the population mean, starting with a set of experimental replicate measurements.

To do that we need to be able to calculate the *t* (or *z*) value corresponding to a specified probability, rather than the other way round.

This requires two new subroutines, which you should now incorporate in your program by typing the following new lines.

```
  30  READ G1,G2,G3,G4
  40  DATA 2.30753,0.27061,0.99229,0.04481
1100  REM Inverse normal          prob --> z
1110  IF (P<0 OR P>1) THEN 9600
1120  P0=.5-SGN(.5-P)*(.5-P)
1130  Z1=SQR(-2*LN(P0))
1140  Z=Z1-(G1+G2*Z1)/(1+Z1*(G3+G4*Z1))
1150  Z=Z*SGN(.5-P)
1160  RETURN
1300  REM Inverse t            prob --> t
1310  GOSUB 1100
1320  T=SGN(Z)*SQR(K*(EXP((Z*(8*K+3)/(8*K+1))↑2/K)-1))
1330  IF ABS(T)<3 THEN 1350
1340  T=T/(1+.05*LN(ABS(T))↑2/(2*K-1))
1350  RETURN
9600  PRINT"STOP  *  prob not between 0.0 and 1.0!"
```

Lines 1100–1150 provide a new subroutine to perform the inverse of our previous normal distribution calculation. From a specified probability P it calculates the corresponding z-value. The formula to achieve this involves parameters labelled G1–G4 whose values are set via lines 30 and 40. Lines 1300–1350 allow t to be calculated, given P and K. K is again used as the programming symbol for ν. Notice that the inverse t subroutine first finds the z-value for the specified P. This value is then used in a formula to estimate the corresponding t for the appropriate number of degrees of freedom, K. Yet again we make no effort to justify the formulae used.

We must still type in some further programming so that we can make use of the new subroutines. We shall take the opportunity to introduce a new 'menu' section which will allow you to make full use of any of the four subroutines which we now have for the two statistics z and t. In doing this we shall overwrite the previous instructions on lines 400 to 500. You are getting quite a lot of typing to do in this Section! Don't despair; in this step we are making our biggest single stride towards our aim of obtaining a flexible and quite powerful package. Type the following lines.

```
400  PRINT"1= prob -> z     2= z -> prob"
410  PRINT"3= prob -> t     4= t -> prob"
420  PRINT"5=stop"
430  INPUT I
440  IF I>1 THEN 500
450  PRINT"Give one-tail prob ";
460  INPUT P
470  GOSUB 1100
480  PRINT"Corresponding z-value = ";INT(Z*100+.5)/100
490  GOTO 390
500  IF I>2 THEN 560
560  IF I>3 THEN 620
570  PRINT"Give nu, one-tail prob ";
580  INPUT K,P
590  GOSUB 1300
600  PRINT"Corresponding t-value = ";INT(T*1000+.5)/1000
610  GOTO 390
620  IF I>4 THEN 670
670  GOTO 9999
```

When you RUN the revised program you should be presented with the menu below.

$$1 = \text{prob} \rightarrow z \quad 2 = z \rightarrow \text{prob}$$
$$3 = \text{prob} \rightarrow t \quad 4 = t \rightarrow \text{prob}$$
$$5 = \text{stop}$$

First let's test the new z subroutine. Select option 1, then give 0.025 when asked to specify the required one-tail probability. You should get the response:

Corresponding z-value = 1.96

If there is a discrepancy in your z-value, then search for a mistake in lines 1110–1150 or in 30–40. Take care for instance that you have not typed letters I or O in place of numbers 1 or 0. Once your program agrees with the above value try option 1 for other one-tail probability levels. Cross-check the quoted z-values either with the z-table (Fig. 2.3g) or with the bottom row of the t-table (Fig. 3.4b). (Remember that when ν becomes large t becomes identical with z.)

Now we can check the new t subroutine. Select option 3 and, when asked for ν and the one-tail probability enter 1, 0.025. The computer's response should be:

Corresponding t-value = 12.483

If your result does not agree then you will need to check for a programming error. The most likely place to have made a slip is in lines 1310–1340. Having managed to get your program to agree with ours now consult once more the t-table (Fig. 3.4b). The value of t which corresponds to P = 0.025 and ν = 1 should be 12.706. The program obtains 12.483, a result on the low side by 1.8%. For any situation we are likely to meet, this is too small a discrepancy to matter: it is the price we must pay for keeping our programming reasonably short. Test the program against other entries in the t-table. You should notice that as the number of degrees of freedom is increased the program results become very close to the exact values.

Quite a lot has been achieved in this Section. The program you now have is growing quite substantial, and you have built it brick by brick! We shall make a further addition in the next Section so that with the program we can calculate confidence limits for the population mean from sample measurements. If you intend to take a well-deserved rest before you continue, be sure to save the latest version of the program! Use any name for it you like.

3.4.5.   Programming to find Confidence Intervals

After the last Section we are most of the way through what will be the biggest single extension of our program. So far we have simply been adding new features and using each of them independently. Now we will use the very earliest segments of the program to help us to complete our task of deriving confidence limits for the population mean, given a particular set of experimental measurements. LIST lines 4000 and 4010 of your program. You should find that they still contain DATA for the twelve replicate determinations which we first introduced in SAQ 1.5b and which we also examined in Section 2.5.

4000 DATA 12
4010 DATA 27.86, 27.68, 28.02, 28.05, 27.60, 27.55, 27.03, 27.76,
27.73, 27.75, 27.77, 27.53

These values, you may remember, represented a random sample from a population whose true mean was 27.66 cm^3 and whose population standard deviation was 0.32 cm^3. You will also find these two values within the program if you LIST line 3900.

3900 DATA 27.66,0.32

You should make sure that the DATA statements in the current version of your program agree with the above. Edit the lines if necessary.

We wish to add to our program a subroutine to calculate confidence intervals for the mean derived from the array of replicate values held in the program (ie in line 4010). To achieve this enter the following new lines.

```
1690 RETURN
1700 PRINT"Bracketing the mean"
1710 PRINT"... give req'd confidence (prob<1.00)";
1720 INPUT P
1730 IF P>.99999 THEN 1710
1740 P=(1-P)/2
1750 PRINT"Which st dev:  1=sample   2=population";
1760 INPUT J
1770 GOSUB 1640
1780 IF J=2 THEN 1830
1790 K=N-1
1800 GOSUB 1300
1810 W=T*S/SQR(N)
1820 GOTO 1860
1830 PRINT"Population st. dev. = ";INT(S9*1000+.5)/1000
1840 GOSUB 1100
1850 W=Z*S9/SQR(N)
1860 PRINT"It is predicted with confidence ";(1-2*P)
1870 PRINT"that the mean lies in the range:"
```

1880 PRINT"   ";INT((M-W)*100+.5)/100;" to ";

INT((M+W)*100+.5)/100

1890 RETURN

The RETURN instruction on line 1690 simply converts the preceding segment of the program into yet another subroutine. That earlier segment starts on line 1640 and is in fact the very earliest part of the program you wrote, away back in Part 1, to calculate the mean and standard deviation of the set of measurements contained in the DATA line 4010. Notice the GOSUB 1640 on our new line 1770; our new segment actually uses that earlier work! We will not discuss in detail how our new segment does its job. (In fact its operation is not all that complicated, it's just that we want to move on.)

So that we can direct the program actually to perform our new calculation we have to extend the menu section. Lines to do this are as follows.

360 PRINT"1=tables   2=sample   3=stop";
370 INPUT I
380 IF I>1 THEN 780
780 IF I>2 THEN 9999
790 PRINT"- -"
800 PRINT"1=output mean & st.dev"
810 PRINT"2=bracket population mean"
840 INPUT I
850 IF I<1 OR I>2 THEN 9999
860 ON I GOSUB 1640,1700
870 GOTO 790

Line 860 is of a structure which was not explained in the Appendix. Notice that this line is only reached if I (input on line 840) is either 1 or 2. If I=1 line 860 directs the program to the subroutine beginning on line 1640 (which is the one required for option 1). If I=2 then we go to the new subroutine beginning on line 1700. The "ON I GOSUB" line is an efficient way of achieving this result, which saves several lines of typing.

RUN the program. The early lines (below 300) ensure that the values on the DATA line 4010 are read in, and that their mean and standard deviation are calculated (but not printed). The program has in fact been doing this in previous runs also, but this will be the first occasion for some time on which we shall make use of the fact. The first obvious response of the program occurs when it reaches line 360. It prints the following.

> 1=tables   2=sample   3=stop?

If we take option 1 the program will not be deflected by line 380 and will run through to line 390 where the previously written menu for finding z- and t-values will be obtained. The option we want just now is option 2, which will deal with the sample data from line 4010 (contained in array X).

Selecting option 2 means that the program's action jumps (from line 380) to 780 then 790, where a new menu is offered as follows.

> 1=output mean & st.dev
> 2=bracket population mean

If you type 1 now you will meet a long lost friend! The sample mean and standard deviation will be printed out, just as you arranged for it to happen in Part 1 (for the set of 12 values from DATA line 4010). After that you will be returned to the present menu. Clearly what interests us now is option 2. Give 2 as your input and our new subroutine should swing into operation.

> Bracketing the mean
> ... give req'd confidence (prob<1.00)?

Let us find the 95% confidence interval. You should therefore enter 0.95 as the 'required confidence'. This 95% confidence allows for an area of 0.025 (2.5%) in each tail. In fact the program works out the t- or z-value corresponding to the upper-tail area[*].

---

[*] The program actually works out the upper-tail area from the specified confidence level in line 1740. Here it will obtain $P = 0.025$.

The next response to the screen is given below.

which st dev   1 = sample   2 = population?

Let us suppose that the population standard deviation may be unknown. Enter 1 as your choice. The program will now carry out a t-calculation, based on the sample standard deviation of our 12 values. The output is below.

No of values   = 12
Sample mean = 27.694
Est. st. dev.   = 0.2655
% Rel.st.dev. = 0.957
It is predicted with confidence 0.95
that the mean lies in the range:
  27.53 to 27.86

The true population mean, which we happen to know in this instance, is 27.66. This value lies comfortably within the predicted range.

We also happen to know the true population standard deviation in this case, so we could have taken choice 2 above instead of 1. Let us try that now. Take option 2 again ('bracket population mean'). Again give 0.95 as the required confidence level but this time enter 2 as the choice for the standard deviation. The program holds the value of the population standard deviation (as variable S9), and now uses this in a z-calculation, with results below.

No. of values   = 12
Sample mean   = 27.694
Est. st. dev.   = 0.265
% Rel.st.dev.   = 0.957
Population st. dev. = 0.32
It is predicted with confidence 0.95
that the mean lies in the range:
  27.51 to 27.88

Note that the predicted range is almost the same as before. This does not always happen – quite often in practice the range based on σ turns out significantly narrower than one based on s.

Before you continue, save your new program. We shall refer to it as "STAT3".

What if we had had fewer than 12 measurements to go on? We can investigate the consequences of this quite easily. Simply retype line 4000.

      4000 DATA 2

Now when we RUN the program only the first two measurements on line 4010 will be considered. Check what happens to the 95% confidence limits. By relying on the sample standard deviation estimated from these two values, the program predicts that the mean lies in the range 26.65 to 28.89. If on the other hand, the population standard deviation is used the reported range is 27.33 to 28.21.

Notice that here the range based on the sample standard deviation is more than twice as broad as that based on σ. This is because of the very large value of t appropriate to this example. We have only one degree of freedom ($n = 2$ means $\nu = 1$), so for a tail-area of 0.025 we have $t = 12.706$. On the other hand the sample standard deviation is considerably underestimated.

      Est. st. dev. = 0.127

The first two measurements happen to be relatively close together. (If you check line 4010 you can confirm that they are 27.86 and 27.68.) The value of s happens to come out almost a factor of three smaller than the true population standard deviation of 0.32. As we have said before this sort of thing can happen to us from time to time, and it is in order to cope with this possibility that the appropriate t-value is so high in the first place. The final outcome is quite satisfactory to the extent that the predicted range does indeed encompass the true population mean of 27.66.

SAQ 3.4b

Use program "STAT3" to obtain the 95% confidence range for the mean derived from the first three replicate measurements noted in DATA line 4010. Compare the results obtained by using the sample standard deviation with those based on the population standard deviation. Repeat the process considering in turn the first 4, 6, 8, and finally, 10 measurements.

SAQ 3.4c

The level of manganese in a sample of steel was investigated by atomic absorption spectroscopy. Ten replicate determinations gave the following results.

% Mn: 0.957, 0.922, 0.839, 0.803, 0.724, 0.857, 0.816, 0.918, 0.767, 0.747

Use program "STAT3" to derive the 99% confidence interval for the mean Mn level in the steel. Examine also the 90%, 95% and the 99.9% confidence intervals.

SAQ 3.4c

During the last two sections we have substantially developed our program. So that if the need arises you can cross-check details, we give below a full listing of the new version 'STAT3'. We shall develop the package further in Part 4 of the Unit, to its final form. A number of REM statements have been added to the listing below so that it is easier to sort out which part of the now rather large program is doing what.

```
 9  REM Parameters used in z subroutines
10  READ D1,D2,D3,D4,D5,D6
20  DATA 4.98673470E-2,2.11410061E-2,3.2776263E-3,
                   3.80036E-5,4.88906E-5,5.3830E-6
30  READ G1,G2,G3,G4
40  DATA 2.30753,0.27061,0.99229,0.04481
49  REM M9 and S9 are pop mean and s.d. for data in array X
50  READ M9,S9
80  DIM X(30)
99  REM Read N values into array X, calc sample mean and s.d.
```

```
100  READ N
110  A=0
120  FOR I=1 TO N
130     READ X(I)
140     A=A+X(I)
150     NEXT I
160  M=A/N
170  A=0
180  FOR I=1 TO N
190     A=A+(X(I)-M)↑2
200     NEXT I
210  S=SQR(A/(N-1))
220  R=100*S/M
359  REM Main menu ************************
360  PRINT"1=tables   2=sample   3=stop";
370  INPUT I
380  IF I>1 THEN 780
389  REM Menu for statistical tables *******
390  PRINT"- - - - - - - - - - - - - - - - - - - - - - - - - - - - - - - -"
400  PRINT"1= prob -> z      2= z -> prob"
410  PRINT"3= prob -> t      4= t -> prob"
420  PRINT"5=stop"
430  INPUT I
440  IF I>1 THEN 500
450  PRINT"Give one-tail prob ";
460  INPUT P
470  GOSUB 1100
480  PRINT"Corresponding z-value = ";INT(Z*100+.5)/100
490  GOTO 390
500  IF I>2 THEN 560
510  PRINT"Give z ";
520  INPUT Z
530  GOSUB 1000
540  PRINT"Prob of EXCEEDING this value is ";
                                   INT(P*100000+.5)/1000;"%"
550  GOTO 390
560  IF I>3 THEN 620
570  PRINT"Give nu, one-tail prob ";
580  INPUT K,P
590  GOSUB 1300
```

```
600  PRINT"Corresponding t-value = ";INT(T*1000+.5)/1000
610  GOTO 390
620  IF I>4 THEN 670
630  PRINT"Give nu, t";
640  INPUT K,T
650  GOSUB 1200
660  GOTO 540
670  GOTO 9999
780  IF I>2 THEN 9999
789  REM Menu for analysis of sample *********
790  PRINT"- - - - - - - - - - - - - - - - - - - - - - - - - - - - - - - - - -"
800  PRINT"1=output mean & st.dev"
810  PRINT'2=bracket population mean"
840  INPUT I
850  IF I<1 OR I>2 THEN 9999
860  ON I GOSUB 1640,1700
870  GOTO 790
1000 REM One-tail normal: z --> prob
1010 Z9=ABS(Z)
1020 P=1+Z9*(D1+Z9*(D2+Z9*(D3+Z9*(D4+Z9*
                                    (D5+Z9*D6)))))
1030 P=.5/P↑16
1040 P=SGN(Z)*(P-.5)+.5
1050 RETURN
1100 REM Inverse normal   prob --> z
1110 IF (P<0 OR P>1) THEN 9600
1120 P0=.5-SGN(.5-P)*(.5-P)
1130 Z1=SQR(-2*LN(P0))
1140 Z=Z1-(G1+G2*Z1)/(1+Z1*(G3+G4*Z1))
1150 Z=Z*SGN(.5-P)
1160 RETURN
1200 REM t       t --> prob
1210 IF ABS(T)<3 THEN 1230
1220 T=T*(1+.037*LN(ABS(T))↑2.5/(2*K-1))
1230 Z=SGN(T)*(8*K+1)*SQR(K*LN(1+T*T/K))/(8*K+3)
1240 GOSUB 1000
1250 RETURN
1300 REM Inverse t      prob --> t
1310 GOSUB 1100
1320 T=SGN(Z)*SQR(K*(EXP((Z*(8*K+3)/(8*K+1))↑2/K)-1))
```

```
1330  IF ABS(T)<3 THEN 1350
1340  T=T/(1+.05*LN(ABS(T))↑2/(2*K-1))
1350  RETURN
1640  REM Show sample mean and st dev
1650  PRINT"No of values = ";N
1660  PRINT"Sample mean = ";INT(M*1000+.5)/1000
1670  PRINT"Est. st. dev.= ";INT(S*1000+.5)/1000
1680  PRINT"% Rel.st.dev.= ";INT(R*1000+.5)/1000
1690  RETURN
1700  PRINT"Bracketing the mean"
1710  PRINT"... give req'd confidence (prob<1.00)";
1720  INPUT P
1730  IF P>.99999 THEN 1710
1740  P=(1-P)/2
1750  PRINT"Which st dev:  1=sample   2=population";
1760  INPUT J
1770  GOSUB 1640
1780  IF J=2 THEN 1830
1790  K=N-1
1800  GOSUB 1300
1810  W=T*S/SQR(N)
1820  GOTO 1860
1830  PRINT"Population st. dev. = ";INT(S9*1000+.5)/1000
1840  GOSUB 1100
1850  W=Z*S9/SQR(N)
1860  PRINT"It is predicted with confidence ";(1-2*P)
1870  PRINT"that the mean lies in the range:"
1880  PRINT"     ";INT((M-W)*100+.5)/100;" to ";
                                    INT((M+W)*100+.5)/100
1890  RETURN
3900  DATA 27.66,0.32
4000  DATA 12
4010  DATA 27.86, 27.68, 28.02, 28.05, 27.60, 27.55, 27.03, 27.76,
                                    27.73, 27.75, 27.77, 27.53
9600  PRINT"STOP   * prob not between 0.0 and 1.0!"
9999  END
```

3.5. HYPOTHESIS TESTING

We have been discussing the problem of estimation. We have, in the light of knowledge of a sample mean and standard deviation, been able to estimate a population mean. We now turn our attention to another type of problem; it is *deciding whether or not a population mean has a particular stated value*. The process we carry out in order to make such decisions is known as *hypothesis testing*. Before we start our formal treatment of this topic, let us remind ourselves that statistics and probability are interrelated. When we use phrases such as 'we are $(1-\alpha) \times 100\%$ confident that the population mean lies in the range $\bar{x} \pm z_{\alpha/2}\sigma/\sqrt{n}$', we are making use of our knowledge of the probability distribution function for the variable $\bar{x}$. The truth or falsity of such a statement is never known with absolute certainty; we are never 100% confident. What we are doing when we make such statements is what professional gamblers call 'playing the odds'. On $(1-\alpha) \times 100\%$ of the occasions, the population mean will be embraced by the interval $\bar{x} \pm z_{\alpha/2}\sigma/\sqrt{n}$; however, on $100\alpha\%$ of the occasions it will not.

Keep this in mind as we develop our arguments about hypothesis testing.

3.5.1. Testing Assertions

Every chemist is confronted daily with assertions about materials: the sodium hydroxide solution is 50.0% w/v; the propan-1-ol is 99.8% pure; the EDTA solution is 0.0105 M. He is confronted too with assertions about methods: a given spectrophotometric method gives the same result for the lead content of a material as a given potentiometric method; for determinations of calcium at the 2 ppm level, the precision obtained with flame atomic absorption is the same as that obtained with flame emission spectrophotometry. It is often necessary to test the truth of such assertions. For example, in the bulk buying of sodium hydroxide solution, the amount paid depends upon the agreed assay for the material. A tank car of a caustic soda solution might, on delivery, be labelled 50.0% (w/v) sodium hydroxide. Is this statement true? We, as chemists, know how to determine the NaOH content of a solution: we take a sample and carry out several replicate acid-base titrations on the material. It is unlikely that we would obtain a result that was exactly 50.0% (w/v) NaOH. What sort of leeway should we permit the supplier? Should we reject his assertion about the caustic soda solution if we obtain a result 49.6% (w/v) NaOH? Would 49.2% (w/v) NaOH be considered good enough? Obviously we must have previously agreed upon a decision-making criterion for accepting or rejecting the supplier's assertion about his material. For any student of statistics, it is equally obvious that the criterion must have a statistical basis. (Just in case there is one reader who still wonders why this should be so, let us drop a hint – means of random samples are normally distributed!) There are several procedures which could be used as tests of the supplier's assertion that his material is 50.0% (w/v) NaOH. We shall use the formal procedure known as *hypothesis testing*. This procedure is quite general. A little effort is required to become familiar with it. However, once mastered, it can be carried out almost without thinking, well, without too much thinking!

3.5.2. Hypothesis Testing – the Null Hypothesis

In statistics, a hypothesis is an assertion, statement, or conjecture about a measurable property or event. The supplier's assertion that the NaOH content of a solution is 50.0% (w/v) is thus a statistical hypothesis. So is the assertion that the NaOH content of a solution is not equal to 50.0% (w/v). Since it is the supplier's assertion that is to be tested, *our starting point is the hypothesis that the NaOH content is 50.0% (w/v). This is called stating the null hypothesis*. The use of the word 'null' is rather interesting. In the past, any hypothesis which was made with the hope that it would be rejected was called the null hypothesis. Nowadays, the *null hypothesis* refers to any hypothesis that we wish to test. It is denoted by the symbol H_0. In our example we have H_0: $\mu = 50.0\%$. If on testing it, we decide to reject the null hypothesis, we must be ready to accept an *alternative hypothesis*. In our example, a reasonable alternative hypothesis would be that the NaOH content is *not* 50.0% (w/v). The alternative hypothesis is denoted as H_1; we thus have H_1: $\mu \neq 50.0\%$. The next step in our procedure for testing the truth of the null hypothesis is to assume that it is true! This might seem like a silly thing to do but there are very logical reasons underlying our action. If the null hypothesis is true, we can immediately make statements about samples which are drawn from the population. For example, 95% of all samples of size n drawn from the population will have a mean NaOH content in the range below.

$$\text{Range} = 50.0 \pm 1.96 \, \sigma/\sqrt{n}$$

Equivalently we could say that 5% of such samples would fall outside of the range $50.0 \pm 1.96 \, \sigma/\sqrt{n}$. Let us assume that we know from past experience that $\sigma = 0.30\%$. The consequences of this for samples of size 4 drawn from the NaOH solution are illustrated in Fig. 3.5a.

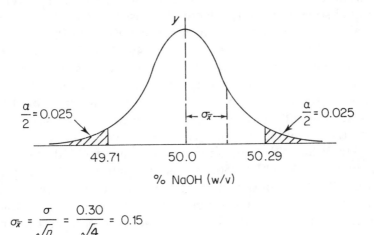

$$\sigma_{\bar{x}} = \frac{\sigma}{\sqrt{n}} = \frac{0.30}{\sqrt{4}} = 0.15$$

Fig. 3.5a. *The distribution of means for samples of size 4 drawn from a population with $\mu = 50.0\%$ and $\sigma = 0.30\%$*

The distribution-of-means diagram shows that 95% of all samples of size 4 will have means which lie in the range 49.71–50.29% or, equivalently that 5% of such samples have means which fall outside this range. Now, here comes the part of our procedure which students find tricky! We are going to lay down a *decision-making criterion*: it is that if a sample of size 4 has a mean which lies in the range 49.71–50.29%, then we are willing to accept that the sample mean is compatible with a population mean of 50.0%; ie we are willing to accept the null hypothesis H_0: $\mu = 50.0\%$. If, however, the sample has a mean which falls outside of this stated range, then we are not willing to accept that the sample mean is compatible with a population mean of 50.0%. This is synonomous with saying that we shall reject the null hypothesis and accept the alternative hypothesis, H_1: $\mu \neq 50.0\%$. 'What are they doing?' you are probably asking. 'A sample of size 4 with a mean of say 49.6% could very well have come from a population with a mean of 50.0%.' We agree, but must point out that we do not know what the population mean really is. It might be 50.0% but then again, it might be 49.5%. The only way that we can be sure about the true value of the population mean is to analyse the whole tank car; to carry out a very large number of determinations. Since we are unlikely ever to do something so silly, we have decided to use statistics to help us. We have said that if the population mean is 50% then 95% of samples of size 4 will have means in the range 49.71–50.29%. We then decided to state a decision-making criterion. It was that if the mean of a sample of size 4 lay in this range, we would be willing to accept that there did not appear to be anything incompatible between this mean and the asserted population mean. This does not imply that the true population mean is 50.0%. It could be 49.5%, 49.2%, 50.6% etc. What it does mean is that we have decided that a sample mean in the range 49.71–50.29% is compatible with a population mean of 50.0%. It follows that we have also decided that a sample mean outside of the stated range is not considered compatible with a population mean of 50.0%. But, as we have pointed out on your behalf, a sample of size 4 could have a mean which falls outside of this range. Yes, it could. What we are doing is 'playing the odds'. *If the population mean is 50.0%*, then 5% of all samples of size 4 will fall outside of the stated range and this will result, on 5% of all occasions, in the wrong decision, ie in the incorrect rejection of the null hypothesis, H_0: $\mu = 50.0\%$. *If the population mean is 50.0%*, then on 95% of all occasions, the right decision will be made, ie we shall accept the null hypothesis H_0: $\mu = 50.0\%$. Stated in probability terms, this means that *if the population mean is 50.0%*, then our decision-making criterion results in a probability such that 19 times out of 20 we should make the right decision and only once in twenty times should we make the wrong decision. Gamblers would be happy with such odds. For the time being, let us be too!

3.5.3. Level of Significance and Type I Error

Focus your attention on the 5% of all occasions that a wrong decision could be made. This percentage was decided upon by us when we set up the decision-making criterion. It is called the *level of significance*. *The level of significance is the probability of incorrectly rejecting the null hypothesis, ie of not accepting the null hypothesis when it is true. When we incorrectly reject the null hypothesis, we make what is known as a type I error. The level of significance is thus the probability of making a type I error*.

Is it possible to reduce the probability of making a type I error? Yes, we simply reduce the level of significance. For example, in the hypothesis test, $H_0: \mu \neq 50.0\%$ we could decide that $(1-\alpha)$ should be 99%, or equivalently that $\alpha = 1\%$. This means that we are willing to accept the null hypothesis if the mean of a sample of size 4 falls in range $50.0 \pm 2.576 \times 0.30/(\sqrt{4}) = 50.0 \pm 0.39\%$. This is wider than the previous range $(50.0 \pm 0.29\%)$. If the population mean is 50.0%, then there is a 99% probability that a sample of size 4 will have a mean in the stated range and hence a 99% probability that we shall correctly accept the null hypothesis if it is true. There is, for a population with $\mu = 50.0\%$, a 1% probability that a sample of size 4 will have a mean outside of the stated range and hence a 1% probability of incorrectly rejecting the null hypothesis when it is true, ie of committing type I error. The value for α, the level of significance, determines the probability of committing a type I error. 'Fine', you are probably thinking, 'let us always have a small value for the level of significance.' Hold on! Think about the consequence of your suggestion. A small value for the level of significance results in a wider range for the null-hypothesis acceptance region. If we make α very small, we could end up accepting almost any sample mean as being compatible with a population mean equal to 50.0%! Admittedly, if the population mean was indeed 50.0%, then a very small α would result in a very low probability of making a type I error. However, we do not know that the population mean is 50.0%. That is why we are carrying out the test!

If we insist on having a low probability of committing a type I error, ie of incorrectly rejecting the null hypothesis *when it is true*, we increase the probability of accepting the null hypothesis when it is false, ie of accepting that a sample belongs to a population with the stated mean, when it in fact belongs to a population with a different mean. *Accepting the null hypothesis when it is false is termed committing a type II error.* This matter requires some further discussion.

3.5.4. Type I and Type II Error

A type I error arises when we incorrectly reject the null hypothesis, a type II error arises when we incorrectly accept the null hypothesis. In Fig. 3.5b below we have represented the distribution of means for samples of size 4 taken from a population with mean 50.0% and standard deviation 0.30%. This is the distribution which exists if the assertion that $\mu = 50.0\%$ is true. We have shown on this distribution the *critical values* are 49.71% and 50.29%. These values are termed critical because we have decided to reject the null hypothesis H_0: $\mu = 50.0\%$ if a sample mean falls outside of the range defined by the values. The areas of the distribution curve which fall outside of the critical values have been cross-hatched with vertical lines. The total vertically cross-hatched area represents the level of significance, which, as you will recall, is 5%. The level of significance defines the probability of making a type I error, ie of falsely rejecting the null hypothesis *if it is true*.

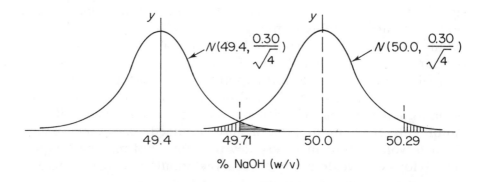

% NaOH (w/v)

Fig. 3.5b. *The distribution of means for samples of size 4 taken from: (i) the population with $\mu = 50.0\%$ and $\sigma = 0.30\%$, (ii) the population with $\mu = 49.4\%$ and $\sigma = 0.30\%$. The critical values at the 5% level of significance for (i) are indicated*

If a sample of size 4, taken from the population with $\mu = 50.0\%$ and $\sigma = 0.30\%$, had a mean which fell outside the critical values we would conclude that the sample mean was incompatible with a population mean of 50.0% and would thus make a type I error.

Now consider the second distribution of sample means shown on the left in Fig. 3.5b. It represents the distribution of means for samples of size 4 taken from a population with mean 49.4% and standard deviation 0.30%. Let us now suppose that this population is actually the true one, ie that the material which is labelled 50.0% (w/v) NaOH is actually 49.4% (w/v) NaOH. As can be seen from Fig. 3.5b, the distribution of sample means for the true population ($\mu = 49.4\%$) overlaps the distribution of sample means for the assumed population ($\mu = 50.0\%$). The consequences of this are very interesting. In setting up the test for the assertion that $\mu = 50.0\%$ we set out critical values at 49.71% and 50.29%. Now a sample of size 4 drawn from a population of mean 49.4% could very well have a value which fell in the acceptance region for the null hypothesis H_0: $\mu = 50.0\%$. We can easily calculate the probability of this occurring. It will occur for any value above 49.71. The probability is represented by the tail in the curve $N(49.4, 0.30/\sqrt{4})$: this tail is cross-hatched with horizontal lines. The area in this tail is obtainable in the usual way from the relationship

$$z = \frac{\bar{x} - \mu}{\sigma/\sqrt{n}}$$

When $\bar{x} = 49.71$, $\quad z = \dfrac{49.71 - 49.4}{0.30/\sqrt{4}} = 2.067$

When $z = 2.067$ the area in the tail is 0.019 or 1.9%. Therefore, there is a probability of 1.9% that a sample of size 4 drawn from a population of mean 49.4 and standard deviation 0.30 will have a value which falls within the acceptance region for the null hypothesis, H_0: $\mu = 50.0\%$, when the level of significance for the test is 5%. We will say this again in a different way. At the 5% level of significance, there is a 1.9% probability that we shall falsely accept the null hypothesis, H_0: $\mu = 50.0\%$ when we draw samples of size 4 from a population of mean 49.4%. *When we incorrectly accept a null hypothesis we make what is known as a type II error.* We accept a batch which should have been rejected.

We will now examine the consequence of reducing the level of significance, ie of reducing the probability of making a type I error. In Fig. 3.5c we have once again represented the distribution of means for samples of size 4 taken from a population of mean 50.0% and standard deviation 0.30% (the asserted population) and the distribution of means taken from a population with mean 49.4% and standard deviation 0.30% (the actual population in the example). We have indicated on the first of these distributions the critical values which result from choosing a level of significance of 1% for testing the null hypothesis, H_0: $\mu = 50.0\%$. When the area in both tails is 1% the area in one tail is 0.5% or 0.005. The z-value for an area of 0.005 in a tail is 2.576. The acceptance range is therefore

$$50.0 \pm 2.576 \frac{(0.30)}{\sqrt{4}}$$

The critical values work out as 49.61 and 50.39. We have not bothered to cross-hatch the areas of the distribution curve which fall outside these critical values. We have, however, cross-hatched the area in the tail of the distribution $N(49.4, 0.30/\sqrt{4})$ which corresponds to the probability of a sample of size 4 (taken from the population $N(49.4, 0.30)$) having a value greater than 49.61%. This probability is calculated below.

$$z = \frac{\bar{x} - \mu}{\sigma/\sqrt{n}} = \frac{49.61 - 49.4}{0.30/\sqrt{4}} = 1.4$$

when $z = 1.4$, the area in the tail is 0.081 or 8.1%.

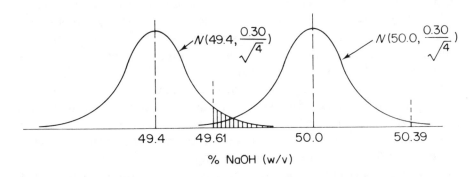

% NaOH (w/v)

Fig. 3.5c. *The distribution of means of samples of size 4 taken from: (i) the population with $\mu = 50.0\%$ and $\sigma = 0.30\%$, (ii) the population with $\mu = 49.4\%$ and $\sigma = 0.30\%$. The critical values at the 1% level of significance for (i) are indicated*

This means that at the 1% level of significance there is a probability of 8% that we shall falsely accept the null hypothesis, H_0: $\mu = 50.0\%$ when we take samples of size 4 from a population of mean 49.4%.

We are now in a position to put all of the above together! We have been involved in the testing of the null hypothesis H_0: $\mu = 50.0\%$ (w/v) NaOH. We decided in the first instance upon a level of significance for our test of 5%. There was thus a 5% probability that we should falsely reject the null hypothesis *if it was true*, ie a 5% probability of making a type I error. We supposed that we might actually be drawing our samples of size 4 from a population with mean 49.4%. (In practice we should not know the true population mean.) We calculated that if the population mean was in fact 49.4%, then there was a 1.9% probability that we should falsely accept a sample as being compatible with the null hypothesis H_0: $\mu = 50.0\%$. There was therefore a 1.9% probability of making a type II error. When we reduced the level of significance (the probability of making a type I error) to 1%, the probability of making a type II error increased to 8%. We can therefore say that in our attempt to reduce the probability of falsely rejecting the null hypothesis, we increased the probability of falsely accepting the null hypothesis. This result is quite general. Type I and type II errors are obviously interrelated. As we extend the range within which we are willing to accept that a sample mean is compatible with a stated population mean, we unavoidably increase the probability of accepting, as being compatible with the stated population mean, samples which belong to populations having means other than the stated one! This leaves us with a problem; should we start out with a level of significance equal to 5% or to 1%? The answer is, it depends! If, in our example, it would be very costly to reject the assertion that $\mu = 50.0\%$, when in fact the assertion was true, then we should be more inclined to adopt a lower level of significance such as 1%. If, on the other hand, it would be very costly to accept the assertion that $\mu = 50.0\%$, when in fact the assertion was false, then we should be more inclined to adopt a higher level of significance such as 5%. This lowers the probability of our making a type II error, ie of false acceptance of the null hypothesis. What do you think we should do? Yes, we need more information about the cost of the caustic soda solution and perhaps the uses for which it is intended. Let us leave the matter for the time being, and turn our attention to a fresh example.

∏   A manufacturer claims that his crisps are bagged with a mean weight of 30.0 g and a standard deviation of 0.24 g. A sample of nine bags is taken from the production line and the mean weight is found to be 30.17 g. Test the manufacturer's assertion at (i) 5% level of significance and (ii) 1% level of significance.

(i)   The Null Hypothesis is H_0: $\mu = 30.0$ g. The alternative hypothesis is H_1: $\mu \neq 30.0$ g. At the 5% level of significance, the decision making criterion is that we shall accept the null hypothesis if a sample of 9 bags has a mean weight in the range

$$30.00 \pm \frac{1.96(0.24)}{\sqrt{9}} = 30.00 \pm 0.16 \text{ g}$$

The critical values are therefore 29.84 g and 30.16 g. Since the mean weight of 9 bags was found to be 30.17 g, we reject the null hypothesis H_0: $\mu = 30.0$ g, and accept the alternative hypothesis H_1: $\mu \neq 30.0$ g.

(ii) At the 1% level of significance the decision-making criterion is that we should accept the null hypothesis if a sample of 9 bags has a mean weight in the range:

$$30.00 \pm \frac{2.576(0.24)}{\sqrt{9}} = 30.00 \pm 0.21 \text{ g}$$

The critical values are now 29.79 g and 30.21 g. The mean weight for the sample tested was 30.17 g. This lies inside the critical range, so under this criterion we accept the manufacturer's assertion, $H_0: \mu = 30.0$ g.

We apply two different criteria (5% or 1% level of significance) and come to two different conclusions. First, we decided to reject the null hypothesis, then we decided to accept it! On which occasion did we make the right decision? Strange as it may seem, *we were right on both occasions!* Right or wrong in statistical decision-making simply refers to going along with the decision criterion previously agreed.

Clearly in actual fact the manufacturer's claim is either valid or it is not. *The trouble is that we do not know which,* otherwise we would not need to carry out the test! Unfortunately our test fails, in this awkward instance, to give us a fully convincing answer. Given the known standard deviation for the result of our sample measurement, a value of 30.17 g is not a very reassuring one from the manufacturer's point of view. If the null hypothesis is valid, if the true mean is in fact 30.00 g, then a result so far removed as this one would only be expected to occur, because of variability in the weight of bags of crisps, 3.4% of the time.*

* If you wished you could confirm this value of 3.4% by using program "STAT2" (notice not STAT3). The normal distribution corresponding to our null hypothesis is $N(30.0, 0.24/\sqrt{9})$, so line 3900 would have to be edited to read: 3900 DATA 30,.08. A RUN of the program would then tell you that the probability of EXCEEDING the value 30.17 is 1.7%. There would be an equal probability of deviating by as much in the opposite direction, hence the total probability quoted is $2 \times 1.7\%$. (This calculation uses the one feature of STAT2 which we have overwritten in building the more powerful program STAT3.)

Hence if we adopt a 5% level of significance we must reject the null hypothesis whilst if instead we adopt a 1% level of significance we must accept it.

You choose your criterion then you make the decision. At a 5% level of significance we reject the manufacturer's claim. We might be making a type I error in so doing, but on the other hand we might not. In the long run adopting a 5% level of significance means that five times in a hundred we shall make a type I error; we cannot know whether or not we are actually doing so in this instance.

At a 1% level of significance we accept the manufacturer's claim. We *know* this time that we are not making a type I error, we have not rejected the null hypothesis, so we cannot have rejected if falsely! If we are wrong this time our error is of type II. We cannot predict the probability that a test will lead us into making a type II error, not at least without knowing what the true population mean actually is. And if we knew that there would be no need for our test! What we can say is that in reducing the level of significance for our test we reduced the probability of making a type I error. In increasing the range of test results that lead us to acceptance, we inevitably increased the probability of false acceptance, ie of type II error. In an uncertain world we must live with the possibility of error, whichever way we play it. What is important is that we choose our criterion appropriately, in line with the relative costs to us of type I and type II error. We choose our criterion, then we must live with the consequences of our choice!

3.5.5. Sample Size and Hypothesis Testing

Let us start this topic by posing a question, 'Does sample size have any effect on either type I or type II errors?' In order to provide answers to this question, we shall return to our example of the bagging of crisps. You will recall that the manufacturer claims that crisps are bagged with a mean weight of 30.0 g and a standard deviation of 0.24 g. In testing this assertion we shall state the null hypothesis as H_0: μ = 30.0 g and the alternative hypothesis as H_1: $\mu \neq$ 30.0 g. We shall test at the 5% level of significance and consider the consequences of having (a) a sample size of 9, (b) a sample size of 16. Further, in order to make our point, we shall suppose that the true population mean is actually 29.76%. (Keep in mind the fact that in 'real life' we could not know the true population mean without weighing all the bags. Notice that the value we have chosen is one standard deviation, 0.24 g, below the manufacturer's specified value of 30.0 g.) All of the above information makes it possible for us to sketch, for sample size 9 and for sample size 16, a distribution of means curve for the asserted population, with the critical values indicated, and a distribution of means curve for the true population. This has been done in Figs. 3.5d and 3.5e.

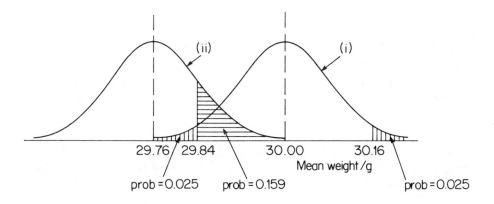

Fig. 3.5d. *The distribution of means curves for a sample of size 9 taken from: (i) the population with $\mu = 30.00$ g and $\sigma = 0.24$ g, (ii) the population with $\mu = 29.76$ g and $\sigma = 0.24$ g. The critical values at the 5% level of significance for (i) are indicated*

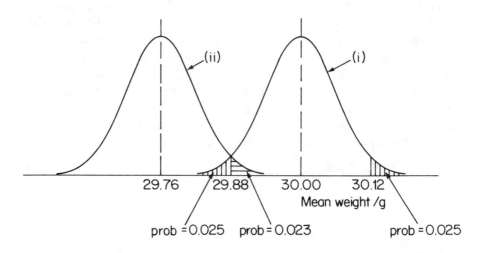

Fig. 3.5e.   *The distribution of means curves for a sample of size 16 taken from: (i) the population with μ = 30.00 g and σ = 0.24 g, (ii) the population with μ = 29.76 g and σ = 0.24 g. The critical values at the 5% level of significance for (i) are indicated*

First consider Fig. 3.5d. We can see that the null hypothesis (H_0: μ = 30.0 g) will be accepted if a sample of size 9 has a mean weight in the range

$$30.00 \pm 1.96(0.24)/\sqrt{9} \approx 30.00 \pm 0.16 \text{ g}$$

This follows from our decision to test at the 5% significance level ($z_{.025}$ = 1.96). The probability of a sample of size 9 taken from the asserted population falling outside of this range is 5%: this is the probability of a type I error occurring if the null hypothesis is true. It can be seen from the figure that the distribution $N(29.76, 0.24/\sqrt{9})$ overlaps the distribution $N(30.00, 0.24/\sqrt{9})$. The percentage of samples of size 9 taken from the true population (μ = 29.76) which would have a mean weight of 29.84 g or above is calculable from the area in the tail of the distribution $N(29.76, 0.24/\sqrt{9})$.

We have $\quad z = \dfrac{\bar{x} - \mu}{\sigma/\sqrt{n}} = \dfrac{29.84 - 29.76}{0.24/\sqrt{9}} = 1.0$

When $z = 1.0$, the area in the tail is 0.159 or 15.9%. This area is the probability of making a type II error, ie of falsely accepting the null hypothesis if the true mean is 29.76 g. When we falsely accept the null hypothesis, we are accepting that a sample mean is compatible with the asserted population mean, when it actually is a mean of a sample taken from a population with a mean different from the one asserted.

Let us denote the probability of making a type II error as β. In our example $\beta = 15.9\%$. We shall now introduce a new statistical idea. *If β is the probability of accepting the null hypothesis when it is false, then $1 - \beta$ is the probability of rejecting the null hypothesis when it is false.* (We either accept or reject: total probability $= \beta + (1 - \beta) = 1$.) In our example $1 - \beta$ is the area under the curve $N(29.76, 0.24/\sqrt{9})$ to the left of the value 29.84: it has value of $(100 - 15.9\%) = 84.1\%$. The probability of rejecting the null hypothesis when it is false is termed the *Power of the Test*. In our example the power of the test is 84.1%.

Now let us consider the case represented in Fig. 3.5e where the sample size is increased to 16.

Again, the null hypothesis is H_0: $\mu = 30.0$ g and the alternative hypothesis is H_1: $\mu \neq 30.0$ g. The decision to work at the 5% level of significance has now fixed the critical values at approximately 29.88 and 30.12. (Check this!) Samples of size 16 having mean weight outside of these values will be considered to be incompatible with null hypothesis H_0: $\mu = 30.0$ g. The acceptance region $(30.0 \pm 0.12$ g) is narrower than that for means of samples of size 9. Were the null hypothesis true, then the probability of a type I error remains at 5%, the level of significance at which we have chosen to work. Let us suppose that the true distribution is in fact curve (ii) in the figures. Let us calculate therefore the probability of a type II error – the probability of falsely accepting the null hypothesis. For a population with mean 29.76 g, the probability of a sample of size 16 having a mean weight >29.88 g is obtained by calculating the area in the tail of the distribution $N(29.76, 0.24/\sqrt{16})$.

This time,

$$z = \frac{\bar{x} - \mu}{\sigma/\sqrt{n}} = \frac{29.88 - 29.76}{0.24/\sqrt{16}} = 2.0$$

When $z = 2.0$ the area in the tail is 0.023, or 2.3%. This gives the probability, β, of making a type II error, ie of falsely accepting the null hypothesis. The probability of correctly rejecting the null hypothesis, the power of the test, is give by $(1 - \beta) = 0.977$, or 97.7%[*].

Let us review our discussion. We have considered tests at the 5% level of significance of the null hypothesis H_0: $\mu = 30.0$ g. We supposed that in reality (unknown to the tester) the true mean is 29.76 g, ie that it is below specification by one standard deviation, 0.24 g. We calculated the probability, β, that our test would fail us, ie that we would obtain a result which would lead us to accept the null hypothesis and thus make a type II error. We also looked at things the other way round and considered the probability $(1 - \beta)$ that our test would succeed, ie that it would lead us, correctly, to reject the null hypothesis. Our results were as follows.

---

[*] Since the critical value 29.88 g is exactly half way between the means of the curves in Fig. 3.5e(i) and Fig. 3.5e(ii), you might just be wondering why the tail area of curve (i) (below 29.88) is quoted as 0.025 whilst that of curve (ii) (above 29.88) is found to be 0.023. If you noticed this you are to be congratulated! Both tail areas are in fact 0.023. This minor discrepancy occurs because we rounded off the 5% critical values for curve (i). To three places of decimals the critical value would be 29.882 g, or just a shade above the half-way point between the means of the two curves, so the two areas would indeed be slightly different. We do not apologise for rounding, it is standard and reasonable practice. In this example it seems reasonable to work to two decimal places. Having decided that, the stated critical values are as close as possible to the exact 5% significance values. We shall have more to say about significant figures in the final Part of the unit.

Sample size taken	9	16
Probability of type II error, β	15.9%	2.3%
Power of test, $(1 - \beta)$	84.1%	97.7%

We can now see clearly the advantage of taking a larger sample size. We have not changed the probability of making a type I error (where the test is applied to a batch which satisfies the null hypothesis). We set that probability at 5% by our decision to apply our test at that level of significance. But we have substantially increased the power of the test correctly to reject the null hypothesis if the true mean is awry by the stated amount. We have reduced the chance of making a type II error.

We must stress again that in practice we shall not know the value of the true mean, so we shall be unable to evaluate the actual value of β for a given case. Nevertheless numerical calculations such as the above can be of direct quantitative value to us. We can present our conclusions as follows:

> 'If the mean weight of bags of crisps being produced happens to be 0.24 g below the stated weight of 30.0 g, then a test at the 5% level of significance with a sample size of 16 has a 97.7% chance of rejecting the batch.'

Consider what would happen to the curve (*ii*) in Fig. 3.5e if the mean weight happened to be even lower than 29.76 g. It would be moved further to the left and would overlap even less with curve (*i*) in Fig. 3.5e. The chance of rejection would be higher still. We can use a calculation of the power of a test to give ourselves a kind of benchmark. We calculate the probability that we shall correctly reject a batch which is off-specification by some given margin. Batches which are worse than that will have a still higher chance of rejection.

We shall return to this topic in later work, where we shall apply it to a familiar problem in analytical chemistry. We shall not set you a numerical problem at this point. If you simply have a feeling for what we have been discussing, that will be enough meantime.

3.6. ONE AND TWO-TAIL TESTS

When the manufacturer of crisps tests whether or not the weight of his bags is on specification ($\mu = 30.0$ g, $\sigma = 0.24$ g) he is interested in both the possibility of putting too much in each bag (a decreased profit) and the possibility of putting too little in each bag (a fine, and some bad publicity). It therefore makes sense for him to have *an upper and a lower critical region for his test*. The fact that there is to be an upper and a lower critical region in the test is reflected in the way the alternative hypothesis is stated. You will recall that the null hypothesis is H_0: $\mu = 30.0$ g and the alternative hypothesis is H_1: $\mu \neq 30.0$ g. The statement of the null hypothesis tells us that we are to start with the assumption that the population mean is 30.0 g. The statement of the alternative hypothesis tells us that the reason for rejecting the null hypothesis can be either evidence that the population mean is greater than 30.0 g or evidence that it is less than 30.0 g, ie any evidence that the population mean is not 30.0 g. Such evidence makes itself known when the mean weight of a sample of size n falls outside of the range $30.0 \pm z_{\alpha/2}(0.24)/\sqrt{n}$ where α is the level of significance. When this happens a sample mean lies in the region of one of the tails. Since the test involves two-tails, it is known as a *Two-Tail test*.

Now it might occur that the authorities receive a complaint about the weight held by bags of crisps produced by the manufacturer. The complaint will obviously suggest that the average weight is less than 30.0 g. (It is an almost unheard of complaint that a consumer is receiving more than he has paid for!) In the light of the complaint, the authorities decide to carry out a test. They state the null hypothesis, it is H_0: $\mu = 30.0$ g. They now need to state the alternative hypothesis. It obviously needs to be stated in terms of the conclusion which will be reached if the null hypothesis is rejected. What is that conclusion? Well it surely has to be related to the purpose of their test. They are not testing to see whether the population mean is greater than 30.0 g: they are interested in whether the population mean is *less* than 30.0 g. They consequently state the alternative hypothesis as H_1: $\mu < 30.0$ g. This is the conclusion which will be reached if H_0 is rejected. The testing procedure now requires a level of significance. Let us assume that their testing protocol calls for a 5% significance level. This means that there is a 5% probability of falsely rejecting the null hypothesis when it is true. Now, if we reject the null hypothesis, we draw the conclusion that $\mu < 30.0$ g. To reach such a conclusion we must surely have witnessed a sample whose mean weight was less than the population mean! How much less? The answer is given by the critical value which defines a left-side tail, whose area is 5% of the whole, in the distribution $N(30.0, 0.24/\sqrt{n})$. What is this critical value? Easy! For an area in the tail of a normal distribution to be equal to 5%, the z value must be 1.645 (check it). The critical value is therefore $30.0 - 1.645(0.24)/\sqrt{n}$. If you are slightly confused, refer to the graphical representation in Fig. 3.6a.

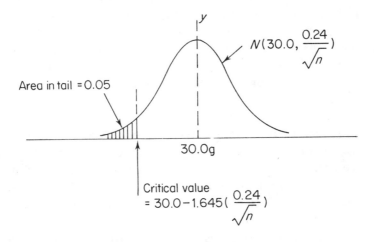

Fig. 3.6a. *The distribution of means for samples of size n drawn from the assumed population N(30.0,0.24) showing the critical value for rejection of the null hypothesis $H_0: \mu = 30.0$ g, and acceptance of the alternative hypothesis, $H_1: \mu < 30.0$ g; the level of significance is 5%*

What we have here is a *one-tail test*. Rejection of the null hypothesis will occur if a sample of size n has a mean weight less than $30.0 - 1.645(0.24)/\sqrt{n}$, ie if it falls in the region of the tail of the distribution presented in Fig. 3.6a. *One-tail tests are appropriate whenever an alternative hypothesis is stated with either a less-than or greater-than sign. If the alternative hypothesis is stated with a less-than sign*, then we have a *'lower tail'* or *'lower one-sided'* test.

An '*upper tail*' or '*upper one-sided*' test is appropriate to an *alternative hypothesis stated with a greater-than sign*. How do we know when to state the alternative hypothesis with either a greater-than or a less-than sign? It is actually quite easy; one simply refers to the question which is either explicitly asked or is implicit in the purpose of the test. Is the population mean 30.0 g? A two-tailed test is implied. Is the population mean less than 30.0 g? That question requires a one-tailed test.

Π   A supplier asserts that the sodium hydroxide content of a solution is 50.0% (w/v). You, as purchaser, are suspicious that the material might not be up to specification. State the null and alternative hypotheses you would apply in a test.

H_0: $\mu = 50.0\%$. (That is easy enough.) Since you are testing to ascertain whether you are being supplied with a weaker solution than you are paying for, the only sensible alternative hypothesis is H_1: $\mu < 50.0\%$.

Π   The Health and Safety Executive quote a threshold level for airborne dirt of 10 mg m^{-3}. (Actually an 8-hour time-weighted average figure.) There is some concern that a certain work environment exceeds this figure. Tests are to be carried out. State a null and an alternative hypothesis.

Here our interest is in the possibility that the threshold levels are being exceeded. Hence we have:

$$H_0: \mu = 10 \text{ mg m}^{-3} \text{ (8-hour time-weighted average)}$$
$$H_1: \mu > 10 \text{ mg m}^{-3} \text{ (8-hour time-weighted average)}$$

Assess your progress by attempting the following SAQ. It should help to reinforce your understanding of one- and two-tail tests.

SAQ 3.6a   A manufacturer produces dry cells which under a standardised test are claimed to have a lifetime of 28.0 hours with a standard deviation of 1.50 hours. $\longrightarrow$

SAQ 3.6a
(cont.)

(*i*) Test the manufacturer's claim in the light of the knowledge that a random sample of 25 dry cells was found to have a mean life-time of 27.45 hours. Use a 5% level of significance for the test.

(*ii*) The manufacturers's research team believe that a slight change in the production procedure will result in dry cells with an improved life-time. The changes in the production procedure are effected and a random sample of 25 dry cells was found to have a mean life-time of 28.60 hours. Carry out an appropriate test at both the 5% and 1% level of significance.

3.7. TAKING STOCK

In this Part of our Unit we have covered a good deal. We started with a discussion of the properties of random samples drawn from a population of possible measurements. The central-limit theorem told us that the mean result derived from a sample of a given size, n, should conform to a normal distribution whose mean coincides with the population mean, μ, and whose standard deviation equals $\sigma/\sqrt{n}$, where σ is the population standard deviation. We went on to discuss how, from our sample mean, we can estimate that the value of the population mean lies within a specified confidence interval. Initially, we did this on the basis of an assumption that we knew the value of σ. In this case the $(1-\alpha) \times 100\%$ confidence range is given by $\mu = \bar{x} \pm z_{\alpha/2}\sigma/\sqrt{n}$. We then had to discuss long and hard the situation when the population standard deviation is unknown, and we were forced to use the standard deviation, s, from our sample as an estimate of σ. We were led to introduce the t-distribution. Eventually we emerged with a revised formula to give the $(1-\alpha) \times 100\%$ confidence range for the mean as $\mu = \bar{x} \pm t_{\alpha/2}s/\sqrt{n}$, where the t-value chosen was that appropriate to $(n-1)$ degrees of freedom. Section 3.4.2 gives a fuller review of the story to this point.

We then made substantial additions to our programming package, to allow it to do the jobs of the z- and t-tables, and then to automate the calculation of confidence intervals for the mean derived from any given sample of individual measurements.

Section 3.5 and Section 3.6 have introduced the subject of hypothesis testing. The procedures described are particularly applicable to routine quality control checking. We are supplied with a material claimed to conform to a certain specification. We make measurements on a sample. These are subject to random error (or to normal random variations in the in the materials sampled). What results will lead us to accept that the material is as specified, and what results will lead us to declare that it is not? This general issue led us to introduce quite a number of concepts as follows.

Section 3.5.2: null hypothesis, alternative hypothesis, decision-making criterion.

Section 3.5.3: level of significance, type I error.

Section 3.5.4: type II error, critical values.

Section 3.5.5: power of a test.

Section 3.6: two-tail test, one-tail test, upper one-sided test, lower one-sided test.

SAQ 3.7a may allow you quickly to check whether you have grasped the meanings of the more awkward of these.

SAQ 3.7a   Associate each of the terms (i) to (v) below with the most appropriate item from the list (a) to (f).

(i) null hypothesis,
(ii) power of a test,
(iii) significance level,
(iv) type II error,
(v) type I error.

(a) accepting a batch which is off specification,
(b) rejecting a batch which is off specification,
(c) accepting a batch which conforms to specification,
(d) rejecting a batch which conforms to specification,
(e) the assumption that a batch conforms to specification,
(f) the assumption that a batch is off specification.

SAQ 3.7a

The mechanics of calculating critical values should have been familiar. The crucial step was to obtain the value of the statistic z, corresponding to the tail area we decided to allow. The critical values for the sample mean were then given by $\bar{x} = \mu \pm z\sigma/\sqrt{n}$.

3.7.1.   The t-distribution in Hypothesis Testing

It has probably not escaped your notice that so far we have been testing assertions about population means in situations where the population standard deviation is known. We shall now turn our attention to situations where σ is not known. This, as was mentioned when discussing estimation, is the more common state of affairs in analytical chemistry and therefore deserves some considerable attention. We judge the matter so important that nearly all of the next Part of the unit will be dedicated to it. All we are going to do now is to introduce the topic. In our work on estimation we saw that when σ was not known, we switched from using the normal distribution variable to employing the t distribution variable. In practical terms this meant, for example, that we used the relationship $\mu = \bar{x} \pm t_{\alpha/2}s/\sqrt{n}$ for obtaining the $(1-\alpha) \times 100\%$ confidence interval for the population mean instead of $\mu = \bar{x} \pm z_{\alpha/2}\sigma/\sqrt{n}$, the relationship which is used when σ is known. It was argued that the relationship involving t was necessary because s was not always a good estimate of σ. By analogy, we now argue that when it comes to testing hypotheses, if σ is unknown, then s must be used and this necessitates the use of the t distribution. An example should get us quickly into the swing of things.

∏   A manufacturer claims that his caustic solution is 50.0% NaOH(w/v). Test the manufacturer's assertion in the light of the fact that four replicate determinations on a random sample of the caustic solution yielded a mean NaOH content of 49.70% (w/v) with a sample standard deviation of 0.25%.

The Null Hypothesis is H_0: $\mu = 50.0\%$. We shall settle for a two-tail test, even though we recognise that a one-tail test might be appropriate. Consequently, the alternative hypothesis is H_1: $\mu \neq 50.0\%$. We shall test at the 5% level of significance. For an area of 0.025 in a single tail and for 3 degrees of freedom (recall $\nu = n\text{-}1$) the value for t is 3.182. The acceptance region is given by $\mu \pm ts/\sqrt{n}$ and is therefore $50.0 \pm 3.182(0.25/\sqrt{4}) = 50.0 \pm 0.40\%$ NaOH (w/v).

The critical values are 49.60 and 50.40% (w/v) and since the mean NaOH content of the sample was 49.70% (w/v), the null hypothesis is accepted. The conclusion drawn is that there is no evidence that the sample mean is incompatible with a population mean of 50.0% (w/v).

Easy? Yes! Now try an SAQ.

SAQ 3.7b   A new manufacturer claims that his caustic solution is 51.0 (w/v) NaOH. There is reason to doubt this assertion! Nine replicate determinations are carried out on a random sample: the results are $\bar{x} = 50.75$ (w/v) and $s = 0.30$ (w/v). Test the manufacturer's assertion at the 5% significance level.

SAQ 3.7b

Objectives

As a result of completing Part 3 you should feel able to:

• explain the precise meanings given in statistics to the terms population, random sample and sample size;

• recognise the status of the sample mean and the sample standard deviation as *estimates* of the population mean and population standard deviation, respectively;

• in a case where the population mean and the population standard deviation are known, use the table of z to predict the probability that a measured sample mean would be found to lie within some specified interval;

• in a case where the population standard deviation is known, use a measured value for a sample mean, together with the table of z, to predict a *confidence interval* within which the population mean should lie;

• sketch roughly the shapes of t-distribution curves, illustrating how the t-distribution narrows to coincide with the standardised normal distribution as the *number of degrees of freedom* becomes large;

- use the percentile table of t (Fig. 3.4b) to evaluate one-tail areas for a t-distribution;

- recognise that, if the population standard deviation is unknown, then estimates of the type defined above require use of the percentile table of t in place of the table of z;

- use your program "STAT3" to find:

  — tail areas corresponding to specified z or t values,

  — z or t values corresponding to specified tail areas,

  — estimates of the type defined above;

- define the terms *decision-making criterion, null hypothesis, alternative hypothesis, level of significance, type I error and type II error*;

- perform, as required, either a one-tail or a two-tail significance test on whether a result for the mean of measurements on a sample is consistent with some hypothesised value for the population mean.

4. Comparison of Means and of Standard Deviations

Introduction

Before we embark on this stage of our studies, let us reflect upon the activities of three imaginary chemists. It is likely that you may recognise some of their activities. However, rest assured, the chemists are imaginary!

The first chemist is a final-year student who has been assigned the often difficult project of developing a novel analytical method. The initial stages of the work go well and the time comes when it is necessary to evaluate the method for potential sources of systematic error. The student is aware of the seriousness of the work: there can be no 'cooking of the books' – a not unheard of phenomenon in student practical work! Is there any evidence, no matter how small, of systematic errors? Statistics can help here.

The second chemist is employed in process control. His employer recognises its vital importance. He has to; there is money in it! Is the process under control? Answers are required, and they are often required quickly. The process-control chemist must sometimes work under considerable pressure: not the least of his worries is the fact that his results are subject to a final checking – the finished product is invariably analysed in the quality-control laboratory. Both the process and the quality-control chemist use statistical methods, often slightly hidden from view, but they are very important.

The last of our chemists is employed in an industry in which the raw materials and the finished products offer an almost nightmarish matrix for analytical determinations. This is because not only are the materials under examination a mixture of complex substances but also because the composition of the mixture is always changing. To add to this, our scientific knowledge of the materials is often incomplete. Nevertheless, analytical determinations must be carried out. The methods employed are often highly empirical – a way of saying that the method works but that we do not fully understand why. There is in such situations a marked reluctance to change methods. 'Better the devil you know'! Sometimes, however, for one reason or another a new method is required. Does the new method give the same results as the old method? Does it give consistent results which can be related to those of the old method? If either of these questions can be answered 'yes', then the advantages of the new method may lead to its adoption. Our chemist is given the job of evaluating the new method. Here too, statistics can be of help. The sort of questions asked of our three chemists are typical of those encountered by analytical chemists the world over. To answer them, an analytical chemist must first and foremost be a chemist: but it helps if he also understands the topics which are covered in this section of the unit.

4.1.   HYPOTHESIS TESTING – A MODIFIED PROCEDURE

We have just given in Sections 3.5 and 3.6 an introduction to hypothesis testing, which, incidentally, is also often called *significance testing*. The procedure we used involved the following steps:

(*a*)   stating the null hypothesis

(*b*)   stating the alternative hypothesis

(*c*)   deciding upon a level of significance

(*d*)   calculating the critical value(s)

(*e*) examining the value of the sample mean to ascertain whether it fell within the span of the critical value(s) – if it did, then the null hypothesis was accepted; if it did not, then the null hypothesis was rejected and the alternative hypothesis accepted.

By this time you are probably quite happy with that procedure, so it is with some diffidence that we announce that we are about to modify it! We shall give our reasons for doing so later. Our new procedure still requires us to state the null and the alternative hypotheses and to decide on a level of significance. It is beyond that stage that we introduce our modifications. To illustrate them we shall return to the example of the caustic soda solution, whose sodium hydroxide content is asserted to be 50.0% (w/v). We shall assume that σ is known and is 0.30% (w/v) and that it is our intention to draw a sample of size 4 from the population and use the value of its mean to test the assertion $\mu = 50.0\%$ at the 5% significance level. Finally, we shall assume that the nature of the investigation is such that a two-tail test is in order, ie $H_1 : \mu \neq 50.0\%$. In previous work, this knowledge was used to determine the critical values:

$$\mu \pm z_{0.025}\sigma/\sqrt{n} = 50.0 \pm 1.96 \times 0.30/\sqrt{4} = 50.00 \pm 0.29$$

Fig. 4.1a gives this information graphically.

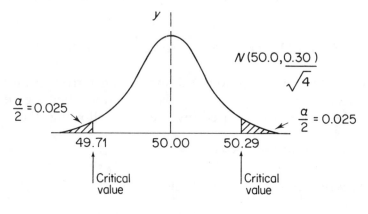

Fig. 4.1a.   *The distribution of means for samples of size 4 drawn from a population with $\mu = 50.0\%$ and $\sigma = 0.30\%$*

Now you will recall that any normal distribution can be transformed into a standardised normal distribution by using an appropriate transformation: here the relevant equation is:

$$z = \frac{\bar{x} - \mu}{\sigma/\sqrt{n}}$$

If we transform the distribution $N(50.0, 0.30/\sqrt{4})$ in this way, we obtain the standardised distribution $N(0,1)$, and the critical values 50.00 ± 0.29 become ± 1.96. This information is presented graphically in Fig. 4.1b.

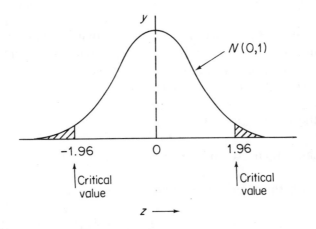

Fig. 4.1b. *A standardised normal distribution showing an area of 0.025 in each tail*

Let us refresh our memory as to the meaning of this distribution in the context of our example. It means that if we were to take the values of samples of size 4 from the population $N(50.0, 0.30)$ and then transform them into z-values by the operation $z = (\bar{x} - \mu)/(\sigma/\sqrt{n})$, we should find that the z-values would be distributed normally with mean 0 and standard deviation 1. We should also discover that 95% of all these z-values fell in the range $z = -1.96$ to $z = +1.96$, or equivalently, that 5% of the z-values were in the tails beyond $z = \pm 1.96$. With this in mind, let us return to our test. The null hypothesis is $H_0 : \mu = 50.0\%$, the alternative hypothesis is $H_1 : \mu \neq 50\%$ (a two-tail test). The significance level is 5%. This means that we shall take as evidence of incompatibility between a sample mean and an asserted population mean any value of a sample mean which puts it in one or other of the two tails whose total area is 5%. Such a sample mean would have a z-value, calculated on the basis that $\mu = 50.0\%$, which was either -1.96 or $+1.96$. For example, let us assume that we took a random sample of size 4 from the caustic soda solution and found that its mean NaOH content was 49.60% (w/v). On the basis that the null hypothesis is true, ie that $\mu = 50.0\%$, the z-value associated with this value of a sample mean is

$$ z = \frac{\bar{x} - \mu}{\sigma/\sqrt{n}} = \frac{49.60 - 50.0}{0.30/\sqrt{4}} = -2.67 $$

Since this z-value is less than -1.96, it must be associated with a sample-mean value which is incompatible with a population mean of 50.0%. This thinking leads us to suggest a new way of testing a hypothesis. The procedure is as follows:

(a) state the null hypothesis,

(b) state the alternative hypothesis,

(c) decide on the significance level,

(d) write down the critical z-value(s) associated with the chosen level of significance.

(e) *Calculate the z-value for the experimental sample-mean* on the basis that the null hypothesis is true. If the z-value falls within the critical range of z-values, then accept the null hypothesis; if it does not, then reject the null hypothesis.

We will immediately put our new procedure into use.

∏   Nine replicate determinations of the NaOH content of a solution which is asserted to be 49.8% (w/v) NaOH are carried out on a random sample; the mean NaOH content is found to be 49.65% (w/v): the population standard deviation is known to be 0.25% (w/v). Test the assertion at the 5% significance level.

(*a*) $H_0 : \mu = 49.8\%$.

(*b*) $H_1 : \mu \neq 49.8\%$.

(*c*) Significance level = 5%.

(*d*) The appropriate z-values (area of 0.025 in each tail) are ± 1.96.

(*e*) On the basis that the null hypothesis is true, the z-value for the experimental sample mean is given by:

$$z = \frac{\bar{x} - \mu}{\sigma/\sqrt{n}} = \frac{49.65 - 49.80}{0.25/\sqrt{9}} = -1.80$$

Since this value is within the critical range, the null hypothesis is accepted. The experimental mean is compatible with a population mean of 49.85 (w/v) NaOH.

We think that you will agree that the new procedure offers the advantage of simplicity; only one simple calculation is required. Let us try another example.

∏   A manufacturer produces dry cells which, under a standardised test, are claimed to have a life-time of 28.0 hours with a standard deviation of 1.50 hours. It is believed that a slight change in production process will lead to dry cells with an enhanced life-time. The process was duly changed and a random sample of 16 dry cells was found to have a mean lifetime of 28.80 hours. Carry out an appropriate test at (*i*) the 5% (*ii*) the 1% level of significance.

(*i*) $H_0 : \mu = 28.0$ hours.
$H_1 : \mu > 28.0$ hours (we are interested in an enhanced lifetime). Significance level = 5%, therefore the appropriate z-value is $z_{0.05}$, ie $+1.645$ (all of the 5% is in the upper tail).

$$z_{calc} = \frac{\bar{x} - \mu}{\sigma/\sqrt{n}} = \frac{28.80 - 28.0}{1.50/\sqrt{16}} = 2.13$$

Since z_{calc} is $> z_{critical}$, the null hypothesis is rejected and the alternative hypothesis, $\mu > 28.0$ hours is accepted. At the 5%

significance level the evidence is that the new process produces dry cells with a longer life-time than those made with the old process.

(*ii*) $H_o : \mu = 28.0$ hours.
$H_1 : \mu > 28.0$ hours.

$z_{critical} = +2.326$. (This is $z_{0.01}$. The plus sign indicates an upper tail).

Now $z_{calc} = 2.13$. Since this is less than the value for $z_{critical}$, we accept the null hypothesis, $H_o : \mu = 28.0$ hours. We therefore are unable to conclude, at the 1% level of significance, that the new process produces dry cells with an improved life-time.

4.1.1.   The *p*-Value

This is an appropriate occasion to make an interesting point. We rejected the null hypothesis at the 5% level of significance but accepted it at the 1% level. At the 5% level, the calculated z-value was *significant*; it resulted in a rejection of the null hypothesis. At the 1% level the calculated z-value was *not significant*; the null hypothesis was accepted. At what level of significance would a change from the acceptance of the null hypothesis to its rejection take place? This is easy to calculate; the cross-over point occurs when $z = 2.13$. The area in the tail for this value of z is 0.01659 or approximately 1.66%. Keeping in mind that this is a one-tail test, we see that 1.66% is the *smallest level of significance which would have allowed the null hypothesis to be accepted*. The smallest level of significance, or the *p*-value as some statisticians call it, is an interesting concept. It permits one to assess just how significant a significant result is! For example, in the test we carried out at the 5% level of significance, a calculated z-value of 1.645 would have resulted in rejection of the null hypothesis. Our calculated value of z was considerably larger than this critical value, but all we were able to say was that the null hypothesis was to be rejected. At the 1% level of significance, a calculated z-value of 2.327 would have resulted in rejection of the null hypothesis. Our calculated value was not very far from this value. Nevertheless, all that we were able to say at the time was that the null hypothesis was to be accepted – the calculated z-value was not significant. Even if the calculated z-value had been 2.325 we would still have accepted the null hypothesis at the 1% level.

If our sample gives a z-value of 2.325 we come to one conclusion, if it gives a z of 2.327 we come to the opposite one! Clearly there have to be borderline cases, but if we simply note the decision reached we lose a lot of information. Nobody will realise that it was a close-run thing. What made us choose a significance level of 1% anyway? Had we chosen a level of 1.01% both $z = 2.325$ and $z = 2.327$ would have been significant. Had we chosen a level of 0.99% neither would have been.

The p-value leaves us much better informed. Our sample gives us a z-value of 2.13, which leads to a p-value of 1.66%. *The significance level of our finding is 1.66%.* If the new production process had made no difference, ie if the null hypothesis holds, then a z-value as large as this would have had only a 1.66% chance of occurring. We still have our decision to make. If we stick to requiring a 1% level of significance before we think it worthwhile to pursue the new process, then we don't make the change. If a 2% level of significance was judged sufficient to act on, then we would adopt the new process. More likely, if either of these two criteria had been our proposed stance, we might well suggest that a further sample be taken! Had we, on the other hand, been prepared to act on evidence at, say, a 10% level, then we would treat our result as adequate encouragement to go ahead with no further ado. Quoting a p-value keeps us informed of the strength of the evidence from our test.

The value of p simply gives the area in the tail(s) of z. It gives the probability that a z-value as extreme as that obtained, or more extreme, would occur by chance if the null hypothesis were valid. Our test above was a one-tail test, so p is a one-tail area. Where the appropriate test is two-tailed, then p will be given by the corresponding two-tail area.

4.1.2. Hypothesis Testing by Using the t-Distribution

We shall now use our modified procedure for hypothesis or significance testing when σ is unknown, ie when we have to use the *t*-distribution. The procedure is so closely analogous to what we have just been doing that we can attempt an example immediately.

∏   A given steel has a manganese specification of 0.820%. A random sample of the steel was analysed by a standard method (systematic error can be assumed to be absent). Four replicate determinations yielded a mean manganese content of 0.830% with sample standard deviation of 0.0080%. Is the steel being produced according to specification? Test at the 5% significance level.

The null hypothesis is $H_o : \mu = 0.820\%$.

A two-tail test is in order: the steel could be out of specification by having either too high or too low a manganese content. Hence the alternative hypothesis is $H_1 : \mu \neq 0.820\%$. The level of significance is 5%. The area in each tail is therefore 2.5% or 0.025. Four measurements were taken, therefore the number of degrees of freedom, ν is 3. For a tail-area of 0.025 and 3 degrees of freedom the *t*-value is 3.182 (see Fig. 3.4b). The critical *t*-values are thus ±3.18.

The calculated *t*-value – calculated on the basis that the null hypothesis is true – is obtained from $t = (\bar{x} - \mu)/(s/\sqrt{n})$. This gives the value below.

$$t_{calc} = \frac{0.830 - 0.820}{0.0080/\sqrt{4}} = 2.50$$

Since this value lies inside the critical range for t, we accept the null hypothesis: the sample mean is compatible. at the 5% significance level, with a population mean of 0.820%. We therefore accept that the steel is being produced to specification. Yes, it is all relatively simple!

What about calculating the *p*-value for this test? This is again, in principle, straightforward; it is a matter of finding what tail area corresponds to a *t*-value of 2.50, for $\nu = 3$. But this is where the shortcomings of the *t*-table become clear. For $\nu = 3$, Fig. 3.4b gives $t_{0.05} = 2.353$ and $t_{0.025} = 3.182$. The *t*-value 2.50 lies between these and so corresponds to a tail area intermediate between 0.025 and

0.050. Since 2.50 is closer to 2.353 than to 3.182 we can see that the area will be closer to 5% than to 2.5%, but it is quite awkward to set a precise figure on it. We used the inverse t-subroutine in our computer program STAT3 (option 4 under 'tables'). For $t = 2.50$ and $\nu = 3$ it gave a tail area of 4.44%. Our test on this occasion was two-tailed, therefore the significance of our result is twice this tail area, $p = 8.9\%$.

The t-table is not suited to finding the precise tail-area for a particular t_{calc}. For this reason p is often not calculated. This is a pity, because the p-value is more informative than a simple recommendation of acceptance or rejection. With a computer package, on the other hand, p can be easily generated.

SAQ 4.1a | A new method for the determination of the aspirin content of analgesic tablets is being developed. The method was applied to tablets which were known to contain 300 mg of aspirin: the results obtained on four tablets were 308 mg, 307 mg, 304 mg and 301 mg. Is there any evidence of a systematic error?

4.2. COMPARISON OF TWO SAMPLE MEANS: CONFIDENCE INTERVALS

In our work so far on estimation and hypothesis testing. we have been concerned with applying statistics to situations involving a single sample mean. If we knew an experimental sample mean and the population standard deviation, we could estimate confidence intervals for the population mean, and could test the likelihood of the sample having come from a population with a stated mean. Our justification for being able to do this was our belief that sample means are normally distributed. When we knew an experimental sample mean, but not the population standard deviation, we used the sample standard deviation and the t-distribution in order to estimate, and to test hypotheses. Our justification for this action was again our belief that sample means are normally distributed. However, we also knew that the behaviour of sample standard deviations was such that it required a relatively large sample size before the application of the normal distribution to problems of estimation and testing was valid. The capricious nature of sample standard deviations forced us to use the more tolerant t-distribution for small samples. Nevertheless, it is worthwhile repeating that even here, our starting point is our belief that sample means are normally distributed. Keep this in mind as we introduce and discuss the implications of the following theorem.

> *Theorem*: If the observed values of two independent random variables x_1 and x_2 come from two normal (or near normal) distributions with mean μ_1 and μ_2 and standard deviation σ_1 and σ_2 respectively, and if $\bar{x}_1$ is the mean of a random sample size n_1 drawn from the first population, and if $\bar{x}_2$ is the mean of a random sample of size n_2 from the second population; then *the distribution of the difference between the sample means*, $\bar{x}_1 - \bar{x}_2$, *is normal with mean* $\mu_{(\bar{x}_1 - \bar{x}_2)} = \mu_1 - \mu_2$ and standard deviation
>
> $$\sigma_{(\bar{x}_1 - \bar{x}_2)} = \sqrt{\sigma_{\bar{x}_1}^2 + \sigma_{\bar{x}_2}^2} = \sqrt{(\sigma_1^2/n_1) + (\sigma_2^2/n_2)}.$$

Did you follow that?(!) It isn't really that hard, though the formula for the standard deviation looks a little complicated at first sight. It helps to take the square of the relationship.

$$\sigma^2_{(\bar{x}_1 - \bar{x}_2)} \;=\; \sigma^2_{\bar{x}_1} \;+\; \sigma^2_{\bar{x}_2} \tag{4.1}$$

Do you remember the name we gave to the square of the standard deviation? You could be excused for forgetting because we introduced the concept only very briefly. The standard deviation squared is known as the *variance*. In the same way that we discuss population standard-deviation, σ, and sample standard-deviation, s, so we shall have occasion to refer to population variance, σ^2, and sample variance, s^2.

Now the values of $\bar{x}_1$ and $\bar{x}_2$ are both subject to random error. You would expect that the value of their difference $\bar{x}_1 - \bar{x}_2$ would be subject to somewhat greater uncertainty than either of the individual values. In fact we have met this problem before, long ago, in section 1.6.2. There we explained that when we took sums or differences of values which were individually subject to error, then the *variance* of the result was equal to the *sum of the individual variances*. Eq. 4.1 above says simply that.

The theorem allows us to calculate the standard deviation in the difference between the means. Let us give it in full.

$$\sigma_{(\bar{x}_1 - \bar{x}_2)} \;=\; \sqrt{\sigma^2_{\bar{x}_1} + \sigma^2_{\bar{x}_2}} \;=\; \sqrt{\sigma^2_1/n_1 + \sigma^2_2/n_2} \tag{4.2}$$

In the final expression on the right the standard deviations, $\sigma_{\bar{x}_1}$ and $\sigma_{\bar{x}_2}$, of the sample means have been expressed in the familiar way in terms of the standard deviations, σ_1 and σ_2, of the individual measurements from which they are derived.

Fig. 4.2a illustrates the theorem. When sample means are drawn respectively from the populations $N(\mu_1, \sigma_1)$ and $N(\mu_2, \sigma_2)$, the *difference between the sample means* is normally distributed, with mean $\mu_1 - \mu_2$ and standard deviation $\sqrt{(\sigma^2_1/n_1) + (\sigma^2_2/n_2)}$.

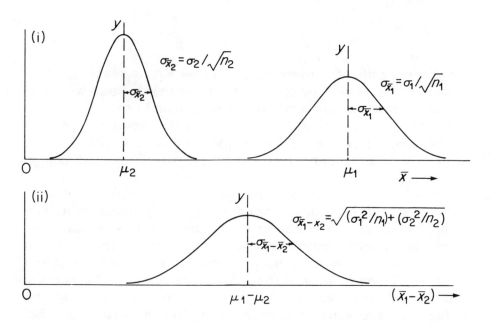

Fig. 4.2a. *(i) Distribution of two sample means $\bar{x}_1$ and $\bar{x}_2$, drawn from two normal populations $N(\mu_1, \sigma_1)$ and $N(\mu_2, \sigma_2)$. (ii) Distribution of the difference $(\bar{x}_1 - \bar{x}_2)$ between the sample means*

The parallel with our previous work is apparent. Let us continue with it. If we wish to transform the distribution of sample means curve into a standard normal curve, we utilise the operation below.

$$z = \frac{\bar{x} - \mu}{\sigma/\sqrt{n}}$$

We can do the same thing with the distribution being discussed. The operation is given below.

$$z = \frac{(\bar{x}_1 - \bar{x}_2) - (\mu_1 - \mu_2)}{\sqrt{\sigma_1^2/n_1 + \sigma_2^2/n_2}} \tag{4.3}$$

In the past, the z-variable made it possible for us to use the normal distribution table and from that we were able to make estimates, and to test hypotheses. The same is true now. Our main concern will be with the testing of hypotheses. However, we can learn a great deal from first considering problems of estimation. This we shall do, mainly by working through a number of examples.

4.2.1.   Case 1: σ_1 and σ_2 are Known

When we were called on to estimate a population mean from knowledge of a sample mean, we used the idea of a *confidence interval*. The $(1-\alpha) \times 100\%$ confidence interval for a population mean was given by the equation below.

$$\mu = \bar{x} \pm z_{\alpha/2}\sigma/\sqrt{n}$$

where $\bar{x}$ is the experimental sample mean, σ the population standard deviation, n the sample size and $z_{\alpha/2}$ the z-value appropriate to the confidence interval required.

The $(1 - \alpha) \times 100\%$ confidence interval for the difference $\mu_1 - \mu_2$ is given analogously.

$$(\mu_1 - \mu_2) = (\bar{x}_1 - \bar{x}_2) \pm z_{\alpha/2}\sqrt{(\sigma_1^2/n_1) + (\sigma_2^2/n_2)} \quad (4.4)$$

We shall immediately apply this equation to a problem of estimation.

∏   Manufacturer A boxes breakfast flakes with a standard deviation of 10 g; manufacturer B boxes the same sort of flakes with a standard deviation of 4 g. A random sample of 20 boxes of flakes from manufacturer A had a mean weight of 215 g; a random sample of 16 boxes of flakes from manufacturer B had a mean weight of 202 g. Give a 95% confidence interval for $(\mu_A - \mu_B)$.

All that is required here is simply to substitute into Eq. 4.4.

$$\mu_A - \mu_B = (\bar{x}_A - \bar{x}_B) \pm z_{\alpha/2}\sqrt{\sigma_A^2/n_A + \sigma_B^2/n_B}$$

$$= (215 - 202) \pm 1.96\sqrt{10^2/20 + 4^2/16}$$

$$= 13 \pm 1.96\,(\sqrt{6})$$

$$= 13 \pm 4.8\,\text{g}$$

The 95% confidence interval for $(\mu_A - \mu_B)$ is thus from 8.2 g to 17.8 g. You might see this question as simply a 'number crunching' exercise! We admit that we wanted you to start this work with a 'plug-into-the-formula' question. However, there are some 'messages' here which are worthy of comment. It is reasonable to suppose that both manufacturers aim to produce boxes which contain close to 200 g of flakes. However, because manufacturer A has more variability in his packaging procedure, he is forced to settle for boxes which contain 8 to 18 g of flakes more than those of his competitor. (This estimate of the difference is at the 95% level of confidence.) We have had to give a relatively large weight range for the estimated difference; however, there can be no doubt that manufacturer A gives away more flakes than manufacturer B. In analytical chemistry we are foolish if we ignore the consequence of a large variability in a measurement. In production, a large variability in a manufacturing process is equally something which cannot be ignored.

4.2.2.   Case 2: σ_1 and σ_2 Unknown – Large Samples

You will recall that if σ was not known, but we had a good estimate of it from the standard deviation, s, of a large sample, we were able to use the relationship

$$\mu = \bar{x} \pm z_{\alpha/2}s/\sqrt{n}$$

as the $(1-\alpha) \times 100\%$ confidence interval for the population mean. With this in mind, it should not surprise you that we can do the same sort of thing when estimating $(\mu_1 - \mu_2)$.

∏   A manufacturer produces dry cells at two different locations. Fifty dry cells from the first location have a mean life-time of 35 hours, with a standard deviation of 2 hours. Forty dry cells from the second location have a mean life-time of 31 hours, with a standard deviation of 3 hours. Give a 95% confidence interval for the difference in the mean life-times of the dry cells.

For population 1: $\bar{x}_1 = 35$ hours, $n_1 = 50$, and $\sigma_1 \approx s_1 = 2$ hours,

For population 2: $\bar{x}_2 = 31$ hours, $n_2 = 40$, and $\sigma_2 \approx s_2 = 3$ hours.

For a 95% confidence interval, we need $z_{.025}$ (which is 1.96.)

$$\therefore \quad (\mu_1 - \mu_2) = (\bar{x}_1 - \bar{x}_2) \pm z_{\alpha/2}\sqrt{(\sigma_1^2/n_1) + (\sigma_2^2/n_2)}$$

$$= (35 - 31) \pm 1.96\sqrt{(2^2/50) + (3^2/40)}$$

$$= 4 \pm 1.96(\sqrt{0.305})$$

$$= 4.0 \pm 1.1 \text{ hours}$$

The 95% confidence interval for the difference in mean life-times is from 2.9 to 5.1 hours.

Yes, it is essentially the same procedure as for Case 1. The point we are making is that when sample sizes are large we can use sample standard deviations as good estimates of population standard deviations.

4.2.3.   Case 3: σ_1 and σ_2 Unknown but Equal – Small Samples

If you are anything like countless other chemists meeting the above heading for the first time, you are probably dumbfounded. 'How can we not know σ_1 and σ_2, yet know that they are equal?' is a reasonable question. Reasonable questions from chemists deserve reasonable answers, preferably in terms of examples appreciated by chemists! Here goes! If we examine two sets of analyses carried out by the same procedure, or two sets of measurements carried out under the same conditions, we often find that the standard deviations of the results are very similar. Think of how often one hears statements such as 'that is the sort of precision one could expect from that sort of titration' or 'I have always found my results for this type of determination to be highly reproducible: the precisions are better than 1%'. These generalisations refer to given methods or given procedures being applied to different determinations. The results of these different determinations will obviously vary but the measurement precisions will be very similar. In order to know the population standard deviations, the σ's, for two sets of measurements it would be necessary to carry out very large numbers of measurements. This is something which we obviously avoid doing. Consequently we do not know σ_1 and σ_2. However, this does not prevent us from saying that σ_1 and σ_2 are equal.

If $\sigma_1 = \sigma_2 = \sigma$, we can re-write Eq. 4.3 as

$$z = \frac{(\bar{x}_1 - \bar{x}_2) - (\mu_1 - \mu_2)}{\sigma\sqrt{(1/n_1) + (1/n_2)}}$$

Furthermore the $(1-\alpha) \times 100\%$ confidence interval for the difference becomes

$$\mu_1 - \mu_2 = (\bar{x}_1 - \bar{x}_2) \pm z_{\alpha/2}\sigma\sqrt{(1/n_1) + (1/n_2)} \qquad (4.6)$$

Now, when we had to write confidence intervals for a population mean in the light of an experimental sample-mean, and had a small sample size to contend with, we used the relationship $\mu = \bar{x} \pm t_{\alpha/2}s/\sqrt{n}$ rather than the relationship $\mu = \bar{x} \pm z_{\alpha/2}\sigma/\sqrt{n}$; ie we switched to using a t-distribution. We do the same sort of thing here. The t variable is given by

$$t = \frac{(\bar{x}_1 - \bar{x}_2) - (\mu_1 - \mu_2)}{s\sqrt{(1/n_1) + (1/n_2)}} \qquad (4.7)$$

and the $(1-\alpha) \times 100\%$ confidence interval for the difference, $\mu_1 - \mu_2$, is

$$(\mu_1 - \mu_2) = (\bar{x}_1 - \bar{x}_2) \pm t_{\alpha/2}s\sqrt{(1/n_1) + (1/n_2)} \qquad (4.8)$$

That is easy enough. However a question arises, 'What is s?' The answer is that it is an estimate of σ. Unfortunately, we do not immediately know its value. Why not? Because all that we have from our two sets of measurements is s_1 and s_2, the respective sample standard deviations.

Can we use these to calculate a single best approximation to σ? Yes, we use the following equation

$$s_p^2 = \frac{(n_1 - 1)s_1^2 + (n_2 - 1)s_2^2}{n_1 + n_2 - 2} \qquad (4.9)$$

where s_p is called the *pooled standard deviation* (s_p^2 is the pooled variance). What do we mean by 'pooled'? It means that we have used both s_1 and s_2, which are both estimates of σ, to obtain a sort of weighted average estimate of σ. Let us explain. Recall that the basic definition of a sample standard deviation. It is given by

$$s = \sqrt{\sum_{i=1}^{n} d_i^2/(n - 1)},$$

where the d_i's are the individual deviations of measurements from their mean ($d_i = x_i - \bar{x}$). Therefore, Eq. 4.10 follows.

$$(n - 1)s^2 = \sum_{i=1}^{n} d_i^2 \qquad (4.10)$$

Hence $(n-1)$ times the sample variance s^2 equals the sum of the squares of the deviations from the sample mean. The numerator of Eq. 4.9 therefore gives the sum of the squares of the deviations for the first sample plus that for the second sample (each calculated with respect to its own sample mean), ie the numerator gives a global sum of squared deviations.

What of the denominator in Eq. 4.9? To obtain a sample variance we always divide $\sum d_i^2$ by the number of degrees of freedom ($n-1$). The first sample is of size n_1 and so has ($n_1 - 1$) degrees of freedom.

The second sample has $(n_2 - 1)$ degrees of freedom. Notice that $(n_1 - 1) + (n_2 - 1) = (n_1 + n_2 - 2)$. The denominator therefore gives the total number of degrees of freedom.

So the name 'pooled variance' for s_p^2 is an apt one. It is given by the total of the sums of squared deviations divided by the total number of degrees of freedom. Let us now make use of Eq. 4.8 and 4.9. Before we do, see if you can demonstrate the fact that when $n_1 = n_2$, the pooled variance is given by $(s_1^2 + s_2^2)/2$.

It is quite easy — it's only algebra! (Put $n_1 = n_2 = n$ in Eq. 4.9, then cancel a factor of $(n\text{-}1)$ from the numerator and denominator).

Now, let us return to the 's' in Eq. 4.8. It is an estimate of σ and the best experimental estimate we have of σ is s_p. Therefore, from now, we will use s_p for s in these equations.

∏   In a certain batch process, the efficacies of two catalysts are being compared. The procedure is to quench the reaction after 15 min and to determine the mainproduct as a percentage of the 'theoretical yield'. With catalyst A, ten batches were found to have a mean yield of 45.0% with a standard deviation of 1.5%. With catalyst B, twelve batches were found to have a mean yield of 41.5% with a standard deviation of 1.2%. Give a 95% confidence interval for the difference $\mu_A - \mu_B$.

It is highly likely that $\sigma_1 = \sigma_2$. We are using essentially the same procedure for our reaction – only the catalysts differ – and the measurement system is undoubtedly identical. Consequently, we will start by obtaining a value for the pooled standard deviation.

$$s_p^2 = \frac{(n_A - 1)s_A^2 + (n_B - 1)s_B^2}{n_A + n_B - 2}$$

$$s_p^2 = \frac{(10 - 1)(1.5)^2 + (12 - 1)(1.2)^2}{10 + 12 - 2}$$

$$= \frac{9(2.25) + 11(1.44)}{20}$$

$$= 1.8045$$

$$\therefore \qquad s_p = 1.34\%$$

Note that the total number of degrees of freedom is 20. The *t*-value for 2.5% in a tail and $\nu = 20$ is 2.086 (check this!). The 95% confidence interval is thus obtained from Eq. 4.8.

$$(\mu_A - \mu_B) = (\bar{x}_A - \bar{x}_B) \pm t_{\alpha/2}s\sqrt{(1/n_A) + (1/n_B)}$$

$$= (45.0 - 41.5) \pm 2.086(1.34)\sqrt{1/12 + 1/10}$$

$$= 3.5 \pm 1.2\%$$

Yes, once one knows what one is doing it is relatively easy. A few comments are appropriate here. We were able to carry out this calculation becuase we believed that $\sigma_1 = \sigma_2$. Sometimes it is not so easy to see that this equality holds. For example, let us say that we wished to estimate the difference $\mu_1 - \mu_2$ in the light of two sets of determinations, the first of which had been carried out by a spectrophotometric method and the second by a spectrofluorimetric method. Have we any right to assume that $\sigma_1 = \sigma_2$? The answer is; perhaps. Often, different instrumental methods do have approximately the same measurement precision. If you were aware that this was true in the example above, then it would be reasonable to assume that $\sigma_1 \approx \sigma_2$.

A quick way of comparing measurement precisions is to examine the sample standard deviations to hand. Are they approximately the same? If so, you might be safe in your assumption that $\sigma_1 \approx \sigma_2$. Of course, it is best when one compares sample standard deviations to have $n_1 = n_2$. It is usually possible to arrange this. Even so, for small sample sizes, s_1 and s_2 might appear to be quite different even though $n_1 = n_2$. Is there any way in which we can test whether s_1 and s_2 are compatible with the idea that $\sigma_1 = \sigma_2$? The anwer is yes. However, we shall delay introducing you to this test. In the same way, we shall step back from introducing you to the procedure which must be employed when we are certain that $\sigma_1 \neq \sigma_2$. Patience!

SAQ 4.2a   In each of the following examples information is given for two samples. For each derive a 95% and a 99% confidence interval for $(\mu_1 - \mu_2)$, the difference between the means of the populations sampled.

(*i*) $\bar{x}_1 = 550, \bar{x}_2 = 480, n_1 = 8, n_2 = 6$, and $\sigma_1 = \sigma_2 = 10$.

(*ii*) $\bar{x}_1 = 155$ ppm, $\bar{x}_2 = 148$ ppm, $n_1 = 32, n_2 = 36, s_1 = 8$ ppm, and $s_2 = 6$ ppm.

(*iii*) $\bar{x}_1 = 13.50\%, \bar{x}_2 = 13.00\%, n_1 = 10, n_2 = 10, s_1 = 0.10\%$ and $s_2 = 0.08\%$

(*iv*) $\bar{x}_1 = 0.1015M, \bar{x}_2 = 0.01010M, n_1 = 15, n_2 = 10, s_1 = 0.0002M$, and $s_2 = 0.0003M$.

4.3.   COMPARISON OF TWO SAMPLE MEANS: HYPOTHESIS TESTING

Here are two questions which are of a type occurring frequently in chemical science and technology. Do two alternative chemical processes give the same yield? Do two alternative analytical methods give the same result? To answer questions of this type, one would carry out replicate experiments and then compare sample means. There are essentially two statistical procedures for the comparison of sample means. The first, which has already been discussed, is the estimation of the difference in population means by using the confidence interval. The second is that of hypothesis (or significance) testing. When we discussed hypothesis testing in the context of a single sample mean we outlined the following procedure:

(*a*) state the null hypothesis,

(*b*) state the alternative hypothesis,

(*c*) decide on the significance level,

(*d*) write down the critical z (or t)-values,

(*e*) calculate the z (or t)-value on the basis that the null hypothesis is true. If the calculated z (or t)-value falls within the range of the critical z (or t)-value(s), then accept the null hypothesis; if it does not then reject the null hypothesis and accept the alternative hypothesis.

We shall use exactly the same procedure for hypothesis testing for situations involving two sample means. The equations for calculating z- and t-values are as follows.

Case 1. When σ_1 and σ_2 are known we use Eq. 4.3.

$$z = \frac{(\bar{x}_1 - \bar{x}_2) - (\mu_1 - \mu_2)}{\sqrt{(\sigma_1^2/n_1) + (\sigma_2^2/n_2)}} \qquad (4.3)$$

If $\sigma_1 = \sigma_2 = \sigma$, this equation becomes

$$z = \frac{(\bar{x}_1 - \bar{x}_2) - (\mu_1 - \mu_2)}{\sigma\sqrt{(1/n_1) + (1/n_2)}}$$

Case 2. When σ_1 and σ_2 are unknown but the sample sizes are large we simply substitute s_1 and s_2 for σ_1 and σ_2 in Eq. 4.3. Hence:

$$z = \frac{(\bar{x}_1 - \bar{x}_2) - (\mu_1 - \mu_2)}{\sqrt{(s_1^2/n_1) + (s_2^2/n_2)}}$$

Case 3. When σ_1 and σ_2 are unknown but are equal, and when the sample sizes are small we use the pooled standard deviation, and Eq. 4.7.

$$t = \frac{(\bar{x}_1 - \bar{x}_2) - (\mu_1 - \mu_2)}{s_p\sqrt{(1/n_1) + (1/n_2)}}$$

where $$s_p = \sqrt{\frac{(n_1 - 1)\,s_1^2 + (n_2 - 1)s_2^2}{n_1 + n_2 - 2}}$$

If $n_1 = n_2$, $s_p = \sqrt{(s_1^2 + s_2^2)/2}$

The situation when $\sigma_1 \neq \sigma_2$ and when sample sizes are small will be referred to later.

4.3.1. The Null Hypothesis

When we considered situations involving a single sample mean, the question asked was essentially, 'Is the sample mean compatible with the asserted population mean?' In order to obtain the null hypothesis, we answered this question in the affirmative, ie we adopted as our null hypothesis the statement that the population mean was as asserted. In comparing two sample means, we need to be clear about just what questions are being asked. They are either, 'Are the sample means from the same population', or, 'Are the two sample means from two separate populations whose means are equal?' Again, in obtaining our null hypothesis, we shall answer these questions in the affirmative, ie we shall say, as appropriate, 'Yes, the two sample means could have come from the same population', or 'Yes, the two sample means are compatible with two populations which have the same mean'. The consequence of both of these answers is that the null hypothesis is $\mu_1 = \mu_2$. Think about that! If the two sample means come from the same population then μ_1 must equal μ_2. Equally obviously, μ_1 must equal μ_2 when the population means are equal! Now it follows that when we come to calculate a $\bar{z}$- (or a t-) value on the basis of the truth of the null hypothesis, the difference $\mu_1 - \mu_2$ must be zero. (At last, you are probably thinking, something to make the equations simpler!) This is all the introduction we need. Let us proceed to test some hypotheses.

∏ A laboratory is asked to investigate the possibility that an oil spillage came from a tanker which was close to the scene of the 'crime'. Several lines of investigation are undertaken; one of which is to compare the sulphur content of specimens from both the spillage and the tanker. Past experience with the determination of sulphur in oil indicates a measurement precision, in terms of standard deviation, of 0.02% of sulphur. Six samples from the tanker had a mean sulphur content of 0.21%; an equal number of samples from the spillage had a mean sulphur content of 0.18%. Test, at the 5% significance level, whether it is likely that the two sample means are compatible with having population means which are equal. In the light of this test, comment.

We have phrased this question, like others to follow, in such a way that the statistical test may be seen in its real-world context.

The null hypothesis is $H_O : \mu_1 = \mu_2$.

We shall conduct a two-tail test, consequently the alternative hypothesis is $H_1 : \mu_1 \neq \mu_2$.

We are given the measurement precision as 0.02% of sulphur. Therefore $\sigma = \sigma_1 = \sigma_2 = 0.02\%$. Since σ is known, we can use a normal distribution. For a two-tail test at the 5% significance level the critical values of z are ± 1.96.

For a normal distribution, when σ_1 and σ_2 are known we can calculate the z-value corresponding to the test result from the relationship below.

$$z_{calc} = \frac{(\bar{x}_1 - \bar{x}_2) - (\mu_1 - \mu_2)}{\sqrt{(\sigma_1^2/n_1) + (\sigma_2^2/n_2)}}$$

Now since the null hypothesis is that $\mu_1 = \mu_2$, and since also $\sigma_1 = \sigma_2 = \sigma$, this equation reduces to the following.

$$z_{calc} = \frac{\bar{x}_1 - \bar{x}_2}{\sigma\sqrt{(1/n_1) + (1/n_2)}}$$

In our example, $\bar{x}_1 = 0.21\%, \bar{x}_2 = 0.18\%, n_1 = 6, n_2 = 6$ and $\sigma = 0.02\%$.

Consequently the value for z calculated on the basis that the null hypothesis is true is

$$z_{calc} = \frac{0.21 - 0.18}{0.02\sqrt{(1/6) + (1/6)}} = 2.60.$$

Since z_{calc} (2.60) falls outside the critical range (± 1.96) we reject the null hypothesis and accept the alternative hypothesis, $\mu_1 \neq \mu_2$. What is the meaning of this result in the context of the investigation? In statistical terms it means that at the 5% significance level, the evidence is that the two samples do not come from populations with the same mean. In scientific language, the evidence is that the two samples did not have the same origin. Could they have had the same origin? The answer is: 'Yes, but it is unlikely on the basis of this test'. It might be that other scientific tests, perhaps more definitive in nature, would demonstrate that the spillage oil did come from the tanker. Then again, other tests might supplement the evidence from the sulphur analyses by showing that it was unlikely that the two oil specimens had the same origin. The point being made is that statistics is the servant of chemists, not their master.

∏   A laboratory is concerned with the determination of metals in saline waters, particularly the levels of calcium present.

Several years ago, it switched from determining calcium gravimetrically to using a complexometric titration. It is now suggested that atomic absorption spectrophotometry (AA) might replace the titrimetric method; the main advantage being speed of analysis. Are the two methods equally reliable – do they give the same result? The answer to this question is most important. The chemists like the titrimetric method; they are not familiar with AA! Furthermore, AA instrumentation is costly. (Yes, this is a long story.) A comparison (competition?) of the two methods is arranged. A quantity of water is collected from a salt-lake and divided into two parts. Thirty-two determinations of calcium by each method are carried out. The result for the titrimetric method is: mean calcium content = 420.1 ppm, with a standard deviationof 2.01 ppm. For the AA determinations, the mean calcium content is 418.9 ppm, with a standard deviation of 2.99 ppm. Do the two methods give the same result? Carry out a test at the 5% significance level.

Here, we have an example of a comparison-of-means test where the sample sizes are large enough to permit the approximations $\sigma_1 \approx s_1$ and $\sigma_2 \approx s_2$. We can therefore use the normal distribution for both the critical values and the calculated values of the tested variable, ie we use z.

The null hypothesis is $H_o : \mu_1 = \mu_2$.

The alternative hypothesis is $H_1 : \mu_1 \neq \mu_2$.

The level of significance is 5%, therefore for a two-tail test the critical values of z are ± 1.96.

The z_{calc} value is obtained as follows:

$$z_{calc} = \frac{\bar{x}_1 - \bar{x}_2}{\sqrt{(\sigma_1^2/n_1) + (\sigma_2^2/n_2)}} =$$

$$\frac{420.1 - 418.9}{\sqrt{(2.01)^2/32 + (2.99)^2/32}} = 1.88$$

(Don't forget that $\mu_1 - \mu_2 = 0$, if the null hypothesis is true.)

Since z_{calc} falls within the critical range, we shall accept the null hypothesis $\mu_1 = \mu_2$. This means that we accept the idea that the sample means are from populations whose means are equal. In practical terms, this means that we conclude that the two methods give results which are not significantly different. (Notice, however, that the result of this test was a close-run thing.)

One should be cautious about extrapolation from this statement. It would be a mistake to believe that the two methods will always give results which are not significantly different. Titrimetric methods are generally used at concentration levels of 0.1 to 0.01 M (400 ppm Ca is approximately 0.01 M). AA methods are usually applied to metal determinations at roughly the ppm level. More concentrated solutions can of course be diluted to bring them into the concentration ranges appropriate to AA. Even so, it is generally thought better to carry out determinations at levels appropriate to titrimetric methods by titrimetric methods. When systematic error is absent, the high precision obtainable by titrimetry gives results of high accuracy. Of course, titrimetric methods are not applicable at the ppm level. If a laboratory was concerned with calcium determinations at this level, titrimetry would not come into consideration.

∏   A company is engaged in the fats-and-oils industry. One of its quality control tests is the saponification value determination. This involves treating the fat or oil with an excess of potassium hydroxide and determining the alkali used up in the process (saponification) by back titration with HCl. By definition the saponification value is the number of mg of potassium hydroxide required to saponify 1 g of fat or oil.

The company is concerned that several of its recent production batches of castor oil appear to have low saponification values. A sharp-eyed supervisor notices that those batches with low saponification values were analysed by a new laboratory technician. Of course, it might be that the new technician was given batches to analyse which had low saponification values. Then again, it might also be that he had a bias in his method. It was decided to compare the results from the new employee with those of a long-serving and trusted employee. (The new employee couldn't do determinations on a standard material – the lab standard had gone missing!) Each analyst carried out six determinations on a batch of the same material. However, one of the determinations of the trusted employee was rejected as an outlier (see later)! The results obtained were as follows:

New Employee	Old Employee
$\bar{x}_1 = 177.3$	$\bar{x}_2 = 178.8$
$s_1 = 0.5$	$s_2 = 0.3$
$n_1 = 6$	$n_2 = 5$

Carry out appropriate statistical test on these results at the 5% significance level.

Since our sample sizes are small, the t-distribution must be brought into play. We may assume that $\sigma_1 = \sigma_2$, (Why?) and therefore use both s_1 and s_2 to estimate σ, ie to calculate a pooled standard deviation.

The null hypothesis is $H_o : \mu_1 = \mu_2$.

The alternative hypothesis is $H_1 : \mu_1 \neq \mu_2$. Some will disagree with this and suggest an alternative hypothesis of $H_1 : \mu_1 < \mu_2$. We agree that this would be appropriate and will consider this alternative too.

First though we apply the two-tail test. If we carry out the test at the 5% significance level and for 9 degrees of freedom $(6 + 5 - 2)$, the critical value sought will be $t_{0.025,9}$; this is 2.262, and the critical values are thus ± 2.262.

The calculated t-value is obtained from

$$t_{calc} = \frac{\bar{x}_1 - \bar{x}_2}{s_p \sqrt{(1/n_1) + (1/n_2)}}$$

$$s_p = \sqrt{\frac{(n_1 - 1)s_1^2 + (n_2 - 1)s_2^2}{n_1 + n_2 - 2}}$$

We calculate s_p first and obtain

$$s_p = \sqrt{\frac{(6 - 1)(0.5)^2 + (5 - 1)(0.3)^2}{6 + 5 - 2}} = 0.423.$$

$$t_{calc} = \frac{177.3 - 178.8}{0.423\sqrt{(1/6 + (1/5)}} = -5.86$$

Since t_{calc} is outside the critical range (or equivalently $|t_{calc}| > |t_{crit}|$) we reject the null hypothesis and accept the alternative hypothesis $H_1 : \mu_1 \neq \mu_2$.

Since we know that both chemists were analysing the same material, we interpret this result as evidence that the new chemist does not get the same result as his longer-serving colleague, ie he gets significantly different results (at the 5% level of significance). Actually, the result is so significant that we would have rejected the null hypothesis at the 0.2% significance level (the lowest level for a two-tail test which appears in our table). For a one-tail test at the 5% significance level the result would have been even more convincing; the critical t-value becomes $t_{0.05,9} = 1.833$. The company obviously has a minor problem with their new technician. However, it can be dealt with. We discussed this kind of problem in Section 1.7 of the unit.

SAQ 4.3a

An agricultural research laboratory carries out a large number of protein-nitrogen determinations by the Kjeldahl method. This method involves digesting a substance with concentrated H_2SO_4, which converts the nitrogen into $(NH_4)_2SO_4$, and then adding an excess of NaOH to liberate NH_3. The NH_3 is distilled into stan-

$\longrightarrow$

SAQ 4.3a
(cont.)

dard acid and is determined by means of a back titration of the excess of acid. Knowing the amount of NH_3 produced enables the nitrogen content to be calculated (and the protein content of the sample to be estimated). The laboratory has heard reports that the NH_3 in Kjeldahl digests can be determined by an 'ammonia electrode'. It decides to compare this procedure with the distillation/back-titration procedure.

A batch of barley is used for the test. Ten Kjeldahl digests are prepared; five have their NH_3 content determined by the ammonia-electrode procedure and five by the distillation/back-titration procedure. The results, reported as % N, are below.

Ammonia Electrode	Distillation-Back-Titration
1.20	1.25
1.30	1.24
1.25	1.26
1.25	1.22
1.26	1.27

Can we conclude at the 5% level of significance that the two procedures for nitrogen determination give the same result?

SAQ 4.3a

We hope that you found this SAQ reasonably easy. If you did, then congratulations! We think it is unlikely that you will find anything more difficult in the rest of the unit. Once one knows which test to apply, the rest is arithmetic. Like everyone else, you can make mistakes with arithmetic. However, give yourself nearly full marks if you used the right test and made only one arithmetical error. Two arithmetical errors is carelessness! Of course the computer could help with the donkey work.

Our last worked example and SAQ 4.3a both raised the following question. Had we any right to assume that σ_1 and σ_2 were the same. Without going into a test which could have tested the hypothesis $\sigma_1 = \sigma_2$, let us answer – probably. You will have to wait for a short time before this topic is discussed in more depth. In the meantime, let us introduce and examine a new type of test.

4.4. PAIRED DIFFERENCES

The test which we are about to introduce is, once the associated experiment has been designed and carried out, relatively easy to apply. Unfortunately, it is, at first, somewhat difficult to visualise. With this in mind, we shall start by discussing an issue, which for many, especially in summer-time, is of great importance. The issue is diet and weight loss! Let us assume that we have invented a new method of dieting and that we wish to investigate its efficiency. This is a messy type of investigation, with many variables to be considered. A detailed discussion of the problems which will be met is not necessary here – we will pretend that matters are as simple as we describe them! We start by finding a few volunteers; say seven in number, all mature adults. They are instructed in the new diet, weighed, and told to come back after one week to be weighed again. The volunteers are of the highest integrity – they stick to their diets! The before- and after-dieting weights and the differences in weights are tabulated below

Individual	Weight before/kg	Weight after/kg	Difference d_i/kg
1	54.488	53.287	−1.201
2	57.424	57.015	−0.409
3	60.216	60.216	0.000
4	55.545	55.321	−0.224
5	70.653	70.989	+0.336
6	75.385	74.926	−0.459
7	81.720	82.823	+1.103

Is the diet effective? Maybe! Is there any statistical test which could be applied? Yes. First let us identify the variable which concerns us. It is the difference, d_i between the before- and after-dieting weight. Now some differences are negative (a weight loss) and some positive (a weight gain); one is no change at all. It makes sense to take the mean of these differences: it is $\bar{d} = -0.854/7 = -0.122$. (Check this: do not forget to take account of the sign of each difference.) Does the mean of the seven differences indicate that the diet plan is successful? Well, if we have learned one thing from this Unit, it is that we should never look at a mean without also looking at a standard deviation. In our example the standard deviation of the differences is:

$$s_d = \sqrt{\sum_{i=1}^{n}(d_i - \bar{d})^2/(n - 1)} = \sqrt{3.095/6} = 0.718$$

(Check it. If you use statistics functions on your calculator, take care! One option is to put the data into your computer program!)

Can we read anything into the results $\bar{d} = -0.122$ and $s_d = 0.718$? Yes, it does look rather inconclusive! Is there a way of putting these results into a formal statistical test? Yes, of course we can, provided we make the assumption that the weight changes for the entire population of 'mature adults', who follow the diet, would fit a normal distribution. We refer to the *population of paired differences*. Our set of 7 measured differences is then a sample drawn from this population. Our sample mean, $\bar{d}$, and sample standard deviation, s_d are estimates of the population mean, μ_d, and population standard deviation, σ_d. We do not know the value of σ_d, and so we have to rely on our estimate s_d. Our sample size is small, so we must use t.

We can calculate a confidence range for the population mean μ_d. Our usual equation

$$\mu = \bar{x} \pm t_{\alpha/2}s/\sqrt{n}$$

becomes, in the notation we have been using for the differences:

$$\mu_d = \bar{d} \pm t_{\alpha/2}s_d/\sqrt{n}$$

We have 7 measurements and hence 6 degrees of freedom. For the 95% confidence range we thus use $t_{0.025,6} = 2.447$. This gives

$$\mu_d = -0.122 \pm 2.447 \times 0.718/\sqrt{7} = -0.122 \pm 0.664 \text{ kg}$$

This confidence range is from $+0.54$ kg to -0.79 kg. It is not clear that the diet leads, on average, to a reduction of weight at all!

We can, alternatively, frame a hypothesis test. Take as our null hypothesis the assertion that the diet makes no difference, ie H_o : $\mu_d = 0$. As the alternative hypothesis, since diets are supposed to work, we should take H_1 : $\mu_d < 0$. We require a one-tail test. Let us test at the 5% level of significance. For an area of 0.05 in one tail, we need $t_{0.05,6} = 1.943$. The tail which interests us is the lower tail ($\mu_d < 0$) so the critical t-value is -1.94.

It remains to find t_{calc} from the results for our sample. The usual equation

$$t_{calc} = \frac{\bar{x} - \mu}{s/\sqrt{n}}$$

becomes, in our present terminology,

$$t_{calc} = \frac{\bar{d} - \mu_d}{s_d/\sqrt{n}} = \frac{-0.122 - 0.000}{0.718/\sqrt{7}} = -0.45$$

(recall that we are testing for $\mu_d = 0$). t_{calc} lies a long way inside of the critical value of -1.94, so we accept the null hypothesis. The evidence is that the diet makes no difference.

It can be shown that a t-value of -0.45, for $\nu = 6$, corresponds to a p-value of 33%. (We again used program STAT3 to find this value. For $\nu = 6$, it reported a probability of 66.6% of exceeding $t = -0.45$, ie a probability of 33.4% in the lower tail.) If the null hypothesis is true, ie if the diet makes no difference, then there is a 33% chance that a test of a random sample of 7 individuals will give a mean weight loss of 0.122 kg or more. The author of the diet plan should think again*: undoubtedly he will not!

---

* He does in fact have a little room for manoeuvre. The width of the confidence range for μ_d indicates that to obtain a more definitive answer a much larger sample size should be taken. He might notice that the outcome of the above test would have appeared rather more promising had individual number 7 not participated. Number 7 was the heaviest to begin with, and thus possibly the most weak-willed when it comes to diet! Taking only the first 6 individuals the p-value for the hypothesis test drops to 9%. It still does not reach the 5% threshold, but if our criterion was set at 10% significance then ... Hope springs eternal!

Now, what does all this mean to analytical chemists? Actually, a great deal. For example, there are many situations in which it is desired to compare two methods of analysis on batches of material which have a wide variation in level of the determinand and/or in composition of the matrix. Such a situation would arise in comparing the results obtained with an ammonia electrode with those of the back-titration procedure for the NH_3 present in Kjeldahl digests from a wide range of foods. The agricultural research laboratory, mentioned in SAQ 4.3a, might wish to compare determinations by the two methods not only on digests from barley, but also for example from cheese, eggs, fish, beef. In such a situation a paired difference test would be considered. The following might be typical results:

Substance	%N NH$_3$-electrode	%N Back-titration	d_i Difference
1	1.24	1.24	0.00
2	1.23	1.24	-0.01
3	4.09	4.06	$+0.03$
4	13.67	13.64	$+0.03$
5	4.22	4.17	$+0.05$
6	3.28	3.26	$+0.02$

Now we can apply a *t*-test on the differences provided we assume that they represent a sample drawn from a normal distribution. Is this valid in a case like this? Yes, probably.

It can be shown that the assumption is valid provided that each of the measurement procedures is subject to a constant systematic error (if any) and to random errors characterised by a constant standard deviation. By 'constant' we mean independent of the concentration level of analyte actually present, and unaffected by differences in the matrix composition. We do not need to know specific values for the bias or for the standard deviation, only that they do not vary, at least across the range of substances compared. This must hold for both measurement procedures, though the values of the 'constant' involved will usually be different. We must be aware of this assumption, but we may suppose that it holds for the nitrogen analysis results quoted above.

Let us then apply a paired-difference test to compare the two procedures for nitrogen determination. The mean of the tabulated differences is $\bar{d} = +0.020$ and the sample standard deviation is $s_d = 0.022$. (Check for yourself.) For the null hypothesis we assert that the two methods agree, ie that the population mean difference is zero. $H_0 : \mu_d = 0$. For the alternative hypothesis we take $\mu_d \neq 0$, implying a two-tail test. We have 6 determinations, therefore $\nu = 5$.

If we test at the 5% significance level we need $t_{0.025,5} = 2.571$. Our critical values are therefore $t \pm 2.57$. Finally we compute:

$$t_{calc} = \frac{\bar{d} - \mu_d}{s_d/\sqrt{n}} = \frac{0.020 - 0.000}{0.022/\sqrt{6}} = 2.23$$

t_{calc} lies within the critical range, though not by a wide margin.

At the 5% level of significance we accept the null hypothesis that the two methods give the same results. That should please the laboratory! However, it would not hurt to do some comparison-of-means tests, perhaps selecting substance 5 for a series of replicate determinations by each method.

The application of a t-test to paired comparisons is widely used in science and technology. Let us think why this is so. First, consider the diet-plan investigation. We are investigating whether or not human beings will lose weight if they stick to the diet. Now human beings vary by sex, by age, by weight, by activity, by metabolic rate, etc. Trying to define a typical human being is impossible. The diet-plan investigation thus presents us with a situation that physical scientists and engineers try to avoid like the plague. Why? Because our approach to an investigation is usually to hold constant all variables but one. This is clearly impossible in the diet-plan investigation. The paired comparison test suggests a way round this. It says, 'Measure the difference in weight for a single human being from the beginning to the end of the investigation, ie obtain a difference. This is the same human being in all respects but one – he is on a diet. Repeat this for all the human beings in the investigation, and we will have a sample of differences. Now analyse the differences. Some will be big and negative, others small, some might actually be big and positive. However, we can calculate the mean difference and the standard deviation of the differences. These, once known, permit us to use statistics to see whether the differences are significant'.

With regard to the diet investigation, this approach seems to be common sense. Now let us view the test in the light of our paired comparisons of the two ways of determining ammonia (nitrogen) in a Kjeldahl digest. Agricultural products vary widely in their composition and in their nitrogen content. We could analyse each given product many times in order to compare the two methods. This would necessitate many experiments. The paired-difference *t*-test permits us to compare the two methods for a wide range of materials with what is essentially a single trial. Of course, this does not mean that all investigation would stop there. However, it is not naive to believe that the number of trials is considerably reduced by this procedure.

Let us conclude this discussion by once again showing the common-sense nature of a paired difference test. Let us say that we wished to compare the durability of two brands of tyres. Let us think of as many variables as we can: slight variations in the quality of a given brand of tyre, variability between different sizes of the same brand of tyre, variability depending upon the type of car to which the tyres are attached, variability according to whether the tyres are attached to the front or the back wheels, variability due to different driving habits, variability according to the environmental conditions, variability due to different road surfaces, etc. What sort of test would you apply here? If you cannot answer this question, we have failed you! The experiment is a hard one to design, but it is differences that would be measured and analysed statistically.

4.5.   COMPARING TWO VARIANCES

When we were finding confidence intervals and testing hypotheses for situations involving comparison of means, we were forced to contend with the fact that population standard deviations, σ_1 and σ_2, were often unknown. When sample sizes were large we found little difficulty: we simply assumed that each sample standard-deviation, s, was a good estimate of the corresponding σ-value. However, when we dealt with small sample sizes, this approximation was not considered valid. But, if we made the assumption that the population standard deviations for the two samples were equal, ie that $\sigma_1 = \sigma_2 = \sigma$, then we were able to apply a t-test in which the pooled sample standard deviation s_p was taken as an estimate of σ. It is all very well to make an assumption about equal σ-values, but can we justify doing so? Can we test whether the two sample-standard deviation values, s_1 and s_2, are consistent with the proposition $\sigma_1 = \sigma_2$? You will have guessed it, we can! The test which we are interested in deals explicitly with values of s^2 rather than of s. For that reason, what we are about is referred to as *comparison of variance*.

Comparing variances (or, what is equivalent, comparing standard deviations) is often of interest in its own right. We have referred to examples when the population standard deviation, σ, is known. A typical situation for this would be when we use a method which is well-tried and tested, and the quoted value of σ has been obtained on the basis of long experience. But there may well be occasions when we wish to check whether the sample standard deviation, s, derived from a particular set of results, is consistent with the quoted σ-value. We might, for instance, wish to check whether a new and inexperienced technician is achieving the level of precision generally expected for the method. Then again, we might be concerned that less precise results are being obtained for samples in a new concentration range, or for samples drawn from a different kind of source. On such an occasion we ask ourselves whether a particular s-value is consistent with the assertion that our sample was a random one drawn from a population of standard deviation, σ.

Again we might be comparing a new method with an old one. Certainly we want an answer to the question: 'Do the methods give the same answer (on average)?' That question we have already considered. We know how to compare the means obtained when both methods are used for replicate determinations on the same material. We also know how to assess paired differences, when the methods have each been applied to a range of different materials. To question whether both methods give the same answer (on average) is tantamount to asking whether there is any difference in *systematic error* between the methods. A difference in systematic error will show up in a difference in the population mean for any determination. If our two methods pass the comparison-of-means test of Section 4.3, then we can assert, at the appropriate level of significance, that there is no difference between the methods as regards systematic error. Either both are free of systematic error or both suffer from it to the same extent.

There is, however, a second question which we shall wish to ask when comparing two methods. 'Do the methods differ in precision?' We want to know whether one or other is subject to a greater *random error*. Our measure of random error is the population standard deviation, σ. If we know the σ-values for both methods then we can give an immediate answer – the method with the smaller σ-value is the more precise. But often in practice we have only s-values, and s-values, for small samples, can be misleading guides to σ-values. How can we test, by using the s-values obtained from measurements by both methods, whether there is evidence, at a specified level of significance, that one method is more precise than the other?

Tests have from the start involved us with distribution functions, and areas in their tails. We have met z and we have met t. We hope you are beginning grudgingly to accept them as old friends. We are now going to meet a new character, with a similarly short name but with yet more extensive tables to its credit! With suitable respect we designate it with a capital letter, *F*. *F* will help us to compare variances.

4.5.1. The F-distribution

So we are going to introduce a new statistic, *F*. Past experience should cause you to take a deep breath. It is always a bit uphill at first, but the pattern should be familiar. First we describe the shape of the distribution. Next we fix our attention on areas in its tail; and we introduce you to tables giving values of the statistic corresponding to particular areas. Then we turn to applying this information to practical problems. Once we get to that stage it is easy enough.

Well what is the *F* statistic all about? For our purposes *F represents a ratio of variances*. Let us explain. If a sample of size n_1 is drawn from a normal population with standard deviation σ_1, and a sample of size n_2 is drawn from a population of standard deviation σ_2, then the *F*-value corresponding to this pair of samples is given by

$$F = \frac{s_1^2/\sigma_1^2}{s_2^2/\sigma_2^2} \tag{4.11}$$

This is a general definition. In practice, we shall usually be investigating the significance of a given *F*-value in the context of the assumption that $\sigma_1 = \sigma_2$. (Often we shall test whether that is a reasonable assumption.) This gives us the simplification below.

If $\sigma_1 = \sigma_2$, $\qquad\qquad F = s_1^2/s_2^2 \tag{4.12}$

Now s_1 and s_2 are sample standard deviations. If the samples are small we know that their values may vary substantially from one random sample to the next, so a wide range of different values of F may arise. What is the relative frequency with which particular F-values should occur? In other words, what is the form of the distribution of F?

We can answer that question, but first we must point out that the answer depends on the size of each of the two samples. For a small sample the s-value varies quite widely from one occasion to the next: for a large enough sample we can almost guarantee that s will be very close to σ. *The distribution of F depends on the number of degrees of freedom ν_1 and ν_2 of the two samples.* We always give an equation for our distribution functions, so here goes:

$$y = \frac{CF^{(\nu_1 - 2)/2}}{(\nu_2 + \nu_1 F)^{(\nu_1 - \nu_2)/2}} \qquad (4.13)$$

where y gives the relative frequency of occurrence of a particular value of F. C is a constant whose value is such that the total area under the curve is 1.00 unit of area. As we always say on these occasions, this is not an equation which chemists try to remember! We wrote it down to prove to you that an equation does exist and that it depends in quite a complicated way on the values of ν_1 and ν_2. It would allow you, if you really felt like it, to check the shape of the graphs which we now discuss.

Fig. 4.5a shows the shape of each of two F-distribution curves. Curve (i) applies if our first sample is of size 11 and our second sample of size 25, ie if $\nu_1 = 10$ and $\nu_2 = 24$. Notice first that the graph starts at $F = 0$, this is the smallest conceivable value for F, as variances s_1^2 and s_1^2 can never be negative.

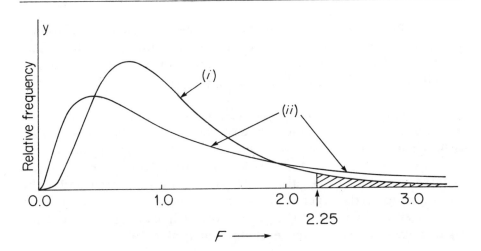

Fig. 4.5a. *Graphs of the F-distribution function when: (i) $\nu_1 = 10, \nu_2 = 24$, (ii) $\nu_1 = 6, \nu_2 = 4$. The upper tail of curve (i) corresponding to an area of 0.05 has been indicated by hatching*

Now our previous curves, for z and for t, have always been symmetrically shaped (about $z = 0$ and $t = 0$ respectively). Half of the area under the curve for z or for t lies to the right of the axis of symmetry and half to the left. By contrast curves for F are *skewed*. The maximum, of the curve occurs at an F-value below 1.0 (for curve (i) at $F = 0.74$), but the curves tail away more slowly to the right than to the left. The total area under each curve is 1.00, but more than half of this area lies to the right of the maximum.

As usual we shall be concerned with tail areas. The lack of symmetry means that the upper and the lower tail have radically different shapes. We shall get by, however, by considering areas only for upper tails. For example, the upper tail for curve (i), having an area of 0.05, has been indicated. The F-value corresponding to this tail is 2.25. If we take two samples from populations with the same σ, the first of size 11 and the second of size 25, and if we compute $F = s_1^2/s_2^2$, then there is a 5% chance of obtaining a value of 2.25 or larger. Obviously we shall need to know how to look up F-values. Patience!

Now turn your attention to curve (ii) in Fig. 4.5a. This shows the F-distribution when the samples are smaller: $\nu_1 = 6$ and $\nu_2 = 4$. This is much more heavily skewed than curve (i). Its maximum comes at a lower F-value ($F = 0.44$) and it tails off much more slowly at large values of F. A substantial part of the total area under the curve lies towards larger values of F. The upper tail corresponding to an area of 0.05 is not shown because it lies entirely beyond the right-hand edge of the figure. It starts at $F = 6.16$!

The shape of the F distribution is always skewed, but the precise shape depends on ν_1 and ν_2. Small ν-values are associated with more pronounced skewing, particularly so if ν_2 is small. High values for both ν_1 and ν_2 lead to sharper peaking of the curve, with the maximum moving closer to $F = 1.0$ and the curve dropping more steeply and more symmetrically on either side of its maximum.

One final point we should emphasise. We must not confuse ν_1 with ν_2. Curve (ii) was drawn for $\nu_1 = 6$ and $\nu_2 = 4$. The curve for $\nu_1 = 4$ and $\nu_2 = 6$ is different. We must be careful: ν_1 is the number of degrees of freedom for the sample whose variance has been put in the numerator when we calculate $F(= s_1^2/s_2^2)$. ν_2 relates to the sample consigned to the denominator.

4.5.2. Percentile Tables of F

Our statistical functions are becoming more awkward. We met z first: it is represented by a single curve, the standardised normal distribution. We were able to give a very detailed table for the tail area as a function of the value of z. When we came to t, we had to face the fact that there is a separate curve for every value of the number of degrees of freedom, ν. This meant we had to compromise in constructing our table: we listed t-values, for each ν, for only seven different chosen values for the tail area. Now, for F there is a different distribution curve for every combination of values for ν_1 and ν_2, the two separate numbers of degrees of freedom. The critical F-value defining a tail area therefore depends on three parameters; the size of the tail area required, α, and the value of ν_1 and of ν_2; we shall use the notation $F_{\alpha(\nu_1,\nu_2)}$. It is necessary to lay out a separate F-table for each chosen tail-area. Three such tables are given in Figs. 4.5b, 4.5c, and 4.5d, corresponding to upper-tail areas of 0.05, 0.025 and 0.01 respectively.

Area in tail = α

$F_{\alpha(\nu_1,\nu_2)}$

$F \longrightarrow$

ν_1	1	2	3	4	5	6	7	8	9	10	12	15	20	24	30	∞
ν_2																
1	161.45	199.50	215.71	224.58	230.16	233.99	236.77	238.88	240.54	241.88	243.91	245.95	248.01	249.05	250.10	254.30
2	18.51	19.00	19.16	19.25	19.30	19.33	19.35	19.37	19.38	19.40	19.41	19.43	19.45	19.45	19.46	19.50
3	10.13	9.55	9.28	9.12	9.01	8.94	8.89	8.85	8.81	8.79	8.74	8.70	8.66	8.64	8.62	8.53
4	7.71	6.94	6.59	6.39	6.26	6.16	6.09	6.04	6.00	5.96	5.91	5.86	5.80	5.77	5.75	5.63
5	6.61	5.79	5.41	5.19	5.05	4.95	4.88	4.82	4.77	4.74	4.68	4.62	4.56	4.53	4.50	4.36
6	5.99	5.14	4.76	4.53	4.39	4.28	4.21	4.15	4.10	4.06	4.00	3.94	3.87	3.84	3.81	3.67
7	5.59	4.74	4.35	4.12	3.97	3.87	3.79	3.73	3.68	3.64	3.57	3.51	3.44	3.41	3.38	3.23
8	5.32	4.46	4.07	3.84	3.69	3.58	3.50	3.44	3.39	3.35	3.28	3.22	3.15	3.12	3.08	2.93
9	5.12	4.25	3.86	3.63	3.48	3.37	3.29	3.23	3.18	3.14	3.07	3.01	2.94	2.90	2.86	2.71
10	4.96	4.10	3.71	3.48	3.33	3.22	3.14	3.07	3.02	2.98	2.91	2.84	2.77	2.74	2.70	2.54

11	4.84	3.98	3.59	3.36	3.20	3.09	3.01	2.95	2.90	2.85	2.79	2.72	2.65	2.61	2.57	2.40
12	4.75	3.89	3.49	3.26	3.11	3.00	2.91	2.85	2.80	2.75	2.69	2.62	2.54	2.51	2.47	2.30
13	4.67	3.81	3.41	3.18	3.03	2.92	2.83	2.77	2.71	2.67	2.60	2.53	2.46	2.42	2.38	2.21
14	4.60	3.74	3.34	3.11	2.96	2.85	2.76	2.70	2.65	2.60	2.53	2.46	2.39	2.35	2.31	2.13
15	4.54	3.68	3.29	3.06	2.90	2.79	2.71	2.64	2.59	2.54	2.48	2.40	2.33	2.29	2.25	2.07
16	4.49	3.63	3.24	3.01	2.85	2.74	2.66	2.59	2.54	2.49	2.42	2.35	2.28	2.24	2.19	2.01
17	4.45	3.59	3.20	2.96	2.81	2.70	2.61	2.55	2.49	2.45	2.38	2.31	2.23	2.19	2.15	1.96
18	4.41	3.55	3.16	2.93	2.77	2.66	2.58	2.51	2.46	2.41	2.34	2.27	2.19	2.15	2.11	1.92
19	4.38	3.52	3.13	2.90	2.74	2.63	2.54	2.48	2.42	2.38	2.31	2.23	2.16	2.11	2.07	1.88
20	4.35	3.49	3.10	2.87	2.71	2.60	2.51	2.45	2.39	2.35	2.28	2.20	2.12	2.08	2.04	1.84
21	4.32	3.47	3.07	2.84	2.68	2.57	2.49	2.42	2.37	2.32	2.25	2.18	2.10	2.05	2.01	1.81
22	4.30	3.44	3.05	2.82	2.66	2.55	2.46	2.40	2.34	2.30	2.23	2.15	2.07	2.03	1.98	1.78
23	4.28	3.42	3.03	2.80	2.64	2.53	2.44	2.37	2.32	2.27	2.20	2.13	2.05	2.01	1.96	1.76
24	4.26	3.40	3.01	2.78	2.62	2.51	2.42	2.36	2.30	2.25	2.18	2.11	2.03	1.98	1.94	1.73
25	4.24	3.39	2.99	2.76	2.60	2.49	2.40	2.34	2.28	2.24	2.16	2.09	2.01	1.96	1.92	1.71
26	4.23	3.37	2.98	2.74	2.59	2.47	2.39	2.32	2.27	2.22	2.15	2.07	1.99	1.95	1.90	1.69
27	4.21	3.35	2.96	2.73	2.57	2.46	2.37	2.31	2.25	2.20	2.13	2.06	1.97	1.93	1.88	1.67
28	4.20	3.34	2.95	2.71	2.56	2.45	2.36	2.29	2.24	2.19	2.12	2.04	1.96	1.91	1.87	1.65
29	4.18	3.33	2.93	2.70	2.55	2.43	2.35	2.28	2.22	2.18	2.10	2.03	1.94	1.90	1.85	1.64
30	4.17	3.32	2.92	2.69	2.53	2.42	2.33	2.27	2.21	2.16	2.09	2.01	1.93	1.89	1.84	1.62
40	4.08	3.23	2.84	2.61	2.45	2.34	2.25	2.18	2.12	2.08	2.00	1.92	1.84	1.79	1.74	1.51
60	4.00	3.15	2.76	2.53	2.37	2.25	2.17	2.10	2.04	1.99	1.92	1.84	1.75	1.70	1.65	1.39
120	3.92	3.07	2.68	2.45	2.29	2.18	2.09	2.02	1.96	1.91	1.83	1.75	1.66	1.61	1.55	1.25
∞	3.84	3.00	2.60	2.37	2.21	2.10	2.01	1.94	1.88	1.83	1.75	1.67	1.57	1.52	1.46	1.00

Fig. 4.5b. *The value of F for a tail-area, $\alpha = 0.05$*

ν_1	1	2	3	4	5	6	7	8
ν_2								
1	647.79	799.50	864.16	899.58	921.85	937.11	948.22	956.66
2	38.51	39.00	39.17	39.25	39.30	39.33	39.36	39.37
3	17.44	16.04	15.44	15.10	14.88	14.73	14.62	14.54
4	12.22	10.65	9.98	9.60	9.36	9.20	9.07	8.98
5	10.01	8.43	7.76	7.39	7.15	6.98	6.85	6.76
6	8.81	7.26	6.60	6.23	5.99	5.82	5.70	5.60
7	8.07	6.54	5.89	5.52	5.29	5.12	4.99	4.90
8	7.57	6.06	5.42	5.05	4.82	4.65	4.53	4.43
9	7.21	5.71	5.08	4.72	4.48	4.32	4.20	4.10
10	6.94	5.46	4.83	4.47	4.24	4.07	3.95	3.85
11	6.72	5.26	4.63	4.28	4.04	3.88	3.76	3.66
12	6.55	5.10	4.47	4.12	3.89	3.73	3.61	3.51
13	6.41	4.97	4.35	4.00	3.77	3.60	3.48	3.39
14	6.30	4.86	4.24	3.89	3.66	3.50	3.38	3.29
15	6.20	4.77	4.15	3.80	3.58	3.41	3.29	3.20
16	6.12	4.69	4.08	3.73	3.50	3.34	3.22	3.12
17	6.04	4.62	4.01	3.66	3.44	3.28	3.16	3.06
18	5.98	4.56	3.95	3.61	3.38	3.22	3.10	3.01
19	5.92	4.51	3.90	3.56	3.33	3.17	3.05	2.96
20	5.87	4.46	3.86	3.51	3.29	3.13	3.01	2.91
21	5.83	4.42	3.82	3.48	3.25	3.09	2.97	2.87
22	5.79	4.38	3.78	3.44	3.22	3.05	2.93	2.84
23	5.75	4.35	3.75	3.41	3.18	3.02	2.90	2.81
24	5.72	4.32	3.72	3.38	3.15	2.99	2.87	2.78
25	5.69	4.29	3.69	3.35	3.13	2.97	2.85	2.75
26	5.66	4.27	3.67	3.33	3.10	2.94	2.82	2.73
27	5.63	4.24	3.65	3.31	3.08	2.92	2.80	2.71
28	5.61	4.22	3.63	3.29	3.06	2.90	2.78	2.69
29	5.59	4.20	3.61	3.27	3.04	2.88	2.76	2.67
30	5.57	4.18	3.59	3.25	3.03	2.87	2.75	2.65
40	5.42	4.05	3.46	3.13	2.90	2.74	2.62	2.53
60	5.29	3.93	3.34	3.01	2.79	2.63	2.51	2.41
120	5.15	3.80	3.23	2.89	2.67	2.52	2.39	2.30
∞	5.02	3.69	3.12	2.79	2.57	2.41	2.29	2.19

Fig. 4.5c. *The value of F for*

9	10	12	15	20	24	30	∞
963.28	968.63	976.71	984.87	993.10	997.25	1001.4	1018
39.39	39.40	39.41	39.43	39.45	39.46	39.46	39.50
14.47	14.42	14.34	14.25	14.17	14.12	14.08	13.90
8.90	8.84	8.75	8.66	8.56	8.51	8.46	8.26
6.68	6.62	6.52	6.43	6.33	6.28	6.23	6.02
5.52	5.46	5.37	5.27	5.17	5.12	5.07	4.85
4.82	4.76	4.67	4.57	4.47	4.41	4.36	4.14
4.36	4.30	4.20	4.10	4.00	3.95	3.89	3.67
4.03	3.96	3.87	3.77	3.67	3.61	3.56	3.33
3.78	3.72	3.62	3.52	3.42	3.37	3.31	3.08
3.59	3.53	3.43	3.33	3.23	3.17	3.12	2.88
3.44	3.37	3.28	3.18	3.07	3.02	2.96	2.72
3.31	3.25	3.15	3.05	2.95	2.89	2.84	2.60
3.21	3.15	3.05	2.95	2.84	2.79	2.73	2.49
3.12	3.06	2.96	2.86	2.76	2.70	2.64	2.40
3.05	2.99	2.89	2.79	2.68	2.63	2.57	2.32
2.98	2.92	2.82	2.72	2.62	2.56	2.50	2.25
2.93	2.87	2.77	2.67	2.56	2.50	2.44	2.19
2.88	2.82	2.72	2.62	2.51	2.45	2.39	2.13
2.84	2.77	2.68	2.57	2.46	2.41	2.35	2.09
2.80	2.73	2.64	2.53	2.42	2.37	2.31	2.04
2.76	2.70	2.60	2.50	2.39	2.33	2.27	2.00
2.73	2.67	2.57	2.47	2.36	2.30	2.24	1.97
2.70	2.64	2.54	2.44	2.33	2.27	2.21	1.94
2.68	2.61	2.51	2.41	2.30	2.24	2.18	1.91
2.65	2.59	2.49	2.39	2.28	2.22	2.16	1.88
2.63	2.57	2.47	2.36	2.25	2.19	2.13	1.85
2.61	2.55	2.45	2.34	2.23	2.17	2.11	1.83
2.59	2.53	2.43	2.32	2.21	2.15	2.09	1.81
2.57	2.51	2.41	2.31	2.20	2.14	2.07	1.79
2.45	2.39	2.29	2.18	2.07	2.01	1.94	1.64
2.33	2.27	2.17	2.06	1.94	1.88	1.82	1.48
2.22	2.16	2.05	1.94	1.82	1.76	1.69	1.31
2.11	2.05	1.94	1.83	1.71	1.64	1.57	1.00

a tail-area, α = 0.025

ν_1	1	2	3	4	5	6	7	8
ν_2								
1	4052.2	4999.5	5403.4	5624.6	5763.6	5859.0	5928.4	5981.1
2	98.50	99.00	99.17	99.25	99.30	99.33	99.36	99.37
3	34.12	30.82	29.46	28.71	28.24	27.91	27.67	27.49
4	21.20	18.00	16.69	15.98	15.52	15.21	14.98	14.80
5	16.26	13.27	12.06	11.39	10.97	10.67	10.46	10.29
6	13.75	10.92	9.78	9.15	8.75	8.47	8.26	8.10
7	12.25	9.55	8.45	7.85	7.46	7.19	6.99	6.84
8	11.26	8.65	7.59	7.01	6.63	6.37	6.18	6.03
9	10.56	8.02	6.99	6.42	6.06	5.80	5.61	5.47
10	10.04	7.56	6.55	5.99	5.64	5.39	5.20	5.06
11	9.65	7.21	6.22	5.67	5.32	5.07	4.89	4.74
12	9.33	6.93	5.95	5.41	5.06	4.82	4.64	4.50
13	9.07	6.70	5.74	5.21	4.86	4.62	4.44	4.30
14	8.86	6.51	5.56	5.04	4.69	4.46	4.28	4.14
15	8.68	6.36	5.42	4.89	4.56	4.32	4.14	4.00
16	8.53	6.23	5.29	4.77	4.44	4.20	4.03	3.89
17	8.40	6.11	5.18	4.67	4.34	4.10	3.93	3.79
18	8.29	6.01	5.09	4.58	4.25	4.01	3.84	3.71
19	8.18	5.93	5.01	4.50	4.17	3.94	3.77	3.63
20	8.10	5.85	4.94	4.43	4.10	3.87	3.70	3.56
21	8.02	5.78	4.87	4.37	4.04	3.81	3.64	3.51
22	7.95	5.72	4.82	4.31	3.99	3.76	3.59	3.45
23	7.88	5.66	4.76	4.26	3.94	3.71	3.54	3.41
24	7.82	5.61	4.72	4.22	3.90	3.67	3.50	3.36
25	7.77	5.57	4.68	4.18	3.85	3.63	3.46	3.32
26	7.72	5.53	4.64	4.14	3.82	3.59	3.42	3.29
27	7.68	5.49	4.60	4.11	3.78	3.56	3.39	3.26
28	7.64	5.45	4.57	4.07	3.75	3.53	3.36	3.23
29	7.60	5.42	4.54	4.04	3.73	3.50	3.33	3.20
30	7.56	5.39	4.51	4.02	3.70	3.47	3.30	3.17
40	7.31	5.18	4.31	3.83	3.51	3.29	3.12	2.99
60	7.08	4.98	4.13	3.65	3.34	3.12	2.95	2.82
120	6.85	4.79	3.95	3.48	3.17	2.96	2.79	2.66
∞	6.63	4.61	3.78	3.32	3.02	2.80	2.64	2.51

Fig. 4.5d.   *The value of F for*

9	10	12	15	20	24	30	∞
6022.5	6055.8	6106.3	6157.3	6208.7	6234.6	6260.6	6366
99.39	99.40	99.42	99.43	99.45	99.46	99.47	99.50
27.35	27.23	27.05	26.87	26.69	26.60	26.50	26.12
14.66	14.55	14.37	14.20	14.02	13.93	13.84	13.46
10.16	10.05	9.89	9.72	9.55	9.47	9.38	9.02
7.98	7.87	7.72	7.56	7.40	7.31	7.23	6.88
6.72	6.62	6.47	6.31	6.16	6.07	5.99	5.65
5.91	5.81	5.67	5.52	5.36	5.28	5.20	4.86
5.35	5.26	5.11	4.96	4.81	4.73	4.65	4.31
4.94	4.85	4.71	4.56	4.41	4.33	4.25	3.91
4.63	4.54	4.40	4.25	4.10	4.02	3.94	3.60
4.39	4.30	4.16	4.01	3.86	3.78	3.70	3.36
4.19	4.10	3.96	3.82	3.66	3.59	3.51	3.16
4.03	3.94	3.80	3.66	3.51	3.43	3.35	3.00
3.89	3.80	3.67	3.52	3.37	3.29	3.21	2.87
3.78	3.69	3.55	3.41	3.26	3.18	3.10	2.75
3.68	3.59	3.46	3.31	3.16	3.08	3.00	2.65
3.60	3.51	3.37	3.23	3.08	3.00	2.92	2.57
3.52	3.43	3.30	3.15	3.00	2.92	2.84	2.49
3.46	3.37	3.23	3.09	2.94	2.86	2.78	2.42
3.40	3.31	3.17	3.03	2.88	2.80	2.72	2.36
3.35	3.26	3.12	2.98	2.83	2.75	2.67	2.31
3.30	3.21	3.07	2.93	2.78	2.70	2.62	2.26
3.26	3.17	3.03	2.89	2.74	2.66	2.58	2.21
3.22	3.13	2.99	2.85	2.70	2.62	2.54	2.17
3.18	3.09	2.96	2.81	2.66	2.58	2.50	2.13
3.15	3.06	2.93	2.78	2.63	2.55	2.47	2.10
3.12	3.03	2.90	2.75	2.60	2.52	2.44	2.06
3.09	3.00	2.87	2.73	2.57	2.49	2.41	2.03
3.07	2.98	2.84	2.70	2.55	2.47	2.39	2.01
2.89	2.80	2.66	2.52	2.37	2.29	2.20	1.80
2.72	2.63	2.50	2.35	2.20	2.12	2.03	1.60
2.56	2.47	2.34	2.19	2.03	1.95	1.86	1.38
2.41	2.32	2.18	2.04	1.88	1.79	1.70	1.00

a tail area, $\alpha = 0.01$

Suppose we are interested in the F-value associated with an upper-tail area of 0.025, in a distribution with $\nu_1 = 5$ and $\nu_2 = 6$. In terms of our notation we are seeking $F_{0.025(5,6)}$. We must consult the table for a tail area of 0.025, ie we look at Fig. 4.5c. *Do this now*, and find the value in the column corresponding to $\nu_1 = 5$ and in the row for $\nu_2 = 6$. You should thus be able to confirm that $F_{0.025(5,6)} = 5.99$, ie that there is only a 2.5% probability of occurrence of an F-value of 5.99 or larger. Easy enough, is'nt it?

You just have to be careful to look up the right table, and to locate the correct entry in it! Notice that if we interchange the values of ν_1 and ν_2 we should have to look for $F_{0.025(6,5)}$. This would give a different answer, $F = 6.98$.

Try it again. In the previous section, in Fig. 4.5a, a tail area was shown. Look back at the figure and concentrate on curve *(i)*; convince yourself that Fig. 4.5a implies that $F_{0.05(10,24)} = 2.25$. Check that this is consistent with the F-table for $\alpha = 0.05$, given in Fig. 4.5b.

We are now in a position to do one-tail tests involving F. Having read off that $F_{0.025(5,6)} = 5.99$ we can immediately say that if we take two random samples, one of size 6 (ie $\nu_1 = 5$) and the other of size 7 (ie $\nu_2 = 6$), then there is only a 2.5% probability of obtaining an F-value exceeding 5.99. If we are testing whether two such samples are drawn from populations of the same 'true' standard deviation, ie if our null hypothesis is $H_0 : \sigma_1 = \sigma_2$, then the F-value for our samples is given by s_1^2/s_2^2 (Eq. 4.12). If the calculated value in fact exceeds 5.99 then, at a 2.5% level of significance, we reject the null hypothesis and accept the alternative hypothesis, ie $H_1 : \sigma_1 > \sigma_2$. The argument, as we said earlier, follows a familiar pattern.

We have one further point to consider, however. The test we have just described is an upper one-tail test – the tables we have given all refer to the upper tail of F. Now previously, with z and t, it has been a simple matter to deal also with the lower tail; the symmetry of the z- and t-distributions has ensured that. When our table told us, for instance, that $z = +1.96$ was the critical value for an upper tail area of 0.025 we understood immediately that $z = -1.96$ was the critical value for a lower tail area of 0.025. Now F lacks this symmetry, and this raises the question of how we can estimate the lower tail, or perform a two-tail test. Do we need yet a further set of tables?

For once the answer is a convenient one! No, we do not need separate tables, in fact we generally do not even need to bother to work out the critical F-value for a lower tail. Yet we can still do two-tail tests! This can be done by a simple device. F is calculated as s_1^2/s_2^2, ie it is the variance for sample 1 divided by the variance for sample 2. But who decides which of our samples is to be labelled 'sample 1' and which as 'sample 2'? We do! We can exploit this freedom so that we always obtain an F-value greater than 1.0 (we simply choose the sample with the larger variance as 'sample 1'). Now a value of F greater than 1.0 could breach only the upper critical value; we know without consulting any table that it lies clear of the lower tail, so an upper-tail test will suffice to clinch things.

Again that needs explanation! Suppose we want to apply a two-tail test at a significance level of α. Let us for a minute call our samples A and B. We ask whether the sample variances, s_A^2 and s_B^2, are compatible with the hypothesis $\sigma_A = \sigma_B$. There are two circumstances under which we shall reject this hypothesis:

(*a*) s_A turns out larger than s_B to an extent which has less than a probability of $\alpha/2$ of occurring by chance;

(*b*) s_A turns out smaller than s_B to an extent which has less than a probability of $\alpha/2$ of occurring by chance.

Every two-tail test is a combination of two one-tail tests, one for each tail at half the overall two-tail significance level. In carrying out an F-test we need only check the tail which is threatened by our results, and we arrange that this is an upper tail. Thus, in (*a*) above we calculate $F = s_A^2/s_B^2$, and we look up the critical value $F_{\alpha/2(\nu_A, \nu_B)}$; if the calculated F exceeds the critical F then we reject the hypothesis $\sigma_A = \sigma_B$.

If (*b*) arises then we calculate $F = s_B^2/s_A^2$, and we now look up the critical value $F_{\alpha/2(\nu_B \nu_A)}$ again, if the calculated F exceeds the critical

F we reject the hypothesis $\sigma_A = \sigma_B$. Notice carefully that the critical F-value will be different when ν_1 and ν_2 are interchanged*.

We are now ready to consider a number of practical applications of F tests.

SAQ 4.5a

> Measurements were made on two samples. On sample A we made 8 replicate measurements and this yielded a sample standard deviation $s_A = 0.12$. On sample B we made 4 replicate measurements and this gave $s_B = 0.24$. Is this result consistent, at a 5% significance level, with both samples conforming to the same population standard deviation (ie is $\sigma_A = \sigma_B$)?

* If we do ever wish to find a lower critical value for F it is in fact quite simply done. It is obtained from the formula below.

$$F_{1-\alpha(\nu_A,\nu_B)} = 1/F_{\alpha(\nu_B,\nu_A)}$$

The notation needs explanation. On the left we have an F-value with an *upper tail* of area $(1 - \alpha)$, it is therefore the value which has a *lower tail* of area α. (Remember the total area under the distribution is 1.00.) The F-value on the right is that for an *upper tail* of area α, but for the case where ν_A and ν_B are interchanged. This value is read from the tables. The formula holds because a lower-tail situation for $F = s_A^2/s_B^2$ corresponds to an equivalent upper-tail situation for the inverted ratio $F = s_B^2/s_A^2$.

SAQ 4.5a

4.5.3. Applying F-tests – Some Examples

Having just learned how to do F-tests it is important to practise applying them to a number of problems. Without any more ado, we can simply launch ourselves into examples involving the comparison of two variances. First let us review the example we have already discussed in SAQ 4.3a.

∏ An agricultural research laboratory is comparing two methods, an ammonia electrode and a titrimetric procedure, for the determination of NH_3 in Kjeldahl digests. Five digests had their ammonia determined by use of the NH_3 electrode, and five by the titrimetric method. The results of the determinations were as follows.

For the ammonia-electrode method: $\bar{x}_1 = 1.252$, $s_1 = 0.0356$.

For the titrimetric method: $\bar{x}_2 = 1.248$, $s_2 = 0.0192$.

In SAQ 4.3a we applied a simple t-test to the difference between the sample means. This test was based on the assumption that $\sigma_1 = \sigma_2$. Was the use of this t-test justified? Test at the 5% level of significance.

We are asked if it is reasonable to assume that $\sigma_1 = \sigma_2$. We shall therefore apply a two-tail test. Our null hypothesis is: H_0: $\sigma_1 = \sigma_2$, and our alternative hypothesis is H_1: $\sigma_1 \neq \sigma_2$. The level of significance is 5%, ie 2.5% for each tail.

Now, it has turned out that $s_1 > s_2$, so we calculate

$$F = s_1^2/s_2^2 = (0.0356/0.0192)^2 = 3.44$$

Both samples were of size five, so $\nu_1 = \nu_2 = 4$. Hence our critical F-value is $F_{0.025(4,4)}$. Look at Fig. 4.5c and check that this equals 9.60. Hence, the calculated F-value from our samples (ie 3.44) lies well below the critical F-value (9.60). We therefore accept the null hypothesis. It was perfectly reasonable in SAQ 4.3a to apply the t-test to the difference of means. The assumption that $\sigma_1 = \sigma_2$ is tenable. This, of course, does not mean that the two population standard deviations are exactly equal. It means rather that *the evidence of the test is not inconsistent with their being equal*. Once again, the arithmetic of testing is easy, it is getting to appreciate what a test tells us which is more difficult!

SAQ 4.5b	In each of the following examples, test the null hypothesis H_o: $\sigma_1 = \sigma_2$ against the stated alternative hypothesis at the given level of significance.

        (*i*)   Sample A: $n_A = 6, s_A = 0.0715$.
                Sample B: $n_B = 4, s_B = 0.0220$.
                Level of significance = 5%: one-tail test.

        (*ii*)   Sample A: $n_A = 6, s_A = 0.0715$.
                Sample B: $n_A = 4, s_B = 0.0220$.
                Level of significance = 5%: two-tail test.

        (*iii*) Sample A: $n_A = 11, s_A = 0.0220$.
                Sample B: $n_B = 16, s_B = 0.0715$.
                Level of significance = 5%: two-tail test.

        (*iv*) Sample A: $n_A = 11, s_A = 0.0220$.
                Sample B: $n_B = 16, s_B = 0.0715$.
                Level of significance = 2%: two-tail test.

SAQ 4.5b

We have so far discussed the *F*-test in a context in which the variances of two separate samples are being compared. There is another common situation where the same procedure can be applied, but where only one sample is involved. Let us go straight into an example.

A firm making dry cells is proud of its ability to manufacture the cells both with a long life-time, and with little variability in life-time: the claim is $\sigma = 1.0$ hour. When a sample of 10 dry cells is tested, a sample standard deviation of 2.0 hours is found. Is the firm's assertion that $\sigma = 1.0$ hour sustainable? Can we test this claim at, say, a 5% level of significance?

Previously we have been presented with data from two separate samples, typically sets of measurements obtained in different ways: here on the face of it we have only a single sample. But we are still comparing variances. Is the observed (sample) standard deviation of 2.0 hours compatible with the claimed (population) standard deviation of 1.0 hour? We are comparing a particular sample with a presumed population.

It is this population which gives us the second sample for our F-test. Imagine we had taken a sample of *infinite size* from that population: it would have $s = \sigma = 1.0$ hour. So we set up our F-test quite happily as follows:

Sample A.   Size, $n_A = 10$, $s_A = 2.0$ hour.
Sample B.   Size, $n_B = \infty$, $s_B = \sigma_B = 1.0$ hour.

What claim are we testing? It is whether the population from which sample A was drawn could reasonably have a standard deviation, $\sigma_A = 1.0$ hour. In other words, we are testing whether $\sigma_A = \sigma_B$ (= 1.0 hour).

We therefore have our null hypothesis, H_0: $\sigma_A = \sigma_B$. Now we must be clear about our alternative hypothesis, and it seems that here a one-tail test might be most appropriate. We are suspicious that our manufacturer is over-stating the consistency of his dry cells. We shall pose the alternative hypothesis, H_1: $\sigma_A > \sigma_B$.

Since $s_A > s_B$, we define $F = s_A^2/s_B^2 = (2.0/1.0)^2 = 4.0$.

The critical value of F will be $F_{0.05(9,\infty)}$. Do you agree? Sample A is of size 10: this is 'sample 1' and so $\nu_1 = 10 - 1 = 9$. Sample B is a population! This is 'sample 2' and so $\nu_2 = \infty$. Look up Fig. 4.5b and confirm that $F_{0.05(9,\infty)} = 1.88$.

The calculated F-value is larger than the critical F-value so we reject the null hypothesis. The variability of the life-times of the dry cells is greater than claimed!

Examples of a similar nature to this are quite often met. A product is claimed to have a certain consistent quality or a method is believed to be of a certain precision. We wish to check whether an actual set of results obtained for some sample conforms with the claim. We have now seen that we can apply a standard F-test – we ask whether it is tenable, at a specified level of significance, that the true population standard deviation corresponding to our sample is equal to the claimed value. We perform the F-test precisely as before, only one of our numbers of degrees of freedom is infinite.

Let us conclude this section with a few comments about the significance of a conclusion. We chose to test the manufacturer's claim about his dry cells at a 5% level of significance. Perhaps our manufacturer would come back to us and say: 'In applying a 5% level of significance you are carrying out a test which has a 5% probability of resulting in a type I error. One time in twenty you will accuse me of producing batteries below specification when in fact there is no problem. I am sure this is just such an occasion. I believe my batteries are OK'. He is a prickly character, and he knows that the best form of defence is attack!

Well we have got him this time! 'Certainly', we can say, 'your first two sentences are perfectly fair; it was a test at 5% significance we chose and this did imply a 1 in 20 chance of rejecting your claim even if it was valid. However, this particular sample has failed our 5% test by a wide margin. If we consult Fig. 4.5d we find $F_{0.01(9,\infty)} = 2.41$. That would be the critical F-value at a 1% level of significance, and that still lies well below the calculated F of 4.0. So the chance that we are wrongly rejecting your claims, far from being 1 in 20, is in fact less than 1 in 100.'

1% is the lowest level of significance for which we have given a table of F-values, but the margin by which the calculated F-value exceeds even the 1% critical value indicates that the probability that we may be wrongly rejecting the manufacturer's claim is a great deal less than 1 in 100.

We have gone into this discussion to encourage you, whenever you apply a test, to apply some attention to the margin by which it is passed or failed. Certainly, if we are working on day-to-day routine quality-control, we may well have to stick to a hard-and-fast significance level. If a sample passes it passes, if it fails we 'blow the whistle' on it. Nevertheless, if we have a sample which nearly fails we might whisper to someone that we are a little worried about it. If we have a string of such results we should be all the more ready to sound a note of caution. Equally, a sample which our test just leads us to reject, whilst we must indeed reject it, should cause us less alarm than a sample which fails by a massive margin. Then we really should tell someone that it looks as though something drastic has gone wrong with the production process. We shall return to some of these issues in Part 7 of this Unit.

SAQ 4.5c

A new technician started a month ago in the quality-control lab. Since he arrived things seem to have gone swimmingly. His results indicate consistently that the plant's production is almost exactly on specification. He seems to be very efficient too, having plenty of time between samples to pursue his interest in reading the fine print on page 3 of his newspaper. Suddenly a rash of complaints start coming in from customers. Their quality checks have led them to refuse delivery of certain batches. Is something going wrong during the transport of the product to the customers? Or could it be that the technician is 'cooking the books'?

The analytical method being used has, from long experience, a standard deviation of 0.05%. The last ten results obtained by the technician, from this morning's production, are found to have a sample standard deviation of 0.015%. Is there evidence of 'cookery'? Test the data at a 1% level of significance.

SAQ 4.5c

4.6. COMPLETION OF PROGRAM STAT

In Parts 1, 2, and 3 of the unit we have painstakingly put together more and more segments of a statistics programming package. If this unit has been your first introduction to computing you were probably quite surprised when we gave a full listing of STAT3 at the end of Section 3.4.5. The program now contains over a hundred lines, some of them complicated. It was starting to be quite flexible; it has taken a lot of effort to produce, but it allows you to do an analysis on a new set of measurements very easily; it is just a matter of inserting revised DATA lines. You could also use the package in place of the statistical tables which give z and t. The program is more flexible than the tables, especially for t, though in this case it is a little less accurate.

We now propose to add further to the package, and so to increase its power substantially. We shall do this in a number of stages. First, we shall introduce programming to obtain approximate tail areas for the F-statistic (and also to obtain F-values corresponding to specified tail areas). We shall then automate tests of the mean and standard deviation for a sample, against the mean and standard deviation of the population whence, it is surmised, the sample may have been drawn. Finally we shall extend the program to allow the comparison of two samples – comparison of their means, paired-difference tests, and comparison of their sample standard deviations. This process will complete our main package, which you will then be in a position to apply very readily to a great variety of analytical problems.

4.6.1.   Programming *F*

Program STAT3 already has a 'tables' menu:

> 1= prob −> z   2= z −> prob
>
> 3= prob −> t   4= t −> prob

We shall now add:

> 5= prob −> F   6= F −> prob

Load program STAT3 into your computer, then add the following new lines.

```
 420  PRINT"5= prob −> F      6= F −> prob   7=stop"
 670  IF I>5 THEN 730
 680  PRINT"Give nu1, nu2, one-tail prob ";
 690  INPUT K1,K2,P
 700  GOSUB 1500
 710  PRINT"Corresponding F-value = "; INT(F * 100+.5)/100
 720  GOTO 390
 730  IF I>6 THEN 9999
 740  PRINT"Give F-Value, nu1, nu2 ";
 750  INPUT F,K1,K2
 760  GOSUB 1400
 770  GOTO 540
1400  REM F      ......F      −> prob
1410  F3=F↑(1/3)
1420  K8=2/(9*K1)
1430  K9=2/(9*K2)
1440  Z=((1-K9)*F3-1+K8)/SQR(K9*F3↑2+K8)
```

```
1450 IF K2>3 THEN 1470
1460 Z=Z*(1+.08*Z*(Z/K2)↑3)
1470 GOSUB 1000
1500 REM Inverse F  .. prob −> F
1510 GOSUB 1100
1520 IF K2>3 THEN 1550
1530 K4=Z/K2↑.75
1540 Z=Z/(.64+.36*(K4+(1-K4+.25*K4↑4)/(1+K4↑8)))
1550 K8=2/(9*K1)
1560 K9=2/(9* K2)
1570 K5=(1-K9)↑2-K9*Z↑2
1580 K6=-(1-K8)*(1-K9)
1590 K7=(1-K8)↑2-K8*Z↑2
1600 F=((SQR(K6↑2-K5*K7)-K6)/K5)↑3
1610 RETURN
```

The two new sub-routines accomplishing 'F −> prob' and ' prob −> F' start on lines 1400 and 1500 respectively. As usual we shall decline to explain in detail how these work though we make two comments:

(*a*) As before the upper-tail area (probability) is represented by symbol P. The *F*-value is represented by F, and ν_1 and ν_2 by K1 and K2 respectively.

(*b*) As for the *t*-subroutines, the *F* calculation works through approximate formulae which relate the particular *F* under consideration (for the given number of degrees of freedom) with a corresponding *z*-value. The *z*-subroutines are therefore called to establish the link between *z*-value and tail area.

As with the *t*-subroutines, we have to make compromises to keep the programming reasonably short. As a result our *F* calculations will be approximate. Nevertheless, the results will be accurate enough for all normal purposes.

Line 420 adds the promised new choices to our menu, and lines 670 to 770 arrange for the new subroutines to be brought into action, and for the results obtained to be printed.

RUN the program. Select the 'tables' option, then within that take our new option 5. You will be asked:

'Give nu1, nu2, one-tail prob?'

It's a pity the computer cannot easily handle the Greek nu (ν)! For your reply we suggest you give:

5, 6, 0.025.

The computer should respond:

'Corresponding F-value = 6.10'.

If your program does not produce this answer check your newly-typed lines carefully, particularly lines 1530-1600. Check that you have put brackets in all the right places, that minus signs appear where they should do, and that the right things have been raised to the right powers!

Now let's check against the tables. We have just asked the computer for $F_{0.025(5,6)}$. Confirm, from Fig. 4.5c, that the proper value is 5.99. The value obtained by the program is slightly too large, it is in error by 1.8%. For our purposes such a discrepancy will be of no importance.

As further tests use your program to compute $F_{0.05(20,2)}$, $F_{0.05(2,20)}$ and $F_{0.01(10,10)}$. See if you can reproduce the values below.

	Value from program	Value from table
$F_{0.05(20,2)}$	20.91	19.45
$F_{0.05(2,20)}$	3.47	3.49
$F_{0.01(10,10)}$	4.91	4.85

For the first value our program makes a rather large error (+ 7.5%). This is a feature which occurs for small ν_2 values. Small ν_2 (the top two or three rows of the F-tables) are associated with large values for F. In these cases the F-distribution has a very long tail. It is difficult for relatively simple formulae (such as those used in the sub routines) to calculate the values. Nevertheless, we still insist that these errors are seldom of practical significance.

As ν_2 becomes larger, F is calculated with rather good accuracy as is demonstrated by the last two values above. If you need reminding of the importance of not getting the order of ν_1 and ν_2 mixed up then you need do no more than notice that $F_{0.05(20,2)}$ is six times larger than $F_{0.05(2,20)}$!

Now test option 6 of your program. This will instruct you:

'Give F-value, nu1, nu2 ?'

From the values you specify the program will compute the corresponding probability (or tail area). Three examples are given below: check that your program can reproduce the quoted probabilities.

F	ν_1	ν_2	Prob from program(%)
3.46	5	5	10.008
3.43	10000	20	0.100
3.43	20	10000	0.000

If your program does not produce the above, you should check your typing, particularly of lines 1410 to 1460.

You cannot check these results against the tables we gave for F. The first example gives a probability, P, of 10%, ie it corresponds to $\alpha = 0.10$, and the second gives $P = 0.1\%$, ie $\alpha = 0.001$. Now published F-tables are available for α-values of 0.10 and 0.001; we could have given you them. But we already gave three full pages of tabulation for F and we felt we had to stop somewhere. One of the advantages of using the computer is that it can replace a bookful of tables. You can obtain the F-value corresponding to any probability level whatsoever and, *vice versa*, you can obtain the probability level corresponding to any value of F which turns up. Granted, our program does not give completely accurate answers, but they are good enough for our purposes. We could in fact make the computer's results fully accurate if we simply used an alternative (but rather longer) program.

If we are using an F-test to compare a sample variance, s^2, with a population variance, σ^2 then we should assign an infinite number of degrees of freedom for the population. On the computer, setting $\nu = 10000$, or any similarly large value, will suffice to mimic $\nu \to \infty$. The third example given above is not very useful as a test: it is included to remind you yet again that interchanging ν_1 and ν_2 gives a quite different answer. Here the probability is 0.000% when truncated to three places of decimals (without truncation it would give $\alpha = 4.6 \times 10^{-7}$).

In practice we shall find option 6 very useful in significance testing. For the particular measurements being tested we shall compute F_{calc} (as you have already been used to doing). Option 6 then gives us directly the probability (ie the tail area) corresponding to F_{calc}. This value represents the (one-tail) significance level of the results. We shall say more about this later!

SAQ 4.6a

(*i*) Use your computer program to calculate the F-value, at a significance level $\alpha = 0.025$, when $\nu_1 = 8$ and $\nu_2 = 1$. Repeat this calculation for the same values of α and ν_1, but for successively larger ν_2 values, viz for $\nu_2 = 2, 3, 5, 10, 20, 30$ and 10000. Compare your values with those tabulated in Fig. 4.5c. Comment.

(*ii*) For the same series of values of ν_1 and ν_2 ie for $\nu_1 = 8$ and $\nu_2 = 1, 2, 3, 5, 10, 20, 30,$ and 10000, use the F-values tabulated in Fig. 4.5c (viz $F = 956.66, 39.37, ...$etc) to obtain corresponding one-tail probability values from the program. Comment on the accuracy of these values.

SAQ 4.6a

4.6.2. Comparison of a Sample with its Presumed Parent Distribution

At last we are ready to incorporate hypothesis testing within our program. We shall deal first with tests involving a single sample:

(*a*) is the sample mean, $\bar{x}$ consistent with the sample having been drawn from a population with 'true' mean, μ?

(*b*) is the sample standard-deviation, s, compatible with the measurements belonging to a population with a standard deviation, σ?

Consideration of tests in which two separate samples are compared will be postponed until the next section.

Testing a sample mean against a claimed population value was the very first topic which concerned us when we introduced hypothesis testing long ago in section 3.5. In fact our attention was focused exclusively on that test until the end of section 4.1, so we return to it now rather as to an old and trusted friend. Testing a sample standard deviation, on the other hand, was one of our more recent ventures, which we met in section 4.5.3.

The initial menu presented when we run our program is:

'1 = tables   2 = sample   3 = stop?'

If you select the 'sample' option you are next presented with:

'1 = output mean and st.dev

2 = bracket population mean'.

What we are about to do here is to add new options:

'3 = test sample mean vs pop mean

4 = test sample st dev vs pop st dev'.

We can do this by typing in the following new lines.

```
 820  PRINT"3=test sample mean vs pop mean"
 830  PRINT"4=test sample st dev vs pop st dev"
 850  IF I<1 OR I>4 THEN 9999
 860  ON I GOSUB 1640,1700,1900,2100
1900  PRINT"Which st dev?   1=sample   2=population";
1910  INPUT J
1920  PRINT"   Sample mean = ";INT(M*1000+.5)/1000
1930  PRINT"Population mean = ";INT(M9*1000+.5)/1000
1940  IF J=2 THEN 2000
1950  T=ABS(M-M9)/(S/SQR(N))
1960  K=N-1
1970  GOSUB 1200
1980  PRINT"   Sample st dev = ";INT(S*1000+.5)/1000
1990  GOTO 2030
2000  Z=ABS(M-M9)/(S9/SQR(N))
2010  GOSUB 1000
2020  PRINT"Population st dev = ";INT(S9*1000+.5)/1000
2030  PRINT"One-tail significance = ";INT(P*100000+.5)/1000;"%"
```

```
2040  RETURN
2100  PRINT"    Sample st dev = ";INT(S*1000+.5)/1000
2110  PRINT"Population st dev = ";INT(S9*1000+.5)/1000
2120  IF S<S9 THEN 2170
2130  K1=N-1
2140  K2=10000
2150  F=(S/S9)↑2
2160  GOTO 2200
2170  K1=10000
2180  K2=N-1
2190  F=(S9/S)↑2
2200  GOSUB 1400
2210  PRINT"Calculated F-value = ";INT(F*100+.5)/100
2220  PRINT"One-tail significance = ;INT(P*100000+.5)/1000;"%"
2230  RETURN
```

If you LIST lines 789 to 870 you will see how the new options have been fitted into the original menu. The INPUT I statement on line 840 allows you to select which procedure to carry out. If the test on the sample mean is selected (ie $I=3$) then line 860 directs the program to line 1900. If instead the sample standard deviation is to be tested (ie $I=4$) then the action is directed to line 2100. The bulk of the new programming above supplies the necessary new lines running from 1900 and from 2100.

Having added the new program options it remains to see how we can make use of them. Consider the following example.

∏   A new method for the determination of the aspirin content of analgesic tablets has been proposed. It is claimed that the method is free from systematic error and that measurements made by using it are subject to a standard deviation of 1.0%. The method was applied to tablets which were known to contain 300 mg of aspirin: the results obtained on ten tablets are given below.

302 mg	295 mg	301 mg	299 mg	305 mg
306 mg	309 mg	305 mg	311 mg	295 mg

Assess the claims made for the new method.

We have two judgements to make. We shall conclude that the method is free of systematic error if the mean of our set of measurements is sufficiently close to the true result of 300 mg, ie if there is no systematic error we should be able to assert that the above measurements are consistent with being a random sample from a population with a mean of 300 mg. Our second judgement must be whether the measurements are consistent with the stated precision for the new method. The asserted figure of 1.0% would lead to a population standard deviation of 3.0 mg for measurements on tablets containing 300 mg of aspirin. In short, therefore, we are to test whether the above measurements could constitute a random sample from a population of mean 300 mg and standard deviation 3.0 mg

We have to enter the data for this example in our program. LIST lines 3900 to 4010. You will find that these contain the values appropriate to the last occasion when you ran one of the 'sample' options. Now you should replace them as follows.

```
3900 DATA 300,3.0
4000 DATA 10
4010 DATA 302,295,301,299,305,306,309,305,311,295
```

Line 3900 contains the population mean and standard deviation. Line 4010 lists the sample measurements the total number of which is recorded on line 4000.

Now RUN the program and select the 'sample' option. Within that take option 3 to 'test sample mean vs pop mean'. You should then find that you are immediately confronted with a further question!

'Which st dev?   1=sample   2=population?'

At this point it is brought home to us that the conclusion to our first query (whether the method is free from systematic error) might depend on our answer to the second question (whether the precision of the method is as claimed). If we can trust the reported precision of the method we can use the population standard deviation ($\sigma = 3.0$ mg) and then a z-test is in order. If we cannot use σ, then we must use the sample standard deviation, s, and the test must be based on the *t*-statistic. We shall hedge our bets for a time and try out both possibilities. First let's take option 2, and accept $\sigma = 3.0$ mg. The computer's response should be that given below.*

Sample mean $= 302.8$
Population mean $= 300.0$
Population st dev $= 3.0$
One-tail significance $= 0.158\%$

The sample mean comes out almost one standard deviation above the true value of 300 mg. With a sample size of 10, the *z*-test leads to the conclusion that this difference is highly significant. If the population mean for the analytical measurements was indeed 300 mg (and assuming that the standard deviation is 3.0 mg) then a sample mean as high as 302.8 mg has only a probability of 0.158% of occurring by chance. The method appears to have a positive systematic error. Let us put things in terms of our normal language for hypothesis testing. Our null hypothesis was $H_0: \mu = 300$ mg. We shall take as the alternative hypothesis the assertion that there is a positive systematic error. i.e. $H_1: \mu > 300$ mg. Our conclusion is that if we adopt any significance level for our test greater than 0.16% then we shall reject the null hypothesis in favour of the alternative hypothesis.

Notice the way in which our answer is framed. In the past we have usually had a set significance level in our mind as we conducted a hypothesis test (usually 5%, 2.5% or 1%). We would come to the conclusion (either to accept or to reject the null hypothesis) which was valid for the chosen level of significance. But we were often left quite unsure whether we would have come to the same conclusion if we had chosen a somewhat different significance level. This state of affairs arose because we depended on printed tables; we had values of our statistical parameters tabulated for only a limited number of set probability levels. Using the computer frees us from that limitation. What we have produced in this instance is the '*p*-value' of section 4.1.1. This is *the largest value for the level of significance which would allow the null hypothesis to be accepted*.

---

* If your answers do not agree with ours check your DATA statements and program lines 1910–1940 and 2000–2040.

Our value tells us *how significant* our conclusion is. In this instance the evidence is very significant indeed in indicating the existence of a positive systematic error in the method. All of our computer-based hypothesis testing will be done on this basis, i.e. giving *p*-values.

But we were unsure whether we were justified in using σ, the claimed population standard deviation! Let us now see what conclusion results if instead we use the sample standard deviation, *s*. Select 'sample' option 3 to test the sample mean again, but this time when confronted with

'Which st dev? 1=sample 2=population?'

enter 1, so that *t*-test is done. The output you should get is below*.

$$
\begin{aligned}
\text{Sample mean} &= 302.8 \\
\text{Population mean} &= 300.0 \\
\text{Sample st dev} &= 5.432 \\
\text{One-tail significance} &= 6.877\%
\end{aligned}
$$

The sample standard deviation is rather larger than 3.0 mg. This, together with the fact that the program now has to use a *t*-test, results in less significant evidence for the existence of a systematic error. Our null hypothesis is $H_0: \mu = 300$ mg. If our alternative hypothesis is that there is a positive systematic error, ie $H_1: \mu > 300$ mg then the null hypothesis is rejected only if the significance level of the test is set higher than 6.9%. On the basis of the sample standard deviation, we conclude that if the method had zero systematic error then a random sample of ten measurements would give a mean aspirin content as large as 302.8 mg almost 7 times in every hundred trials. Indeed, testing the null hypothesis on a two-tail basis, almost 14 times in a hundred one would expect the sample mean to differ from the true value (300 mg) by as much as has occurred this time, ie for the sample mean to be either as low as 297.2 mg or as high as 302.8 mg. (297.2 = 300.0 - 2.8.)

* If your answers do not agree with those given the most likely source of error is in the typing of lines 1950–1990.

For the time being we are left in a state of uncertainty: if we accept the claimed population standard deviation, σ, there is strong evidence for the existence of a positive systematic error: if we do not accept the σ-value then the evidence is rather unconvincing. Clearly it is time we checked whether it is reasonable or not to accept the claim that $\sigma = 3.0$ mg.

RUN the program again, this time selecting 'sample' option 4, to test the sample standard deviation, s, against the population value, σ. The output should be as below.

$$
\begin{aligned}
\text{Sample st dev} &= 5.432 \\
\text{Population st dev} &= 3.0 \\
\text{Calculated } F\text{-value} &= 3.28 \\
\text{One-tail significance} &= 0.059\%
\end{aligned}
$$

There we have it, a very significant conclusion, if the true standard deviation of our measurement process was 3.0 mg then only 0.059% of the time (around 6 times in ten thousand) should a sample of ten measurements give a sample standard deviation as large as was obtained here (5.43 mg).

This clarifies our earlier uncertainty. Our results give very strong evidence that the method (at least as practised by ourselves!) is less precise than was claimed. In testing the mean we are driven to rely on the t-test, and the consequences of this are that we have only very weak evidence for the existence of any systematic error. If there is a systematic error it seems that it must be relatively small, of the same order as or smaller than random errors. If one wanted to pursue further the existence of systematic error it would be a matter of analysing a much larger sample.

By simply modifying the DATA lines in the program, it is now possible quite painlessly to tackle other problems concerning the consistency of sample measurements with a presumed population. As

---

* If your output disagrees with ours you probably have a typing error somewhere in lines 2100–2230.

always the program has been written so as to minimise the amount of typing required. You might, optionally, like to try your hand at improving the program. For instance, you might have noticed that whereas in the F-test the value of F was output (line 2210), there was no corresponding output of z or of t during the tests on the mean. You might like to add some PRINT lines of your own to rectify this (immediately after line 2000 for z and after line 1950 for t). You could also have numbers of degrees of freedom output. (In the t-test ν is represented by K in the program, and its value is set on line 1960. In the F-test ν_1 and ν_2 are represented by K1 and K2 respectively, and their values would be most easily printed out after line 2200). If you do not feel inclined to attempt these refinements they are by no means essential for our purposes.

SAQ 4.6b

A manufacturer produces dry cells which under a standardised test have an average life-time of 28.0 hours with a standard deviation of 1.50 hours. It is suggested that a slight change in the production process will lead both to an increase in mean life-time and to reduced variability in performance. The production process is duly changed and a random sample of 16 dry cells on testing gives the following life-times (in hours).

29.8 28.3 27.4 29.2 29.5 29.2 27.5 29.2

27.9 30.3 27.5 30.3 28.7 28.4 29.2 29.7

Use your computer program to assess the claims made for the modified process.

SAQ 4.6b

4.6.3. Comparison of Two Sample Means and Paired-differences Tests

In the previous two sections we have substantially increased the power of our program by expanding the options available under both the 'tables' and 'sample' headings. We added the facility of the F-table, and its inverse, and we have introduced significance testing of the compatibility of the mean or the standard deviation of sample measurements with a proposed parent distribution.

Now, before we returned to programming, we had just devoted a great deal of attention, from Section 4.2 till Section 4.5, to comparing measurements from *two separate samples*. The time has come to incorporate this facility in our program. We shall extend the initial choice presented in a run to the following.

"1-tables   2 = one sample   3 = two samples ?"

From the start our program has been designed to cater for only a single sample at a time, with a single corresponding 'shadow' population. The sample data are read-in at the beginning of each run and the sample mean and sample standard deviation are calculated. We shall now have to duplicate these features for a second sample and its potential parent population. We propose initially to introduce two options, each comparing the two samples; these will allow a significance test to be made on the difference between the sample means, or for a paired-differences test to be done.

We need to take a deep breath at this point, there are about 60 lines of new typing involved. That is the bad news! The good news is that this will all but complete our package "STAT". There will be a mere dozen lines to add in the next section to incorporate a test of the difference in sample standard deviations, and then that will be that. So, with light showing at the end of the tunnel, we ask you to plod through the following.

```
  60  READ M8,S8
  80  DIM X(30),Y(30),D(30)
 230  READ N1
 240  A=0
 250  FOR 1=1 TO N1
 260     READ Y(I)
 270     A=A+Y(I)
 280     NEXT I
 290  M1=A/N1
 300  A=O
 310  FOR I=1 TO N1
 320     A=A+(Y(I)-M1)↑2
 330     NEXT I
 340  S1=SQR(A/(N1-1))
 360  PRINT"1=tables    2=one sample    3=two samples ";
 780  IF I>2 THEN 890
 890  PRINT"- - - - - - - - - - - - - - - - - - - - - - - - - - - - - - - - - - - - - - -"
 900  PRINT"1=test difference of means"
 910  PRINT"2=paired difference test"
 940  INPUT I
 950  IF I<1 OR I>3 THEN 9999
 960  ON I GOSUB 2300,2500,2700
 970  GOTO 890
2300  PRINT"Which st devs?    1=pop    2=pooled";
2310  INPUT J
```

```
2320  PRINT"Sample sizes    ";N;"       ";N1
2330  PRINT"Sample means    ";
                      INT(M*1000+.5)/1000;"     ";INT(M1*1000+.5)/1000
2340  IF J=2 THEN 2400
2350  PRINT"Pop st devs    ";INT(S9*1000+.5)/1000;"    ";
                                        INT(S8*1000+.5)/1000
2360  Z=ABS(M-M1)/SQR(S9↑2/N+S8↑2/N1)
2370  PRINT"Corresponding z-value = ";INT(Z*100+.5)/100
2380  GOSUB 1000
2390  GOTO 2470
2400  PRINT"Sample st devs    ";INT(S*1000+.5)/1000;" ";
                                        INT(S1*1000+.5)/1000
2410  K=N+N1-2
2420  S5=((N-1)*S↑2+(N1-1)+S1↑2)/K
2430  PRINT"Pooled st dev ........... ";INT(SQR(S5)*1000+5)/1000
2440  T=ABS(M-M1)/SQR(S5/N+S5/N1)
2450  PRINT"t = ";INT(T*100+.5)/100;" with ";K;" deg of freedom"
2460  GOSUB 1200
2470  PRINT"One-tail significance = ";INT(P*100000+.5)/1000;"%"
2480  RETURN
2500  PRINT"Differences: ";
2510  A=O
2520  FOR I=1 TO N
2530     D(I)=X(I)-Y(I)
2540     A=A+D(I)
2550     PRINT INT(D(I)*1000+.5)/1000;"     ";
2560     NEXT I
2570  M5=A/N
2580  A=O
2590  FOR I=1 TO N
2600     A=A+(D(I)-M5)↑2
2610     NEXT I
2620  K=N-1
2630  S5=SQR(A/K)
2640  PRINT
2650  PRINT"Mean difference = ";INT(M5*1000+.5)/1000
2660  PRINT"St dev of diffs = ";INT(S5*1000+.5)/1000
2670  T=ABS(M5)*SQR(N)/S5
2680  GOTO 2450
2700  GOTO 9999
```

New lines 60–340 duplicate the work of the pre-existing lines 50–220, but for the new second sample. It is perhaps helpful to identify the symbols used, and the DATA statements which will be relevant to each sample.

	Previously programmed		Newly Added	
	Sample 1	Pop.1	Sample 2	Pop.2
Array of values	X		Y	
Sample size	N		N1	
Mean	M	M9	M1	M8
Standard deviation	S	S9	S1	S8
DATA line(s)	4000–4010	3900	4020–4030	3910

Note that DATA lines have yet to be entered.

Line 360 introduces the newly expanded initial menu, and line 780 ensures that if the new choice 'two samples' is selected (ie $I = 3$) the program is directed to line 890 where the second stage of the menu lays out the options:

'1 = test difference of means
2 = paired difference test'

The difference-of-means test is carried out through lines 2300–2480 and the paired-difference test through lines 2500–2680. Line 2700 is a temporary one; in the next section we shall insert our final option there, to compare the two sample standard deviations.

We cannot RUN the program yet, until we provide appropriate DATA. Let us therefore move on to consider an example. As in the last section we shall choose problems which should be familiar from earlier work.

∏   An investigation is undertaken to test whether an oil spillage could have come from a 'suspect' tanker. The sulphur content of oil samples from both the tanker and the spillage is measured by a procedure which is known, from experience, to have a standard deviation of 0.020% Six replicate determinations on oil from the tanker yield the results below.

%S: 0.217   0.202   0.188   0.231   0.232   0.207

Six replicate determinations from the spillage give the following.

%S: 0.202   0.154   0.204   0.181   0.192   0.151

Are these results compatible with the tanker being the source of the spillage?

Well, our null hypothesis must be that the two oil samples come from the same population. We are not told what the mean of this population is, but we shall have to test whether the two samples are compatible with the *same* population mean. We are told what the population standard deviation is for the measurements; it is 0.020% S. We can feed the information into the computer program through the following DATA statements.

```
3900 DATA 0,0.020
3910 DATA 0,0.020
4000 DATA 6
4010 DATA 0.217,0.202,0.188,0.231,0.232,0.207
4020 DATA 6
4030 DATA 0.202,0.154,0.204,0.181,0.192,0.151
```

The first two statements specify the mean and standard deviation for each parent population. The population means are unknown to us and have each been set at zero; these values will not in fact be used by the program. (Doing this sort of thing is not regarded as good programming practice, our excuse as usual is that it helps us keep a reasonably general-purpose program shorter than it would otherwise be.) The last four DATA statements give the size and the measurements for each of the two samples.

RUN the program, select 'two samples', and then take option 1 to 'test difference of means'. The program should come up with the further question:

"Which st devs?   1=pop   2=pooled?"

Well, in this example we are assured that we know the population standard deviation, so you should select 1. The output then rolls out.*

Sample sizes   6   6
Sample means   0.213   0.181
Pop st devs   0.020   0.020
Corresponding z-value = 2.79
One-tail significance = 0.267%

After all the effort in typing it is reassuring how effortlessly an actual example can be handled. Here the conclusion is very clear cut; the difference in the sample means is very significant. The oil from the tanker seems to have truly the higher sulphur content. If the null hypothesis held, ie if the two samples corresponded to the same population mean H_0: $\mu_1 = \mu_2$ then there would be a probability of only just over 0.5% (strictly 2 × 0.267%) that the sample means would differ by as much as has occurred. (Notice that we have doubled the significance level given by the program, so that it may apply to a two-tail test.) This would be regarded as very strong evidence, particularly by the counsel for the defence of the tanker owners, that the null hypothesis should be rejected. The test appears to exonerate the tanker from blame for the oil slick.

Given that the relevant data are in our program, let us for the fun of it see what happens if we suppose that we did not know the value for σ, the population standard deviation for the analysis. We should have to depend then on the sample standard-deviations, and in fact on the *pooled* standard deviation. Select the option 'test difference

---

* If you have trouble duplicating our results, first check that your DATA lines are correct, then check the typing especially of lines 240–290 (if the mean for sample 2 is wrong) and lines 2330–2390.

of means' again, but this time when asked 'Which standard deviations?' select 'pooled'. The output should be as follows.*

$$
\begin{array}{lcc}
\text{Sample sizes} & 6 & 6 \\
\text{Sample means} & 0.213 & 0.181 \\
\text{Sample st devs} & 0.017 & 0.023 \\
\end{array}
$$

Pooled st dev 0.020
$t = 2.72$ with 10 deg of freedom
One-tail significance $= 1.086\%$

In fact the pooled standard deviation here agrees exactly with the
claimed population value. The fact that the program now has to
resort to a t-test, however, leads to the difference in means being
judged somewhat less significant. Nevertheless the null hypothesis
is still rejected in a two-tail test down to a significance level of just
above 2% (strictly 2 × 1.086%).

Now you may remember that, back in Sections 4.2 and 4.3, we described three cases for tests of the difference between sample means.
Our program offers only two options, viz 'case 1' when the population standard deviations σ_1 and σ_2 are known, and 'case 3' when
the pooled sample standard deviation is used. To hold down the size
of the program we have omitted 'case 2' which was applied when
we had large samples but the population standard deviations were
unknown. Then we simply used the sample standard deviations s_1
and s_2 as approximations for σ_1 and σ_2. This case can be dealt with
by the program by the following device:

(*a*) enter the sample measurements (DATA lines 4000–4030),

(*b*) test the difference of means by using the pooled s option,

(*c*) read off the output sample standard deviations,

* If you have problems here check lines 300–340 (for the standard
deviation of sample 2), and 2400–2460.

(*d*) put these as the population σ-values in DATA lines 3900 and 3910,

(*e*) RUN the test selecting the 'population' option.

A little messy, but it works. The alternative is to compose a half dozen or so additional lines of programming to cope with this case.

Having operated the test options for the significance of the dif-ference between sample means it remains to try out the paired-difference test facility. You may recall that earlier we discussed a comparison of the Kjeldahl method with a proposed new 'NH_3 elec-trode' procedure for determining the percentage of nitrogen in agri-cultural products. Results (for % N) were quoted for six substances.

NH_3 -electrode: 1.24   1.23   4.09   13.67   4.22   3.28
Kjeldahl: 1.24   1.24   4.06   13.64   4.17   3.26

We described a paired-difference test on these measurements. We urged you: 'check for yourself!' We hope you did so then, but now use your program to tackle it. Enter the measurement DATA below.

```
4000 DATA 6
4010 DATA 1.24,1.23,4.09.13.67,4.22,3.28
4020 DATA 6
4030 DATA 1.24,1.24,4.06,13.64,4.17,3.26
```

RUN the program, selecting 'two samples' and then option 2 ('paired difference test'). The individual differences will be output, then the following*.

Mean difference   = 0.02
St dev of diffs   = 0.022
$t = 2.24$ with 5 deg of freedom
One-tail significance = 3.794%

* If your output disagrees with the above check the listed individual differ-ences (a cross check on the DATA) and also lines 2510–2680.

You should find that these results agree with those quoted previously in Section 4.4, with the trivial exception that t was there given as 2.23 rather than 2.24 (the result of rounding the value of s in the previous hand calculation).

Previously we made a two-tail test at a 5% level of significance. Our conclusion was that the null hypothesis, $H_0: \mu_d = 0$, should be accepted, ie at 5% significance there was no demonstrated systematic bias between the two methods.

Our computer has given a one-tail significance level of 3.8%. If the null hypothesis holds, ie if there is no systematic bias, the the evidence is that there was a 3.8% chance that the results from the NH_3-electrode would on average be found to exceed those from the Kjeldahl method by as much as has happened. If we judge the results on the basis of a two-tail test, then there would be a 7.6% chance ($2 \times 3.8\%$) of the results differing on average, one way round or the other, by as large a margin as has occurred. So if we take as our criterion for assessing the evidence any level of significance less than 7.6% (for instance the 5% level) we shall accept the null hypothesis that the two methods agree. As we commented in Section 4.4, the evidence is not all that reassuring, indicating that additional tests might be worthwhile.

SAQ 4.6c

The cadmium content of various tissue samples was investigated by two procedures using atomic absorption spectrometry. Comparative results (for Cd in ppm) are given below. $\longrightarrow$

SAQ 4.6c (cont.)	Procedure 1	Procedure 2
	1.6	1.3
	4.1	4.4
	2.8	3.3
	5.7	6.5
	1.9	1.4
	2.4	2.5
	4.2	3.5
	10.3	10.2
	11.3	12.8
	10.4	10.7
	3.5	4.5

Procedure 2 gives a larger result for 7 of the 11 samples reported. Is there evidence that this due to a systematic discrepancy between the two methods?

4.6.4. Comparison of Two Sample Standard Deviations

We promised you: just a dozen lines more! The final component for our program 'STAT' will allow an F-test to be run to assess the significance of the difference in the standard deviations of our two samples. We introduced this test in Section 4.5.3, and we practised applying it there to a number of examples. The necessary programming is given below.

```
 920  PRINT"3=test difference of st devs"
2700  PRINT"Sample A:   st dev = ";INT(S*1000+.5)/1000;" size = ";N
2710  PRINT"Sample B:   st dev = ";INT(S1*1000+.5)/1000;" size = ";N1
2720  IF S<S1 THEN 2770
2730  F=(S/S1)↑2
2740  K1=N-1
2750  K2=N1-1
2760  GOTO 2200
2770  F=(S1/S)↑2
2780  K1=N1-1
2790  K2=N-1
2800  GOTO 2200
```

Line 920 adds the new option to the 'two samples' menu. Lines 2700-2800 calculate the value of F and assign the numbers of degrees of freedom ν_1 and ν_2. Note that the program follows the practice we established earlier by ensuring that $F > 1$. If the sample standard deviation for sample A, s_A, is larger than that for sample B, s_B, then $F = s_A^2/s_B^2$, whereas if $s_B > s_A$ we set $F = s_B^2/s_A^2$. ν_1 and ν_2 (K1 and K2 in the program) have to be assigned so that ν_1 always equals the number of degrees of freedom for the sample allocated to the numerator in F. Once these values are assigned earlier parts of the program are used to evaluate and output the corresponding tail area. Let us then try out the new option on an example.

∏ A comparative test is run on the performance of an x-ray fluorescence technique (method A) and flame atomic absorption spectrometry (method B) to determine trace levels of Zn in an aqueous solution. Ten replicate determinations are performed by each method on a single solution. The following are the results (in μg l^{-1} Zn).

Method A: 138	117	108	127	116	138	120
117	112	110				
Method B: 130	140	119	125	135	132	138
131	127	122				

Is one method superior to the other in terms of precision?

The two sets of data need to be substituted into the relevant DATA lines:

    4000 DATA 10
    4010 DATA 138,117,108,127,116,138,120,117,112,110
    4020 DATA 10
    4030 DATA 130,140,119,125,135,132,138,131,127,122

RUN the program choosing 'two samples'. Then simply select the new option 3 ('test difference of st devs') and the following results should spill out.[*]

> Sample A: st dev = 10.740   size = 10
> Sample B: st dev = 6.773   size = 10
> Calculated F-value = 2.51
> One-tail significance = 9.292%

Method A comes up with a noticeably greater standard deviation, but this result is not very significant. In the absence of any difference in precision between the methods there would be a 9.3% chance that s_A would turn out to exceed s_B by this much. Put in two-tail terms, we accept the null hypothesis H_0: $\sigma_A = \sigma_B$ at any level of significance smaller than 18.6%. Such a discrepancy as this, one way round or the other, would be expected to occur by chance almost one time in five such tests.

So, there is no strong evidence that either method is to be preferred to the other in terms of precision, but is there a *systematic difference* between the results of the two methods? To test this is now as easy as pie! At the end of the test of the standard deviations our program leaves us sitting faced with the 'two-samples' menu. Select option 1 ('test difference of means') and, supported by the respectability of

---

[*] If your output disagrees with our standard deviations the most likely cause is mis-typed data. If you have difficulty with F or the significance level you should careful check lines 2720–2800.

the conclusion that $\sigma_1 = \sigma_2$, ask for the pooled standard deviation to be used. The resulting output is given below.

Sample sizes   10   10
Sample means   120.3   129.9
Sample st devs   10.74   6.773
Pooled st dev 8.978
$t = 2.39$ with 18 of freedom
One-tail significance = 1.398%

There is rather strong evidence that method A gives systematically a lower result than method B. If there was no bias as between the methods such a large discrepancy in the mean result would have only a 1.4% probability of occurrence.

We conclude that there is strong evidence for the existence of a systematic difference between the methods but no convincing evidence that there is anything to choose between them in terms of precision. There is one general comment which we should make on this outcome, that is to reassert our old point about the variability of the sample standard deviation. It is quite difficult to establish convincing evidence that two samples correspond to different population standard deviations, except where the difference is a particularly marked one. We must continually be aware that a small sample frequently gives an s-value noticeably removed from its true population σ-value. We can combat this problem only by taking large sample sizes.

SAQ 4.6d

> Two laboratories were asked to apply a standard method to determine the total phosphate concentration in a sample of river water. The following replicate results were reported, on a basis of $\mu g\ l^{-1}$ of phosphate.
>
> Lab A: 20.7   27.5   30.4   23.9   18.1   24.1
>   24.8   28.9
> Lab B: 20.9   21.4   24.9   20.5   19.7 $\longrightarrow$

SAQ 4.6d
(cont.)

(*i*) Are the results from the two labs consistent with one another?

(*ii*) How does it affect your conclusions if it is known from long experience that the analytical method, properly applied to such samples, has a standard deviation of 2.0 μg 1^{-1}?

4.6.5. Program STAT: Taking Stock

Our main program package is complete at last. We shall refer to it simply as 'STAT', and you will find a complete reference listing of it at the end of this section. This is an appropriate point at which to remind ourselves that we have gone to all the effort of building this package in pursuit of two main aims:

(*a*) to develop your practical experience of programming,

(*b*) to reinforce your 'feeling' for statistics.

As far as the first of these aims is concerned you have put together, in stages, a quite complicated program and you have built up quite a bit of experience in applying it in various ways. We have not asked you (except optionally) to devise your own programming steps, nor have we explained in detail how some of our program segments work. Nevertheless we shall be disappointed if you do not feel that you have some grasp, in broad terms, of the way in which the program functions. You will almost certainly by now have had considerable experience of debugging the program to get rid of typing errors. You may even find that you have become a somewhat better typist!

As far as programming is concerned we earnestly suggest that you set aside some time for trying to add some new refinements of your own. Start perhaps in a modest way, tidying up the presentation of output, and having more information printed. More ambitiously you might try adding further options. The possibilities are numerous. We suggest this not so much so that you can end up with a more useful package, though that is certainly a plus point! We are more concerned that you gain practice, and confidence, in successfully making changes to a program which is non-trivial in structure. We deliberately refrain from making specific suggestions as to what you should try; we think you will gain more if you follow your own initiative. We do, however, suggest it might be wise to store a reserve copy of the program in its present form. Any new options you write should be tested on an exercise for which you know in advance what the answers should be. If it happens you are already an experienced programmer, then it is less important that you do further development work on the package.

Turning to the second aim, that of reinforcing your 'feeling' for statistics, again we would like to encourage you to devote some time doing some further investigations of your own. You might use the package to analyse results you yourself obtain in the lab, or information from elsewhere in your studies, or data which you dream up out of thin air! Gaining a wide experience of significance testing is particularly valuable; the conclusions reached are often not what one might have guessed from a first glance at the data. Again we leave you to follow your own initiative; we think you are at a point where you will learn more if you do so. We would, however, suggest that you experiment with changing sample sizes, something which it is generally very easy to do with the program*.

We know that you are probably busy, but we do hope you take our advice. Experience of taking the driving seat, both in programming and in statistical studies, will help you make rapid progress, both in competence and in confidence. Below, as promised, we give a complete listing of program 'STAT', with just a few further REM statements included to help identify what its various segments do.

```
 9 REM Parameters used in z subroutines
10 READ D1,D2,D3,D4,D5,D6
20 DATA 4.98673470E-2,2.11410061E-2,3.2776263E-3,3.80036E-5,
                                  4.88906E-5,5.3830E-6
```

---

* There is one small programming point of which we should warn you. When we worked with only a single sample in the program we often changed sample size simply by changing DATA line 4000. This changing *4000 DATA 8* to *4000 DATA 4*, for instance, means that only the first four entries in DATA line 4010 are read into array X as sample A. Now that a second sample has been added to the program this change will mean that the next item read, which happens to be the size of sample B, will be set equal to the fifth item in DATA line 4010. The way round this is to add a line *223 RESTORE 4020* which ought to ensure that the size of sample B is read from the correct place (line 4020). If your computer will not accept that new statement you will have to remove the extra values from line 4010.

```
 30  READ G1,G2,G3,G4
 40  DATA 2.30753,0.27061,0.99229,0.04481
 49  REM M9 and S9 are pop mean and s.d. for sample A (as are M8
                                                    & S8 for sample B)
 50  READ M9,S9
 60  READ M8,S8
 80  DIM X(30),Y(30),D(30)
 99  REM Read N values into array X, calc sample mean and s.d.
100  READ N
110  A=0
120  FOR I=1 TO N
130      READ X(I)
140      A=A+X(I)
150      NEXT I
160  M=A/N
170  A=0
180  FOR I=1 TO N
190      A=A+(X(I)-M)↑2
200      NEXT I
210  S=SQR(A/(N-1))
220  R=100*S/M
229  REM Read N1 values into array Y, calc sample mean and s.d.
230  READ N1
240  A=0
250  FOR I=1 TO N1
260      READ Y(I)
270      A=A+Y(I)
280      NEXT I
290  M1=A/N1
300  A=0
310  FOR I=1 TO N1
320      A=A+(Y(I)-M1)↑2
330      NEXT I
340  S1=SQR(A/(N1-1))
350  R1=100*S1/M1
359  REM Main menu **************************
360  PRINT"1=tables   2=one sample   3=two samples ";
370  INPUT I
380  IF I>1 THEN 780
389  REM Menu for statistical tables **********
```

```
390  PRINT"- - - - - - - - - - - - - - - - - - - - - - - - - - - - - - - - - - - - - -"
400  PRINT"1= prob -> z      2= z -> prob"
410  PRINT"3= prob -> t      4= t -> prob"
420  PRINT"5= prob -> F      6= F -> prob       7=stop"
430  INPUT I
440  IF I>1 THEN 500
450  PRINT"Give one-tail prob ";
460  INPUT P
470  GOSUB 1100
480  PRINT"Corresponding z-value = ";INT(Z * 100+.5)/100
490  GOTO 390
500  IF I>2 THEN 560
510  PRINT"Give z ";
520  INPUT Z
530  GOSUB 1000
540  PRINT"Prob of EXCEEDING this value is ";
                                        INT(P*100000+.5)/1000;"%"
550  GOTO 390
560  IF I>3 THEN 620
570  PRINT"Give nu, one-tail prob ";
580  INPUT K,P
590  GOSUB 1300
600  PRINT"Corresponding t-value = ";INT(T*1000+.5)/1000
610  GOTO 390
620  IF I>4 THEN 670
630  PRINT"Give nu, t";
640  INPUT K,T
650  GOSUB 1200
660  GOTO 540
670  IF I>5 THEN 730
680  PRINT"Give nu1, nu2, one-tail prob ";
690  INPUT K1,K2,P
700  GOSUB 1500
710  PRINT"Corresponding F-value = ";INT(F*100+.5)/100
720  GOTO 390
730  IF I>6 THEN 9999
740  PRINT"Give F-value, nu1, nu2 ";
750  INPUT F,K1,K2
760  GOSUB 1400
770  GOTO 540
```

```
 780  IF I>2 THEN 890
 789  REM Menu for analysis of single sample **********
 790  PRINT"- - - - - - - - - - - - - - - - - - - - - - - - - - - - - - - - - - - - -"
 800  PRINT"1=output mean & st.dev"
 810  PRINT"2=bracket population mean"
 820  PRINT"3=test sample mean vs pop mean"
 830  PRINT"4=test sample st dev vs pop st dev"
 840  INPUT I
 850  IF I<1 OR I>4 THEN 9999
 860  ON I GOSUB 1640,1700,1900,2100
 870  GOTO 790
 889  REM Menu for comparison of two samples **********
 890  PRINT"- - - - - - - - - - - - - - - - - - - - - - - - - - - - - - - - - - -"
 900  PRINT"1=test difference of means"
 910  PRINT"2=paired difference test"
 920  PRINT"3=test difference of st devs"
 940  INPUT I
 950  IF I<1 OR I>3 THEN 9999
 960  ON I GOSUB 2300,2500,2700
 970  GOTO 890
1000  REM One-tail normal:   z --> prob
1010  Z9=ABS(Z)
1020  P=1+Z9*(D1+Z9*(D2+Z9*(D3+Z9*(D4+Z9*
                                     (D5+Z9*D6)))))
1030  P=.5/P↑16
1040  P=SGN(Z)*(P-.5)+.5
1050  RETURN
1100  REM Inverse normal   prob --> z
1110  IF (P<0 OR P>1) THEN 9600
1120  P0=.5-SGN(.5-P)*(.5-P)
1130  Z1=SQR(-2*LN(P0))
1140  Z=Z1-(G1+G2*Z1)/(1+Z1*(G3+G4*Z1))
1150  Z=Z*SGN(.5-P)
1160  RETURN
1200  REM t       t --> prob
1210  IF ABS(T)<3 THEN 1230
1220  T=T*(1+.037*LN(ABS(T))↑2.5/(2*K-1))
1230  Z=SGN(T)*(8*K+1)*SQR(K*LN(1+T*T/K))/(8*K+3)
1240  GOSUB 1000
1250  RETURN
```

```
1300  REM Inverse t       prob --> t
1310  GOSUB 1100
1320  T=SGN(Z)+SQR(K*(EXP((Z*(8*K+3)/(8*K+1))↑2/K)-1))
1330  IF ABS(T)<3 THEN 1350
1340  T=T/(1+.05*LN(ABS(T))↑2/(2*K-1))
1350  RETURN
1400  REM F    ......   F   -> prob
1410  F3=F↑(1/3)
1420  K8=2/(9*K1)
1430  K9=2/(9*K2)
1440  Z=((1-K9)*F3-1+K8)/SQR(K9*F3↑2+K8)
1450  IF K2>3 THEN 1470
1460  Z=Z*(1+.08*Z*(Z/K2)↑3)
1470  GOSUB 1000
1500  REM Inverse F    .. prob -> F
1510  GOSUB 1100
1520  IF K2>3 THEN 1550
1530  K4=Z/K2↑.75
1540  Z=Z/(.64+.36* (K4+(1-K4+.25*K4↑4)/(1+K4↑8)))
1550  K8=2/(9*K1)
1560  K9=2/(9*K2)
1570  K5=(1-K9)↑2-K9*Z↑2
1580  K6=-(1-K8)*(1-K9)
1590  K7=(1-K8)↑2-K8*Z↑2
1600  F=((SQR(K6↑2-K5*K7)-K6)/K5)↑3
1610  RETURN
1640  REM Show sample mean and st dev
1650  PRINT "No of values = ";N
1660  PRINT "Sample mean = ";INT(M*1000+.5)/1000
1670  PRINT "Est. st. dev.= ";INT(S*1000+.5)/1000
1680  PRINT "% Rel.st.dev.= ";INT(R*1000+.5)/1000
1690  RETURN
1700  PRINT"Bracketing the mean"
1710  PRINT"...give req'd confidence (prob<1.00)";
1720  INPUT P
1730  IF P>.99999 THEN 1710
1740  P=(1-P)/2
1750  PRINT"Which st dev   1=sample   2=population";
1760  INPUT J
1770  GOSUB 1640
```

```
1780  IF J=2 THEN 1830
1790  K=N-1
1800  GOSUB 1300
1810  W=T*S/SQR(N)
1820  GOTO 1860
1830  PRINT"Population st. dev. = ";INT(S9*1000+.5)/1000
1840  GOSUB 1100
1850  W=Z*S9/SQR(N)
1860  PRINT"It is predicted with confidence ";(1-2*P)
1870  PRINT"that the mean lies in the range:"
1880  PRINT"   ";INT((M-W)*100+.5)/100;" to ";
                              INT((M+W)*100+.5)/100
1890  RETURN
1899  REM Test difference of sample mean from supposed population value
1900  PRINT"Which st dev?   1=sample   2=population";
1910  INPUT J
1920  PRINT" Sample mean = ";INT(M*1000+.5)/1000
1930  PRINT"Population mean = ";INT(M9*1000+.5)/1000
1940  IF J=2 THEN 2000
1950  T=ABS(M-M9)/(S/SQR(N))
1960  K=N-1
1970  GOSUB 1200
1980  PRINT" Sample st dev = ";INT(S*1000+.5)/1000
1990  GOTO 2030
2000  Z=ABS(M-M9)/(S9/SQR(N))
2010  GOSUB 1000
2020  PRINT"Population st dev = ";INT(S9*1000+.5)/1000
2030  PRINT"One-tail significance = ";INT(P*100000+.5)/1000;"%"
2040  RETURN
2099  REM Test sample st dev vs supposed population value
2100  PRINT" Sample st dev = ";INT(S*1000+.5)/1000
2110  PRINT"Population st dev = ";INT(S9*1000+.5)/1000
2120  IF S<S9 THEN 2170
2130  K1=N-1
2140  K2=10000
2150  F=(S/S9)↑2
2160  GOTO 2200
2170  K1=10000
2180  K2=N-1
2190  F=(S9/S)↑2
```

```
2200  GOSUB 1400
2210  PRINT"Calculated F-value  = ";INT(F*100+.5)/100
2220  PRINT"One-tail significance = ";INT(P*100000+.5)/1000;"%"
2230  RETURN
2299  REM Test difference between sample means
2300  PRINT"Which st devs?   1=pop    2=pooled";
2310  INPUT J
2320  PRINT"Sample sizes    ";N;"    ";N1
2330  PRINT"Sample means    ";INT(M*1000+.5)/1000;"    ";
                                 INT(M1*1000+.5)/1000
2340  IF J=2 THEN 2400
2350  PRINT"Pop st devs    ";INT(S9*1000+.5)/1000;"    ";
                                 INT(S8*1000+.5)/1000
2360  Z=ABS(M-M1)/SQR(S9↑2/N+S8↑2/N1)
2370  PRINT"Corresponding z-value = ";INT(Z*100+.5)/100
2380  GOSUB 1000
2390  GOTO 2470
2400  PRINT"Sample st devs    ";INT(S*1000+.5)/1000;"    ";
                                 INT(S1*1000+.5)/1000
2410  K=N+N1-2
2420  S5=((N-1)*S↑2+(N1-1)*S1↑2)/K
2430  PRINT"Pooled st dev .......... ";INT(SQR(S5)*1000+.5)/1000
2440  T=ABS(M-M1)/SQR(S5/N+S5/N1)
2450  PRINT"t = ";INT(T*100+.5)/100;" with ";K;" deg of freedom"
2460  GOSUB 1200
2470  PRINT"One-tail significance = ";INT(P*100000+.5)/1000;"%"
2480  RETURN
2499  REM Paired difference test
2500  PRINT"Differences: ";
2510  A=0
2520  FOR I=1 TO N
2530      D(I)=X(I)-Y(I)
2540      A=A+D(I)
2550      PRINT INT(D(I)*1000+.5)/1000;"    ";
2560      NEXT I
2570  M5=A/N
2580  A=0
2590  FOR I=1 TO N
2600      A=A+(D(I)-M5)↑2
2610      NEXT I
```

```
2620  K=N-1
2630  S5=SQR(A/K)
2640  PRINT
2650  PRINT"Mean difference = ";INT(M5*1000+.5)/1000
2660  PRINT"St dev of diffs = ";INT(S5*1000+.5)/1000
2670  T=ABS(M5)*SQR(N)/S5
2680  GOTO 2450
2699  REM Test difference between sample standard deviations
2700  PRINT"Sample A: st dev = ";INT(S*1000+.5)/1000;" size = ";N
2710  PRINT"Sample B: st dev = ";INT(S1*1000+.5)/1000;" size = ";N1
2720  IF S<S1 THEN 2770
2730  F=(S/S1)↑2
2740  K1=N-1
2750  K2=N1-1
2760  GOTO 2200
2770  F=(S1/S)↑2
2780  K1=N1-1
2790  K2=N-1
2800  GOTO 2200
3899  REM Mean and st dev for pops A & B
3900  DATA 0,2.0
3910  DATA 0,2.0
3999  REM Size and measurements - samples A & B
4000  DATA 8
4010  DATA 20.7,27.5,30.4,23.9,18.1,24.1,24.8,28.9
4020  DATA 5
4030  DATA 20.9,21.4,24.9,20.5,19.7
9600  PRINT "STOP * prob not between 0.0 and 1.0!"
9999  END
```

4.7. THE ANALYSIS OF VARIANCE: AN INTRODUCTION

We shall now spend a short time introducing a topic of great importance to all users of statistics. It is known as the *analysis of variance*. We say 'introducing' because that is all that we can do here. If our introduction convinces you of the importance of this topic, then there are many textbooks which should enable you to take your studies of the analysis of variance much further.

In our work on hypothesis testing, we have so progressed that we can now test for the equality of two population means and for the equality of two population variances. What do we mean by this? Well, with regard to testing the equality of population means, we compare the sample means to ascertain whether or not they are compatible with having come from two populations with equal means. Our tests involve a null hypothesis, H_0: $\mu_1 = \mu_2$, and an appropriate form of an alternative hypothesis, eg H_1: $\mu_1 \neq \mu_2$. In testing for the equality of two population variances, we compare two sample variances to determine whether or not they are compatible with having come from two populations with equal variances. The tests involve a null hypothesis, H_0: $\sigma_1 = \sigma_2$ and an appropriate alternative hypothesis, eg H_1: $\sigma_1 \neq \sigma_2$.

It would be easy if all testing involved no more than two sample means or two sample variances. Unfortunately, the real world often throws up situations which are far more complicated. Some examples follow. We might wish to compare three analytical methods. Why? For the same reason that we compared two methods; to ascertain whether or not they give essentially the same result. Here, we would need to answer the question, 'Are the *three* sample means compatible with having come from *three* populations with equal means?' The null hypothesis for this test would be H_0: $\mu_1 = \mu_2 = \mu_3$; the alternative hypothesis H_1: at least two of the population means are not equal (think about this statement).

A second example arises in the need to compare the analytical results obtained by different laboratories. This is a most important type of comparison: it is known as an *interlaboratory comparison*. If the results from four laboratories are to be compared, then the question which would have to be answered is 'Are the *four* sample means compatible with having come from *four* populations with equal means?' The null hypothesis for a statistical test would be H_0: $\mu_1 = \mu_2 = \mu_3 = \mu_4$; the alternative hypothesis H_1: at least two of the population means are not equal. Yes, the issues being discussed are increasingly complex! We shall introduce you to the analysis of variance by considering a situation which involves *five* sample means. Read on!

A situation which analytical chemists often confront involves the sampling of bulk material. Such sampling is carried out to ascertain the composition of the material, or relatedly its homogeneity. We shall focus our attention on the problem of determining whether or not a material is homogeneous. Random samples are taken from the bulk material and replicate determinations are carried out on each sample. The replicate results for each sample are scattered about their mean. Why? Because of the random error of measurement. Random error causes variability in the results of replicate determinations carried out on a sample. It is also the sole source of variability in the mean values from samples taken from a homogeneous materials. (In the unlikely event that you have forgotten the Central-Limit Theorem, it is worthwhile repeating the fact that sample means are normally distributed!) If a material is heterogeneous, then there are two sources of the variability in the mean values of samples. The first is the ever-present random error of measurement. The second is, obviously, the heterogeneity itself: each sample in a heterogeneous material has a different composition! Now, we are interested in testing whether or not a material is homogeneous. *If a material is homogeneous, then the variability in the means of samples can be explained in terms of the random error of measurement. If it is not homogeneous, then the variability in sample means will have to be rationalised in terms of both the random error of measurement and the differing composition of samples.*

4.7.1. Within-sample and Between-sample Variances

Let us examine the consequences of carrying out replicate determinations on samples of a homogeneous material, ie consider a situation where the only source of variability in the mean values of samples is random error. If we were to make an infinite number of measurements on a homogeneous material, we should expect the measurements to be normally distributed with mean μ and standard deviation σ. Another way of saying this is that we would have a population of measurements with mean μ and standard deviation σ. *The population standard deviation, σ, is a measure of the random error present in the measurement system.* If samples of size n were to be taken from such a population, the sample means would be distributed with mean μ and standard deviation $\sigma_{\bar{x}} = \sigma/\sqrt{n}$.

Let us assume that we take k samples, each of size n, from a *ho-*

mogeneous material. We shall label an individual measurement on a given sample as x_{ij}; this means that we are dealing with the jth measurement on the ith sample. Thus, in order to obtain the mean for the first sample, we calculate

$$\bar{x}_1 = \frac{\sum\limits_{j=1}^{n} x_{1j}}{n}$$

The sample means are thus termed $\bar{x}_1, \bar{x}_2, \bar{x}_3, \ldots\ldots \bar{x}_k$.

The sample standard deviations are denoted $s_1, s_2, s_3, \ldots\ldots s_k$, and the sample variances $s_1^2, s_2^2, s_3^2, \ldots\ldots s_k^2$. To obtain the variance for say the first sample we calculate s_1^2 as below.

$$s_1^2 = \frac{\sum\limits_{j=1}^{n} (x_{1j} - \bar{x}_1)^2}{n - 1}$$

The standard deviation for a sample is obtained by taking the square root of the sample variance.

For a homogeneous material, each of the sample means is an estimate of the population mean. (Yes, for chemists this is a statement of the obvious!) Each of the sample standard deviations (the s_i's) is an estimate of the population standard deviation, σ. (This too, by now, is a statement of the obvious. However, we shall nevertheless, explain this statement!)

The origin of the population standard deviation (σ) is the random error of measurement; the origin of a sample standard deviation (s_i is, again, the random error of measurement.

When the sample size is very large, the value for a sample standard deviation is very close to that of the population standard deviation. For normal sample sizes, a sample standard deviation may be considerably different from the population standard-deviation; it is nevertheless an *estimate* of it. It follows that each of the sample variances are *estimates* of the population variance, σ^2. (We shall, from now on, talk about variances rather than standard deviations – this topic is, after all, called the analysis of variance!) We have learned that when s_1^2 and s_2^2 are estimates of σ_1^2 and σ_2^2 respectively and when $\sigma_1^2 = \sigma_2^2 = \sigma^2$, then the *pooled sample variance* is an even better estimate of σ^2. If we have k sample variances, then the pooled sample variance is given by Eq. (4.14).

$$s_p^2 = \frac{(n_1 - 1)s_1^2 + (n_2 - 1)s_2^2 + (n_3 - 1)s_3^2 + \ldots\ldots + (n_k - 1)s_k^2}{n_1 + n_2 + n_3 + \ldots\ldots + n_k - k}$$

(4.14)

(The value of the denominator in Eq. 4.14 defines the number of degrees of freedom in the estimate of σ^2 by s_p^2.)

The pooled variance s_p^2 is an estimate of σ^2. There is another way in which we could estimate σ or σ^2. We have k sample means. We could calculate either the standard deviation or the variance of these means. Why should we wish to do this? First recall that sample means are distributed with mean μ and standard deviation $\sigma_{\bar{x}} = \sigma/\sqrt{n}$. The standard deviation of the sample means, we shall call it $s_{\bar{x}}$, is an estimate of $\sigma_{\bar{x}}$. If we multiplied $s_{\bar{x}}$ by $\sqrt{n}$ then we would have an estimate of σ, ie $\sqrt{n}s_{\bar{x}}$ is an estimate of σ. Alternatively, *we say that $ns_{\bar{x}}^2$ is an estimate of σ^2*.

The calculation of $s_{\bar{x}}^2$ for k sample means is quite easy: see Eq. 4.15:

$$s_{\bar{x}}^2 = \frac{\sum\limits_{i=1}^{k}(\bar{x}_i - \bar{x})^2}{k - 1}$$

(4.15)

where $\bar{x}$ is the mean of the k means. (The value of the denominator in this expression defines the number of degrees of freedom in the estimate of σ^2 by $s_{\bar{x}}^2$.) Of course, if we wished to know $s_{\bar{x}}$, we would simply take the square root of the right-hand side of the equation. We now have two methods for estimating σ^2. The first method uses a pooled variance approach. *The variances within samples are used in this estimate:* the origin of each sample variance is the random error of measurement. The second method utilises the variance of the means of samples. *This is called the between-sample estimate of the population variance.* Since we are dealing with a homogeneous material the variability in the sample means is, like the variability of measurements within a sample, due solely to the random error of measurement. Let us summarise all of the above in the form of a diagram (Fig. 4.7a).

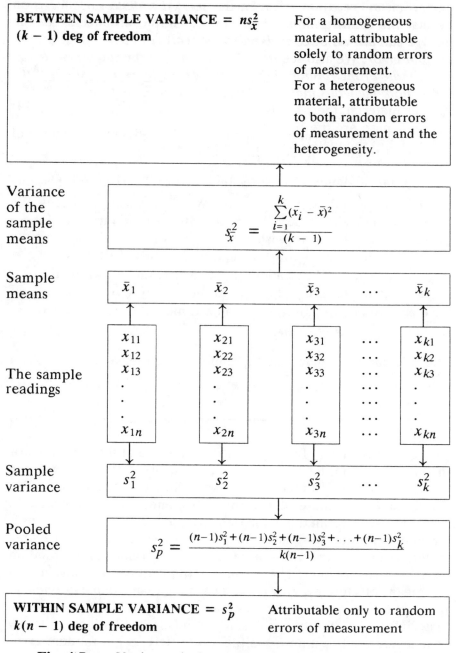

BETWEEN SAMPLE VARIANCE = $ns_{\bar{x}}^2$
$(k-1)$ deg of freedom

For a homogeneous material, attributable solely to random errors of measurement.
For a heterogeneous material, attributable to both random errors of measurement and the heterogeneity.

Variance of the sample means

$$s_{\bar{x}}^2 = \frac{\sum\limits_{i=1}^{k}(\bar{x}_i - \bar{x})^2}{(k-1)}$$

Sample means

| $\bar{x}_1$ | $\bar{x}_2$ | $\bar{x}_3$ | ... | $\bar{x}_k$ |

The sample readings

x_{11}	x_{21}	x_{31}	...	x_{k1}
x_{12}	x_{22}	x_{32}	...	x_{k2}
x_{13}	x_{23}	x_{33}	...	x_{k3}
.	.	.	...	.
x_{1n}	x_{2n}	x_{3n}	...	x_{kn}

Sample variance

| s_1^2 | s_2^2 | s_3^2 | ... | s_k^2 |

Pooled variance

$$s_p^2 = \frac{(n-1)s_1^2 + (n-1)s_2^2 + (n-1)s_3^2 + \ldots + (n-1)s_k^2}{k(n-1)}$$

WITHIN SAMPLE VARIANCE = s_p^2
$k(n-1)$ **deg of freedom**

Attributable only to random errors of measurement

Fig. 4.7a. *Variance in homogeneous and heterogeneous materials*

From a careful reading of the information presented in Fig. 4.7a, we see that *the within-sample variances and their weighted or pooled average can always be considered as estimates of the variance due to random error*. We also see that *the variance of the sample means is an estimate of the random-error variance only when we are dealing with a homogeneous material*. When we are dealing with a heterogeneous material, there are two causes of variability in the means of samples; these are the random error of measurement and the heterogeneity of the material. Let us modify our symbolism in order to discuss both sources of variability in the means of samples. We will term the variance due to random error σ_0^2. This can *always* be estimated from *within-sample* variances. For a homogeneous material it can also be estimated by the variance between sample means.

The variance due to the heterogeneity of a material will be termed σ_1^2. If we carry out a very large number of measurements on a heterogeneous material, the total variance will be $\sigma_0^2 + \sigma_1^2$. *Note that we have summed the separate variances. This additive property obtains when the sources of variation are independent.**

If a sample of size n is taken from a population of measurements when the two sources of variance are present, then the sample mean belongs to a population with variance $\sigma_0^2/n + \sigma_1^2$. When we calculate the variance between sample means (ie when we calculate $s_{\bar{x}}^2$) for a heterogeneous material, we have an estimate of the combined effect of both sources of variability, ie $s_{\bar{x}}^2$ is an estimate of $(\sigma_0^2/n + \sigma_1^2)$. It follows that $ns_{\bar{x}}^2$ is an estimate of $(\sigma_0^2 + n\sigma_1^2)$. Believe it or not, we are almost there!

Let us summarise, making use of our new nomenclature, all that we have learned so far: this is done below.

---

* We have met this point before. It was the basis of the theorem we set down in Section 4.2 where we were discussing the variance of the difference between two sample means. That was given, in equation (4.1), as the sum of the separate variances for each individual sample mean. (Each sample mean gave an independent contribution to the variance of the difference).

Sample Variance	Population – Variance Estimated	
	Homogeneous material	Heterogeneous material
Within Sample: s_p^2	σ_0^2	σ_0^2
Between Sample: $ns_{\bar{x}}^2$	σ_0^2	$\sigma_0^2 + n\sigma_1^2$

It is easily inferred from the estimates which obtain when we deal with a heterogeneous material, that an estimate for the variance due to the heterogeneity of the material can be obtained from $\sigma_1^2 = (ns_{\bar{x}}^2 - s_p^2)/n$.

Two independent sources of variability exist, and we can estimate the contribution of each to the total variance:

(*a*) s_p^2 is an estimate of σ_0^2 – the variance which has its origin in random error;

(*b*) $(ns_{\bar{x}}^2 - s_p^2)/n$ is an estimate of σ_1^2 – the variance which has its origin in the heterogeneity of the material.

Incidentally, chemists often call σ_0^2 the variance due to analysis and σ_1^2 the variance due to sampling.

4.7.2. Applying an F-Test

Now for a very important question. How can one distinguish between a homogeneous and a heterogeneous material? Well, in terms of variances, it is quite easy; it is when σ_1^2 is zero! This might seem to be an unhelpful comment but have patience! If $\sigma_1^2 = 0$, then we are dealing with a homogeneous material. This is when only the random error of measurement contributes to the variability in measurements. Here s_p^2 and $ns_{\bar{x}}^2$ are both estimates of the same thing, ie σ_0^2. Now, we know enough about statistics to know that we can test whether s_p^2 and $ns_{\bar{x}}^2$ are compatible with having their origins in a population with a common variance. We utilise an *F*-test.

The calculated F-value is obtained from

$$F_{calc} = \frac{ns_{\bar{x}}^2}{s_p^2}$$

The numerator has $(k-1)$ degrees of freedom, where k is the number of samples. The denominator has $k(n-1)$ degrees of freedom, since each sample is of size n.

The critical value for F depends upon the level of significance and, of course, the respective degrees of freedom. For a level of significance α and degrees of freedom $(k-1)$ and $k(n-1)$ we have

$$F_{crit} = F_{\alpha[k-1,k(n-1)]}$$

This, as you will recall, implies a one-tail test. You might ask; 'Why a one-tail test and why, in the calculated value of F, is $ns_{\bar{x}}^2$ in the numerator?' The questions are related.

The null hypothesis (in words) is that $ns_{\bar{x}}^2$ and s_p^2 are both estimates of σ_0^2. The conclusion which is drawn if this hypothesis is accepted is that the material is homogeneous. If the null hypothesis is rejected, ie if $F_{calc} > F_{crit}$ the material is assumed to be heterogeneous. This is synonomous with saying that $ns_{\bar{x}}^2$ is an estimate not of σ_0^2 but of $\sigma_0^2 + n\sigma_1^2$. In this situation $ns_{\bar{x}}^2$ would overestimate the magnitude of σ_0^2. In other words $ns_{\bar{x}}^2$ would thus be larger than s_p^2 and our interest consequently lies in the critical region defined by the right-hand tail of the F-distribution. A quick way of remembering which estimate of σ_0^2 to place in the numerator is to keep in mind that between-sample estimates of variance will, for rejection of the null hypothesis, always be larger than within-sample estimates.

There is one further point to be mentioned before we start to apply our knowledge to specific examples. It is that we can, through our latest analysis-of-variance test, assess the equality of any number of sample means. The null hypothesis, that the between-sample and the within-sample variances are equal, is tantamount to the hypothesis

$$H_0 : \mu_1 = \mu_2 = \mu_3 = \ldots = \mu_k = \mu,$$

ie that all the sample means are compatible with having come from populations with equal means. This is the conclusion to be drawn, for instance, when all the samples are withdrawn from a homogeneous material. The rejection of the null hypothesis leads to acceptance of the alternative hypothesis H_1: at least two of the sample means are not compatible with having come from populations with equal means.

If you are still with us, you are undoubtedly ready for an example!

∏ A manufacturer prepares ready-to-use fertilisers by mixing various salts in predetermined ratios. He is concerned that the mixing and the subsequent storage might result in a product which is not homogeneous with respect to the various salts. An investigation on a large batch of a given fertiliser was initiated. It was decided to base the investigation on the water-insoluble 'phosporus' in the fertilisers : such results are usually reported as % water-insoluble P_2O_5. Five random samples of the fertiliser were taken and 4 replicate determinations made on each sample. The results of the investigation are listed below, along with the calculated sample means, standard deviations, and variances.

% insoluble P_2O_5				
Sample 1	Sample 2	Sample 3	Sample 4	Sample 5
2.05	2.00	1.99	2.03	1.96
2.03	2.00	1.96	2.05	1.93
2.06	1.98	1.96	2.06	1.98
2.04	1.98	2.01	2.04	1.93
$\bar{x}_1 = 2.045$	$\bar{x}_2 = 1.990$	$\bar{x}_3 = 1.980$	$\bar{x}_4 = 2.045$	$\bar{x}_5 = 1.950$
$s_1 = 1.29 \times 10^{-2}$	$s_2 = 1.15 \times 10^{-2}$	$s_3 = 2.45 \times 10^{-2}$	$s_4 = 1.29 \times 10^{-2}$	$s_5 = 2.45 \times 10^{-2}$
$s_1^2 = 1.67 \times 10^{-4}$	$s_2^2 = 1.33 \times 10^{-4}$	$s_3^2 = 6.00 \times 10^{-4}$	$s_4^2 = 1.67 \times 10^{-4}$	$s_5^2 = 6.00 \times 10^{-4}$

The mean of the means, $\bar{x} = 2.002$

Armed with these results, the manufacturer asked two questions. Is the material homogeneous? If it is not, what are the relative contributions of random error and heterogeneity to the variability of results? Let us answer these questions.

Assuming that the material is homogeneous, the *between-sample estimate of* σ_0^2 is obtained from $ns_{\bar{x}}^2$; where

$$s_{\bar{x}}^2 = \frac{\sum\limits_{i=1}^{k} (\bar{x}_1 - \bar{x})^2}{k - 1}$$

This is our Eq. 4.15. It requires taking the sum of the squares of the differences between the sample means and the mean of the means, $\bar{x}$.

The calculation is shown below.

Sample	$(\bar{x}_i - \bar{x})$	$(\bar{x}_i - \bar{x})^2$
1	$2.045 - 2.002 = 4.3 \times 10^{-2}$	18.49×10^{-4}
2	$1.990 - 2.002 = -1.2 \times 10^{-2}$	1.44×10^{-4}
3	$1.980 - 2.002 = -2.2 \times 10^{-2}$	4.84×10^{-4}
4	$2.045 - 2.002 = 4.3 \times 10^{-2}$	18.49×10^{-4}
5	$1.950 - 2.002 = -5.2 \times 10^{-2}$	27.04×10^{-4}

$$\sum_{i=1}^{k} (\bar{x}_1 - \bar{x})^2 = 70.3 \times 10^{-4}$$

The number of samples, k, is 5, hence the number of degrees of freedom, $k - 1$, is 4. Now we need $ns_{\bar{x}}^2$, and n, the sample size, is also 4.

$$ns_{\bar{x}}^2 = \frac{4(70.3 \times 10^{-4})}{4} = 70.3 \times 10^{-4}$$

Keep in mind that this estimate has 4 degrees of freedom.

The within-sample estimate of σ_0^2 is obtained from the pooled variance s_p^2, which is calculated according to Eq. 4.14. This, in our example, gives

$$\frac{3(1.67 \times 10^{-4} + 1.33 \times 10^{-4} + 6.00 \times 10^{-4} + 1.67 \times 10^{-4} + 6.00 \times 10^{-4})}{5(4-1)}$$

viz $s_p^2 = 3.33 \times 10^{-4}$

There are 15 degrees of freedom in this estimate. If we assume that the material is homogeneous, then both $ns_{\bar{x}}^2$ and s_p^2 are estimates of σ_0^2. The calculated F-value is obtained from

$$F_{calc} = \frac{ns_{\bar{x}}^2}{s_p} = \frac{7.03 \times 10^{-3}}{3.33 \times 10^{-4}} = 21.1$$

Now, we need to compare this calculated F-value with a critical F-value. Let us work at the 5% level of significance and hence look up the value for $F_{0.05(4,15)}$, ie the F-value associated with 4 and 15 degrees of freedom respectively. We see from the table that F_{crit} is 3.06. Since $F_{calc} > F_{crit}$, we conclude that $ns_{\bar{x}}^2$ and s_p^2 are not compatible with having come from populations whose variances are equal. Since we know that s_p^2 is always an estimate of σ_0^2, we must conclude that $ns_{\bar{x}}^2$ is not an estimate of σ_0^2, that there are two sources of variability in the between-sample variance, that consequently the material is heterogeneous in nature and the $ns_{\bar{x}}^2$ is an estimate of the sum of the variance due to random error and the variance due to heterogeneity, ie that $ns_{\bar{x}}^2$ is an estimate of $\sigma_0^2 + n\sigma_1^2$.

We can estimate the variance due to the heterogeneity σ_1^2 from

$$\sigma_1^2 \approx \frac{ns_{\bar{x}}^2 - s_p^2}{n} = \frac{70.3 \times 10^{-4} - 3.33 \times 10^{-4}}{4}$$

$$\approx 16.7 \times 10^{-4}$$

Now, the estimate of the variance due to random error is obtained from the within-sample variance, s_p^2.

$$\sigma_0^2 \approx s_p^2 = 3.33 \times 10^{-4}$$

Comparing σ_0^2 with σ_1^2 demonstrates that the major source of variability in the sample means is the heterogeneity of the material. The manufacturer's suspicions are well founded, either the mixing or storage is resulting in a product which is *not* homogeneous with respect to the various salts. The method of analysis of variance has permitted us to separate and to estimate the different contributions to the variability in measurements on samples.

There is quite a bit of labour in doing analysis-of-variance calculations, and there are plenty of opportunities to make numerical errors. If you were doing these calculations on a regular basis it would be very worthwhile to tackle them by computer. We do not, however, propose to describe a program to do this. It would involve a fair amount of work, and we are in any case presenting you only with an introduction to this whole topic. As we said at the outset, our main purpose is simply to convince you that this topic is important. We are discussing some specific examples to demonstrate the issues which can be addressed. One thing, though, which we can easily do with our program, is to find out *how significant* our F-value is. If you run 'tables' option 6, you will find that the calculated F-value above has a one-tail significance-level of 0.001%. The evidence of the test is very significant indeed. If the fertiliser was in fact homogeneous, results such as these would have only one chance in 100,000 of occurring.

As far as the manufacturer is concerned, the investigation has yielded answers to the questions asked. Obviously other work will be needed before the problem of lack of uniformity in his fertiliser mixture is solved. We have done our bit for the time being, though.

For our own purposes, there is one other aspect of the example of the fertiliser mixture which is worthy of comment. We can use the data obtained during the investigation to estimate the insoluble P_2O_5 content of the fertiliser. (The manufacturer will not be interested in this estimate; he already knows the composition of his fertiliser!) We are quite familiar with estimating the population mean, μ, from a knowledge of a sample mean, $\bar{x}$, and its standard deviation, either the population value, $\sigma_{\bar{x}}$, or an estimate, $s_{\bar{x}}$. We dealt with this in detail in Sections 3.2 and 3.3, and we built the facility into our program in Section 3.4.5. Here, although we do not know $\sigma_{\bar{x}}$ exactly, we do have the estimate $s_{\bar{x}}$, obtained from our k sample means.

Above we quoted $ns_{\bar{x}}^2 = 70.3 \times 10^{-4}$. Recalling that $n = 4$, this gives

$$s_{\bar{x}} = \sqrt{(70.3 \times 10^{-4})/4} = 0.0419$$

The number of degrees of freedom corresponding to this estimate is k-1, ie $\nu = 4$. The 95% confidence interval for the population mean can be obtained from $\mu = \bar{x} \pm t_{0.025,4}s_{\bar{x}}$, where $\bar{x}_i$ is any one of our sample means. From our t-table, Fig. 3.4b, $t_{0.025,4} = 2.78$ (alternatively, our program gives $t = 2.79$). By using, for example, the first sample mean $\bar{x}_1$, we get

$$\mu = 2.045 \pm 2.78 \times 0.0419$$

$$= 2.045 \pm 0.116\%$$

You might at this juncture think it would be better to estimate the population mean from the mean of the five samples. You are right! The mean, of the k sample means or, as some call it, the grand mean, is $\bar{x}$. It belongs to a population with mean μ and population standard deviation $\sigma_{\bar{x}}/\sqrt{k}$. Our estimate of this standard deviation is $s_{\bar{x}}/\sqrt{k}$, based on k-1 degrees of freedom. Hence a better estimate of the 95% confidence interval for the population mean is:

$$\mu = \bar{x} \pm t_{0.025,4}s_{\bar{x}}/\sqrt{k}$$

$$= 2.002 \pm 2.78 \times 0.0419/\sqrt{5}$$

$$= 2.002 \pm 0.052$$

We now consider another example, which introduces another very important application of analysis of variance. It involves comparing results obtained by different laboratories.

Π   A company is concerned with the determination of copper at concentration levels ranging from 5 to 100 ppm. The company has four analytical laboratories and compares their work on a regular basis. A given copper-containing material is supplied to each of the laboratories at the start of each month and the laboratories are requested to carry out, by the company's standard method, 5 replicate determinations of copper and to report the sample mean and sample variance. The results of one months' joint test are reported below (in terms of ppm Cu).

Lab 1	Lab 2	Lab 3	Lab 4
$\bar{x}_1 = 10.1$	$\bar{x}_2 = 10.6$	$\bar{x}_3 = 11.0$	$\bar{x}_4 = 9.5$
$s_1^2 = 9.19 \times 10^{-2}$	$s_2^2 = 28.09 \times 10^{-2}$	$s_3^2 = 1.44 \times 10^{-2}$	$s_4^2 = 36.0 \times 10^{-2}$

Carry out an analysis of variance on these results. *Hint.* Use the within-sample variances, which we can call here the *within-laboratory* variances, to obtain a pooled-variance estimate of σ_0^2. Then use the sample means to obtain a *between-laboratory* estimate of σ_0^2. Compare the two estimates by use of an *F*-test. If the *F*-test suggests more than one source of variability is present, then you assume that $ns_{\bar{x}}^2$ is an estimate of $\sigma_0^2 + n\sigma_1^2$; where σ_0^2 is the variance due to random error and σ_1^2 is the variance due *to systematic differences between laboratories* (more will be said of this later). Give estimates of both σ_0^2 and σ_1^2, and be prepared for a discussion on repeatability and reproducibility!

The *within-laboratories* estimate of σ_0^2 is obtained from the pooled variance, when the sample sizes are the same and equal to n, Eq. 4.14 gives:

$$s_p^2 = \frac{(n-1)(s_1^2 + s_2^2 + s_3^2 + \ldots + s_k^2)}{k(n-1)}$$

where k is the number of laboratories; $(n-1)$ cancels.

This, in our example becomes:

$$s_p^2 = \frac{(0.0919 + 0.2809 + 0.0144 + 0.3600)}{4} = 0.1868.$$

(Note that there are 16 degrees of freedom in this estimate.)

The between-laboratories estimate of σ_0^2 is obtained from $ns_{\bar{x}}^2$.

Eq. 4.15 defines $s_{\bar{x}}^2$ as below,

$$s_{\bar{x}}^2 = \sum_{i=1}^{k}(\bar{x}_i - \bar{x})^2/(k-1)$$

We now need an ordinary standard deviation calculation with the $\bar{x}_i$'s. This gives the result below.

$$\bar{x} = 10.3 \text{ ppm}$$

$$s_{\bar{x}} = 0.648 \text{ ppm}$$

$$s_{\bar{x}}^2 = 0.420$$

and hence $\qquad ns_{\bar{x}}^2 = 5 \times 0.420 = 2.100$

There are four laboratories, hence $\nu = 3$ for this estimate.

Our test now is: '*Are* s_p^2 and $ns_{\bar{x}}^2$ estimates of the same thing, ie are they both estimates of σ_0^2?' We test this by calculating F.

$$F_{calc} = \frac{ns_{\bar{x}}^2}{s_p^2} = \frac{2.100}{0.1868} = 11.24$$

Suppose we wish to test at the 5% level of significance. Our test is to be a one-tail one, so the critical value is $F_{0.05(3,16)}$. From our tables, $F_{crit} = 3.24$ (alternatively, our program estimates F_{crit} as 3.23). Since the calculated F-value exceeds the critical F-value we reject the null hypothesis. In fact our program will tell us that F_{calc} is significant right down to a level of 0.036% !

We conclude that $ns_{\bar{x}}^2$ and s_p^2 are not estimating the same thing. Whilst s_p^2 estimates σ_0^2, $ns_{\bar{x}}^2$ estimates $\sigma_0^2 + n\sigma_1^2$. We obtain the separate estimates below:

$$\sigma_0^2 \approx s_p^2 = 0.187.$$

$$\sigma_1^2 \approx (ns_{\bar{x}}^2 - s_p^2)/n = (2.100 - 0.187)/5 = 0.383.$$

The variance due to systematic differences between laboratories is larger than the within-laboratory variance, roughly by a factor of two. This is not an uncommon occurrence. We are ready for a new look at the concepts of repeatability and reproducibility! We shall refer back to these results in the course of doing so.

4.7.3. Repeatability and Reproducibility Revisited

We know that σ_0^2 has its origin in the random error of measurement. What is the origin of σ_1^2 ? We have advised you to call it the variance due to systematic differences between laboratories. What does this mean? Let us reflect upon what went on during the inter-laboratory comparison. The same material was analysed in each of the laboratories and the same method was used. However, there were differences. Each of the laboratories used different instruments. They have been made by the same manufacturer but they were undoubtedly different instruments! The ambient temperatures in each of the laboratories were unlikely to have been exactly the same. The reagents used were unlikely to have had exactly the same purity. The procedures used, even though they were based on the same standard procedure, probably varied slightly. Chemists rarely have exactly the same work habits or, for that matter, the same skills. In the light of these and other factors, we have present in our inter-laboratory comparison a source of variability other than that of the random error of measurement. Let us discuss the matter by using some familiar terms.

Within each of the laboratories, random error will cause the results of replicate measurements to be scattered about the mean. We can assume that all the chemists work carefully and therefore are likely to obtain similar sample variances. These sample variances are all estimates of the random error of measurement σ_0^2 (by pooling these variances we get a better estimate of σ_0^2). The individual sample variances or their pooled variance are measures of what is known as the *repeatability* of the results: s_P^2, which provides an estimate of σ_0^2, is a measure of *repeatability*. Repeatability refers to *within-laboratory* or *within-run precision*.

When the same procedure is carried out in different laboratories, there are systematic differences between laboratories in many of the factors which govern results. This means that when comparing results obtained in different laboratories, two sources of variability are operative, ie random error and systematic differences. The variance due to systematic differences between laboratories has been denoted by σ_1^2. Since the two sources of error are independent their variances may be added. The total variance affecting single measurements in different laboratories is thus $\sigma_0^2 + \sigma_1^2$. This is a measure of the *between-laboratory* or *between-run precision*, ie a measure of what is known as the *reproducibility of results*.

In our example where four laboratories analysed a given material for copper, we obtained estimates $\sigma_0^2 \approx 0.187$ and $\sigma_1^2 \approx 0.383$. Thus the estimate of repeatability, as measured by σ_0^2, is 0.187 and the estimate of reproducibility, as measured by $(\sigma_0^2 + \sigma_1^2)$ is 0.570. If one prefers to discuss precision in terms of standard deviations we must take square roots.

Repeatability: $\sigma_0 \approx 0.432$ ppm Cu

Reproducibility: $\sqrt{(\sigma_0^2 + \sigma_1^2)} \approx 0.755$ ppm Cu

You might recall that it is customary to express precisions in terms of percentage relative standard deviations. To obtain % RSD we divide the appropriate standard deviation by the mean and multiply by 100. In our case the best estimate of the mean is the grand mean, $\bar{x}$, which has a value of 10.3 ppm Cu. We thus obtain:

$$\% \text{ RSD for repeatability} \quad = \frac{0.432}{10.3} \times 100\% = 4.19\%$$

$$\% \text{ RSD for reproducibility} \quad = \frac{0.755}{10.3} \times 100\% = 7.33\%$$

As one might expect, the value for the measure of reproducibility is greater than the value for the measure of repeatability. This result can be taken as a particular example of a generalisation: between-laboratory precision is always worse than within-laboratory precision. Actually, in the example given, the between-laboratory precision is not all that much larger than the within-laboratory precision. Someone who is not aware of this, might be shocked at how large the difference between repeatability and reproducibility can be. The question arises: 'Can we do anything to minimise the systematic differences between laboratories – ie to make the between-laboratories precision similar in magnitude to the within-laboratory precision?' The answer is yes, there are two possible approaches. In the first of these, the protocol for the method would be tightened up, ie the tolerances usually permitted with the method would be made even more stringent. As a result of this, temperature variations between laboratories would be reduced, the reagents used in each laboratory would have little difference and the calibrations used in each laboratory would be almost identical. This is easier said than done and it still leaves us with the differences in work habits and skills of the chemists.

The second approach to minimising between-laboratory variability is by far the more practicable. This involves the use of standard reference materials. A standard reference material (SRM) has a certified value for its composition or for one or more of its components. Each laboratory involved in determinations when between-laboratory precision is considered inadequate would calibrate its method with the SRM. This offers the opportunity to reduce systematic error within a laboratory and consequently to reduce between-laboratory variability.

4.7.4. The Youden Plot

Before leaving the topic of repeatability and reproducibility, it is illuminating to consider a graphical procedure, introduced by W. J. Youden, for the analysis of sources of variability in inter-laboratory studies. The procedure starts with each of several laboratories analysing two similar yet different materials, we shall call them A and B. The same method is used in all the laboratories and a single determination on each material is carried out. The pair of results reported by each laboratory constitutes a data point in a correlation diagram which is plotted so that a result for material A is represented on the x-axis and the corresponding result for material B is represented on the y-axis.

Consider an example. The results of determinations, made in ten laboratories, on two similar aspirin preparations are given below.

| | Wt % Aspirin | | |
Laboratory	Material A	Material B	Difference
1	50.45	52.55	2.10
2	49.89	52.00	2.11
3	49.60	51.70	2.10
4	50.26	52.11	1.85
5	49.78	51.79	2.01
6	49.92	51.81	1.89
7	50.22	52.35	2.13
8	50.40	52.26	1.86
9	50.17	52.24	2.07
10	49.85	51.87	2.02
Means	50.054	52.068	2.014

From the data we see, both for material A and for material B, that the ten laboratories report results having considerable variation: there is obviously between-laboratory variability. It is also seen that for each of the laboratories, the observed difference between the aspirin content of the two materials is approximately the same. Keep these comments in mind as we proceed with our argument.

The average values for the materials (50.054% for A and 52.068% for B) are used as the co-ordinates for a point on the Youden plot which is known as the centroid, labelled point 'O'. A Youden plot for the data appears in Fig. 4.7b. From an examination of the diagram, we see that it is divided into four quadrants by the dashed lines drawn through the centroid. It is also seen that the data-points on the diagram fall in either the first or the third quadrant. This is a consequence of the fact that laboratories which obtained 'high' values for material A also obtained 'high' values for material B. The opposite is also true; laboratories reporting 'low' values for A also report low values for B. If a line at 45° to the x-axis is drawn through the centroid and a second line drawn perpendicular to it, we have two new axes, ox' and oy'. The data points tend to be scattered along the ox' axis. *This scatter is a result of and in a sense a measure of the systematic between-laboratory variability.* The Youden plot is such that it effectively adds the between-laboratory variability observed for material A to that observed for material B.

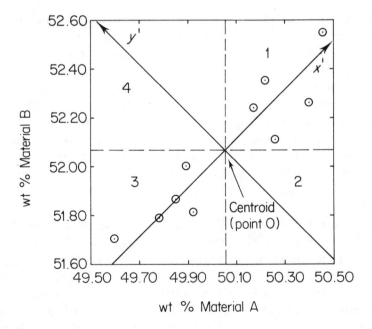

Fig. 4.7b. *The Youden plot for the aspirin analysis data*

We can also see from the plot that the data points are also scattered in the oy' direction. This scatter is, in our example, not nearly as pronounced as the scatter along ox'; most of the data-points lie close to the ox'-axis. What is the origin of the scatter along the oy' axis? First, let us state what this scatter measures; *it is the variability in the differences between the two materials as reported by the ten laboratories.* You will recall that these differences were all close to about 2% aspirin by weight. What caused these differences to be approximately the same? Perhaps it would be better to ask, 'What could cause the differences to be much more varied?' *There are two factors which influence the variability of the differences. These are the random error of measurement and the heterogeneity of the materials.* Let us deal with heterogeneity first. It is easy to infer that if the materials were relatively heterogeneous then the reported differences would show considerable variability. That this is not so in our example is evidence that the laboratories were dealing with relatively homogeneous materials, ie the composition of the samples of material A and of material B received by each laboratory was essentially the same.

We know the larger the random error of measurement associated with an analytical procedure the more varied the results of replicate measurements. A relatively large random error of measurement will also cause the differences between the results of determinations on different materials to vary considerably. In our example, observed differences are approximately the same. This suggests that the random error experienced at each laboratory is relatively small. The scatter along the oy' axis is, for homogeneous materials, a reflection of the within-laboratory variability. In our example, the within-laboratory variabilities are small, compared with the systematic between-laboratory variabilities.

We now see the utility of a Youden plot. If the data points from an inter-laboratory study lie within a thin band along the ox'-axis then the systematic between-laboratory variability is the main cause of the variability in the results reported by the laboratories. Since one of the objectives of an inter-laboratory study is to find out why different laboratories get different results and to rectify the situation, this information is most useful. If given the task of minimising systematic between-laboratory variability, we already have some ideas on how to proceed. For example, we could 'tighten-up' the protocol for the method. If, as a result, a further inter-laboratory study yields data points which result in a Youden plot in which the scatter along ox' and oy' are more similar, then we have evidence that we are succeeding in our task. When dealing with homogeneous materials, the less elongated the pattern, the less the systematic between-laboratory variability is as compared with the repeatability of the method.

If you need more convincing as to the value of a Youden plot, please spend a few moments looking up the results of the American National Cholesterol Survey [published in *Anal. Chem.*, **43**(6), 28A (1971)]. This has become regarded as a classical case which should serve as a warning to all analytical chemists. What is this warning? It is undoubtedly of the form 'Lest we Forget'! Do look it up, it makes very worthwhile reading.

Let us make one final point before leaving this topic. It is sometimes believed that the centroid in a Youden diagram gives us a good estimate of the true values of the two materials. This is often not so. The scatter of the results obtained by several laboratories around the mean value is a result of both random within-laboratory and systematic between-laboratory variability. Averaging may result in the mean value being close to the true value. It may also result in a mean value which is, for example, much higher than the true value. Why is this so? Well, suppose that all the laboratories, besides the many other systematic differences between them, all had a similar positive bias in one of the steps of a procedure. The result? Scatter about a mean value which was higher than the true value. In spite of all our efforts, the truth may still be somewhat elusive!

4.8.   SUMMARY

At one level, you can track our progress by the succession of statistical parameters we have met and by their increasing sophistication. In Part 2 we met z, then in Part 3 came t, now in this current Part 4 we have encounted F. But this Part has been more than just another step up a statistical escalator; it marks the climax of our efforts from their gentle beginnings in Part 1. It can be regarded as the heart of the whole Unit.

We have reached a point where we have at our disposal a powerful armoury for evaluating the evidence given us by sample measurements. And we have practised using this weaponry to solve a wide range of practical problems. We can calculate confidence intervals and we can reach decisions about whether or not to accept a given hypothesis. The presence of errors of measurement means that we can never completely eliminate uncertainty, but we have learned to quantify the 'gambling odds'.

Hypothesis testing had already been introduced in Part 3, in order to assess the compatibility of sample measurements with a suggested value for the population mean. In Section 4.1, however, we refined our approach to such tests, carefully distinguishing two cases. If we know the value of the population standard deviation, σ, we could perform a z-test, whilst if we had to rely on the sample standard deviation, s, we were led to a t-test. We learned to go beyond simply answering yes or no to a test at a preset level of significance. We introduced the p-value, which tells us exactly *how significant* our evidence is. Dealing with p is awkward when we are forced to look up statistical tables, but we were able to make full use of this approach when later we computerised our tests.

Next we turned our attention to comparing the results from measurements on two separate samples. We were interested in drawing conclusions about the evidence they gave about any difference between the respective population means. In Section 4.2 we were able to calculate confidence intervals for this difference whilst in Section 4.3 we dealt with significance testing. Again we discovered that if the σ's were known (or reliably estimated) we could apply a z-test. When the samples were small and we were forced to use s-values, but when we had grounds to believe that the two σ-values were almost equal, then we could apply a t-test based on the pooled sample standard deviation.

In Section 4.4 we described the paired-difference t-test, which can be used to judge whether there is a systematic bias between two methods when they are each applied across a range of different materials.

In Section 4.5 we learned how to evaluate evidence on sample standard deviations. This was a very important further step forward in that it allowed us for the first time to form quantitative judgements on precision. We learned how to test the compatibility of a sample standard deviation with an anticipated population value, or with the s-value for another sample. This involved us in discovering F.

Section 4.6 saw us complete the computer package STAT, components of which we had steadily been building together in each of the first three Parts. Through the program we can deal more flexibly with the statistical parameters z, t, and F, and we can analyse and test data for a single sample or for two samples. The final program is reviewed in Section 4.6.5. A full listing is also given there.

Finally, in Section 4.7, we turned to the problem of comparing several samples. This section gave a glimpse of the important subject known as analysis of variance. We discussed within-sample and between-sample variance and devised an F-test which allowed us to examine the homogeneity of a material. The same test in another context distinguished within-laboratory and between-laboratory variance and led us into a discussion, in more depth than before, of the concepts of repeatability and reproducibility. A Youden plot gave us a graphical way of displaying these effects.

We said above that this Part could be viewed as the heart of the whole Unit. This is so both from the viewpoint of statistics and from that of computing. All that we have done, up to the end of Section 4.6, can be seen as one continuous development aiming to build the basis of an understanding of the statistics of measurement and to give experience in applying the basic ideas in contexts relevant for an analytical chemist, whilst introducing the computer as a valuable tool. There is much ground still to cover, but Section 4.7 marks the start of a new approach; we shall use the background we have so painstakingly built up as a basis for a series of brief incursions into a number of important topics. Our aim will always be to obtain an introductory insight, not to pursue the subject in depth. Analysis of variance was the first such venture. Ahead are calibration, regression, correlation, and quality control.

Objectives

As a result of completing Part 4 you should feel able to:

- recognise that the variance of the sum or difference of two independently measured quantities is equal to the sum of the individual variances (see for example Eq. 4.1);

- compute the *pooled standard deviation* as the best estimate of the population standard deviation based on evidence from measurements on two or more small samples (see Eqs. 4.9 and 4.14);

- recognise the status of the F-distribution in assessing sampling estimates of standard deviations;

- make use of percentile tables of F (Fig. 5.4b–d), or your computer, to obtain F-values corresponding to specific tail areas, or to find tail areas corresponding to specific F-values;

- estimate confidence intervals and perform hypothesis tests of the types reviewed in Section 4.8, either by hand using tables of z, t and F, or with the help of your computer;

- use *analysis of variance* F-tests to compare:
  - within-sample and between-sample variances,
  - within-laboratory and between-laboratory variances,
  - within-run and between-run variances;

- recognise the usefulness of a Youden plot in investigating inter-laboratory comparisons;

- feel a sense of achievement at having constructed and debugged a quite substantial statistical computer package!

4.8.1. Comparison of Means – Case 4: Small Samples with Unequal σ's

We cannot leave this Part without appending a brief additional comment on tests comparing two sample means. Remember that in Sections 4.2 and 4.3 we discussed three cases:

Case 1: σ_1 and σ_2 known

Case 2: σ_1 and σ_2 unknown – large samples

Case 3: σ_1 and σ_2 unknown but equal – small samples

Repeatedly, we exhorted: 'Patience!'", whenever it occurred to us that there must be a 'Case 4' where σ_1 and σ_2 were unknown, where our sample sizes were small, and where it was not reasonable to assume that σ_1 and σ_2 were equal. In Case 3 we used the pooled sample standard deviation s_p, but that could no longer be appropriate. Instead there is an alternative t-test.

We define parameters u_1 and u_2 as follows:

$$u_1 = s_1^2/n_1 \text{ and } u_2 = s_2^2/n_2$$

We then obtain a calculated t-value by applying the formula:

$$t = (\bar{x}_1 - \bar{x}_2)/\sqrt{(u_1 + u_2)}.$$

Finally we calculate an appropriate value for the number of degrees of freedom, using the formula:

$$\nu = \frac{(u_1 + u_2)^2}{\dfrac{u_1^2}{n_1 - 1} + \dfrac{u_2^2}{n_2 - 1}}$$

We give these formulae only for the sake of completeness, so that you can refer to them if in the future you meet such a case. We shall make no direct use of them in this Unit. For those of you interested in digging deeper, look up *Biometrics*, **20**, 191 (1964).

5. The Elementary Statistics of Calibration

Many modern analytical procedures depend on the use of calibration curves. A sample containing an unknown quantity of analyte is prepared in some standard way and is then subjected to instrumental measurement. The reading from the instrument is the information from which we hope to infer the concentration of analyte in the sample. What is being measured is some physical property of the prepared analytical sample. We may be using any one of many instrumental methods, perhaps spectroscopic, perhaps electrochemical, perhaps chromatographic. The primary measurement obtained from the instrument may be as a deflection of a needle over a scale, as a digital readout, as a trace on a recorder chart or, increasingly, as an electronic signal transmitted directly to a computer and converted into a number stored in its memory. The important property of the measurement, however and whenever obtained, is that its value varies with the concentration of the analyte.

The relationship between the signal from the instrument and the analyte concentration can be presented as a calibration curve. Fig. 5.0a illustrates the use of a calibration curve. The instrumental reading is conventionally plotted along the y-axis and the concentration of analyte along the x-axis. The curve is obtained by applying the instrumental method to a number of samples of known concentration. The results are plotted as points on the diagram and these are used as a basis for drawing the calibration curve. The instrumental method can then be applied to an unknown material and the analyte concentration is 'read off' from the point on the curve which corresponds to the instrument reading for the unknown.

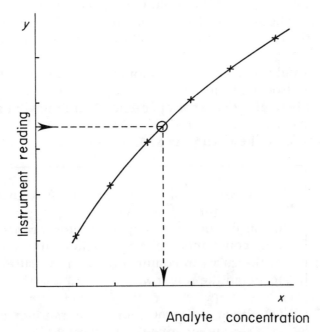

Fig. 5.0a. *Using a calibration graph*
x ... *results obtained for standard samples.*
→– – – O ... *'reading off' a concentration from a measurement on an unknown sample*

We shall be concerned with how best to construct a calibration curve from measurements on known samples, given that experimental errors are inevitable. Fig. 5.0a represents an ideal situation where the standard points all appear to lie perfectly on a smooth curve. We do indeed expect the instrument's response to vary smoothly with concentration, but we can anticipate that because of random measurement errors the calibration points will in practice exhibit a degree of scatter about the 'true' curve. There is a problem, therefore, of deciding what exactly is the 'best' curve to adopt in the light of a set of imperfect measurements. Having once decided upon a particular curve, there is then the question of evaluating just how reliable it is. Finally, we must assess the margin of error to be expected in the determination of an unknown by using the curve. The concept of calibration curves tends to be indelibly associated in our minds with drawings on graph paper. We should make the point that in careful work we shall be interested in the mathematical equation for the curve, and the calculations will be followed through arithmetically.

It will be helpful if you give a little thought in advance to the nature of the errors associated with the use of calibration curves. The SAQ below is deliberately somewhat open-ended, and may be answered in a variety of ways. Please simply jot down your own first thoughts, and then compare these with the SAQ response.

SAQ 5.0a

> An instrumental analysis of an unknown involves a number of stages, including measurements on standard samples of known concentration, construction of a calibration curve, and use of the curve in conjunction with measurements made on the unknown.
>
> (*i*) What different sources of random errors can be identified?
>
> (*ii*) Can you think of any reasons why systematic errors may be avoided?

SAQ 5.0a

Fitting a calibration curve to experimental results can often be a complex exercise, particularly if we wish also to derive from the results an assessment of the size of random experimental errors. Most of our attention will be devoted to examples for which the calibration curve is a *straight line*. This is of wide applicability, though curved calibration graphs also arise quite frequently and they must be recognised when they do. The principles which we develop for the straight line graph can be readily extended to other situations, though we shall not pursue that in detail. We shall, however, be concerned with looking for evidence as to whether a particular set of results is or is not consistent with a linear relationship.

5.1. FITTING A STRAIGHT LINE TO CALIBRATION DATA

We shall use a specific example to bring out the issues involved in drawing a calibration line, and in drawing quantitative conclusions from it. You will be asked to carry out some of the necessary numerical manipulations as our 'case study' unfolds. Some of these are set as SAQs. You are strongly urged to tackle these immediately you come to them; that should help you greatly to follow what we are doing, and to appreciate some of the slightly awkward subtleties which arise from time to time.

We shall consider some calibration results for the spectroscopic determination of Cl^- at the mg dm^{-3} level. The chemistry of the method involves displacement of CNS^- from mercury(II) thiocyanate by Cl^-. The CNS^- released forms a coloured complex with an excess of Fe^{3+} present in the solution. After a standard volumetric procedure has been completed, the spectroscopic absorbance at 465 nm is recorded. Fig. 5.1a shows some typical results. Four points only have been given so as to minimise the labour involved in manipulating the results.

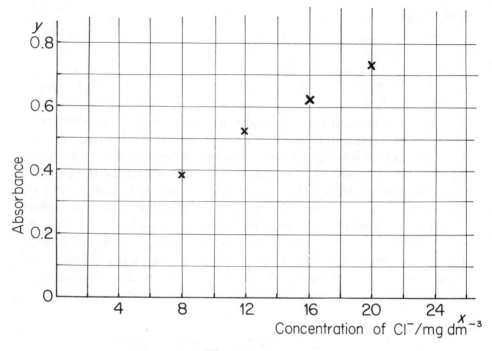

Fig. 5.1a

We shall develop a number of formulae which you will be able to apply generally to other calibration problems. For this reason we shall use the common cartesian-coordinate symbols y and x to represent, respectively, the spectroscopic absorbance and the chloride concentration. When we quote actual values of x, or of a function of x (eg x^2) strictly we ought to quote units, *viz* mg dm^{-3} for x, or mg^2 dm^{-6} for x^2, etc. For ease of presentation, however, we shall generally omit units, leaving them to be understood.

The data-points appear to be roughly linear, so we shall attempt to draw a straight line through the points as closely as possible. Now the general equation for a straight line is of the form below,

$$y = mx + c$$

where c is the value of the intercept made on the y-axis (ie c is the value of y when $x = 0$). The gradient, m, is given by:

$$m = (y_2 - y_1)/(x_2 - x_1)$$

where (x_2, y_2) and (x_1, y_1) are the coordinates of any two points on the line (though to determine m as accurately as possible two points should be chosen which are far apart).

SAQ 5.1a   With the aid of a ruler, draw a straight line to conform with the data in Fig. 5.1a. Read off the gradient and intercept for your line. Check to see whether your values lie within the ranges specified in the SAQ responses section.

If we draw a line 'by eye', there is clearly some scope for subjectivity. Different workers will draw the line differently. Even the same person, on repeating the exercise, would usually draw a slightly different line. The values obtained for the parameters m and c would vary accordingly; it is not at all unusual for different estimates made in this way to vary by 10% or more.

The first problems which confront us are now revealed.

(*a*) What systematic criteria should we use in deciding precisely which line to draw?

(*b*) How can we obtain an estimate of the accuracy with which our line has been determined?

It is clearly important that we have some objective and widely accepted method for selecting the 'best' line fitting our data. The criterion used should give equal weight to all the data points available, unless we have evidence that one or more points have been less reliably determined than others. If the data points all lay exactly on one straight line, then that line would clearly be the one to draw. In practice, however, our line will inevitably miss all, or nearly all, the points. The *distances* between the points and the line will give some measure of the experimental scatter.

Let us presume that the concentrations of our standard solutions of chloride, our x-values, are all known exactly, so that the uncertainty in our data arises entirely from the experimental determination of the absorbance values, y. If the true relationship between x and y is indeed a straight line, it follows that the calibration points would all lie on a single line were it not for the errors in y. The vertical distances d_i in Fig. 5.1b are thus presumed to arise from random errors of measurement. These distances are known as *residuals*. The 'best' line to draw will be the one which keeps the residuals in some sense small.

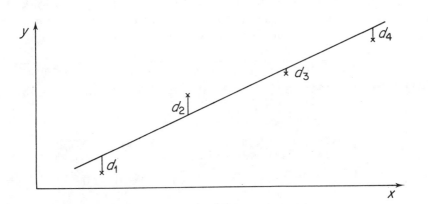

Fig. 5.1b. *y-residuals, d_i, between data-points and a calibration line*

The accepted procedure to determine the best-fit line is known as the *method of least squares*. This involves choosing a line such that the sum of the squares of the residuals (here $d_1^2 + d_2^2 + d_3^2 + d_4^2$) is as small as possible. It is relatively easy to derive general equations giving the values of m and c which define the least-squares line. (The derivation is given as an appendix to Section 5.6). The equations are given below.

$$m = (nS_{xy} - S_x S_y)/(nS_{xx} - S_x^2) \qquad (5.1)$$

$$c = (S_y S_{xx} - S_x S_{xy})/(nS_{xx} - S_x^2) \qquad (5.2)$$

In these formulae n is the total number of data-points used, and the other symbols all represent various sums over all the data points.

$$S_x = \sum_{i=1}^{n} x_i \qquad S_y = \sum_{i=1}^{n} y_i \qquad S_{xy} = \sum_{i=1}^{n} x_i y_i \qquad S_{xx} = \sum_{i=1}^{n} x_i^2$$

These formulae are actually much easier to use than they look! Let us obtain the least-squares fit for the data in Fig. 5.1a. The points from the chloride analysis are given in SAQ 5.1b.

SAQ 5.1b   Derive the least-squares straight line to fit the data below.

x	8.00	12.00	16.00	20.00
y	0.388	0.523	0.624	0.734

Obtain the values of m and of c using Eqs. 5.1 and 5.2. Plot the resulting line on Fig. 5.1a.

The equation for the line is

$$y = 0.02848\, x + 0.1686$$

If you have any difficulty in obtaining these values, or in plotting the resulting line, please consult the SAQ response before reading on. Notice that we have retained one more significant figure in the quoted values for m and c than was available in the y data. This tactic will ensure that we do not introduce significant additional random errors when we use our equation.

The straight-line calibration which we have derived is known as *the regression line of y on x*.

5.1.1.   The Choice of the Regression Line: Some Comments

Note carefully the assumptions we have made in deriving our calibration line.

(*a*) We have assumed that the 'true' relationship between absorbance and concentration in this case is a straight line, ie we have assumed that the only reason why the observed points do not quite fit a straight line is because of random errors of measurement.

(*b*) We have further assumed that the experimental 'scatter' arises entirely in the measurement of y.

We shall have more to say about assumption (*a*) later, but for the time being we shall continue to accept this as reasonable. After all, inspection of Fig. 5.1a does show that our four points are rather close to being collinear. Assumption (*b*) is worth probing in a little more detail because, when we fit experimental data to curves, we may be unsure whether errors in measurement are more associated with our x-values than with our y-values.

In obtaining the calibration measurements, the situation is certainly not symmetrical as between x and y. At one level we can regard x as the independent variable and y as the dependent variable. What we mean in that regard is that we choose the x-value independently by selecting a standard whose known concentration suits us. The y-value is not directly selected by us, it is the instrument's response to our chosen x; it depends on x. Now the value of x will be error-free if our standard solutions are sufficiently reliable. On the other hand, the chemical preparation of our spectroscopic sample (release of thiocyanate, etc) and making the instrumental measurement itself, ie determining y, seems to be more prone to error. Nevertheless it need not always be so. One might, in a particular example, have reservations about the reliability of the standard materials. How were they themselves prepared? In the chloride example they were probably obtained by dilution of a single standard stock solution. This opens the possibility of volumetric errors and, if the prepared standards were inadequately mixed, there could be subsequent sampling errors due to inhomogeneity. So the assumption that the only source of error is in measuring y, is potentially questionable.

Let us for a moment go to the other extreme and suppose that any errors are entirely related to uncertainties in measuring x, whilst our y-values are known exactly. Then we should seek the 'best-fit' straight line which minimises the sum of the *horizontal* residuals h_i as indicated in Fig. 5.1c. In so doing we would be determining the *regression line* of x on y. It can be shown that this is a new line given by $y = m'x + c'$, where

$$m' = (nS_{yy} - S_y^2)/(nS_{xy} - S_xS_y) \tag{5.3}$$

and
$$c' = (S_yS_{xy} - S_xS_{yy})/(nS_{xy} - S_xS_y) \tag{5.4}$$

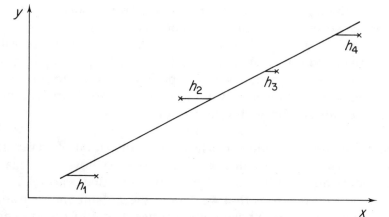

Fig. 5.1c.   *x-residuals h_i between data points and a calibration line*

SAQ 5.1c	Determine the x-on-y regression line for the chloride analysis data given in SAQ 5.1b. Use Eqs. 5.3 and 5.4 to obtain the value of m' and c', and then add this further line to Fig. 5.1a.

Our two regression equations for the measurements made on chloride solutions are given below.

Regression of y on x: $y = 0.02848\ x + 0.1686$

Regression of x on y: $y = 0.02858\ x + 0.1671$

Although the two lines are distinct the difference between them is very small indeed. Here the choice of which line to adopt is of little practical significance. However, if errors of measurement are proportionately larger then the 'y-on-x' and the 'x-on-y' regression may differ more seriously. It will be helpful to explore the relation between them just a little further.

The two regression lines cross at the mean point of the calibration data. This is the point $(\bar{x}, \bar{y})$ whose coordinates are the respective means of the x's and y's of the data ($\bar{x} = S_x/n$ and $\bar{y} = S_y/n$).

SAQ 5.1d

The *y-on-x* and the *x-on-y* regression equations obtained in the previous two SAQs for the chloride analysis data are given below.

$$y = 0.028475\,x + 0.16860$$

$$\text{and } y = 0.028584\,x + 0.16707$$

Solve these as two simultaneous equations to find the *x*- and the *y*-value at their crossing-point. Hence confirm that the crossing point is indeed the mean point $(\bar{x}, \bar{y})$ of the data. (You may notice that a fifth significant figure has been added in quoting the *m* and *c* values. These two equations are so similar that this is needed in order to obtain an answer with sufficient precision.)

The regression of y on x and that of x on y always pass through the mean point, but the lines do have somewhat different gradients so they progressively diverge as we move further away from the mean point. There are yet other possible choices of regression line. If the scatter in the data points is believed to be due to appreciable errors in the measurement of *both* the x- and the y-values, it can be shown that the 'best' least-squares line is an intermediate one. This line would also pass through the mean point, with a gradient intermediate between m and m'.

We shall have occasion later, in a somewhat different context, to return to the distinction between the 'y-on-x' and the 'x-on-y' regression. From the point of view of our current discussion of calibration lines, however, we want you to have gathered only two qualitative points from this discussion.

(a) Strictly specking, the 'correct' regression line depends on the relative magnitudes of random errors in the two variables x and y.

(b) Any uncertainty about the location of the calibration line is least near the mean point of the calibration data.

Point (b) will be reinforced below as we look in more detail at the precision of our line. As to point (a), when we construct a calibration line we shall more often than not expect to know the concentrations of our standard samples (ie our x-values) quite accurately. Therefore it is conventional to work with our first regression equation, that of y on x. When we are dealing with a reasonably reliable experimental technique, as in our chloride example, we can in any event, rely on the fact that an alternative choice of regression line is unlikely to make any significant difference.

5.2. MAKING ESTIMATES BY USING A REGRESSION LINE

We can use our regression line to read off an estimate of the concentration of an unknown sample. The question which we still need to ask is how to derive a quantitative assessment of the errors inherent in such a determination. Let us give a brief overview of how the story will unfold.

The basic evidence we have of the presence of random errors is apparent in the scatter of the calibration points about the least-squares line. In our earliest studies of statistics, when we were concerned with a simple measurement, as for instance in a series of replicate titrations, scatter about the mean provided the fundamental measure of dispersion. In the more complicated situation which we now face, with a series of individual points spread out over a graph, the scatter of the points relative to the regression line gives the corresponding information. From this we obtain a measure which plays a role analogous to that of the standard deviation in our earlier work. That is our starting point.

Given a basic measure of dispersion, we can give an estimate of the standard deviation for each of the various values which characterise the regression line itself, and for measurements made by using it. Thus, for example, we can estimate the standard deviation for the gradient of the line, m, or for the concentration read from the line for a particular unknown. Having values for standard deviations we can follow the path we have used so often before. We can derive confidence intervals, and we can carry out significance tests. We have to bear in mind that our line is determined by using a finite number of points. Our standard deviations are thus not population values, σ, but estimates based on finite samples. They are sample standard deviations, s. Thus we shall find ourselves back in the arms of our old friend t.

So much for the preview. As you may have gathered estimation by using a regression line will have strong parallels with what we have done before for simpler situations. Try to keep a lookout for these parallels as we proceed. First we must obtain our new measure of dispersion which, it turns out, is called *the standard error of the estimate*.

5.2.1. The Standard Error of the Estimate

We have already pointed out that the measure of dispersion for our calibration is given by the scatter of the standard points relative to the regression line. By how much does a typical point deviate from the line? In adopting the *y-on-x* regression we are presuming that errors arise solely from uncertainties in our *y*-values. Hence the natural measure of dispersion is the magnitude of the deviation between a given *y*-value and the *y*-value read off the line for the corresponding value of *x*. These are just the *residuals* d_i sketched in Fig. 5.1b. For the four points of our chloride determination example, the calculation of the residuals is laid out in Fig. 5.2a.

x_i	8.00	12.00	16.00	20.00
y_i	0.388	0.523	0.624	0.734
$y_{est} = 0.02848\, x_i + 0.1686$	0.3964	0.5104	0.6243	0.7382
Residual $d_i = y_i - y_{est}$	-0.0084	0.0126	-0.0003	-0.0042

Fig. 5.2a. *Calculation of residuals for the chloride example*

Some residuals are positive, others are negative. The natural choice of statistic to represent the dispersion about the line is the root-mean-square value. We name this quantity the *standard error of the estimate of y*. We shall use the symbol, σ_y, to represent this quantity, thus emphasising its direct analogy with the standard deviation in the measurement of a single quantity.

If we had a sufficiently large number of readings ($n \rightarrow \infty$) we should put

$$\sigma_y = \sqrt{\sum_{i=1}^{n} d_i^2 / n}$$

In practice, however, a finite set of data points is used for the two purposes of first determining the regression line itself, and then obtaining the standard error of the estimate. This means that we obtain only an estimate of σ_y. Not surprisingly we label this estimate s_y; thus s_y gives a *sample* standard error, whereas σ_y is the *population* value.

The appropriate expression to obtain s_y is

$$s_y = \sqrt{\sum_{i=1}^{n} d_i^2/(n-2)} \qquad (5.5)$$

SAQ 5.2a

Can you think of the reason why we use $(n-2)$ in the denominator of the expression defining s_y, the standard error of the estimate of y?

(Hint. Recall that we divide by $(n-1)$ when we estimate the standard deviation from a set of repeated measurements of a single quantity when the population mean is unknown).

SAQ 5.2b

Calculate the value of s_y for the chloride-ion example. (The necessary data are tabulated in Fig. 5.2a).

SAQ 5.2b

The standard error of the estimate of y does indeed play a role analogous to that of a standard deviation. If the true line is known and if the only reason why experimental points do not lie on it is that there are random errors in measuring y, then, if n is large, 68% of all points will lie within a vertical distance σ_y above or below the regression line. 95.5% will lie within twice this vertical distance, and 99.7%. within $3\sigma_y$, etc (see Fig. 5.2b).

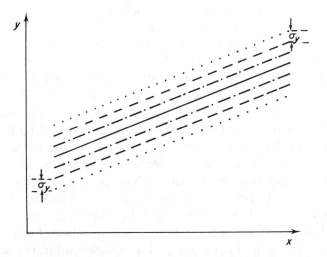

Fig. 5.2b. *Dispersion of points about a y-on-x regression line. The lines drawn are all parallel, each separated from its neighbour by a vertical distance σ_y*

On average: as many points should be above as below the regression line:

68% of points should lie between lines   — · — · — · — · —

95.5% of points should lie between lines   — — — — — — — —

99.7% of points should lie between lines   · · · · · · · · · · · · · · · · · ·

The term s_y is the simplest indicator of the precision of the fit given by our regression line. However, we have already had an indication that the reliability of our calibration varies along its length, and becomes less certain as we move away from the mean-point towards or past the ends of our calibration data. Obtaining s_y is a first step. In Section 5.2.2 below we shall present formulae which allow us to estimate the standard deviation for any point in the calibration, and for the concentration of an unknown determined by measuring its 'y'-value and then reading off 'x' from the calibration line.

5.2.2.   Precision in the Use of a Straight-line Calibration

> So, naturalists observe, a flea
> Hath smaller fleas that on him prey;
> And these have smaller still to bite 'em
> And so proceed ad infinitum.
>   (Swift. 'On Poetry: A Rhapsody', 1733)

Here we go again! Statistical analysis has something about it which is reminiscent of Swift's famous epigram. We first use statistics with our data to obtain a best estimate of the value of some quantity, or at present, a best estimate of the y-value corresponding to every x-value (as given by our regression line). Next we estimate how large our error may be; in the previous section we obtained an estimate of the standard deviation for residuals. Yes, as you may have noticed, in s_y we have an estimate of an estimate! (It is a sample estimate of the standard error of the estimate of y.) In due course we move on to estimate how confident we can be in the reliability of our estimates. Uncertainty constantly preys on us. It does seem a bit like fleas biting fleas (some who are less poetic and more uncharitable may tend to think more in terms of nits picking nits!). It certainly requires a little patience!

Our calibration graph is constructed on the basis of measurements made on a set of standard samples. It is subsequently used with an 'unknown' sample, and a measurement of property y is used to 'read-off' the corresponding concentration value, x. We must again remind ourselves that there are two distinct sources of random error in this process:

(*a*) errors in estimating the regression line itself,
(*b*) errors arising in measuring y for the unknown and hence determining the concentration value x for the unknown.

We shall consider (*a*) first.

We have used experimental results to determine a regression line, and then we have used the spread of our data about the line to obtain the standard error of the estimate, s_y. But, given that we have a limited number of data points, it follows that all our results are liable to the uncertainties of sampling. Hence you may safely conclude that we are embarked upon a voyage in which, in due course, we shall again meet the statistic, t. We can reinforce our argument by considering the extreme situation which arises if we are forced to construct a calibration line on the basis of just two points. The 'best' line could then be only the straight line passing exactly through these points. We would be left with no measure at all of the magnitude of dispersion. At other extreme, if we had the patience to measure an indefinitely large number of calibration points, we should be able to determine the line itself, and the value of σ_y, with complete confidence.

SAQ 5.2c   A 'regression' line is drawn when there are only two calibration points.

(*i*) What will the values of the residuals be?

(*ii*) What result would you obtain for the standard error of the estimate?

SAQ 5.2c

In practical situations we shall find ourselves somewhere between these two extremes. Using a limited number of points (but more than two) allows us to compute an estimate s_y of the value for σ_y. This is an estimate obtained with $(n - 2)$ degrees of freedom. Because of the danger that s_y may be an underestimate of σ_y we have to use t instead of z when we attempt to specify confidence ranges, or to pursue tests of significance. The statistic t leaves us with broader ranges of uncertainty than z would.

But so far we have discussed only s_y, which is effectively the sample standard deviation for the residuals. The residuals are members of a population whose mean is zero and whose estimated standard deviation is s_y. We could use s_y to estimate the proportion of residuals falling in a given range (see, for instance, Fig. 5.2b). We are generally less interested in the residuals however, than in the accuracy of estimates of other quantities, such as the parameters m and c which define the regression line, or the concentration of an unknown obtained by using the line. Happily it turns our that we can estimate the standard deviations for all the quantities of interest once we have our estimate, s_y. Here we shall simply quote the relevant formulae without any attempt to explain how they are derived. For fuller explanations you would have to consult a more advanced text on statistics (some pointers to suitable books are given in our bibliography). The expressions of interest are tabulated in Fig. 5.2c.

Quantity	Value	Standard deviation
(a) Residual	$d_i = y_i - y_{est}$	$s_y \sqrt{= \sum_{i=1}^{n} d_i^2/(n-2)}$
(b) Gradient	$m = \dfrac{nS_{xy} - S_xS_y}{nS_{xx} - S_x^2}$	$\dfrac{s_y}{\sqrt{S_{xx} - S_x^2/n}}$
(c) Intercept	$c = \dfrac{S_yS_{xx} - S_xS_{xy}}{nS_{xx} - S_x^2}$	$s_y\sqrt{\dfrac{S_{xx}}{nS_{xx} - S_x^2}}$
(d) Regression	$y = mx_o + c$	$s_y\sqrt{\dfrac{1}{n} + \dfrac{(x_o - \bar{x})^2}{S_{xx} - S_x^2/n}}$
(e) Predict x (N repeats)	$x = \dfrac{y_o - c}{m}$	$\dfrac{s_y}{m}\sqrt{\dfrac{1}{N} + \dfrac{1}{n} + \dfrac{(y_o - \bar{y})^2}{m^2(S_{xx} - S_x^2/n)}}$

Fig. 5.2c.   *Linear regression of y on x: a summary*

Fig. 5.2c deserves a detailed review. On the first line the y-residual, d_i is first defined, then the formula for obtaining s_y is reiterated. Notice that s_y appears in the formulae for all the other standard deviations listed.

Row (b) and row (c) deal with the parameters m and c which define the y-on-x regression line. The formulae for obtaining these (Eqs. 5.1 and 5.2) are repeated in the middle column. The new information, giving estimated standard deviations, is in the final column. These standard deviations, which we could label s_m and s_c, involve, as well as s_y, the values of the sums S_x and S_{xx}. These latter were required anyway for determining m and c themselves.

Row (d) is labelled, perhaps somewhat dauntingly, 'regression'. Here we are concerned with the position of the line itself, ie with the y-value of the regression line for whatever x-value we choose. Suppose we are interested in an x-value, x_o. The value estimated for

y by using our line is $mx_o + c$. Now we have already indicated that the reliability of our straight line decreases as we move away from the mean-point $(\bar{x}, \bar{y})$ of the calibration data. The standard deviation for y, as a function of x, is generally referred to as the *standard error of the regression line*. Let us look in some detail at the expression given for it. Notice that the second term within the square root gives zero at the mean-point (*viz* at $x_o = \bar{x}$). At this point the standard error of the regression line is at a minimum, given by $s_y/\sqrt{n}$. That result has a very familiar ring to it; remember how the standard deviation of a sample mean depends on the number of measurements $\sigma_{\bar{x}} = \sigma_x/\sqrt{n}$. The second term in the square root is always positive for $x_o \neq \bar{x}$, and its value increases with the square of the difference $(x_o - \bar{x})$. It follows, as we have argued before, that the standard error of the regression line increases steadily as we move away from the mean point of the line.

With row (e) we at last reach our major goal, *viz* the determination of the concentration of an unknown. Suppose the unknown gives a measured y-value of y_o. The middle column gives the corresponding x-value obtained from the regression equation (solving $y_o = mx + c$ for x). The estimated standard deviation for this determined x-value is given in the final column. We shall call this *the standard error of a prediction of x*. Unfortunately the formula is the most complicated of all. One new parameter appears. This is N, the number of repeat determination of y for the unknown. The y_o quoted should therefore be taken as the mean of N repeat determinations. (If only one determination is done we simply take $N = 1$). There are now three terms within the square root. Only the first $(1/N)$ arises from errors in the reading for the unknown itself, but notice that this term may dominate if N is small (eg if only a single measurement is made on the unknown). The third term is zero at the mean point of the regression (ie at $y_o = \bar{y}$), but this term grows steadily in importance as the measured y-value departs from the mean of the original calibration standards. To calculate the standard deviation for the regression and for a predicted x, we need the values for $\bar{x}$ and $\bar{y}$. these do not appear explicitly in the original regression calculation of the m and c values, but they are very easily obtained since

$$\bar{x} = \sum_{i=1}^{n} x_i/n = S_x/n$$

and

$$\bar{y} = \sum_{i=1}^{n} y_i/n = S_y/n$$

SAQ 5.2d Substitute the value $x_o = 0$ into the formula given in Fig. 5.2c for the standard error of the regression line. Show that for this value of x the formula reduces to that given for the standard deviation of the intercept c.

SAQ 5.2e   Consider once more the calibration line for chloride determination. (The data are tabulated in Fig. 5.2a.) An unknown sample, when treated in the standard way, gives an absorbance, $y = 0.660$. Estimate the concentration, x, and evaluate the standard error of this prediction supposing:

(*i*)   the value of y is the result of a single determination,

(*ii*) $y = 0.660$ is the mean of four replicate determinations.

5.2.3. Confidence Intervals and Significance Tests

Now that we have expressions for standard deviations the remaining step, to derive confidence intervals or to test hypotheses, is a by now familiar process. We are retracing the ground of Sections 4.1 to 4.3. Since our standard deviations are sample estimates, and not population values, we must use the methodology based on the t-statistic. The appropriate number of degrees of freedom, we have decided, is $n - 2$. Suppose for instance we wish to establish a 95% confidence interval for some quantity. Each tail, therefore, will carry a probability of 2.5%. So we need to look up the value of $t_{0.025(n-2)}$. If for the quantity we have an estimated value r and standard deviation s, then the confidence interval for the true result is

$$r \pm t_{0.025,(n-2)} \times s.$$

As we have said, it is a matter of using familiar methods in a new context. The way to get on top of things is simply to let yourself loose on some SAQs.

SAQ 5.2f

(*i*) For the example discussed in SAQ 5.2e, in which four determinations of an unknown chloride solution gave a mean absorbance of 0.660, give the 95% confidence interval for the chloride concentration.

(*ii*) Give the 95% confidence interval for the intercept c for our chloride regression line.

(*iii*) Can you, at a 1% level of significance, assert that the true value of the intercept is greater than zero?

SAQ 5.2f

Fig. 5.2d shows the 95% confidence interval for the *regression line*, obtained by using the formulae in row (*d*) of Fig. 5.2c. Check as closely as you can that the range of *y* at *x* = 0 agrees with your answer to part (*ii*) of SAQ 5.2f. Notice on the other hand that the range *for x* at *y* = 0.660 is a little narrower than that from the answer to part (*i*). The latter range is also marked on the figure. This is consistent with what we should expect, as the limits in Fig. 5.2d reflect only the uncertainties in the regression line itself, whereas part (*i*) of the SAQ was concerned also with the effects of uncertainties in the measurement of the unknown.

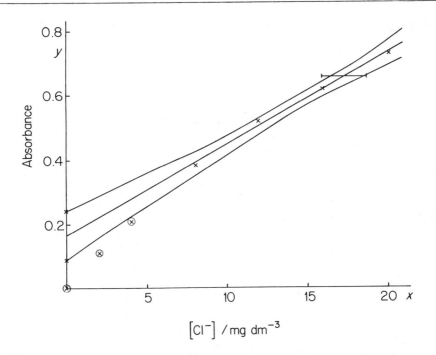

Fig. 5.2d. *The y-on-x regression line, bracketed by its 95% confidence range, for the chloride analysis data (four standard calibration points over the range 8–20 mg dm⁻³). See text for discussion of the horizontal bar at y = 0.660, and for the ringed points at x = 0, 2 and 4 mg dm⁻³*

| SAQ 5.2g | Evaluate the 95% confidence limits of the chloride regression line at $x = 14$ mg dm⁻³ (ie at $x = \bar{x}$). How much smaller is this range than that at $x = 0$? |

SAQ 5.2g

We have mentioned on a number of occasions that the precision of a regression line becomes progressively poorer as we move further away from the mean point of the calibration data. Fig. 5.2d illustrates this feature quantitatively.

5.3.   DANGERS OF EXTRAPOLATION

It is widely acknowledged that interpolation is a more reliable process than extrapolation. It is the object of this section to emphasise the extent to which this is so. *Extrapolation is very unreliable indeed.*

As we noted above from Fig. 5.2d, the uncertainty in the regression line increases steadily as we move away, in either direction, from the mean point of the calibration data. This effect becomes pronounced for an extrapolation, where we are venturing so far from the mean that we are beyond the extreme ends of the calibration points. This is, however, only one reason why it is more error prone to use a calibration line for extrapolation than for interpolation. A further three experimental points have been added to Fig. 5.2d indicating measurements obtained for standard solutions with chloride concentrations of 0.00, 2.00, and 4.00 mg dm^{-3}. These upset the applecart in no uncertain fashion! Our straight line calibration, which fits the data in the range 8–20 mg dm^{-3} quite convincingly, does not at all describe the behaviour of the absorbance at lower chloride concentrations.

From Fig. 5.2d, if we take full account of its 95% confidence lines, we predict very clearly a positive intercept. In SAQ 5.2f (*iii*), you should have found that, in a one-tail test at a 1% level of significance, the hypothesis that the true intercept might be zero had to be rejected. Yet manifestly the experimental data indicate that the true intercept is zero. As you may have suspected from the beginning, this result is inherent in the experimental technique, which rests on observing the absorbance of the test solution relative to a 'blank' containing zero chloride. How do we explain this situation? Have we simply been unlucky in this example, happening on an occurrence which we might expect to arise less frequently than once in a hundred times? Not at all. Our regression line itself is not to blame, *what is wrong is our assumption that the true relationship between x and y is linear.* That assumption seems to be a reasonable approximation to the truth over the range 8–20 mg dm^{-3}, but it is certainly not valid over the whole range 0–20 mg dm^{-3}. Had we in fact included several more calibration points, but still all in the range 8–20 mg dm^{-3}, then our calculated regression line would have been unlikely to change much. The confidence intervals, however, would almost certainly have narrowed considerably. The true experimental result at $x = 0$ would have seemed even more dramatically out of line.

SAQ 5.3a   The regression line in Fig. 5.2d has been calculated on the basis of four calibration points for chloride concentrations in the range 8–20

$\longrightarrow$

SAQ 5.3a
(cont.)

mg dm^{-3}. It has been argued that, has many more calibration points been included, all in the range 8–20 mg dm^{-2}, then a very similar regression line would probably have resulted, but that its 95% confidence limits would have been considerably narrowed. Give arguments for and/or against these claims.

There is a very important general lesson to learn here. *Predictions which involve extrapolation beyond the range of the calibration data must always be viewed with a very healthy scepticism.* To have any reasonable level of confidence in such predictions, we must first be thoroughly satisfied that the true relationship between the measured property and the sample concentration (ie between y and x) fits the same mathematical expression in the extrapolated region as that which applies in the region of the original calibration measurements. If we cannot convince ourselves on that score we cannot give a reliable assessment of the likely error of a prediction made by extrapolation. That is the strongest reason why extrapolation has a bad name. If we *do* have convincing theoretical reasons to believe that the same relationship between y and x will hold in the extrapolated region, then we can give confidence ranges for a result obtained by extrapolation, though such ranges will be wider than for a prediction involving interpolation. (See, for example, the answer to SAQ 5.2g.) For a straight line, the ranges are calculated by using row (e) of Fig. 5.2c.

In many real-life situations we are often forced to make extrapolations. In particular, nearly all predictions about the future rely on extrapolating past trends. We should not therefore be too surprised at the blunders made in forecasting tomorrow's weather, or next year's economic growth. The analytical chemist, on the other hand, can usually avoid the need for extrapolation if a wide enough range of standard samples is used when determining the calibration line.

The siren message of this Section is: whenever practicable, *it is a very good idea to avoid the need for extrapolation.*

5.4.   PROGRAMMING THE REGRESSION CALCULATION

Our regression 'case study' has used only four calibration points. This was a deliberate choice made to minimise the work involved in calculations. In practice, one would normally have many more calibration measurements; the more points the better as far as the reliability of the analysis is concerned. The regression calculation then becomes rather wearisome and vulnerable to mistakes. It is on the other hand quite easily programmed.

It would be perfectly feasible to add this facility into programme "STAT", but we did promise you, in Part 4, that we were prepared to regard that as a finished product! Instead we give a separate program in which the amount of typing required has been kept to a bare minimum. Type in the following lines.

```
10  DIM X(30),Y(30),D(30)
20  READ N
70  FOR I=1 TO N
80      READ X(I),Y(I)
90      C1=C1+X(I)
```

```
100      C2=C2+X(I)↑2
110      C3=C3+Y(I)
120      C4=C4+X(I)*Y(I)
130      NEXT I
140  G=(N*C4-C1*C3)/(N*C2-C1↑2)
150  C=(C3*C2-C1*C4)/(N*C2-C1↑2)
160 PRINT"     x        y      residual"
180 FOR I=1 TO N
190      D(I)=Y(I)-G*X(I)-C
200      A=A+D(I)↑2
210      PRINT X(I),Y(I),INT(D(I)*10000+.5)/10000
220      NEXT I
230 S5=SQR(A/(N-2))
240 G5=S5/SQR(C2-C1↑2/N)
250 C5=S5*SQR(C2/(N*C2-C1↑2))
270 PRINT"     Gradient    = ";G
280 PRINT"     Intercept    = ";C
290 PRINT"St err of estimate = ";S5
300 PRINT"St dev of gradient = ";G5
310 PRINT"St dev of intercept = ";C5
999 END
1000 DATA 4
1010 DATA 8.0,0.388,12.0,0.523,16.0,0.624,20.0,0.734
```

Arrays X and Y hold the calibration x- and y-values, and the first item read, N, is the number of points, n. The DATA inserted on line 1000 and 1010 are just those for our chloride example (check against Fig. 5.2a). Notice that the data on line 1010 come in the order x_1, y_1, x_2, y_2, etc. This is a change from previous practice but it saves us from having to write an extra loop in the program. Every time line 80 is passed one point (x_i, y_i) is read. Parameters C1, C2, C3, and C4 accumulate the sums S_x, S_{xx}, S_y and S_{xy} respectively.*

---

* We have cut more corners than usual in this program. The variables C1–C4 must have zero values when the loop from line 70 is entered for the first time, Most machines set all values to zero at the beginning of a RUN. If you have any doubts about whether your machine does this it would be wise to add lines 30–60 setting C1=0, C2=0, etc. Similarly, you would also need a line 170 A=0.

Lines 140 and 150 apply Eqs. 5.1 and 5.2 to get the gradient, G, and the intercept C. The second loop calculates the residuals (array D, see line 190) and the sum of their squares (as A). Then formulae from Fig. 5.2c are used to obtain the standard error of the estimate (S5) and the standard deviations for the gradient and the intercept (G5 and C5). Finally, the results are printed out.

RUN the program. We get the results below.

x	y	residual
8	0.388	-8.4E-3
12	0.523	1.27E-2
16	0.624	-2E-4
20	0.734	-4.1E-3
Gradient		= 2.8475E-2
Intercept		= 0.1686
St err of estimate		= 1.11512332E-2
St dev of gradient		= 1.24674577E-3
St dev of intercept		= 1.83233458E-2

Your computer may present some of the numbers differently, but the values should agree. If not, then by carefully noting just where disagreement sets in you should be able to locate any typing errors. The results do not agree precisely with those we obtained earlier by hand! The discrepancies are very small and arise because previously we rounded the value of the gradient to 0.02848. This led us to report the residuals as $-0.0084, 0.0126, -0.0003$ and -0.0042, respectively (Fig. 5.2a). We also obtained $s_y = 0.0111$ (SAQ 5.2b) and $s_c = 0.0182$ (SAQ 5.2f). We should always be aware of the possiblity of rounding errors. None of the discrepancies is important, but they would have begun to be so had we rounded the gradient further, to 0.285. The program gives a compact presentation of the results which is quite useful. Having the residuals listed is particularly valuable. It is always worthwhile looking for peculiarities in the list; we shall have more to say about that later.

So far so good! The program quite painlessly rolls out the parameters for the line, and the basic standard deviations. But what about the standard errors of regression and of prediction? These values depend on the point on the line in which one is interested. To cope with them add the following lines.

```
400 PRINT"Err of Regress: Give x (-ve to skip)";
410 INPUT X0
420 IF X0<0 THEN 500
430 S6=S5*SQR(1/N+(X0-C1/N)↑2/(C2-C1↑2/N))
435 PRINT"    . . . . . gives y =    ";G*X0+C
440 PRINT"St err of regression =    ";S6
450 GOTO 400
500 PRINT"Err of Predict: Give y (-ve to stop)";
510 INPUT Y0
520 IF Y0<0 THEN 999
530 PRINT"Give no repeats run on unknown";
540 INPUT N0
545 PRINT"    . . . . . gives x    = ";(Y0-C)/G
550 S7=S5*SQR(1/N0+1/N+(Y0-C3/N)↑2/(G↑2*(C2-C1↑2/N)))/G
560 PRINT"St err of prediction = ";S7
570 GOTO 500
```

Line 430 and line 550 need careful typing; they are based on the rather messy formulae on rows (*d*) and (*e*) of Fig. 5.5c. The effort is, however, worthwhile, as the new values which the program can now rattle out are very tedious to generate manually. Notice that calculation of the error of regression (line 400–450) is repeated again and again till you specify a negative value for *x*. When you finally do that, the program moves on to find errors of prediction (lines 500–570), which in turn is repeated till a negative *y* is specified. If you were using calibration data where negative *y*-values (ie negative instrument readings) were expected, then you would have to use a different *y*-value to signal a wish to stop (lines 500 and 520).

RUN the program. The previous results will be output, followed by:

"Err of Regress: Give x (-ve to skip)?"

Remember Fig. 5.2d, where we showed in detail how the uncertainty in the regression line grew steadily as we got further from the mean-point $\bar{x}$? Well try this out now; put in a succession of x values and see how the reported standard error of regression varies. Check in particular that you can agree with the following results.

x	0.0	4.0	8.0	12.0	14.0
St err of regression	0.0183	0.0137	0.0093	0.0061	0.0056

You might like to check how the first and the last value quoted agree with your previous calculations by hand in SAQ's 5.2f(ii) and 5.2g. You should also check that, as you go beyond the mean point ($\bar{x} = 14.0$), the standard error grows again, in a symmetrical way. thus at $x = 20.0$ (6.0 above $\bar{x}$) the standard error is the same as at $x = 8.0$ (6.0 below $\bar{x}$).

What about Fig. 5.2d in detail? Well that showed the 95% confidence lines. The width of the confidence interval, on each side of the line, was given by $t_{0.025,2}$ (ie 4.30) times the standard error of regression. You might now like to check that the figure has been properly drawn. A price we pay for having written a separate program for regression is that the t-value is not readily available to the program. You could if you chose modify the program so that it gives confidence intervals (feeding it t either as data or by supplying the appropriate subroutine from "STAT"). That is up to you. As far as we are concerned our regression program can now be regarded as effectively complete (apart from half-a-dozen lines which we shall add in Part 6 of this Unit).

Supplying a negative x-value will cause the program to jump to its final stage, that of calculating the standard error of prediction, given a y-value measured for an unknown. You are now asked:

"Err of Predict: Give y (-ve to stop)?"

Well, let us check our answers to SAQ 5.2e, where we supposed that we had, for an unknown, measured an absorbance of 0.660. Feed that value in. The result is another question!

"Give no of repeats run on unknown?"

In the first part of the SAQ we assumed that we had done a single determination. So give 1. This time you should get an answer that the standard error of prediction is 0.4605. This agrees with the answer found previously.

In the SAQ you were asked to do the calculation supposing that 0.660 was the mean of four replicate determinations on the unknown. Try that next and check that you can agree with the SAQ answer that the standard error decreases to 0.31 mg dm^{-3}. Try also supposing that the result was based on even larger numbers of results. The standard error decreases further towards a limit of about 0.24 mg dm^{-3}. This limit is due to uncertainty in the regression line: random errors in the reading for the unknown itself are eventually eliminated by taking the mean of a very large number of replicate measurements.

Investigate how moving further from the mean point increases the standard error of prediction. Some results we obtained are below.

y-value (single measurement):	1.00	0.30	0.20	0.10
Predicted x (from the line) :	29.20	4.61	1.10	(-2.41!!)
Standard error of prediction :	0.80	0.60	0.71	0.84

The mean point of the calibration data comes at $y = 0.55$. the calculated standard error increases steadily the further the y-value strays from this mean. However, we chose these examples to remind you once again that, no matter how sophisticated our programming is, its results are of little value if the treatment being applied is not valid. All the y-values chosen above represent extrapolations beyond the four calibration points. We have seen in Section 5.3 that this is not valid for $x \rightarrow 0$. From Fig. 5.2d you can see that the regression is reasonably applicable at $y = 0.3$, but for $y = 0.2$ the true x-value is just a little below 4.0, far larger than the regression, for all its predicted uncertainty, would allow. It is the *assumption of linearity* which is at fault, not the mathematics of the regression calculation itself – the results are fine where the assumption of linearity is reasonable. The last result, for $y = 0.1$, indicates just how daft an unjustified extrapolation can be! We have not looked at results for extrapolation at the high absorbance side, but the general curvature of the data points in Fig. 5.2d should make us very suspicious also of the regression's prediction for $y = 1.00$.

This would be a good point at which to SAVE your program. We shall refer to it as program "LINE".

Let us apply our program to a new example.

∏ Specially coated quartz piezoelectric crystals can be used as detectors to monitor the SO_2 level in gaseous emissions. Adsorption of SO_2 causes a decrease in the crystal's frequency of vibration. Following a standard measurement procedure the frequency shift, f, was obtained for air samples of known SO_2 concentrations as follows.

SO_2/ppm	f/Hz
5	45
10	55
15	67
20	75
25	90
30	101
35	113
40	122
45	143
50	150
55	159
60	177
65	192
70	195
75	215

There are 15 calibration points here and, on inspection, the variation appears at least roughly linear. Let us see what a linear regression fit to the data looks like.

The DATA lines of the program need to be changed thus.

```
1000 DATA 15
1010 DATA 5,45,10,55,15,67,20,75,25,90,30,101,35,113
1020 DATA 40,122,45,143,50,150,55,159,60,177,65,192
1030 DATA 70,195,75,215
```

The SO_2 concentration is our x-value and the frequency response is our y-value. You could type all the x, y data onto line 1010, or you can split it between two or more lines, as we have done for purposes of presentation.

If you RUN the program you should get the results below.

x	y	Residual
5	45	3.35
10	55	1.2143
15	67	1.0786
20	75	-3.0571
25	90	-0.1929
30	101	-1.3286
35	113	-1.4643
40	122	-4.6
45	143	-4.2643
50	150	-0.8714
55	159	-4.0071
60	177	1.8571
65	192	4.7214
70	195	-4.4143
75	215	3.45

Gradient	= 2.42714286
Intercept	= 29.5142857
St err of estimate	= 3.28683688
St dev of gradient	= 3.92852147E-2
St dev of intercept	= 1.78592966

The fit appears to be good. The residuals do show some degree of scatter but this appears to be fairly random. The standard error of the estimate, at 3.29, is quite small relative to the typical change in y

from point to point. We conclude that it seems perfectly reasonable to accept the line as a good representation of the data.

Investigate the standard error of the regression, as a function of x. The mean-point of the data is at $\bar{x} = 40$. Show that the standard error of the regression is under 0.9 there, but that it grows to around 1.6 at the edges of the calibration data.

SAQ 5.4a	

(i) Use the piezoelectric crystal data for SO_2 analysis to construct a linear regression graph, including 95% confidence lines (ie construct the analogue of Fig. 5.2d for this system). The data given were as follows.

SO_2/ppm	f/Hz
5	45
10	55
15	67
20	75
25	90
30	101
35	113
40	122
45	143
50	150
55	159
60	177
65	192
70	195
75	215

(ii) An emission is required to conform to a standard which imposes a upper limit of 50 ppm of SO_2. Four measurements are made, giving readings of 154, 159, 162 and 155 Hz. Is the limit being breached? Test at a 5% level of significance.

SAQ 5.4a

SAQ 5.4b   *This SAQ should be regarded as optional!*

Incorporate the regression line calculation as a facility within program "STAT". Note that all variables in program "LINE" have been named so as to avoid clashes with others in "STAT". Also note that some gaps in line numbers have been left in "STAT", which might prove convenient for this task.

SAQ 5.4b

The SO_2 example is much more appropriate for linear regression than our case study on the spectroscopic determination of chloride. Notice that the first thing we urged you to do after determining the regression line was to look carefully at the residuals. They give an important indication of how reasonable a linear fit is. The residuals should not only be reasonably small compared to the point-to-point variation in y, they should vary in a random fashion. Any clear trend in the values of the residuals should act as a warning signal that the reason that the points do not lie on the line is not simply due to random errors of measurement. A linear relationship may not be valid.

We say 'look carefully at the residuals,' but nothing in the above paragraph is quantitative. Sometimes, as with the SO_2 example, it is clear that a linear regression is reasonable. Sometimes there is clear evidence of curvature. Certainly we must look for these things. Sometimes we may be left somewhat in doubt, the evidence may not stare us in the face too starkly. Are there quantitative tests we can apply? Yes there are; we shall introduce one to you after the next section.

5.5. FITTING NON-LINEAR CALIBRATION CURVES

Now we have a fuller set of points for our calibration problem for the determination of chloride in water, it is apparent that the data conform distinctly to a curve. The straight line fit is quite inappropriate to use outside the range 8-20 mg dm^{-3}.

It is in principle reasonably straightforward to derive the least-squares 'best' fit to a given set of data for any desired algebraic (or similar) expression. It is beyond our scope here to go into such procedures in any depth, but it will be helpful to discuss briefly some improved fits to our particular data. Fig. 5.5a shows our calibration points with a series of different 'best-fit' lines. The solid line (*a*) is our original linear regression fit to the data in the 8–20 mg dm^{-3} range.

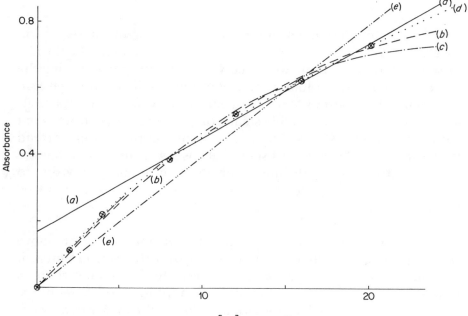

Fig. 5.5a.   *Different fits to the chloride analysis data*

(*a*) ——— linear regression fit in range 8-20 mg dm^{-3}
(*b*) – – – – $y = ax + bx^2$: least-squares fit to all data
(*c*) – · — · – $y/x = a + bx$: linear regression of 'y/x on x'
(*d*) ·········· $x = ay + by^2$: least-squares fit to all data
(*e*) – · · — · · – $y = mx$: linear regression through origin

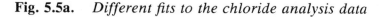

The *true* calibration curve must go through the origin. The simplest expression which would ensure this but would allow for some curvature in the line is of the form below.

$$y = ax + bx^2$$

To find expressions for the optimum values of parameters a and b requires a similar derivation to that for the linear relationship (given as an appendix in Section 5.6). To cut a long story short the result is

$$y = 0.0556\,x - 0.000971\,x^2 \quad [\text{line } (b)]$$

This equation gives (b) in Fig. 5.5a. You can see that it is a moderately good approximation over the whole range 0–20 mg dm^{-3}.

The equation $y = ax + bx^2$, like $y = mx + c$, involves *two* parameters. Where it is believed that such a two-parameter expression might give a good fit to the calibration data, there is a useful trick which can often be played which allows us to make use of the standard procedures which we have already set out for fitting straight lines. We *linearise* the equation. By 'linearise', we mean that the equation should be algebraicly manipulated into the form below,

$$\text{variable} = \text{constant} \times \text{variable} + \text{constant}$$

analogous to the straight-line expression $y = mx + c$. For the expression introduced above, *viz* $y = ax + bx^2$ this task is easily performed. We simply divide throughout by x to give the following equation.

$$y/x = a + bx.$$

We can therefore use our standard results from Fig. 5.2c to produce the best straight-line fit to a graph of y/x *versus* x (its intercept is a and its gradient b). This plot is shown in Fig. 5.5b. The equation is:

$$y/x = 0.0588 - 0.001182\,x$$

Notice that having found the equation we can easily multiply both sides by x to give:

$$y = 0.0588\,x - 0.001182\,x^2 \quad \text{[line (c)]}$$

This is plotted as line (c) in Fig. 5.5a. The linearising procedure has led to values for a and b slightly different from those for line (b).

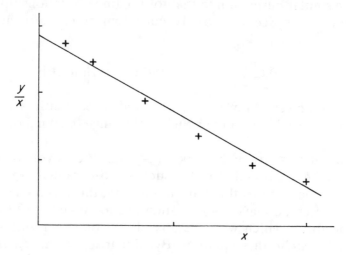

Fig. 5.5b.   *Linear regression of y/x on x for the chloride data*

SAQ 5.5a	Look carefully at the different lines (a), (b) and (c) in Fig. 5.5a. Note some brief critical comments that occur to you, then compare your observations with those we give in our response.

SAQ 5.5a

It is instructive that the two lines (b) and (c) are significantly different from one another, and yet are rival attempts to fit the data to the same general expression $y = ax + bx^2$. This suggests that such an expression is still not very effective in explaining the variation of absorbance, y, with concentration, x. Extrapolated to large x-values curve (b) and curve (c) would both reach maxima, beyond which the predicted absorbance would fall on further increase in chloride concentration. Experimental results would certainly not be expected to behave in this way.

Better fits could be derived if we were prepared to introduce increasing numbers of parameters. For instance the equation $y = ax + bx^2 + cx^3$ would be capable of giving a good fit over the range

0–20 mg dm^{-3}. It is possible to obtain a better fit, however, without going to such lengths. Thus the expression

$$x = ay + by^2 \quad [\text{line (d)}]$$

can be used to give curve (d) in Fig. 5.5a. This expression will give a curve which continues to rise indefinitely as x increases. It fits the experimental points extremely well.

SAQ 5.5b

The expanded set of calibration data for the chloride determination case study is given below.

x mg dm^{-3}	y
2.0	0.113
4.0	0.218
8.0	0.388
12.0	0.523
16.0	0.624
20.0	0.734

The best fit to these data in Fig. 5.5a is curve (d), which has an equation of the form:

$$x = ay + by^2$$

Obtain values for parameters a and b.

(Hint: put the equation into linear form: $x/y = a + by$. Then use program "LINE".)

SAQ 5.5b

Let us take stock. The main purpose of this Section has been simply to point out that the least-squares method can be extended to fit a variety of expressions for given data. We must be careful in choosing appropriate expressions. One can test a given fit statistically to see whether the experimental scatter about it is consistent with a 'true' relationship of the chosen form. In the next section we shall see how to do this for the straight-line case. But vigilance and critical scrutiny of results often gives sufficient warning. Curve (b) and curve (c) in Fig. 5.5a do not fit the trend of the later points at all well. The difficulty with expression $y = ax + bx^2$ is that it predicts a *constant curvature* (the second derivative, $y'' = 2b$). Our calibration-data plot shows decreasing curvature at higher x-values.

Curve (d) conforms with this trend in the measurements.

There is a general lesson here. We are in a much stronger position if we feel confident that we *know* what theoretical form the calibration curve should have. In such circumstance we should try to fit an expression of the theoretically expected form.

Now spectroscopic absorption is subject to Beer's law. If the absorbance, y, at the wavelength of measurement, is due entirely to absorption by a single chemical species, present at concentration x, our graph should be straight line through the origin, of form:

$$y = mx \quad [\text{line } (e)]$$

The best fit of this type to our chloride data is shown as curve (e) in Fig. 5.5a. It represents the data very poorly indeed. There are substantial apparent deviations from the simple Beer-law form in this example because of equilibrium effects on the concentration of the absorbing species in solution.

Often, however, a straight-line graph passing through the origin is indeed what is required. Fig. 5.5c collects together information on how then to obtain and use the y-on-x regression line. The table can be used in exactly the same way as Fig. 5.2c.

Quantity	Value	Standard deviation [n-1 degrees of freedom]
Residual	$d_i = y_i - y_{est}$	$s_y = \sqrt{\sum_{i=1}^{n} d_i^2 / (n - 1)}$
Gradient	$m = S_{xy}/S_{xx}$	$s_y/\sqrt{S_{xx}}$
Regression	$y = mx_0$	$\dfrac{s_y x_0}{\sqrt{S_{xx}}}$
Predict x (N repeats)	$x = y_0/m$	$\dfrac{s_y}{m}\sqrt{\dfrac{1}{N} + \dfrac{y_0^2}{m^2 S_{xx}}}$

Fig. 5.5c.   *Linear regression through the origin*

Many applications of analytical chemistry are based on simple laws from physical chemistry. Where it is known that a certain simple relationship holds one should make use of it! The expression for the calibration curve should be of the expected mathematical form. We should always be mindful, however, that the situation may not be simple. For whatever reason – instrumental factors, chemical equilibrium effects, interference – the simple theoretical relationship may not pertain. If this happens it is important to ensure that a calibration curve truly reflects the observed behaviour.

5.5.1. An F-test for Non-linearity

In discussing calibration curves we have concentrated rather heavily on straight-line regressions. We have, however, also stressed that in practice curved calibrations often arise. Indeed this turned out to be so for our case study of chloride analysis. When non-linearity arises it is important to recognise the fact.

How can we test for non-linearity? Let us choose a clear-cut example to illustrate the matter. We shall take up once more our chloride example, but this time we shall include all seven calibration points shown in Fig. 5.5a. The set of seven shows clear non-linearity even to the most casual observer but we shall choose to ignore that for the moment and blindly proceed to obtain a straight line fit for the whole set. this gives a rather different result from our original line (*a*), because that was based on only four of the points. Program "LINE" gives the results below.

x	y	Residual
0	0.0	−0.051
2	0.113	−0.010
4	0.218	+0.022
8	0.388	+0.048
12	0.523	+0.038
16	0.624	−0.006
20	0.734	−0.041

Gradient	= 0.0362
Intercept	= 0.0510
St err of estimate	= 0.0414

Just look at the trend in the residuals! They start at a negative value, increase, pass through a positive maximum, then finally decrease again. There is a smooth trend with no obvious evidence of randomness at all. This is telling evidence for non-linearity: the principal reason for variation of the residuals is clearly the curvature of the true relationship between y and x, it is not simply a result of random errors of measurement. It is easy to spot this by eye for such a blatant case. But often the evidence is less obvious; we may spot some signs of a trend but much scatter also, and we may wonder whether the apparent pattern could simply be a fluke. Can we carry out a quantitative statistical test? Yes, quite easily in fact!

The clue lies in the point made above that some factor other than random error is contributing significantly to the variation of the residuals. The standard error of the estimate, s_y, measures the standard deviation of the residuals. If variation of the residuals is solely due to random error of measurement, then the standard deviation of the residuals should be same thing as the standard deviation for a simple measurement of y. If from experience of repeat measurements (for a single concentration) we know that the standard deviation for a simple measurement of y is σ, then, if random error is the only contributing factor to variation of the residuals, s_y should simply be an estimate of σ. In other words σ, the population standard deviation applicable to single measurements of y, should be the population standard deviation for the residuals also. Thus we have the basis for our test: *if a linear relationship holds the standard deviation s_y represents an estimate (with $(n - 2)$ degrees of freedom) of the population value σ.* We shall, therefore, have evidence of non-linearity if we can show that s_y has too large a value to be compatible with a population value, σ. How do we test whether standard deviations are compatible with one another? That is a job for an F-test!

To carry out a test we need to know the standard deviation for a single measurement of y. Let us assume that for the chloride determination this value is given by $\sigma = 0.005$. (Much depends on the instrumentation being used, but certainly the data with which we have been working are compatible with this value.) From the above linear-regression calculation $s_y = 0.0414$. Hence the calculated F-value is:

$$F = s_y^2/\sigma^2 = (0.0414/0.005)^2 = 68.6$$

Now there are 5 degrees of freedom associated with s_y and an infinite number with the population value σ. Let us test at a 1% significance level. Thus we need $F_{0.01(5,\infty)}$, whose value turns out to be 3.02. (Check it!) The evidence is overwhelming (as we knew it would be). The calculated F-value is enormously larger than the 1% critical value.

As we indicated, we have chosen a very clear cut case. The F-test really proves its value when the conclusion is not immediately self-evident; it gives a quantitative measure of the strength of evidence for non-linearity. Note that the test depends on having a value for the standard deviation for a simple measurement of y. It is not essential to have a population value. Repeat determinations at a single concentration can provide a sample estimate, s, in place of the population value σ. We then take $F = s_y^2/s^2$. The number of degrees of freedom, ν_2, associated with the denominator is no longer infinite, it is the number associated with the estimation of s.

SAQ 5.5c

Our original linear regression [line (*a*) in Fig. 5.5a], appeared to be a reasonable fit to the four calibration points for the analysis of chloride in the range 8–20 mg dm^{-3}. The standard error of the estimate, s_y for the line was 0.0111 (SAQ 5.2b). If we take the value 0.005 as the accepted population standard deviation for a simple measurement of y, is there evidence, even over this concentration range, that the calibration curve is non-linear?

SAQ 5.5c

5.6.   SUMMARY

A brief survey of what we have been about is now appropriate. In this Part of the Unit we have learned how to fit straight lines to calibration data, and how to make quantitative estimates by use of a regression line, and to come to decisions based upon such lines. We have also learned to test for non-linearity. Perhaps even more important than the specific techniques we have introduced, however, are the general arguments we have advanced about the need for care in interpretation. A healthy scepticism and continual vigilance are required. Not all calibrations are linear. The y-on-x regression line is only one of various possible least-squares fits for a given set of data, one that assumes that all deviations are due to random error in the measurement of y. A calibration curve is at its most reliable near the mid-point of the calibration data; extrapolation is a procedure fraught with danger.

The business of fitting expressions to data is a highly developed field of study. Although we have concentrated on a small area of it we have tried to leave you with some flavour of the subject as a whole. Fitting expressions to data is not something to be done in a cavalier fashion. There is always a danger of falling into the trap of not looking carefully to see whether the fit is reasonable. We are busy people, and we may have a lot of data to handle. The temptation is simply to let rip the mathematics of the chosen treatment and to assume that all is well. Yield not to temptation!

Objectives

As a result of completing Part 5 you should feel able to:

- explain the general role of calibration curves in analysis;

- estimate a best straight line fit 'by eye', and determine its equation;

- obtain and draw a least squares regression line of y on x to fit given data (using Eqs. 5.1 and 5.2);

- distinguish between y-on-x and x-on-y regression lines;

- recognise that a linear regression fit is most precisely determined at the mean point of the calibration data;

- explain the significance of, and calculate the value of, the standard error of the estimate of y (Eq. 5.5);

- appreciate that if an expression with p parameters is fitted to n items of data then $(n\text{-}p)$ degrees of freedom remain to assess the dispersion of the data;

- compute, for a straight line regression, the estimated standard deviation in its gradient, in its intercept, in the regression (at any x), and in a prediction of x from a measurement of y (refer to Fig. 5.2c);

- for the quantities listed above, obtain confidence intervals and apply significance tests, through use of the statistic t;

- carry out straight line regression analyses by computer;

- recognise the manifold dangers of extrapolation, and in particular be aware that a mathematical expression fitting calibration data in one region may not necessarily be valid elsewhere;

- understand the principles of least squares fitting to a general mathematical expression;

- use the results in Fig. 5.5c to obtain a best y on x regression line constrained to pass through the origin, and analyse the statistics of this case;

- transform appropriate equations involving two parameters into linear form;

- apply an F-test to the residuals in a linear regression fit, and hence comment on evidence of non-linearity.

5.6.1. Appendix: A Least-squares Derivation

The goals of this Unit are very much more directed towards helping you to *use* statistical results than for you to become acquainted with the underlying theory. We make an exception to our normal practice here to show you how, by a straightforward application of differential calculus, the formulae can be obtained which give the parameters m and c for the y-on-x regression line. We ask you to accept so many formulae on trust that we felt it might be helpful to derive at least these. This Appendix can be skipped if you choose.

We use the notation from Section 5.1. We suppose that we have a series of n points (x_i, y_i). The object is to find the straight line, $y = mx + c$, which minimises the sum of the squares of the residuals, d_i illustrated in Fig. 5.1b.

The residual for the ith point is defined by the equation

$$d_i = y_i - (mx_i + c).$$

Let S be the sum of the squares of all the residuals.

$$S = \sum_{i=1}^{n} d_i^2 = \sum_{i=1}^{n} (y_i - mx_i - c)^2$$

We seek a minimum in S with respect to the choice of the parameters m and c. At a minimum the partial derivatives of S with respect to both m and c must be zero.

$$\frac{\partial S}{\partial m} = \sum_{i=1}^{n} \{-2x_i(y_i - mx_i - c)\} = 0$$

$$\frac{\partial S}{\partial c} = \sum_{i=1}^{n} \{-2(y_i - mx_i - c)\} = 0$$

Dividing each equation by -2 and separating the different sums gives

$$\sum_{i=1}^{n} x_i y_i - \sum_{i=1}^{n} mx_i^2 - \sum_{i=1}^{n} cx_i = 0$$

and

$$\sum_{i=1}^{n} y_i - \sum_{i=1}^{n} mx_i - \sum_{i=1}^{n} c = 0$$

By using the notation from Section 5.1 for the sums we obtain the equations below.

$$S_{xy} - mS_{xx} - cS_x = 0$$

and

$$S_y - mS_x - cn = 0$$

These are simultaneous linear equations determining m and c. If we subtract S_x times the second equation from n times the first we eliminate c, giving:

$$(nS_{xx} - S_x^2)m = nS_{xy} - S_x S_y$$

Hence $\quad\quad\quad m = (nS_{xy} - S_xS_y)/(nS_{xx} - S_x^2)$

Substituting this equation for m into either of the simultaneous equations gives, after rearrangement:

$$c = (S_yS_{xx} - S_xS_{xy})/(nS_{xx} - S_x^2)$$

These are Eqs. 5.1 and 5.2 of Section 5.1

6. Correlation

The subject of statistical correlation is important, though it has many pitfalls for the unwary. It is concerned with answering the question as to whether there is firm evidence for the existence of a relationship between two measured quantities.

Perhaps the best known example where overwhelmingly strong evidence of statistical correlation was adduced is in the classic studies of the relationship between cigarette smoking and the incidence of lung cancer. This is a good example to highlight the circumstances in which correlation studies are valuable. Not everyone who smokes heavily will contract lung cancer, and some who never smoke will nevertheless develop the disease. What has been shown beyond all reasonable doubt is that smoking markedly increases the chance that any individual will contract the disease. The chance increases the more heavily he or she smokes, and will diminish if the practice is abandoned. We are here dealing with a quite different type of problem from that of Part 5 where we were concerned with calibration lines. The incidence of lung cancer depends on many different factors; there is no 'true' relationship between the number of cigarettes smoked and the date of incidence of the disease. Smoking is simply one factor, an important one, which affects the chance of incidence. In our method for the determination of chloride at low concentration levels in water, we are convinced that there *is* a true and unique relationship between absorbance and concentration, observed deviations from which occur simply because of unavoidable errors in making the experimental measurements. There correlation studies are not particularly relevant. Yet we shall find a number of parallels between our work on correlation and our earlier study of calibration.

Why exactly do we feel the need to discuss the issue of correlation in a course unit for analytical chemists? There are indeed situations where an analytical measurement will depend on variable factors in addition to the concentration of the particular analyte being determined. This phenomenon is known as 'interference'. Thus the presence of bromide or iodide in a water sample will interfere with the procedure we have discussed for the determination of chloride. Where an analytical measurement is subject to interference, however, the normal response of a good analyst will be to seek a method of eliminating it, by a variation in the procedure, or by use of a separate measurement to determine the concentration of the interfering species. In making an analytical determination, it will always in the end be the aim of the analyst to use a process in which he or she is confident that there exists a unique true relationship between measurement and analyte concentration.

To a practising analyst some knowledge of the principles of correlation analysis is of value in understanding, and in being able to defend or criticise, the wider conclusions which might be reached on the basis of analytical studies. Some typical examples may illustrate the kinds of question addressed.

(*a*) Does the concentration level of a particular minor impurity affect the mechanical strength of a given manufactured alloy?

(*b*) Can measurement of blood cholesterol concentration be used as an early warning of susceptibility to heart attack?

(*c*) Does airborne lead from combustion of petrol affect the IQ of children who live near motorways?

(*d*) To what extent are the sulphur emissions of power stations in the UK responsible for 'acid rain' in Scandinavia?

These are all complex questions, increasingly so as we go down the list. The effects observed (mechanical strength, susceptibility to heart attack, IQ of children, acid rain) always depend on many factors. The analyst has produced detailed and usually rather accurate data on one suspected 'cause'. The object of the exercise, each time, will be to determine whether the evidence supports the existence of a causal correlation and, if so, whether the effect is significant enough to justify action.

Studies of this general kind are becoming ever more common, and analysts are increasingly involved in determining the base data for them, particularly in the medical and environmental fields. Many of the issues involved become the subject of sometimes heated public debate. The analyst should have some understanding of the way in which his or her results are being used. Drawing firm conclusions is difficult, and these difficulties should be borne very much in mind in designing the study from the outset, ie in deciding what exactly the analyst should measure, how the samples to be analysed should be chosen, and what level of accuracy is appropriate. As well as providing many pitfalls for the unwary, this area also provides many opportunities for misrepresentation by the unscrupulous. The analyst should be well informed enough to be able to criticise the blunders and the logical excesses of others.

SAQ 6.0a	We have not yet introduced any method of measuring correlation. Nonetheless you are invited to stick your neck out! Which of the following would be expected to show the stronger evidence for correlation, and for which would a correlation study be more profitable?

(*i*) Data for absorbance *versus* concentration in the standard spectroscopic method for determination of chloride.

(*ii*) Data relating the extent of tooth decay to the concentration of fluoride in drinking water.

Explain your conclusions.

SAQ 6.0a

6.1.   THE SCATTER DIAGRAM – LINEAR REGRESSION REVISITED

Before we attempt to analyse our problem quantitatively, let us look at some typical data. Fig. 6.1a is taken from a report of the British Regional Heart Study. It shows results obtained for the 'Standardised Mortality Ratio' (SMR) for cardiovascular disease, plotted as a function of total hardness in the water supply for 234 towns in the UK, for the years 1969–1973. For our purposes we can gloss over the precise definition of 'Standardised Mortality Ratio'. Suffice it to say that a value of 100 would be the national average figure, and that the derivation of values for different towns is adjusted for any abnormalities in the sex and age distribution of its population. Total water hardness is expressed in terms of calcium carbonate equivalent.

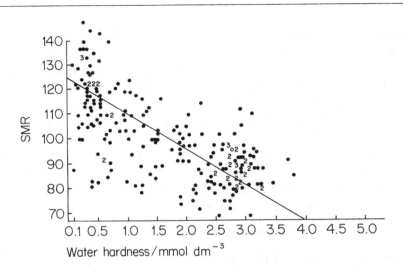

Fig. 6.1a. *The SMR for cardiovascular disease, for all men and women aged 35–74, plotted against water hardness for each of 234 towns in the UK. Water hardness is represented in terms of calcium carbonate equivalent (1 mmol dm^{-3} = 100 mg dm^{-3} CaCO$_3$). The regression line of SMR on water hardness has been added.*
(*British Medical Journal*, 24 May 1980, p 1246 Fig. 3)

Now clearly there is a great deal of scatter in these data. Nevertheless there is an apparent tendency for the mortality ratio to fall as hardness increases, and the authors were able to infer with confidence that this was indeed a significant effect.

The scatter in a diagram such as this is not due appreciably to errors in measurement of either the SMR or water hardness, nor is it due to purely random statistical fluctuations. The authors of the study were able to demonstrate that it was largely explicable in terms of other factors which are also associated with heart disease.

A way to simplify Fig. 6.1a might be to derive values for the *mean* cardiovascular mortality ratio for all towns at each given water hardness level, and then to plot this statistic against hardness. This procedure would in fact give a *regression curve* of SMR on water hardness. Actually to do this would require grouping the towns into bands defined by chosen ranges in water hardness. There would clearly be tactical considerations in deciding how broad to make each water-hardness band. Enough towns would have to be included in each to give a good chance that the main effect of other varying factors affecting SMR would be averaged out within each band. On the other hand, one would want to preserve enough separate bands to hope to be able to see trends in mortality across the whole range. This procedure was actually carried out in the Heart Study, and we shall return to the results obtained later. A simpler method, however, is to derive the *straight-line regression* of SMR on water hardness. This is a process now familiar to us from Part 5. The line has been calculated approximately from the published scatter diagram, and has been superimposed on Fig. 6.1a.

SAQ 6.1a

Study Fig. 6.1a.

(*i*)   From the regression line estimate, on average, the decrease in SMR associated with a 1 mmol dm^{-3} increase in total water hardness.

(*ii*) Comment critically on the validity of your conclusion.

With problems of the type we are now addressing there is no true mathematical equation relating y to x. There is therefore typically a much wider scatter of points about the regression line, due not to errors in the data but to factors other than x which affect the value of y. A natural question to frame in this context is: 'To what extent is the variation in y explained by the variation in x?' It is precisely this question which will allow us to develop a more quantitative treatment. The next Section will be devoted to achieving this at the simplest level of sophistication.

6.2. EXPLAINED AND UNEXPLAINED VARIATION AND THE COEFFICIENT OF CORRELATION

The y-on-x regression line is a 'best fit' in the sense that it minimises the sum of the squares of the y-residuals, d_i, of the data points (see Fig. 5.1b). The line is such that the sum

$$\sum_{i=1}^{n} d_i^2$$

is as small as it is possible to achieve. In an extreme case of perfect correlation, where the data points all fall on a single perfect straight line every residual would equal zero, and so would the sum of their squares. We would then happily claim that the line *completely explains* the variation of y; from any given x-value we would be confident we could predict y precisely.

In practice the sum of the squares of the residuals is not zero. Its magnitude is a measure of the extent to which the regression line fails to fit the data. In recognition of this fact the sum of the squares of the residuals is often referred to as *the unexplained variation of* y.

$$\text{Unexplained variation} = \sum_{i=1}^{n} d_i^2 \tag{6.1}$$

A measure of the overall variation of y is obtained if we adopt a definition of *the total variation of y* as below.

$$\text{Total variation} = \sum_{i=1}^{n}(y_i - \bar{y})^2 \tag{6.2}$$

This is the sum of squares of the deviations from the mean y-value $\bar{y}$, and is equal to $(n-1)$ times the variance of the y-data. It is now natural to define *the explained variation of y* as follows.

$$\frac{\text{Explained}}{\text{variation}} = \frac{\text{Total}}{\text{variation}} - \frac{\text{Unexplained}}{\text{variation}}$$

$$= \sum_{i=1}^{n}(y_i - \bar{y})^2 - \sum_{i=1}^{n} d_i \tag{6.3}$$

There is an alternative expression for the explained variation which can be shown to be equivalent to the above. The explained variation is also given by the sum of the squares of the differences between the estimated y-values obtained by using the regression line, and the mean value of $\bar{y}$.

$$\text{Explained variation} = \sum_{i=1}^{n}(mx_i + c - \bar{y})^2 \tag{6.4}$$

Now the *fraction* of the total variation which is 'explained' by the regression line should give a worthwhile measure of the goodness-of-fit of our line to the data. This quantity is known as the 'coefficient of determination' of the data.

We shall, however, concentrate our attention on the single most widely used statistic which measures the extent to which the variation of the data is explained. This is known as the *coefficient of linear correlation r*, often referred to more concisely as simply the *correlation coefficient*. This is defined by

$$r = \pm\sqrt{(\text{explained variation})/(\text{total variation})} \tag{6.5}$$

The square root is taken for similar reasons to those which cause the standard deviation to be a more commonly used statistic than the variance. The $+$ sign is adopted if m is positive; here the regression line predicts that y increases as x increases, we have *positive correlation*. The -ve sign is taken if m is negative and this case corresponds to *negative correlation*.

There are various ways of calculating the value of the correlation coefficient as defined through the above equations. We shall make use of a single formula given below.

$$r = (nS_{xy} - S_x S_y)/\sqrt{(nS_{xx} - S_x^2)(nS_{yy} - S_y^2)} \qquad (6.6)$$

This equation can be derived from the above definitions though we shall not give the proof. The symbols have the same significance as in Section 5.1.

The above formula automatically gives the r-value with its appropriate sign. Note the symmetry of the expression. It turns out that we obtain exactly the same result for the correlation coefficient if we frame our discussion in terms of the regression line of x on y and in terms of definitions of explained and unexplained variations in x rather than y. Notice also that *r is dimensionless*; its value will be the same whatever units are chosen to express the original x and y values.

We now consider a second case study, which is a little less involved than the heart disease example. Let us look at an early study which provided evidence of the link between fluoride in drinking water and dental decay amongst children. The data below, published by H T Dean in 1943, give the naturally occurring fluoride levels for 21 communities in the USA, along with information of the average number of permanent teeth per child in each location which were affected by dental caries. We denote the fluoride concentration (in mg dm^{-3}) by x, and the average number of teeth affected by y. The data are presented in order of increasing x.

x	y	x	y
0.0	8.2	0.45	4.4
0.0	7.3	0.55	4.1
0.0	6.7	0.9	3.4
0.1	10.4	1.2	3.0
0.1	8.3	1.2	2.7
0.1	7.7	1.2	2.4
0.1	7.0	1.3	3.1
0.2	7.3	1.8	2.4
0.2	7.0	1.9	2.2
0.3	6.4	2.6	2.2
0.35	5.5		

SAQ 6.2a

Evaluate the correlation coefficient between tooth decay and the fluoride concentration in drinking water, by using the data quoted from Dean's study. Use Eq. 6.6.

(i) It takes a little labour to work out the different sums. Check at least one of the following results:

$$S_x = 14.55,$$
$$S_{xx} = 21.27,$$
$$S_y = 111.7,$$
$$S_{yy} = 719.1,$$
$$S_{xy} = 45.31.$$

(ii) Use the values to evaluate r. What conclusion do you draw from your result?

SAQ 6.2a

Dean's observations were the starting point for over 40 years of controversy. There is a strong negative correlation between the fluoride level in drinking water and tooth decay. Thus it is argued that the addition of fluoride to water supplies in areas with a low natural concentration will lead to healthier teeth amongst the local population. Those opposed to such action do not deny this assertion, rather they argue that addition of fluoride could lead to other, and possibly harmful, health effects.

6.3. TESTING THE SIGNIFICANCE OF EVIDENCE OF CORRELATION

The value of r from Dean's data is -0.86. How is it generally accepted that this evidence is *significant* and not just a chance result because the towns surveyed happened to be atypical?

At a simple level the figure of -0.86 is fairly impressive if we remind ourselves that the largest possible magnitude for $|r|$ is 1.0. If $r = +1$ or $r = -1$ we have 100% explained variation, with the points lying precisely on a straight line, with a positive slope if $r = +1$ and a negative slope if $r = -1$. But *how* good a correlation is r

$= -0.86$? Various quantitative tests of significance are possible, but we shall describe just one. Could a given r-value have arisen purely by chance had there in fact been *no* correlation between the properties investigated? We shall apply a one-tail hypothesis test. The coefficient r is an estimate of a 'true' correlation coefficient, r_{true}: it is based on a sample, ie on the evidence which has been collected. It is merely a sample estimate of the 'true' population value. We shall ask whether the value found for r is compatible with the true correlation coefficient being zero.

Our null hypothesis is that there is no correlation, H_0: $r_{true} = 0$. If the calculated r-value is positive, our alternative hypothesis is that there really is at least some positive correlation. H_0: $r_{true} > 0$. If, on the other hand, our calculated r is negative the alternative hypothesis is that there is at least some negative correlation, H_0: $r_{true} < 0$. We are interested in discovering at what level of significance we can reject the null hypothesis in favour of the appropriate alternative hypothesis. At what level of significance can we claim to have established the existence of correlation of the same sign as r?

What statistical parameter can we use for this test? It can be shown that once again we can turn to t. We can apply a conventional t-test, with the critical t-value computed by using the equation.

$$t = r\sqrt{(n - 2)/(1 - r^2)} \tag{6.7}$$

As two degrees of freedom from the n items of data are used in determining the regression line about which the variation is being tested, the t-value has $n - 2$ *degrees of freedom* associated with it. For a test applied at a level of significance of α, the critical t-value is $t_{\alpha,(n-2)}$. If the magnitude of the calculated t-value (from Eq. 6.7) exceeds the critical t-value, then we reject the null hypothesis and assert that there *is* a true correlation between the two properties investigated, a correlation of the same sign as the calculated r-value.

SAQ 6.3a	From the results obtained in SAQ 6.2a, test whether there is truly a negative correlation between fluoride level and tooth decay.

SAQ 6.3a

As always happens when we come to significance testing the argument involved seems at first rather complicated. This is quite a simple process to apply, however, provided we have ready access to the critical t-values. Let us restate the process.

(*a*) We first calculate the correlation coefficient for our data by using Eq. 6.6.

(*b*) Next we compute from r the corresponding t-value for our data by using Eq. 6.7.

(*c*) If this calculated t-value is of a larger magnitude than $t_{\alpha,(n-2)}$, then our evidence can be said, at a significance level of α, to confirm the existence of a correlation.

The result of SAQ 6.3a shows that we can assert with overwhelming confidence that there is indeed a negative correlation between fluoride and tooth decay. A slightly more elaborate test, which we shall not explain here, leads to the conclusion not only that Dean's data confirms that there is *some* degree of negative correlation but that we can assert with over 99% confidence that the true correlation coefficient is more negative than -0.60.

So what? Does the conclusion $r < -0.6$ tell us much more than the conclusion $r < 0$? Look back at Eq. 6.5: r^2 is equal to the ratio of the explained variation to the total variation. The larger the magnitude of r, therefore, the greater the proportion of the variation in y which appears to be 'explained' by the variation in x. Accepting that $r < -0.6$, rather than simply $r < 0$, indicates that the correlation is quite *strong*. We are encouraged to suspect that fluoride levels may have an important influence on tooth decay.

In the fluoride example much further evidence, subsequent to Dean's work, lends support to this conclusion. In general, however, it is important to notice the guarded phrases '*appears* to be explained' and '*may* have an important influence'. Proving the existence of a correlation, even a strong one, is not the same thing as proving that changing one factor *causes* a change in the other. We shall return forcefully to this point in Section 6.6. In the meantime, however, we simply comment that a large value of $|r|$ is consistent at least with the possibility that there is a strong influence. A very small value of $|r|$ inevitably means that any influence must be weak.

SAQ 6.3b

> The data in Fig. 6.1a relating heart disease mortality to water hardness give a correlation coefficient of -0.67. The data are much more widely scattered than for our fluoride example. Show, however, that the evidence for the existence of true negative correlation is very highly significant indeed. (The data cover 243 towns.)

6.4. PROGRAMMING THE CORRELATION CALCULATION

A very minor addition, of five new programming lines, will allow your regression program to automate the calculation of correlation coefficients and their corresponding *t*-values.

Load program "LINE" into your computer and then type in the following lines.

```
125     C6=C6+Y(I)↑2
320 R5=(N*C4-C1*C3)/SQR((N*C2-C1↑2)*(N*C6-C3↑2))
330 PRINT"Correlation coefft    = ";R5
340 T=R5*SQR((N-2)/(1-R5↑2))
350 PRINT"t for correlation     = ";T
```

Line 320 calculates the correlation coefficient (program symbol R5) by using Eq. 6.6:

$$r = (nS_{xy} - S_xS_y)/\sqrt{(nS_{xx} - S_x^2)(nS_{yy} - S_y^2)}$$

The various sums S_x etc, are worked out at the beginning of the program as C1 etc. LIST lines 70 to 130 and you will be able to check that line 320 is a correct translation of the equation for *r*. We have had to add line 125 to the initial loop because the program did not previously calculate S_{yy}. After *r* is calculated, line 340 obtains *t* by applying Eq. 6.7.

$$t = r\sqrt{(n - 2)/(1 - r^2)}$$

Let us check the program by applying it to Dean's study of fluoride *versus* tooth decay. His results, for 21 communities are listed in Section 6.2. All that is needed is to put them into the program's DATA lines. Do so! (Remember that data start on line 1000 with the value of *n*. The (x,y) values follow; you may find that you need to split these between more than one data line.) RUN your program and see whether you can reproduce our results.

x	y	Residual
0	8.2	0.8939
0	7.3	−6.1E-3
0	6.7	−0.6061
0.1	10.4	3.3807
0.1	8.3	1.2807
0.1	7.7	0.6807
0.1	7	−1.93E-2
0.2	7.3	0.5675
0.2	7	0.2675
0.3	6.4	−4.57E-2
0.35	5.5	−0.8023
0.45	4.4	−1.6156
0.55	4.1	−1.6288
0.9	3.4	−1.325
1.2	3	−0.8646
1.2	2.7	−1.1646
1.2	2.4	−1.4646
1.3	3.1	−0.4778
1.8	2.4	0.2562
1.9	2.2	0.343
2.6	2.2	2.3505

gradient	= −2.86795224
intercept	= 7.30612881
St err of estimate	= 1.31674094
St dev of gradient	= 0.393690208
St dev of intercept	= 0.396189705
Correlation coefft	= −0.858114764
t for correlation	= −7.28479448
Err of Regress: Give x (-ve to skip)?	

You can use the initial list of x and y values to check your data. Note just how different this is from a calibration fit; the residuals are very large. The results are given with far too many significant figures, because we did not take the trouble to round them off by using INT, but we need not let that distract us. The points of main interest are the new output lines giving r and t. You will find that the answers agree with those we found in SAQs 6.2a and 6.3a. The program really makes light work of it all. We suggest you opt out of the final program sections on standard errors of regression and prediction; these are of no interest to us in this application.

SAVE your new version of "LINE". As with program "STAT" we hope you will try out some applications on your own initiative. We shall set you just one test for it in the next SAQ. The answer will be of interest to us in the following Section.

SAQ 6.4a

Consider the full set of data for the spectroscopic determination of chloride, discussed at length in Part 5.

$x([Cl-]/mg\ dm-3)$	y (absorbance)
0.0	0.000
2.0	0.113
4.0	0.218
8.0	0.388
12.0	0.523
16.0	0.624
20.0	0.734

Derive the value of the correlation coefficient, r.

SAQ 6.4a

SAQ 6.4b

Draw a scatter diagram for Dean's study relating fluoride levels to tooth decay. The data are listed in Section 6.2. Add the y-on-x regression line to your graph by making use of the fact that our program found that this line has gradient -2.87 and intercept 7.31.

(Hint: look back at Fig. 6.1a; you are being asked to construct the equivalent diagram for Dean's study.)

SAQ 6.4b

6.5. MEASURING 'GOODNESS-OF-FIT' FOR A STRAIGHT-LINE RELATIONSHIP

The fitting of a best least-squares regression line to calibration data was discussed in detail in Part 5. We then have strong theoretical reasons to believe that there is a *true* relationship between y and x. The existence of residuals is put down to errors of measurement. It has become a fairly widespread practice for workers to quote the coefficient of correlation, r, for their data. The assumption is implied that the closer the magnitude of r is to unity, the better is the evidence that the data confirm a linear relationship. Whilst it is certainly true that r can be *exactly* equal to $+1.0$ or -1.0 only when the data lie *exactly* on a straight line, this is of course an extreme case which does not commonly arise in practice. When the data do not conform exactly to a linear relationship, we cannot obtain a reliable measure of 'goodness-of-fit' simply by examining the value of r.

Do you remember the case study in Part 5, concerned with the spectroscopic determination of chloride in a concentration range from 0 to 20 mg dm^{-3}? The full set of data have a very distinctly curved calibration line (see Fig. 5.5a). Now in SAQ 6.4a, we have subjected those data to our program "LINE". The calculated correlation coefficient exceeded $+0.99$! The result seems impressive! It would be an easy matter blandly to compute the equation for the straight-line regression over the full set of the chloride data and to use that to determine the concentrations of unknown samples. We could hope to justify our work by reporting that the value of r for the calibration data is over 0.99, representing an excellent and highly significant level or correlation so close to 1.00 that surely nobody could question that the data were very close to conforming to linearity. Do look back, however, to Fig. 5.5a. The true relationship is most clearly a curve, and we sacrifice accuracy considerably if we represent it by a single straight line. We strongly discourage the use of correlation coefficients in this context. Where the properties concerned are believed to be connected through a true relationship, and deviations from this are due only to errors of measurement, correlation coefficients can provide very misleading evidence about linearity.

In Section 5.5.1 we outlined a much more effective way of checking for linearity through an F-test. When we applied that test to the chloride data it told us very clearly that a straight-line fit was a nonsense. That is the test to use whenever we are working with properties which we believe to be directly related. Correlation coefficients are easily calculated, especially by computer programs, and they are much used by statisticians in other contexts. Unfortunately they are very widely mis-used by scientific workers lazily seeking a justification for assuming that a straight-line relationship fits their data. But now we at least know better!

6.6. FURTHER COMMENTS ON CORRELATION STUDIES

The previous Section represented something of a digression. We thought it important to stress that correlation studies are *not* appropriate in discussing the quality of a fit to measured data when it is believed that a unique exact relationship exists.

We return now to our main current concern, which is with linking analytical evidence with effects which are believed to be associated with a multitude of 'causes'. We shall return to our two case studies, the British Regional Heart Study and Dean's data linking dental decay to the fluoride level in local drinking water. We shall look further at the simpler study, that of Dean, first. We have already concluded that the evidence from it is very highly significant, in that an increase in fluoride concentration is associated with a decreasing incidence of tooth decay. Let us take this conclusion on board. If we wish to improve the dental health of children, we might then advocate that fluoride be added to drinking water in areas where the fluoride content is currently low. But how much should we suggest be added?

Dean's data are plotted in Fig. 6.6a, and the regression line fitting the data is also shown. You should have produced this figure for yourself in SAQ 6.5b.

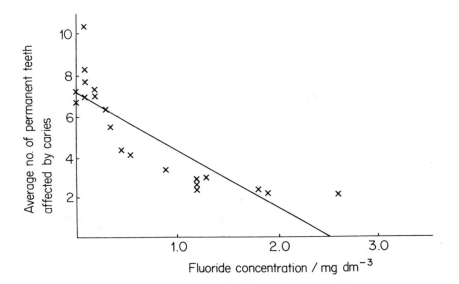

Fig. 6.6a. *Relationship between the incidence of dental caries and the fluoride level in drinking water*

SAQ 6.6a   Three possible prescriptions might be suggested with the aim of decreasing the incidence of tooth decay. They involve increasing the fluoride concentration in fluoride-deficient water supplies:

(i)   to as high a level as is economically feasible,

(ii)   to about 2.5 mg dm^{-3},

(iii)   to about 1 mg dm^{-3}.

Comment critically on all of these suggestions.

SAQ 6.6a emphasises yet again the value of subjecting data to close and careful scrutiny. Two questions still remain to be answered, however, before public health authorities could confidently recommend the addition of fluoride in areas of natural deficiency, to bring the level up to around 1 mg dm^{-3}. The questions are:

(*a*) can we be sure that increasing fluoride will actually *cause* the anticipated improvement?

(*b*) are there other and disadvantageous effects of addition of fluoride?

The first question is a very important one to highlight. We have already seen in Section 6.3 that we can say with extreme confidence that a strong negative correlation between fluoride level and dental caries does exist. *But the existence of a strong correlation is not the same thing as a proof of a causal relationship.* Conceivably, for instance, areas whose natural waters contain significant amounts of fluoride happen also to support the agriculture of some particular crop, which might then feature prominently in the local diet. Perhaps *that* might be the cause of improved dental health. Were that so, simple addition of fluoride to the drinking water would have no effect on local teeth. This kind of danger is always present. Where two properties are strongly correlated it need not be true that one is caused by the other. It is always possible that some different, uninvestigated property is causally connected to both. Then only a change in the unidentified property would be expected to influence things.

This sort of worry can be put to the test through a pilot study. If the fluoride level is increased in the naturally-deficient drinking water of one of the involved communities and a dramatic improvement in dental health is subsequently recorded, then the causal connection suggested by the original correlation study is substantially vindicated. Few now disagree with the claim that increasing fluoride levels to around 1 mg dm^{-3} causes a decrease in dental decay.

The second question identified above was about whether there might be other disadvantageous effects of adding fluoride. This is at the root of the fierce local and international debates about the merits of water fluoridation. No specific evidence has been produced which is sufficiently clearcut to convince supporters of fluoridation that there are such effects. Equally nobody can prove that there may not be any such effect or effects. We do know that, for instance, at very much higher fluoride levels, new dental problems arise. (That was the historical origin of the interest of dentists in the effects of fluoride!)

Now, as Fig. 6.6a indicates, the relationship between fluoride concentration and tooth decay seems pretty direct; at any given fluoride level there is relatively little scatter in the decay-rate data. Not so for our other case study of the relationship between water hardness and cardiac mortality. We are now ready to return to this example. It is a much more sophisticated study statistically; many factors were studied, some of them chemical, others not. In matters of concern to the environment or to health this is a very common situation. A chemist concerned with providing specific analytical evidence should be alive to the possibilities that not only may the concentration of one chemical species be correlated with that of another; they may also be correlated with other factors, climatic, geographical, or even socio-economic.

The data in Fig. 6.1a were in fact shown, in SAQ 6.3b, to provide very firm evidence indeed that increasing water hardness is correlated with a decreased cardiovascular mortality rate. The scatter in the diagram is, however, high, and clearly there must also be other significant factors. Several factors were indeed investigated by the original researchers. In Fig. 6.6b a table taken from their report (published in the *British Medical Journal*, May 1980) is reproduced. It shows highly significant correlations between the mortality rate and each of 24 different properties. Nobody would seriously suggest that all of these factors could be taken as causes of variations in susceptibility to heart disease. Many are interconnected, sometimes in ways that are obvious, sometimes more subtly. There may somewhere be underlying causes which are in turn correlated with many of these factors but which may not have been selected for study.

Factor	Correlation coefficient with SMR	No of towns with data available
Water quality		
Total hardness	−0.67	234
Nitrate	−0.68	229
Calcium	−0.67	163
Langelier index	−0.65	159
Carbonate hardness	−0.65	232
Conductivity	−0.63	178
Silica	−0.58	146
% water from upland sources	+0.69	235
Climate		
Days with >0.2 mm rain	+0.75	
Days with > 1 mm rain	+0.73	
Total annual rainfall	+0.58	253
Mean daily maximum temperature	−0.70	
Mean hours of sunshine	−0.53	
Geographic		
Latitude	+0.74	253
Longitude	+0.68	
Socioeconomic		
% manual workers	+0.64	
% unskilled workers	+0.61	
Mean social class score	+0.63	253
No of cars per household	−0.61	
% large families	+0.56	
% unemployment	+0.53	
Blood group		
% A-gene frequency	−0.59	253
% O-gene frequency	+0.56	
Air pollution		
Mean annual smoke	+0.54	134

Fig. 6.6b. *Correlation between mortality rate and 24 different properties*

SAQ 6.6b   Scan Fig. 6.6b and, from within the main sub-sections, select examples of factors which you might expect to be quite strongly correlated with one another. State whether you would expect such correlations to be positive or negative.

(*i*)   Water quality factors.

(*ii*)   Climatic factors.

(*iii*) Socio-economic factors.

There are substantial problems in analysing this evidence.

(*a*) How can the effect of different factors be separated?

(*b*) How can the most important distinct factors be identified?

(*c*) How can causal relationships be proved?

The way forward is by means of what is known as *multiple regression analysis*. We shall merely give an indication of what is involved in this before we make a few concluding comments about the heart-disease case study.

In looking at the correlation between just two variables x and y we have found the y-on-x regression line:

$$y = mx + c$$

where the parameters m and c are chosen so that the sum of the squares of the y-residual is minimised. In other words m and c are chosen so as to maximise the explained variation of y in terms of variations in x. (The unexplained variation of y equals the sum of the squares of the residuals. This is minimised.) Thus it is with multiple regression. We wish to explain the variation of property y in terms of the variations in a number of other properties x_1, x_2, x_3, say. We thus seek a linear equation of the form,

$$y = c + m_1 x_1 + m_2 x_2 + m_3 x_3$$

where the values of intercept c and 'gradients' (or 'partial derivatives'). m_1, m_2 and m_3 are chosen to minimise the sum of the squares of the y-residuals over the available data. In other words the parameters are chosen to maximise the extent of the explained variation of y. It is not our purpose to explain how to perform multiple regression analysis in practice. You should appreciate however that such a procedure exists and is much used. It represents a natural generalisation of simple regression analysis. Indeed it leads on to the definition of 'coefficients of multiple correlation'.

The research team analysing the British Regional Heart Survey tested many different combinations of factors in multiple regression models. They concluded in the end that five factors collectively had a highly significant effect on the standardised cardiovascular mortally ratio (SMR) and that each of these made a contribution towards 'explaining' the variation in SMR, which was independent of contributions made by the other four chosen factors. Fig. 6.6c. is taken from their report and in it the five key factors are identified. Only one of these is chemical.

	Mean	SD	Minimum		Maximum	
Total water hardness (mmol)	1.70	1.09	0.10	(Colwyn Bay)	5.28	(Hartlepool)
% days with >0.2 mm rain	44.9	4.9	37	(Aylesbury)	58	(Greenock)
Mean daily maximum temperature (°C)	13.11	0.88	11.1	(Aberdeen)	14.6	(Southampton)
% of manual workers	60.5	11.8	30	(Epsom)	83	(Port Glasgow)
No of cars per 100 households	55.2	15.1	26	(Glasgow)	97	(Solihull)

Fig. 6.6c.   *Details of five key factors related to geographic variations in cardiovascular mortality*

SAQ 6.6c

> (*i*) Examine Fig. 6.6c. and note the optimum characteristics of the community you would like to live in if you cared only about trying to minimise your risk of suffering an untimely death due to heart disease.
>
> (*ii*) Is exercise good or bad for your health? Comment.

SAQ 6.6c

We must comment again that proving the existence of a highly significant correlation does not necessarily identify a *casual connection*. For instance the negative correlation between SMR and car ownership is almost certainly a *spurious correlation*. Driving rather than walking is in itself most unlikely to improve the health of your heart. Car ownership was just one of a limited number of socio-economic factors for which the research team were able to get data from the 1971 UK census. It was probably the single factor tested which was most closely related to wealth and would then be expected to be correlated with all sorts of features of life style. Amongst numerous possible underlying causes might, for instance, be some particular difference in typical diet. We can only speculate. If it were that, then it is likely that further detective work pinning down the 'culprit' would involve the services of analytical chemists!

For our final comment let us return to the original data relating SMR to water hardness. When we first displayed these data in Fig. 6.1a. we commented that one way to simplify the diagram with its highly scattered points would be to group towns into bands defined by chosen ranges of water hardness and then to plot simply the average SMR for each band. The research team actually did this. (They happened to calculate the geometric mean rather than the arithmetic mean as the typical SMR value for each band.) The results are shown as the continuous line in Fig. 6.6d.

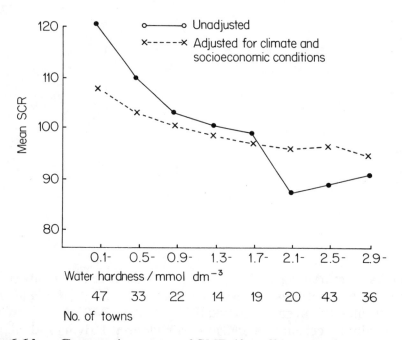

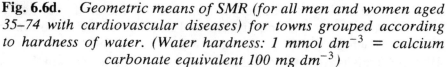

Fig. 6.6d. *Geometric means of SMR (for all men and women aged 35–74 with cardiovascular diseases) for towns grouped according to hardness of water. (Water hardness: 1 mmol dm^{-3} = calcium carbonate equivalent 100 mg dm^{-3})*

Taken at its face value, if a causal connection was involved., the curve would suggest that artificially increasing the hardness of drinking water in a soft-water area from a figure of, say, 0.3 mmol dm^{-3} to, say 2.3 mmol dm^{-3} might decrease local deaths from heart disease by almost 30%.

SAQ 6.6d	How were the figures in the last sentence arrived at? Criticise the argument used.

Soft water tends to occur more frequently in the north together with higher average rainfall, lower temperatures and lower average prosperity. The multiple-regression analysis showed up independent contributions from climatic and socioeconomic factors. When the SMR data for each band are adjusted to take account of differences on these scores the broken curve in Fig. 6.6d. results.

This latter curve gives a rather more reliable picture of the extent of the relationship between SMR and water-hardness. Notice that it eliminates completely the rather surprising and dramatic difference between the 1.7–2.1 mmol dm^{-3} and 2.1–2.5 mmol dm^{-3} bands in the unadjusted data. Note also that the adjusted curve tends to 'level off' at higher hardness levels. Slightly reminiscent of the fluoride *versus* dental-health example, though less dramatic, it seems to suggest that water hardness has a quite significant effect up to about 1.5 mmol dm^{-3} but that it has little impact on SMR at higher levels.

The data presented still do not in any way *prove* the existence of a causal connection, but, if there is one, we now have a more realistic assessment of the likely magnitude of the effect. The sort of evidence which would give an acceptable level of proof would be for the water supply in some soft-water area to be deliberately hardened. If, over the ensuing years, the SMR fell in that area in a way unrelated to any general trends elsewhere, then most sceptics would become convinced of the reality of the effect.

6.7. CONCLUSIONS

For much of this Part of the unit we have been discussing issues and describing methods which cannot strictly be categorised as tools for analytical chemistry. However, we have felt it important to cover this ground because results obtained by analytical chemists are more and more frequently used as the base data for such studies. The issues which arise are often important and nearly always difficult to resolve beyond all reasonable doubt. Conclusions, particularly on topics related to health or to the environment, tend to be the subject of intense public debate. Often such debate is conducted in a superficial and emotive way.

A good analyst should have an appreciation of how his or her results are liable to be used. More important still, an understanding of the questions being addressed should be of considerable importance in deciding exactly what analytical data to obtain. How should samples be selected, how much data do we need and how precise need the analytical measurements be? The design of a programme of measurements should be very strongly influenced by fore-knowledge of the use to which it is intended to put the results.

Our most important aim has been to convey a basic qualitative understanding, about the kinds of problem to which correlation analysis is applicable, about the validity or, more often, the invalidity of particular conclusions, and about the general tactical methodology and the pitfalls involved in interpretation. It is a mistake to view the chemist's role in such studies as limited to providing the technical expertise needed to obtain analyses from samples delivered to the laboratory. The chemist has much to offer to the debate as to which substances are likely to be correlated in their occurrence, which might be suspected of having a causal effect on the behaviour under study and how the occurrence of given chemical species may in turn be connected with other factors such as geography or life-style.

In terms of a quantitative approach we have reminded you how to obtain the regression line, explained how to calculate the coefficient of linear correlation between two properties and described a test to determine the level of significance with which we can claim that a given result implies the existence of a true correlation between the properties studied. The formulae necessary to carry through these tasks are summarised below in Section 6.7.1.

We have strongly recommended, in Section 6.5, that the value of the correlation coefficient should *not* be used as a measure of 'goodness-of-fit' when determining a best line or curve in a calibration problem, or in any other situation where a unique true relationship is believed to exist between two variables.

6.7.1. Simple Linear Correlation Analysis

We suppose that we have n data points, giving corresponding values of properties y and x. We are interested in analysing the extent to which variation in y is explained by variation in x.

(*a*) *The regression equation* of y on x is $y = mx + c$

where $c = (S_{xx}S_y - S_{xy}S_x)/(nS_{xx} - S_x^2)$

and $m = (nS_{xy} - S_xS_y)/(nS_{xx} - S_x^2)$

This information is also given in Fig. 5.2c. and the symbols have the same significance as there. We also give there expressions for the standard errors in c, in m, and in the regression line itself.

(*b*) *The coefficient of (linear) correlation*, r, is given by

$$r = (nS_{xy} - S_xS_y)/\sqrt{(nS_{xx} - S_x^2)(nS_{yy} - S_y^2)}$$

The *explained variation* is given by $100\ r^2\%$

(*c*) *Test of significance.* A test of the level of significance with which a particular observed r value may be taken as indicating that there is indeed a true correlation between y and x involves calculating the t-value as given by

$$t = r\sqrt{(n-2)/(1-r^2)}$$

If the magnitude of this calculated t-value is larger than the critical t-value *for* (n - 2) *degrees of freedom* and a tail-area of α then, at a significance level of α, we can reject the null hypothesis that no correlation exists and assert that the existence of a correlation (of the same sign as r) has been established.

Objectives

As a result of completing Part 6 you should feel able to:

- recognise situations where correlation analysis is appropriate;

- obtain and interpret the regression line (*y*-on-*x*) for a particular set of data;

- distinguish between the existence of correlation and proof of a causal relationship;

- bring a critical approach to bear in assessing correlation data;

- understand the terms *explained variation, unexplained variation, total variation* and *coefficient of determination*;

- calculate the *coefficient of linear correlation, r*, from given data (using Eq. 6.6, or an equivalent formula);

- test the significance with which a given r-value implies the existence of a true correlation (using the critical t-value defined in Eq. 6.7);

- carry out the correlation analyses described above by computer;

- recognise that the correlation coefficient r can be a misleading criterion by which to judge the linearity of a calibration plot, noting that careful examination of the residuals is more appropriate;

- understand, at least superficially, the general principles underlying multiple correlation analysis, and be prepared to discuss, critically, conclusion reached by such means.

7. Statistics in Quality Control

In the very important field of quality control the results of chemical analysis must be judged on a statistical basis. Usually, quality-control work will be a matter of routine. Its purpose may be to monitor the performance of a production process or to determine whether or not a supplied material is up to specification. If the manufacturer or supplier is competent, then it is usually to be expected that all will be in order; the results of the analysis should merely confirm that set standards are indeed being met. Only relatively infrequently should things go wrong. Quality control is concerned with highlighting such occasions.

Since it is a regular and on-going activity, the costs of quality control for an organisation may be quite significant. Exhaustive and detailed studies may be desirable, but they may be uneconomic. So a compromise has to be reached in setting up a programme of analytical work which is thought just sufficient to give adequate assurance that any important deficiencies in a process or a material will be noticed.

When a quality control measurement indicates such a deficiency, decisions will have to be taken. We have dealt quite extensively, in Parts 3 and 4, with the business of setting criteria for such decisions and we practised applying these criteria, given a particular set of measurements. Several of the examples we discussed there had a basis in quality control. You may remember that we dealt with such things as the weight of crisps in a bag, the percentage of NaOH in a solution, and the lifetime of dry cells. Such examples, involving a weight, a chemical concentration, and a lifetime, point up the diversity of properties which may be monitored. The chosen property in any instance is known as the *quality characteristic*. In order to keep a process in check, it is necessary to monitor an appropriate quality characteristic. Typically we shall have in mind situations where a specific chemical analysis is involved. Whatever is measured, the same statistical principles apply when we try to assess the significance of the results.

Because quality-control work is repetitive, over a period of time it will generate a long series of measurements. Clearly it is of benefit to extract as much information as possible from such data. It is mainly in this regard that we shall have to cover new ground. In particular we shall consider some presentational techniques, involving *control charts*. We shall find means whereby we can detect quite sensitively signs of drift in the performance of an established process. In a continuously operating process it is of great value to be able to do this at an early stage, so that corrective action may be taken before deficiences becomes serious.

7.1. BATCH INSPECTION AND PROCESS CONTROL

There are two distinctive reasons for monitoring the output of a process. *The first is to decide whether a batch of the output meets its specification.* This type of quality control is known as *batch inspection*. The procedure for batch inspection is relatively straightforward. A sample of the batch is taken, and some measurements or tests are carried out on the sample. The results of these measurements or tests are then used in a statistical test to ascertain whether or not a batch is up to its specification. A sample from the batch may be inspected in one of two ways. The first way is known as *inspection by attributes*. Here each item in a sample is inspected (the number inspected defines the sample size) and classified as being either good or defective. The proportion of defectives in the sample is used as an indication of the quality of the batch. This is the sort of inspection which would be carried out on light bulbs, toasters, and other items where the product either works satisfactorily or it does not. The second way of inspecting a sample is known as *inspection by variables*. Here, several measurements of a quality characteristic are made - the number of measurements determines the sample size - and the average of the measurements is used as an estimate of the quality of the batch. Inspection by variables is the type of inspection with which we, as chemists, are most familiar. We are almost always concerned with the measurement of continuous variables such as weight, composition, time, etc. This is why we have restricted ourselves in this Unit to the study of statistical tests involving continuous variables and, consequently, why we feel that you would have little difficulty in carrying out batch-inspection procedures involving variables. The hypothesis tests which we dealt with in Parts 3 and 4 are applicable as they stand. Many of the examples we discussed there were direct applications of batch inspection by variables. If you would like to learn something about the statistics of inspection by attributes, you are advised to consult one of the recommended textbooks, particularly one of those addressed to engineers.

Before we leave batch inspection, a few further comments are in order. First, we need to answer the question, 'What constitutes a batch?' Generally, any collection of products which can be sensibly called a batch! The daily output of a process, a single production run, a consignment received by a consumer can all be classified as batches. With consumers and manufacturers in mind, batch inspection is often known as *acceptance sampling*. There is an understanding between a manufacturer and a consumer about the quality of a product: the product's specification is known to both and is an essential part of the contract between them. Consequently, both are concerned that a product meets its specification. If it does, it is accepted – hence the term acceptance sampling. If it does not, it is rejected. Before there is any misunderstanding here, we must realise that the rejection of a batch does not mean that it will be thrown away or destroyed. That may happen. However, what is more likely is that the batch will be sold at a reduced price. With batches of items such as toasters, it is likely that a 'rejected' batch will undergo a 100% inspection, so that items which meet the specification can be retained and be incorporated into a new batch.

The second reason for monitoring the output of a process is concerned with what is known as *process control*. *The aim of process control is to ensure that the quality of the output of a manufacturing process is maintained at a specified level.* A process-control system must therefore have as its objectives the detection of changes in the quality of the output of a process, the identification of the causes of such changes, and the adjustment of the variables which govern the quality of the output, so that the process may, once again have a satisfactory output.

In process control, samples are taken at regular intervals from a production process, measurements or tests are carried out, and the results of these tests, when combined with knowledge of the process, are used to maintain the quality of the output at its specified level. The inspection of samples can be either by attributes or variables and, as with batch inspection, the data obtained are analysed statistically. It is in this analysis that control charts are an important tool and it is to these that we shall now turn our attention.

7.1.1. States of Statistical Control

All manufacturing processes are subject to some degree of variability. This variability may originate in diverse causes. Some likely causes in a chemical manufacturing process are variability in reaction temperatures and pressures, and variability in rates of flow, in efficiencies of mixing, and in other mechanical operations in the process. The manufacturer may do his best to minimise such variability through more stringent specification of raw materials or of aspects of the production process, but it can never be totally eliminated. Slight fluctuations will always remain and the product will thus always exhibit variability in quality. At some level of variability or other, it will be judged that the usefulness of the product is not seriously impaired and that it would be uneconomic to seek further reduction.

Given a production process with its inherent level of variability, if samples for analysis are taken in sequence, these should behave as if they were *statistically random samples drawn from the same population*. The mean and standard deviation obtained from measurements on samples should be estimates of the population mean and standard deviation for the process.*

When all is well, and the results of the analysis conform to the normal pattern, the manufacturing process is said to be in a *state of statistical control*. When a process is in such a state, variability is due to unidentified, but acceptable, *chance fluctuations* in the process conditions.

Now things do not always run smoothly. If they did there would be no need for quality testing! Some aspect of a process may undergo a change which leads to an adverse permanent shift in the average level of quality or in the variability of the product being monitored. It is important to track down such changes and to remedy them. How do we notice their occurrence? It will be when our quality control measurements are so disturbed that *we become convinced that they are no longer behaving as though they were random samples drawn from the previously given population*. The process will no longer be in a state of statistical control. The variability in the quality characteristic will no longer be solely due to chance fluctua-

---

* It is a normal testing procedure for a sample of some chosen size, n, to be collected from the production process over a relatively short time. For the purpose of our present treatment we shall suppose that the inherent variability of the process is such that when a sample is collected over a short time the fluctuations within it are expected to be just as great as they would be if the members of the sample were withdrawn from the process over much longer times. In practice, chance fluctuations may often lead to greater variability between, say, one day and the next than would occur on a shorter time-scale. To deal with such an instance we should have to distinguish carefully between within-sample and between-sample estimates of variability. Our treatment will not contain such complications. In our context, the normal state of the process will be supposed, to be characterised at all times by fixed values for the mean, μ, and for the standard deviation, σ, of the measured quality characteristic.

tions, but will now also be affected by some new cause which can, in principle, be identified and assigned. We say that our measurements have become affected by *an assignable cause*.

It is important to detect the change due to an assignable cause as quickly as possible and to do so for as small an extent of change as possible. Control charts help to do this. *Control charts assist us in testing when variation in a process ceases to be due to chance fluctuations alone and when assignable causes appear.* You are correct if you see an analogy here between random and systematic error on the one hand, and chance fluctuations and assignable causes on the other.

7.2. SHEWHART CONTROL CHARTS

A control chart is, in the first instance, simply a chart on which values of the quality characteristic being utilised are plotted in time sequence. The Shewhart control chart makes use of the fact that when a process is in a state of statistical control, measurements made on physical samples drawn from the production line behave as though they were random samples drawn from the same population. From the Central Limit Theorem we know that the means of random samples are normally distributed with mean μ and standard deviation $\sigma/\sqrt{n}$. The implication of this is that if we were to take physical samples from a production line which was in statistical control and carry out n measurements on them, then the mean of such measurements would be normally distributed with mean μ and standard deviation $\sigma/\sqrt{n}$. All the probability ideas which were introduced in discussing the normal distribution curve are applicable here. For example, 95.45% of all samples drawn from a production line in statistical control will have mean values in the range $\mu \pm 2\sigma/\sqrt{n}$; 99.73% of such samples will have means which lie in the range $\mu \pm 3\sigma/\sqrt{n}$. With this in mind, consider the Shewart chart shown in Fig. 7.2a. As we discuss this chart you may find it helpful to refer also to Fig. 7.2b, which illustrates a normal distribution curve for the sample means, $\bar{x}$, centred on the population mean, μ, and indicates the significance of the tails beyond $\mu \pm 2\sigma/\sqrt{n}$, and beyond $\mu \pm 3\sigma/\sqrt{n}$.

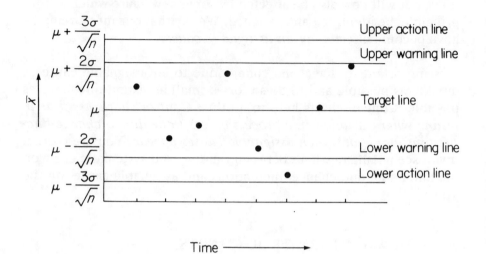

Fig. 7.2a. *A Shewhart control chart showing sample means plotted in sequence*

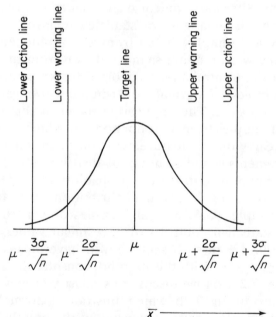

Fig. 7.2b. *A normal distribution curve for the sample mean for samples of size n taken from a production line operating under statistical control and with a process mean of μ*

From an inspection of the chart in Fig. 7.2a we see that time is represented on the horizontal axis and the values of sample means are plotted vertically. Further we see that five horizontal lines have been drawn. The central line represents the value of the population mean, or in production terms, the value for the quality characteristic which is the *specification or target value* for the process. Above and below this target line, at a distance of $2\sigma/\sqrt{n}$, are lines which are called the *warning lines*.

Why should these lines be termed warning lines? Well, if a process is producing material at its specified level (process mean $= \mu$) and if it is *under statistical control*, then the probability that a sample taken from the production line will have a mean value outside of a warning line, ie be either greater than $\mu + 2\sigma/\sqrt{n}$, or less than $\mu - 2\sigma/\sqrt{n}$, is 4.56% or $1 - 0.9544$, ie 4.56% or approximately 1 in 22. The probability that a sample mean will fall outside a *given* warning line is 0.0228 (2.28%) or 1 in 44.

The warning line serves to alert us when this supposedly unlikely event occurs. If a sample mean is outside a warning line, then we are alerted to the possibility that this *might be* because the process mean has changed. Another way of saying this is that although we must expect variability in the values of sample means due to *chance*, when a sample mean has a value outside a warning line we should consider the possibility that an *assignable cause* is present and that the process mean has changed.

Above and below the target line, at distance $3\sigma/\sqrt{n}$ from it, we have two other lines, which are called *action lines*. Why action lines? Again, we must think in probability terms. For a process on target and under statistical control, the probability of a sample mean falling outside a *given* action line is 0.00135 (0.135%) or approximately 1 in 740. The action lines are there to prompt us to respond to an event of such a low probability. They inform us that if a sample mean does have a value which puts it outside of an action line, then, since such events are so rare, we should assume that the process mean has changed its value and some corrective action should be taken. Action lines do more than simply warn us that an assignable cause might be present; they tell us to assume that one is present and to respond to it. If you see an analogy between hypothesis testing, where rejection implies that the population is probably not the one asserted, and control charts, where a sample mean outside an action line implies that the process mean is not the target value, then your efforts to come to terms with the topic of statistics have not been in vain! (The action lines are set at the critical values appropriate to a hypotheses test on the mean at a one-tail significance level of 0.135%.)

∏   If a process is on target, ie has its process mean equal to μ, what is the probability that *two consecutive* sample means will have values which are greater than $\mu + 2\sigma/\sqrt{n}$?

The probability that a sample mean is greater than $\mu + 2\sigma/\sqrt{n}$ is 0.0228. The probability that two consecutive sample means will have values which are greater than $\mu + 2\sigma/\sqrt{n}$ is (0.0228) × (0.0228) = 0.000520 or less than 1 in 1900. This event is of such a low probability that it is used as the basis for an additional criterion for deciding that corrective action is required. *If two consecutive sample means fall outside the same warning line, ie either the upper or the lower warning line, then it is to be assumed that an assignable cause of variability is present and corrective action is warranted.*

Hence we see that both the action lines and the warning lines can be used in deciding whether or not to take corrective action. It is for this reason that *the action lines are called outer control limits, and the warning lines, inner control limits.*

This is an appropriate time to make two comments. The first refers to an issue about which you have probably already been thinking. It is that a process which is on target and under statistical control will, quite naturally, yield samples which have means that fall outside one of the action lines. If you believe in the applicability of statistics to measurement and to control, then you must agree that this will occur, for action lines drawn at $\mu \pm 3\sigma/\sqrt{n}$, once in 370 times on average: for a given action line this occurs once in 740 times on average. By the same token, two consecutive sample means will fall outside a given warning line once in about 1900 times on average, and outside one or the other warning lines (drawn at $\mu \pm 2\sigma/\sqrt{n}$) once in about 950 times on average. The implication of this is that when we respond to an 'out-of-control' signal, we might be making a mistake, ie that we initiate action, even though the process is on target and under statistical control. From the point of view of statistics, this is something that we must simply learn to live with! It is entirely analogous to the situation which exists in hypothesis testing. When we reject a hypothesis, we are aware that there is a possibility that we may be wrong in doing so. (Recall our discussion of type I error.) Nevertheless, we still reject hypotheses when we feel that we have to. We base our decision on the laws of probability and statistical arguments. The same sort of decision-making process is applicable here. 'Out-of-control' decisions have such a low probability of occurrence if the process is actually on target and in a state of statistical control, that when they do occur we feel safe in assuming that the process must be out of control. This necessitates a second comment. Statistics is an aid to control, not its master. The response to an 'out-of-control' signal depends upon a detailed knowledge of the process to which the control chart is applied. To ignore an 'out-of-control' signal would be foolish: almost as foolish would be to start immediately 'twiddling the dials'. The wise action is probably quickly to seek more information, eg from a new sample.

SAQ 7.2a Under statistical control a process produces a powder with a mean ferrous-iron content of 6.70% w/w with a standard deviation of 0.18%. Samples of size 4 are routinely analysed by the quality-control lab. The following sequence of results was obtained for the mean percentage of ferrous iron.

6.68, 6.71, 6.68, 6.72, 6.82, 6.54, 6.59, 6.58, 6.71, 6.48, 6.57, 6.61, 6.64, 6.56, 6.69, 6.64, 6.69, 6.57, 6.54, 6.56, 6.55, 6.50, 6.55, 6.51, 6.42.

Construct a Shewhart control chart for these results.

Comment.

SAQ 7.2a

7.2.1. Average Run-length

We have drawn the warning lines at $\mu \pm 2\sigma/\sqrt{n}$ and action lines at $\mu \pm 3\sigma/\sqrt{n}$.

Others draw them at $\mu \pm 1.96\sigma/\sqrt{n}$ and $\mu \pm 3.09\sigma/\sqrt{n}$ respectively. The advantage of doing this is that the probabilities associated with a sample mean falling outside a *given* warning or action line are easier to remember; the probability is 1 in 40 (2.5%) for a sample mean falling outside a given warning line and 1 in 1000 (0.1%) for a sample mean falling outside a given action line. Check back to our normal distribution table in Fig. 2.3g to see whether you agree with these values! In practical terms it does not matter much which of the two conventions we adopt. It would have made no difference whatsoever to our discussion of the results in SAQ 7.2a. The purpose of a control chart is to enable us to test visually when variability ceases to be due to chance alone and when assignable causes appear. Either convention for control charts affords us this possibility. The question which we must now ask is, 'How good are Shewhart Charts at accomplishing the task for which they are invented?' In order to answer this question, let us imagine a situation where the process mean has changed from μ to $\mu + 3\sigma/\sqrt{n}$, ie it has undergone a unidirectional change of $+3\sigma/\sqrt{n}$ in magnitude because of a permanent change in one (or more) of its governing variables. We now ask the question, 'What is the probability that a sample mean will fall outside of the action line drawn at $\mu + 3\sigma/\sqrt{n}$?' A Shewhart chart helps us to visualise this situation.

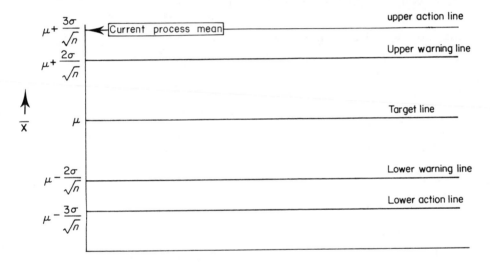

Fig. 7.2c. *A Shewhart chart showing the warning and action lines and illustrating the fact that the current process mean is*
$$\mu + 3\sigma/\sqrt{n}$$

It is easy to see from Fig. 7.2c that if the current process mean is $\mu + 3\sigma/\sqrt{n}$, then the probability that a sample mean will fall outside of the upper action is 0.5, ie it should occur one time in two. In the light of this, let us ask another question. If the process mean changed suddenly from μ to $\mu + 3\sigma/\sqrt{n}$, how many samples would we need to take *on average* before action was signalled? It can be shown that in general, if there is a probability p that any individual sample will give a result outside the action line, then the average number of samples which must be tested before this happens once is given by $1/p$. This average is known as the *average run-length* (ARL). If the process mean moves to $\mu + 3\sigma/\sqrt{n}$ then, as we have said, $p = 0.5$. The average run-length is thus given by $1/0.5$, ie ARL = 2. We should, on average have an indication that the mean has shifted after taking 2 samples. This is a relatively short run-length. Provided that we take samples at reasonably short intervals, we should pick up an out-of-control signal in a short time. Now, however, let us consider a smaller movement of the process mean.

∏   If the process mean changes from μ to $\mu + \sigma/\sqrt{n}$, what is the probability that a sample mean will fall above $\mu + 3\sigma/\sqrt{n}$? What is the average run-length?

Can you work it out? Does Fig. 7.2d help?

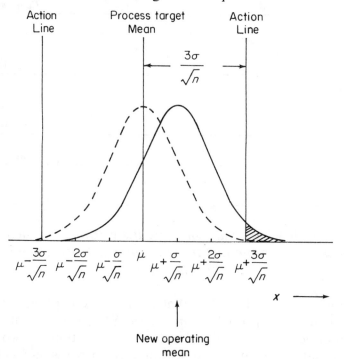

Fig. 7.2d.   *Effect of a shift of one standard deviation in a process operating-mean. The dashed curve shows the distribution of sample means when the process is under statistical control. The continuous curve shows the distribution when the mean increases by one standard deviation*

As the figure illustrates, the upper action-line is set three standard deviations above the process target mean. Now the actual operating mean has shifted to one standard deviation above the target mean, so the upper action-line is now only two standard deviations away. The probability, p, that a sample mean will exceed the limit is given by the shaded area. This is the tail area corresponding to $z = 2$, which is 0.0228 (2.28%). The average run-length is equal to $1/p$, and this gives ARL = 44.

We should have to wait until after an average of 44 samples before we picked up an out-of-control signal. Even with a high frequency of sampling this is a long time to wait! Could we do any better? Well, we could try the 'two consecutive means outside a warning line' rule. The upper warning-line lies two standard deviations above the target mean, so it is just one standard deviation from the new operating mean. The chance that an individual sample will breach this value is given by the tail area for $z = 1$ (convince yourself of this with the help of Fig. 7.2d). This value is 0.1587 (15.87%) (see Fig. 2.3g). The probability that two consecutive means will lie above the warning line is thus $(0.1587)^2$, ie 0.0252. The average run-length is therefore estimated as the reciprocal of this value, which is 40. A little better than before, but not much!

Can we improve upon this situation? Well, we can combine the 'one outside an action line' and the 'two consecutively outside a warning line' rule. The probability of a sample mean falling outside an action line in our example is 0.0228. To add the probability due to the warning-line rule we have to be a little careful; the effect of any value above the action line is already accounted for, it is only values between the warning and the action line which give an added chance that action will be signalled. The probability of a single occurrence of this is $(0.1587 - 0.0228) = 0.1359$, so it will happen twice consecutively with a probability of $(0.1359)^2$, or 0.0185. The combined probability under the two rules is thus estimated as $0.0228 + 0.0185 = 0.0413$. Taking the receiprocal of this value predicts an average run-length of 24[*]. This is a worthwhile improvement, though it is still uncomfortably long.

[*] The statistical argument here is strictly speaking, a little more complicated than we have admitted. The final answer we obtain (as rounded) is still valid, however.

The ARL gives the *average* number of samples taken before an out-of-control signal is obtained. Sometimes one is lucky, the signal may occur sooner than on average. By the same token one may be unlucky and have to go well beyond the ARL. It can be shown that the combined rule in the above example still leaves a 10% chance that more than 54 samples will be required before the out-of-control signal occurs.

The Shewhart chart for sample means is designed to enable us to test when variability in a process is due to chance fluctuations alone and when assignable causes appear. It is a most useful device, but unfortunately it is rather slow in detecting small changes in the process mean. There is another type of chart, known as a Cusum chart, which detects a small drift of the process mean much better. We turn to it in the next section.

SAQ 7.2b

Suppose the operating mean for a process suddenly increases by two standard deviations, ie from μ to $\mu + 2\sigma/\sqrt{n}$. What is the average run-length before a sample mean falls outside an action line?

7.3.   CUSUM CHARTS

Let us assume that we have a process which has a target value of T and that the means of samples taken in sequence from the process are $\bar{x}_1, \bar{x}_2, \bar{x}_3, \ldots, \bar{x}_s$. The differences between the means and the target value are therefore $(\bar{x}_1 - T), (\bar{x}_2 - T), (\bar{x}_3 - T), \ldots, (\bar{x}_s - T)$. The cumulative sums (termed 'cusums') of these differences are given below:

$$S_1 = (\bar{x}_1 - T)$$

$$S_2 = (\bar{x}_1 - T) + (\bar{x}_2 - T) = S_1 + (\bar{x}_2 - T)$$

$$S_3 = (\bar{x}_1 - T) + (\bar{x}_2 - T) + (\bar{x}_3 - T) = S_2 + (\bar{x}_3 - T)$$

.
.
.
.

$$S_v = (\bar{x}_1 - T) + (\bar{x}_2 - T) + (\bar{x}_3 - T) + \ldots + (\bar{x}_v - T)$$
$$= S_{v-1} + (\bar{x}_v - T)$$

It is easy to see that each sum is obtained from its predecessor by simply adding the new difference. *A cusum chart for sample means is a plot for each sample number of the cumulative sum of the differences between the means and the target value.* With this knowledge, we are ready to plot such a chart. All we need are some values to manipulate.

Assume that we have a process whose target value is 50 and which, when under statistical control, has a standard deviation of 8. If we take samples of size 4 from this process, we expect the means to be distributed with standard deviation equal to $\sigma/\sqrt{n} = 8/\sqrt{4} = 4$. (We shall not at present make use of our knowledge of the variability inherent in the process when it is under statistical control. We have given the information here for use later.) Let us assume that samples of size 4 have been taken from the process and that their means are as follows. The cusum associated with each sample has also been calculated.

Sample Number	Sample Mean	Sample Mean – Target Value	Cusum
1	56	+6	6
2	48	−2	4
3	49	−1	3
4	46	−4	−1
5	41	−9	−10
6	53	+3	−7
7	58	+8	+1
8	54	+4	+5
9	50	0	+5
10	49	−1	+4
11	49	−1	+3
12	54	+4	+7
13	56	+6	+13
14	49	−1	+12
15	57	+7	+19
16	57	+7	+26
17	55	+5	+31
18	61	+11	+42
19	60	+10	+52
20	54	+4	+56
21	54	+4	+60
22	50	0	+60
23	49	−1	+59
24	44	−6	+53
25	51	+1	+54
26	58	+8	+62
27	47	−3	+59
28	52	+2	+61
29	41	−9	+52
30	55	+5	+57
31	50	0	+57
32	40	−10	+47
33	46	−4	+43
34	48	−2	+41
35	52	+2	+43
36	40	−10	+33
37	49	−1	+32
38	43	−7	+25
39	50	0	+25
40	42	−8	+17
41	48	−2	+15
42	52	+2	+17
43	55	+5	+22
44	44	−6	+16
45	51	+1	+17

The cusum chart corresponding to these results appears in Fig. 7.3a.

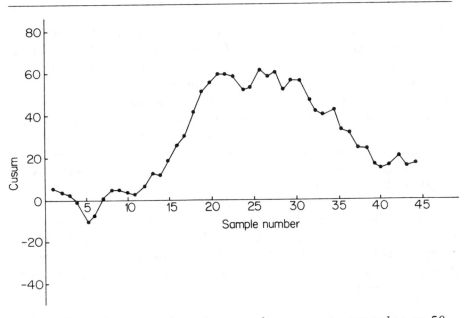

Fig. 7.3a.   *A cusum chart for sample means; target value = 50*

From an examination of the chart, and of the listed values, we can make the following comments.

(*i*)   For the first ten samples the cusum stays relatively close to zero. During this period the average of the sample means is 50.4, quite close to the target value.

(*ii*)   During the period associated with samples 11 to 20 the cusum rises rapidly. Interestingly, the average of the means during this period is 55.2.

(*iii*)   Between samples 21 and 30 the cusum hovers around a value of about 55. The average of the means over this interval is 50.1, very close to the target value.

(*iv*)   The cusum falls rapidly between samples 31 and 40. The average of the means during this time is 46.0.

(*v*)   The last 5 samples sees a return to a relatively constant cusum, and the average of these means is 50.0.

These observations reflect a general relationship between the slope of a cusum chart and the average of the sample means. If the average result for the sample means is close to the target value then the cusum plot follows a broadly horizontal path. In our example this was so in the intervals between samples 1 and 10, between samples 21 and 30, and following sample 40. If the average result for the sample means is above the target value then the cusum plot will have a positive slope (as between samples 11 and 20), whereas if the average sample mean is below the target the cusum will have a negative slope (as between samples 31 and 40).

Let us put this result on a firmer and more quantitative basis. Consider a series of m samples following sample v. The cusum value at sample v is represented by S_v and that at sample $(v + m)$ by S_{v+m}. The average gradient of the cusum plot over the interval between these samples is given by

$$\text{Gradient} = (S_{v+m} - S_v)/m \qquad (7.1)$$

Now the cusum difference $(S_{v+m} - S_v)$ is just the sum of the differences between the sample means and the target value T for each of the m samples following sample v. Hence

$$(S_{v+m} - S_v) = (\bar{x}_{v+1} - T) +$$

$$(\bar{x}_{v+2} - T) + \ldots + (\bar{x}_{v+m} - T)$$

$$= \sum_{i=1}^{m} \bar{x}_{v+i} - mT$$

By dividing both sides by m, and by using Eq. 7.1 we obtain the result:

$$\text{Gradient} = \left(\sum_{i=1}^{m} \bar{x}_{v+i} \right)/m - T$$

The first term on the right is the mean of the sample means in the interval. This is our estimate of the current process mean. Therefore:

Est. process mean = Gradient of cusum + Target value (7.2)

There we have it. If the slope of the cusum plot is close to zero, then it appears that the process operating mean is close to the target value. A positive cusum gradient implies a process mean in excess of target: if the cusum gradient is negative that indicates a process mean below the target value.

For practice, let us apply Eqs. 7.1 and 7.2 to the period between samples 11 and 20 in Fig. 7.3a.

$$\text{Est. process mean} = \frac{S_{20} - S_{10}}{20 - 10} + T$$

$$= \frac{56 - 4}{10} + 50$$

$$= 5.2 + 50 = 55.2$$

We now see that the cusum chart not only gives us a visual signal that a process is moving away from its target value, and the direction of such a change, but it also affords us a quick method of estimating the current process-operating mean during any period of interest.

It is interesting to compare the visual impact of the cusum chart with that of a Shewhart chart. The Shewhart chart for our present example has been constructed in Fig. 7.3b. Yes we agree, it is just possible to detect trends. However, we wonder whether we are being wise after the event: the cusum chart so clearly showed what was happening that we were prepared for trends in the Shewhart chart even before we saw it!

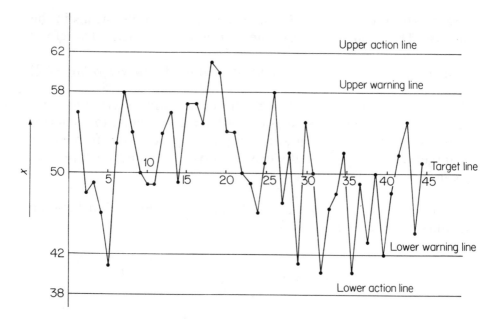

Fig. 7.3b. *The Shewhart chart corresponding to the data in Fig. 7.3a. Target Value = 50: Standard deviation when under statistical control = 8; Sample size = 4*

In the absence of the visual information given by a cusum chart, the Shewhart chart does little to indicate the trends in process mean which we know to be present. There is one indication that an assignable cause may be making an appearance. This occurs with samples 18 and 19: both fall outside of the upper warning line and, hence, according to the 'two outside a warning line' rule, there is a 'signal' that action is warranted. The drift below the target value, which takes place between samples 30 and 40 is not clearly indicated by the Shewhart chart. Although we receive some warning signals (samples 28, 32, and 36), there are no signals which call us to action. This is not surprising. The gradients of the cusum plot between samples 11 and 20, and between samples 31 and 40 suggest that the process mean moved off-target only by $+\sigma/\sqrt{n}$ and $-\sigma/\sqrt{n}$ respectively. For changes of this magnitude, the average length of run based on the 'sample mean outside an action line' rule is 44. There can be no doubt about it, from the point of view of visual impact, the cusum chart is far more effective than the Shewhart chart for indicating small changes in the process mean. A question which now arises is: 'Are there any decision rules which can be used with cusum charts?' We shall shortly introduce such a rule, based on the concept of the V-mask.

SAQ 7.3a

> Draw a cusum chart for the example discussed in SAQ 7.2a where 25 consecutive results are reported for the mean ferrous-iron content in a product. Comment on the chart obtained. Assess quantitatively what is happening to the process mean over the period in which the samples were taken.

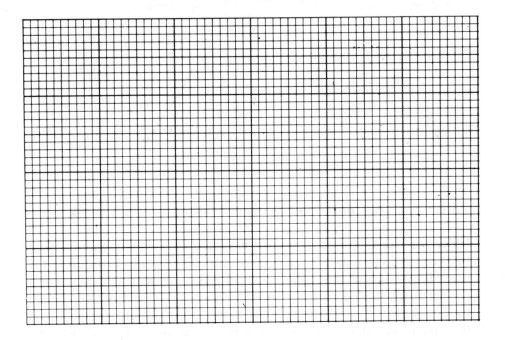

SAQ 7.3a

7.3.1. Cusum Charts and Decision Making with a V-mask

When we use the Shewhart chart we have both warning and action lines to aid us in making decisions. Something similar is required for use with cusum charts. One such aid is known as the V-Mask and its use is illustrated in Fig. 7.3c. The V-mask, which is often made by engraving a V on transparent material, is superimposed upon the cusum in such a way that its axis of symmetry is parallel to the horizontal axis. The vertex of the V is placed at a given distance d ahead of the most recent data-point of the cusum. (The actual value of this distance d, and that of the angle θ made by each arm of the V with the axis of symmetry are most important. We shall discuss the significance of both d and θ later.) As each new data-point for the cusum becomes available the V-mask is moved horizontally so that the lead distance d is maintained.

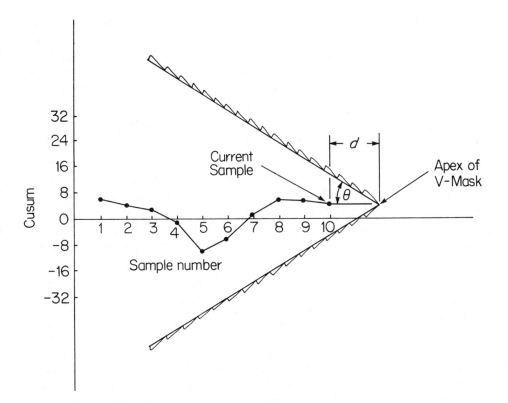

Fig. 7.3c. *The cusum chart for the first ten points from Fig. 7.3a, replotted so that one unit on the sample number axis is of length equal to $2\sigma/\sqrt{n}$ on the cusum axis. In this example $2\sigma/\sqrt{n} = 8$. A V-mask is superimposed, with its apex 2 units to the right of the final data-point. Angle θ equals 30°*

In Fig. 7.3c we have replotted the cusum for the first ten data points from Fig. 7.3a. Superimposed upon the cusum is a V-mask whose apex is positioned a distance d ahead of the current sample point, ie sample number 10. As can be seen from the plot, all the points lie well within the arms of the cusum. This is a typical occurrence for a process which is in a state of statistical control. We can already hear the objections to this statement: 'If different scales for the plot had been used, many of the points would have fallen outside the arms of the V'. You are right! In constructing a cusum chart to be used in conjunction with a V-mask, the scales for the chart and the parameters for the V must both be chosen with great care. We took such care in our constructions. The scaling chosen for the axes was such that the distance between plotted points on the horizontal axis was equal to the distance representing $2\sigma/\sqrt{n}$ on the vertical axis, ie in this example 1 unit on the sample-number axis had the same distance as $2 \times 8/\sqrt{4} = 8$ units on the cusum axis. The parameters for the V-mask are $d = 2$ units on the horizontal axis and $\theta = 30°$. It can be shown that if the scaling is the same as in our plot (ie if the distance representing $2\sigma/\sqrt{n}$ on the vertical axis is the same as the distance representing 1 unit on the horizontal axis) and if $d = 2$ units and $\theta = 30°$, then, for a process under statistical control, the probability of a cusum data-point falling outside the arms of the V-mask is 1 to 500 (the probability of a point falling outside a given arm is 1 in 1000)*. This probability is the same as that for a sample mean falling outside the action lines on a Shewhart Chart, if those are drawn at $T \pm 3.09\sigma/\sqrt{n}$. Yes, it is a little difficult to get a feel for V-masks applied to cusum charts. But practice helps, let us move on.

In Fig. 7.3d we see the replotted cusum for data points 10 to 20 from Fig. 7.3a. Here we see a typical example of a V-mask locating a process whose mean is higher than the target value. The plot straddles the lower arm of the V-mask and actually crosses it on seven occasions. When a data-point crosses the V-mask, ie falls outside one of its arms, then the signal received is equivalent to that when a point falls outside an action line in a Shewhart chart. Action is warranted. In the process under consideration, the V-mask would have shown a data-point outside its lower arm as early as sample no 18 – see Fig. 7.3e. However, the chart gives earlier signals that the current process-mean was running above the target value. Whenever a cusum plot stays close to the lower arm of the V-mask there is a strong indication that the operating mean is above target. It can be argued that this sort of behaviour was evident as early as sample 14.

---

* As usual we opt out of derivations. If you are not happy with statements such as 'it can be shown' you are invited to read R. H. Woodward and P. L. Goldsmith, *Cumulative Sum Techniques*, ICI Monograph No 3, Oliver and Boyd, 1964.

In Fig. 7.3f we see a typical cusum plot for a process where the current mean is running below the target value. Although no data-point actually crosses the V-mask, the points all hug the upper arm. This behaviour, which would have been apparant as early as say sample 35, is rather analogous to successive points crossing a warning line in a Shewhart chart.

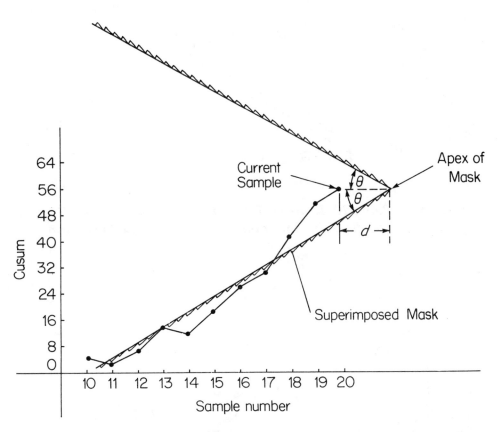

Fig. 7.3d. *The rescaled cusum plot for points 10–20 from Fig. 7.3a. The scaling is the same as in Fig. 7.3c. The V-mask has been aligned after point no. 20*

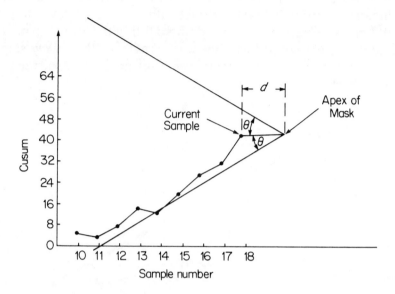

Fig. 7.3e. *Points 10–18 are replotted from Fig. 7.3d, and the V-mask is lined up with point 18. Point 14 just crosses the lower arm*

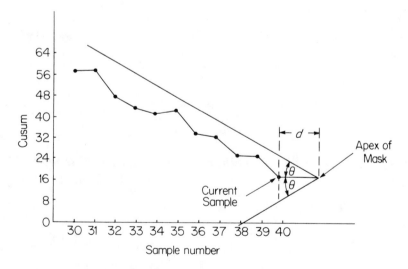

Fig. 7.3f. *The rescaled cusum plot for points 30–40 from Fig. 7.3a. The V-mask is aligned after point 40*

If we simply use cusum charts with a decision rule which signals action only when a point crosses an arm of the V-mask, then the average run-lengths required to detect small changes in the process mean are not much improved on those obtained by using Shewhart charts with the combined action rule. This is because the V-mask decision-rule being imposed still tends to depend on the behaviour of individual points. Thus, for example in Fig. 7.3e, the 'kink' in the cusum line due to the relatively low sample-mean obtained, (probably by chance) for sample 14 is crucial in signalling action after sample 18.

Paying attention to a well-defined trend in the slope of a cusum plot, however, gives a very sensitive indication of even quite small shifts in the process mean. Cusum plots make good use of the information given by a succession of measurements, whereas the Shewhart chart attracts attention only due to the behaviour of an individual point or, at most, two consecutive points. The cusum chart is particularly good at highlighting the kind of problem that frequently occurs in production plants, where there is a gradual drift of the process mean away from the target value. It is often possible to pinpoint the time at which a change in gradient of a plot set in, and this can be of considerable help in identifying the source of the change.

7.3.2. Confidence Intervals for the Operating Process Mean

In our use of control charts we have tended to concentrate on setting decision criteria which signal a shift in the process mean away from its target value. We have been less concerned, on the whole, with finding the actual value of the current operating mean, though we have shown, early in Section 7.3, that the operating mean over any period, can be estimated from the slope of the cusum plot. Eq. 7.2 is restated below.

Est. process mean = Gradient of cusum + Target value

Can we define a confidence range for this estimate?

Well this is old hat to us! If we are concerned with a period over which m samples were withdrawn (each of size n), then our estimate of the process mean is simply the average of the m different sample means. The process standard deviation is σ. All in all, a total of mn measurements are included in this mean of samples means. The standard deviation associated with the estimate is therefore $\sigma/\sqrt{mn}$. So, with a confidence level of $(1-\alpha)$, we can say that the operating process-mean over the interval concerned lies within $z_{\alpha/2}\sigma/\sqrt{mn}$ of the estimated value.

Let us return to our example, whose cusum chart was plotted in Fig. 7.3a. Consider the period of steeply rising cusum between samples 10 and 20. By applying Eq. 7.2, we previously obtained:

$$\text{Est. process mean} = (S_{20} - S_{10})/10 + T = (56 - 4)/10 + 50$$

$$= 55.2.$$

Now, we were told that the process standard deviation was 8. The size of each sample taken was 4. We are considering a period in which 10 samples were taken, so the standard deviation for the mean of sample means is $\sigma/\sqrt{mn} = 8/\sqrt{40} = 1.26$. Let us obtain the 95% confidence interval for the operating mean. We are allowing 2.5% probability for each tail, so we need $z_{0.025} = 1.96$. Hence we have, over the period concerned:

$$\text{Process mean} = 55.2 \pm z_{0.025}\,\sigma/\sqrt{mn} = 55.2 \pm 1.96 \times 1.26$$

$$= 55.2 \pm 2.5$$

We are 95% confident that the actual process-mean over this period was between 52.7 and 57.7. Notice that the target value of 50.0 lies quite clearly outside this interval.

SAQ 7.3b

For the example introduced in SAQ 7.2a, involving measurements of the ferrous-iron content of a product, obtain 99% confidence intervals for the operative process mean over the periods covered by the first ten, and by the last ten samples. Comment. (Note that estimates of the process means for these periods were obtained in SAQ 7.3a, but without any specified confidence intervals.)

We have here the bones of a very useful method for identifying a drift in the process mean. But, as usual, there is a problem! We have accepted throughout, with touching naivety, that we can make use of the quoted value for the process standard deviation, σ! However, disturbances in the smooth running of a process are every bit as likely to affect the variability of the product as they are its mean composition. Indeed, we might expect that variability would be more frequently affected. Now if we cannot rely on the given value for σ, we have to use estimates based on our sample measurements. In the past that has simply meant using a t-value instead of z, but things are not quite so straightforward here. Not only should we have to estimate the actual value for the process standard deviation, but we should have to allow for the possibility that it too might be changing with time.

Although it would be feasible to do so, we shall not follow this line of enquiry any further. We are content simply to have pointed out a topic for possible further study. It is important, however, that we do devote some attention to methods for detecting changes in a process standard deviation.

7.4.   VARIATION IN A PROCESS STANDARD DEVIATION

When a manufacturing process is in a state of statistical control, measurements of some quality characteristic behave as though they are random samples drawn from the same population; the variability in the quality characteristic is due to chance fluctuations in the process. Control charts are devices which enable us to test when the variability in a quality characteristic ceases to be due to chance alone, and when assignable causes make their appearance. So far, we have worked on the assumption that the only type of change in a process which could remove it from a state of statistical control was one which results in the process mean changing. Unfortunately, we now have to recognise that this is not the only type of change which can take place. A process in a state of statistical control is recognised not only by the mean, μ, but also by the standard deviation, σ, of the quality characteristic. Clearly, it is possible for changes to take place within a process which can result in either or both of these parameters changing. Three possibilities are apparent:

(*i*)   the process-mean changes but its standard deviation does not;

(*ii*)   the process standard deviation changes but the mean remains constant;

(*iii*)   both the mean and the standard deviation for the process change.

These situations are presented graphically in Fig. 7.4a. In part (*i*), we see that the process mean has undergone a positive shift in its value but that the variability of the quality-characteristic values, defined by the standard deviation of the distribution, remains the same. This is the only type of change which we have catered for in earlier sections. Part (*ii*) of the figure represents a situation where the process mean remains constant but where the variability in the output changes (for the worse). Part (*iii*) shows a negative shift in the process mean accompanied by an increase in the standard deviation of the quality-characteristic's distribution.

This figure makes it clear that, when a sample mean is found to lie outside an action line, the cause might be drift of the process mean, or of the standard deviation, or of both. If you are thinking that it is also possible to have a situation where both μ and σ change in an irregular fashion, then you are right. We shall not sketch such a case; it only upsets people!

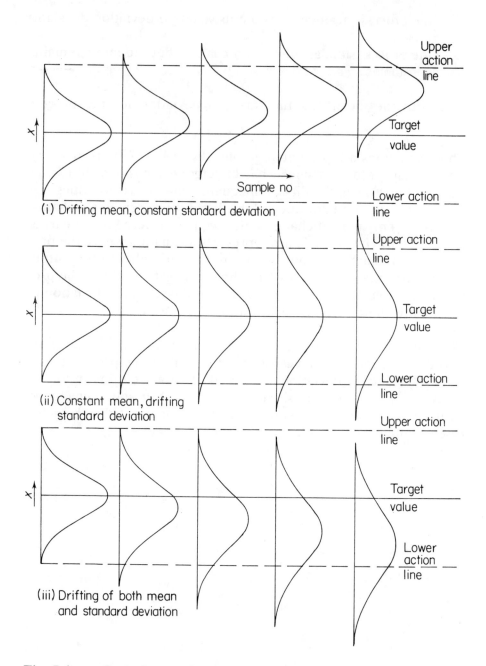

Fig. 7.4a.   *Out of control processes: variation with time of product distribution curve*

7.4.1. A Control Chart for Sample Standard Deviations

Since variability in output can arise from changes in both the process mean and in its standard deviation, we need to have control charts for sample means and for sample standard deviations. It is easy to obtain a theoretical basis for a control chart for sample standard deviations. We used the normal distribution as the basis for the Shewhart control chart for sample means because we expect sample means to be normally distributed. When it comes to sample standard deviations we turn our attention to the F-distribution.

We know the value of σ, the (population) standard deviation, when the process is under statistical control. When we make measurements on each sample, of size n, we can derive the sample standard deviation s. This is an estimate of the current process (population) standard deviation, an estimate with $(n - 1)$ degrees of freedom. In exercising quality control we have to ask whether the sample value, s, is compatible with the target population value, σ.

How do we draw a Shewhart type of control chart for sample standard deviations? From the standard deviation found for each sample we can work out a value of $F = s^2/\sigma^2$, and we can plot this on a diagram of the type sketched in Fig. 7.4b. So far so good, but how do we determine the positions of the horizontal lines on the chart. The target line is easy enough, we can take that at $F = 1$, corresponding to agreement between target and sample values, ie $s = \sigma$.

What about the upper warning line? Let us adopt the convention that this line should be at a level which we would expect to be breached one time in 40 if the process was under statistical control. Once in 40 times is 2.5%, so we need the F-value which has an upper tail area of 0.025. We draw the warning line at $F = F_{0.025(n-1,\infty)}$. Why these numbers of degrees of freedom? Well we are comparing s with σ; s, obtained from a sample of size n, has $n - 1$ degrees of freedom, whilst σ, as a population value, has infinitely many. Note the order in $F_{0.025(n-1,\infty)}$; s is in the numerator when we calculate $F = s^2/\sigma^2$.

The upper control line is obtained in a similar way. If we choose it so that it will be crossed only one time in 1000 if the process remains under statistical control, then we take it at $F = F_{0.001(n-1,\infty)}$. We can stop at that because, although we have sketched lower warning and action lines in Fig. 7.4b, few would actually include them. F-values below 1.0 correspond to situations where the sample standard deviation turns out smaller than the value specified for the process; it is a very rare customer who will complain if the product supplied has a smaller variability in composition than has been claimed!

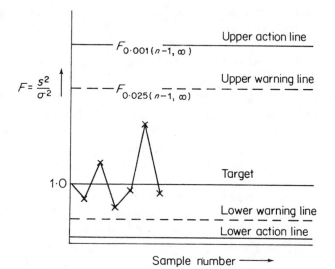

Fig. 7.4b.   *A Shewhart control chart for sample standard deviation*

7.4.2.   A Control Chart for Sample Ranges

Having outlined the theoretical basis for control charts for sample standard deviations, let us make an admission: hardly anyone ever uses them! Why not? Well, calculating sample standard-deviations and F-values involves a little more arithmetic than was needed for a control chart for means. Control charts for sample standard deviations are a bit troublesome to use. With this in mind, an arithmetically simpler control chart has been invented. It is a control chart for *sample ranges. The range for a set of numbers is the difference between the largest and the smallest number. It is thus, like the standard deviation, a measure of the spread of a set of numbers.*

Sample ranges, like sample standard-deviations, have a distribution which can be described mathmatically. Also like sample standard-deviations, the distribution of the sample ranges depends upon the sample size. We shall not go into any of the theory here, we shall simply present a table which enables us to set up a control chart for sample ranges, and demonstrate how to use it.

Sample Size	Lower Limits		Upper Limits		Control Value
	d_2'	d_1'	d_1	d_2	d_n
2	0.00	0.04	3.17	4.65	1.128
3	0.06	0.30	3.68	5.06	1.693
4	0.20	0.59	3.98	5.31	2.059
5	0.37	0.85	4.20	5.48	2.326
6	0.54	1.06	4.36	5.62	2.534
7	0.69	1.25	4.49	5.73	2.704
8	0.83	1.41	4.61	5.82	2.847
9	0.96	1.55	4.70	5.90	2.970
10	1.08	1.67	4.79	5.97	3.078

Fig. 7.4c. *A table for control-chart limits for the sample range*

Each of the values is used to fix the position of one of the horizontal control lines on a Shewhart-type chart for sample ranges. The values must be taken from the row of the table corresponding to the sample size, n, being used. Each value is multiplied by σ, the value for the process standard-deviation under statistical control.

(*i*)   *The lower action limit* for the range of samples of size n is given by $d_2'\sigma$. Under statistical control only 1 sample range in 1000 should lie below this value.

(*ii*) *The lower warning limit* is given by $d_1'\sigma$. Only 1 sample range in 40 should lie below this value.

(*iii*) *The central 'control' value* is given by d_n. This value plays the same role as the target value in earlier Shewhart charts.

(*iv*) *The upper warning limit* is given by d_1. Only one sample range in 40 should exceed this value.

(*v*) *The upper action limit* is given by d_2. Only one sample range in 1000 should exceed this.

Fig. 7.4d gives a sketch of a chart for an example where the process standard deviation under statistical control is 8 units, and where quality-control measurements are made by using samples of size 4. Check that by consulting the table, you can agree with the positioning of the lines. The diagram is of a similar general appearance to the control chart for sample standard deviations sketched in Fig. 7.4b. It performs a similar function and is more popular simply because its use requires marginally less arithmetic. As with standard deviations, the distribution of ranges is skewed. Hence the control lines are not symmetrically spaced on the chart, in contrast to the Shewhart chart for means. The lower warning and action lines are often omitted, as it is not usually regarded as a problem if the process variability is smaller than the specified level.

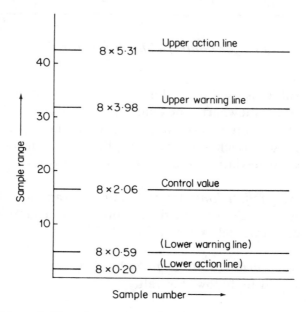

Fig. 7.4d. *A Shewhart chart for sample ranges when $\sigma = 8$ and $n = 4$*

SAQ 7.4a

(*i*) Set up Shewhart Control Charts for sample means and sample ranges for a process whose target value is 100 and which, when under statistical control, has a standard deviation of 6. Assume a sample size of 9 and allow for up to 20 data-points on the x-axis.

(*ii*) Enter the information tabulated below on the appropriate chart and comment on any interesting occurrences.

Sample Number	$\bar{x}$	Range
1	105	19
2	97	13
3	100	25
4	110	40
5	104	36
6	102	10
7	102	27
8	103	20
9	97	16
10	98	10
11	103	12
12	95	20
13	98	20
14	94	23
15	97	12
16	97	19

SAQ 7.4a

7.5. TAKING STOCK

Here we are at the end of our excursion into the statistics of quality control. In some ways this topic appears distinctly different from what has gone before. Control charts certainly represent a new way of working. And once we are used to them they are rather easy to work with. It is much simpler to comprehend information presented in graphical form than it is to follow the rather involved arithmetic which marked earlier topics we have dealt with.

Underlying it all, however, we have been using old methods. The control lines in Shewhart charts are none other than critical values for hypothesis testing. A point falling outside of an action line simply represents a measurement which is incompatible with the established process value (be it a mean or a standard deviation) at some chosen level of significance. If the line is drawn at the '1 in 1000' threshold, then the significance level is 0.1%. The basis for the tests of the sample means and the sample standard deviations is just the same as in Part 3 and in Part 4. The test of sample ranges is a new one, but it follows similar principles to the other tests we have met before. Charts make their appearance because quality control is repetitive work; regular samples are taken, day in and day out, and a chart lets us see any developing pattern much more clearly than if we had to scour through page after page of numbers recorded in a laboratory notebook.

Looking for an emerging pattern in a series of measurements was made somewhat easier by using a cusum type of chart. The accumulation, sample by sample, of discrepancies from the target value means that any drift in behaviour soon shows up by giving the cusum a clearly sloping trend. The V-mask gave us one way of testing for behaviour which would convince us that the process had departed from a state of statistical control. In essence, the V-mask simply represents another way of applying a hypothesis test, with a criterion (any point crossing an arm of the mask) which would lead us to reject the null hypothesis (the assumption of statistical control).

Do we hear you ask: 'What has happened to the computer?' This is the first part of our Unit where we have not explicitly asked you to use it. In compiling simple control charts the computer has relatively little to offer. The one exception to this in what we have covered is in the application of a V-mask to cusum charts. If we were to make a practice of applying that method to a whole range of different examples, then computer manipulation of the data would save a fair bit of bother in scaling our plots, setting up the mask point by point, etc. It would be important, however, to use computer graphics so that the visual impact of trends in the data would be apparent to us. Setting up a program would be a difficult task, and not appropriate to the introductory treatment we have been giving in this Unit.

The fact that we have not introduced the computer in the context of quality control has a very great deal to do with the stopping point in our treatment. Had we gone much further computation would have come into the act in a rather big way. Consider for instance the analysis of cusum plots other than by using a V-mask. We made much of the fact that a non-zero sustained gradient was an indication of a shift in the process mean. But *how steep* need a slope be and *how long* must the trend be sustained before it can be put down to the effect of an assignable cause? In Section 7.2.3 we approached this question by evaluating confidence intervals for the process mean. If the target value lay outside the confidence range then we had grounds for asserting the presence of a shift in performance. But that all relied on our being able to depend on rigid constancy in the process standard-deviation, σ. If we accept, as very often we must, that the process standard-deviation itself might wander then the approach adopted in Section 7.2.3 is of no avail. How then would we test? Well, if we are looking for evidence of a non-zero gradient over a series of readings surely it would be natural to fit *a regression line* to the corresponding section of the cusum plot. We would then wish to test the hypothesis that the gradient of this line was non-zero. Wheel out the methods of Part 5! If one had to do this on a regular basis, as sample data was accumulated, then computer handling would be almost essential.

Remember that one aim of this Unit has been to develop your experience in computation. You will recall that of late we have been progressively encouraging you to take the initiative in this work; to build competence in working with a computer this is a vital transition to make. We suggest that you use program "LINE" to investigate for yourself the significance of the gradients for various sections of the cusum plot for the example which we discussed in Sections 7.3 and 7.3.1. The measured sample means are listed at the beginning of Section 7.3, and the cusum plot is in Fig. 7.3a with sections of it reappearing, rescaled, in Fig. 7.3c–f. See, for instance, the extent to which with program "LINE" you can confirm non-zero gradients before these are detected by a point crossing an arm of the V-mask. We have chosen not to set this as a closely defined SAQ, nor do we give any 'response' in addition to what we already say about the results in Sections 7.3 and 7.3.1. We are setting you off on your own. A possible starting point might be to apply "LINE" to the four points 11–14 to check the comment in Section 7.3.1 that a rising trend might have been identified as early as point 14. Can you assert, with say 95% confidence, that the regression line over these points has a positive slope? What happens if you include point 15?

Computers loom large in the general area of quality control in ways quite distinct from simply carrying out the statistical calculations. By its nature, quality control is a routine and repetitive business. It is very well suited to automation involving the interfacing of computers to equipment. The collection and handling of samples may be automated. Instrumental readings from the analysis may be automatically transmitted into a computer's memory. This information may then be automatically assessed to check the performance of the process. The computer may then send signals which automatically adjust controls governing the production process itself. The principles underlying this whole area would represent a fertile field for further study, though it strays beyond what is generally regarded as the terrain of an analytical chemist into that of the process engineer.

Objectives

As a result of completing Part 7 you should feel able to:

* identify batch inspection and process control as two distinctive objectives for quality control;

* recognise the role of **control charts** as aids to identifying when an **assignable cause** leads to departure from a *state of statistical control*;

- construct a Shewhart chart for successive sample means, interpret data plotted on such a chart, and comment on occurrences of breaches of the inner or outer control lines;

- be aware of the significance of the *average run length* which may elapse before the presence of an assignable cause may be revealed;

- construct a cusum chart for successive sample means and interpret data plotted on such a chart:

  - noting the relation between the cusum gradient and the current process operating mean,

  - if the process standard deviation is known, obtain an estimated confidence interval for the current process operating mean.

  - using an appropriately designed V-mask, identify an 'out of control' condition;

- construct a Shewhart control chart for successive sample standard deviations, noting how to deduce the positions at which to draw the warning and action lines;

- with the aid of Fig. 7.4c, construct and interpret a Shewhart type of control chart for *sample ranges*.

8. Accuracy and Precision in Handling Results

Introduction

We have come a long way and we are within sight of the end of the Unit. We started with a discussion of measurement, random and systematic error, etc, and we shall now return to these same topics. We do so from the particular point of view of considering how results may be presented and how the level of error inherent in them may be clearly appreciated.

Much of what we are about to study could have been discussed earlier. We felt, however, that you would gain more if we dealt with it after covering such topics as populations and samples, estimates of μ and σ etc. The material in this part of the Unit is notorious amongst chemistry teachers! This is synonymous with saying that students do not fare too well with it! Do not let that put you off! In fact the problem is not so much that students find this area difficult to understand, it is rather that they do not pay sufficient attention to it. Some pay no attention at all! The advantage we see in delaying our treatment to this stage, is that you should now be in the best possible position to appreciate why there is a need for care and for judgement. It is the bull-in-a-china-shop syndrome which above all, we must avoid. After all you have been through, you should be able to take the whole business absolutely to heart.

More important, this material is at the very heart of an analytical chemists's function. Let us remind ourselves of what that function is. It is to provide a client with information which is trustworthy. In order to do this, an analytical chemist must understand how errors, both random and systematic, affect results and how to present these results in an honest, tolerant and intelligible manner. This final part of the Unit has, as its aim, the enhancement of your skills in these areas.

8.1. SIGNIFICANT FIGURES

Let us start our work on significant figures by reflecting upon the procedure followed when we take reading from an instrument: we shall use the ubiquitous 50 cm^3 burette as an example of a typical instrument. The burette is marked in 0.1 cm^3 increments from 0 to 50 cm^3. We can therefore easily locate the position of the meniscus of a liquid in the burette as being somewhere between two marks which are, in volume terms, 0.1 cm^3 apart. This means that if we were taking a reading from a burette, we would, at a glance, see that a meniscus was somewhere between say 25.0 cm^3 and 25.1 cm^3. In such circumstances, the procedure we should follow would be to make a mental record of the fact that the reading was greater than 25.0 cm^3 and then, by interpolation, attempt to estimate the actual position of the meniscus. This is the relatively difficult part of reading a burette. We might, for example, say to ourselves that since the meniscus appeared to be halfway between 25.0 cm^3 and 25.1 cm^3, there would be justification in recording the reading as 25.05 cm^3. Before writing this reading down we might decide to take another look at the position of the meniscus. This might lead us to the conclusion that the meniscus was marginally closer to 25.1 cm^3 than it was to 25.0 cm^3. This thinking might lead us to record the result as 25.06 cm^3. Is the reading 25.05 cm^3 or 25.06 cm^3? There is obviously some uncertainty over the last figure in the reading. What should we do? Take another look? Nothing prevents us from doing this. However, the process could go on for ever! The fact remains that there will always be some uncertainty over the last figure, ie the one obtained by interpolation. Let us be decisive! We shall record the reading as 25.06 cm^3. In doing this, we are following a 'rule' known as the *significant figure convention*. This states that we should *record results so that they contain the figures known with certainty plus the first uncertain figure.* In our example, the first three figures in 25.06 cm^3 are known with certainty, the last one is the uncertain figure. We have therefore obeyed the significant figure convention. This seems easy enough! However, there are, especially for novices, continual temptations to break the convention. For example, less experienced people than ourselves might have been tempted to compromise over the uncertainty between 25.05 cm^3 and 25.06 cm^3 and thus to record the result as 25.055 cm^3. This would mean that they would have recorded a result with 3 certain figures and 2 uncertain ones. This would obviously be a departure from the significant figure convention. It would also be a little bit of nonsense! Even the eagle-eyed amongst us cannot read a burette to the nearest 0.005 cm^3.

The reading 25.06 cm^3 has four significant figures; three of them are known with certainty, the last one is the uncertain one. It is important, for reasons which will soon become apparent, for us to know just how many significant figures are present when the result of a measurement is recorded.

As analytical chemists we shall wish to report our measurements in line with this convention by using a number of significant figures such that the last digit quoted is the first one of which we are uncertain. There are, however, circumstances when we deliberately choose to record a value with fewer significant figures than are reliably known. Thus you can probably quote the value of π as 3.1416, that is to 5 significant figures. Or it could be that the value you remember is 3.142, which is correct to 4 significant figures. The appropriate button on your calculator may generate 3.1415927, which is correct to 8 figures. You may have heard of prodigies whom, as a party trick, enunciate the first thousand significant figures. When we round off such a number, say to 3.142, we are making a decision that 4 significant figures (or however many we choose) is *sufficient* for our purposes. By the time you have read on a little you should have a clearer idea of the basis upon which one would decide that a certain number of figures 'is sufficient'.

Let us in the following discussion assume that the right number of significant figures is present: our task is simply to state how many significant figures there are in each example.

1.234   4 Significant figures.

1234   4 Significant figures.

1234.0   5 Significant figures. In 1234, the last figure, the 4, is
               the uncertain one: in 1234.0, the 0 is the uncertain one.

1.234×10^3 This is an interesting one! There are 4 significant fig-
               ures: the uncertain one is the 4. The 10^3 simply places
               the position of the decimal point, ie it tells us that
               we are dealing with the number 1234. When we write
               1.234×10^3, we are aware that we could also write
               12.34×10^2 or 123.4×10^1 or 1234. In each case we
               are making a statement that the 4 is uncertain. If our
               measurement system permits us to report 1234.0, ie 5
               significant figures, then we could write this as 1.2340
               $\times 10^3$ or 12.340×10^2 or 123.40×10^1 or 1234.0.

12340   This is a very interesting one! Why? Because we do
               not really know how many significant figures we are
               dealing with. What? Well, it is not immediately clear
               whether we are, in the example given, dealing with a
               measurement system that yielded 1.234×10^4 (4 signif-
               icant figures) or one that yielded 1.2340×10^4 (5 sig-
               nificant figures)? What lesson do we learn from this? It
               is best to express the results of one's measurements in
               the form 'number multiplied by 10 raised to the appro-
               priate power', where the number written down gives
               the number of significant figures. If you need convinc-
               ing of the wisdom of this procedure, then consider one
               of the best remembered numbers in science, ie 6.022
               $\times 10^{23}$ mol^{-1}. This number, Avogardro's number, is
               usually quoted to 4 significant figures.* If we were to
               meet this number for the first time as 6022 followed
               by nineteen zeros, then we would not have the faintest
               idea of the number of significant figures.

---

* The value of Avogadro's number is actually known to 6 significant figures
as 6.02205×10^{23} mol^{-1}. If we quote it as 6.022×10^{23} mol^{-1}, we are
effectively deciding that 4 significant figures is sufficient for our purposes.

0.1234   Four significant figures. We could write this as 1.234×10^{-1} or 12.34×10^{-3} or 123.4×10^{-3} or 1234×10^{-4}. In each case, the number of significant figures is 4. What about the zero before the decimal point? It is simply a convention to put it in: it is not a significant figure. If you have any temptation to call it a significant figure, then reflect upon the number 0000.1234. Would you be willing to say that this number has 8 significant figures? Of course not!

0.001234   Four significant figures. Why? Consider some alternative ways of expressing this number, ie 1.234×10^{-3} or 1234×10^{-6}. In each case there are 4 significant figures. If we want to express the number as 0.001234, we should keep in mind that the zeros immediately after the decimal point simply tell us the location of the decimal point. They serve the same purpose as 10^{-3} in 1.234×10^{-3}. Do you need convincing? Let us assume that we carry out a titration and find that the titre is $25.01 \ cm^3$. We have here 3 figures of which we are certain and 1 figure which is uncertain, ie we have 4 significant figures. Now let us express the titre value in litres. It is 0.02501 litres or 2.501×10^{-2} litres. How many significant figures? Yes, 4. We would hardly expect the number of significant figures to change when we changed units. Again the message is clear. With significant figures in mind, it is best to express the result of a measurement as a 'number multiplied by 10 raised to the appropriate power' where the number contains the right number of significant figures.

8.1.1. Significant Figures in Addition and Subtraction

There are few of us who, during our formal education, have not experienced a teacher who, on marking an assignment which involved computations, found it necessary to scribble 'significant figures' alongside our answer! If anything, the advent of the pocket calculator has made matters worse. Chemistry teachers find themselves driven to distraction by students who, on reporting the results of a titrimetric procedure, say things such as 'the unknown solution is 0.1010139 molar.'! Let us spend a short time learning how to avoid such horrors!

It is important when we report our results that they still conform to our convention on significant figures. The last digit quoted should be the first one the value of which is uncertain. The values we combine in a calculation may themselves vary in both the number of significant figures each contains and in their relative magnitudes. We need to develop ground rules which will enable us to quote the result of calculations combining such values to an appropriate number of figures.

We shall first deal with calculations involving addition and subtraction.

When we add or subtract numbers, the final result should have no more significant figures after the decimal point than the number being added or subtracted which has the fewest significant figures after the decimal point.

∏   Calculate the relative molar mass of $CaCO_3$.

The relative molar mass of a compound is obtained from the experimentally determined relative atomic masses. The relative atomic masses for the elements in $CaCO_3$ are Ca = 40.08, C = 12.01115 and O = 15.9994.

The relative molar mass of $CaCO_3$ is obtained from:

$$
\begin{array}{llll}
\text{Ca} & = & 40.08 \\
+\,\text{C} & = & 12.01 & 115 \\
+\,3 \times \text{O} & = & 47.99 & 82 \\
\hline
& = & 100.08 & 935
\end{array}
$$

Since there are only 2 significant figures after the decimal point in the relative atomic mass of Ca, there should only be 2 significant figures after the decimal point in the relative molar mass of $CaCO_3$. (In setting out the sum above we left spaces to emphasise this point.) We shall give the result as 100.09. We got this result by carrying out the addition as above and 'rounding-off' to the second place of decimals. In the result 100.09, there is uncertainty associated with the last figure, ie the 9. This is because of the uncertainty in the 8 in 40.08. Notice that the eventual result, 100.09, has five significant figures.

In this example, the same answer would have been obtained by 'rounding off' all the relative atomic masses to 2 places of decimals before carrying out the addition. This is not always true however; if fewer significant figures are retained during the calculation the combined effect may be such as to alter the final result perceptibly. *The best practice is to round off at the end of the calculation.*

∏   Consider the addition: 65.12 + 4.023 + 7.214.

    (*i*)   Add first, then round,

    (*ii*) Round first, then add.

(*i*)   65.12
     4.023
     <u>7.214</u>
   76.357

   Quoted as: 76.36

(*ii*)   65.12
     4.02 (rounded)
     <u>7.21</u> (rounded)
   76.35

   Quoted as: 76.35

This example reinforces our point that it is best to round off at the end of the calculation. Although there is uncertainty in the value of the last place quoted, 76.36 is a better estimate, in the light of the data, than is 76.35.

8.1.2.   Significant Figures in Multiplication and Division

When multiplication or division is involved the situation can be a little more tricky. We shall, however, work to a very straightforward rule of thumb! *When multiplying or dividing, the number of significant figures in the final result generally should equal the smallest number carried by any of the contributing values.*

∏   A solution is 0.1025 M with respect to HCl. How many moles of HCl are present in 250 cm^3 of the solution.

The number of moles is obtained by multiplying the molarity, with units of mol dm^{-3}, by the volume expressed in dm^3, ie moles $= M.V.$ In our example this is as follows.

$$\text{Moles} = 0.1025 \text{ mol dm}^{-3} \times 250 \times 10^{-3} \text{ dm}^3$$

$$= 25.625 \times 10^{-3} \text{ mol}$$

$$= 0.025625 \text{ mol}.$$

Now, the molarity is given with 4 significant figures and the volume with 3 significant figures. This means that we should express the product of the molarity and the volume with no more than 3 significant figures. The answer to the question is therefore 0.0256 mol or 2.56×10^{-2} mol. One can almost hear your objections to this answer. From long experience in titrimetric analysis, you would probably wish to give an answer having 4 significant figures. Let us discuss why you are wrong on this occasion but have probably been right on previous occasions. If we follow the significant figure convention, then we must report our answer with 3 significant figures: there are only 3 significant figures in 250 cm^3. Normally, you are used to quoting volumes with 4 significance figures, eg 23.75 cm^3.

Consequently, since molarities are also often given with 4 significant figures, there is nothing wrong in quoting the product of molarity and volume to 4 significant figures. Had we asked how many moles of HCl were present in 250.0 cm^3 of the solution, then the answer could be given with 4 significant figures. Of course, this may have been what was meant in the first place. We all tend to be lazy at times. If we knew the solution had been competently made up in a 'standard 250 cm^3 flask' then we should be happy that a fourth figure was justified, but we should still tend, in loose expressions, to describe the volume as 250 cm^3. Are we perhaps being a little pedantic? Yes, we admit it, but if we are keeping track of the level of uncertainty then we need to be! If we mean 250.0 cm^3 then that is what we should write. That is our significant number convention, the final '0' is valuable information.

Incidentally, with 4 figures, what is the answer to the question? Does 0.025625 round to 0.02562 or to 0.02563? When the figure being rounded off is a bare 5, then the convention we adopt is to *round to the nearest even figure*. So 0.025625 is rounded to 0.02562. If more than one digit is being rounded away, then when we round a 5, the later figures are taken into account. Thus 0.0256251, rounded to 4 figures, gives 0.02563, though 0.0256250 would still round to 0.02562.

Let us state a reservation about our 'rule of thumb' on significant figures in multiplication or division. Compare two simple examples:

$$3.16 \times 3.16 = 9.9856$$

$$3.17 \times 3.17 = 10.0489$$

Our numbers on the left hand side have three significant figures, hence by our convention so should the answers. Certainly, we should quote the first result as 9.99, but should we really report the second as 10.0? The level of uncertainty in the result has surely not increased by a factor of ten as we go from calculations with 3.17 instead of with 3.16. Clearly not. A change of 0.01 in one of the numbers on the left causes a change of about 0.03 in both answers. It is prefereable to quote the second result as 10.05.

We do not wish to pursue this matter too deeply: our convention does not constitute the last word on significant figures. It is what we said it was, merely a useful rule of thumb. What we are mainly concerned with is that you maintain a keen awareness of significant figures and that you adopt a reasoned view as to how many you quote. This Unit has been about quantifying errors; once you have read Section 8.2 you should be able to carry through this whole process more satisfactorily. The 'rules of thumb' are nonetheless very useful in routine work.

SAQ 8.1a

Each of the following measurements is given to the correct number of significant figures.

$a = 23.70$
$b = 64.82$
$c = 250.0$
$d = 70.77$
$e = 363$

Give the answer to the following calculation to the correct number of significant figures.

$$\frac{(a/b) \times c + d}{e}$$

8.1.3. Significant Figures and Logarithms and Exponentials

We have dealt with the simple arithmetical operations of addition, subtraction, multiplication and division, but what happens to the number of significant figures when we take a logarithm? The rule of thumb is: *when taking the logarithm of a number, quote as many figures after the decimal point as there were significant figures in the original number.* Note 'after the decimal point': this is the part of the logarithm known as its *mantissa.*

∏ Give the logarithm (to base 10) of each of the following:

20, 20.0, 20.00 and 20.000.

The correct answers to quote are 1.30, 1.301, 1.3010 and 1.30103, respectively.

Do you see it? Yes, it is easy!

∏ The stability constant, K, for the EDTA complex of magnesium is 4.9×10^8. What is the value of log K?

The value of log 4.9×10^8 is given by the pocket calculator as 8.6902. There are 2 significant figures in 4.9×10^8. Therefore we quote log $K = 8.69$.

∏ What is the pH of a solution which is 2.0×10^{-3} molar with respect to HCl?

HCl is a strong acid. Hence we can assume that $[H^+] = 2.0 \times 10^{-3}$ mol dm^{-3}

Taking pH $= -\log[H^+]$, we have pH $= -\log 2.0 \times 10^{-3} = 2.69897$

Since there are only 2 significant figures in the number, 2.0×10^{-3} there should only be 2 decimal places in the mantissa. Consequently we report that the pH of the solution is 2.70.

Again, we must emphasise that we are dealing with a rule of thumb. If we apply the rule to take the logarithm of a number whose leading figure is a small one (eg 125.4) then the uncertainty in the last significant figure of the log is substantially larger than in that of the original number. If on the other hand a number has a high leading figure (eg 9.25 × 10^{-6}) then the last figure quoted for the log will be less uncertain than is the last significant digit of the original number. Rules of thumb are only rules of thumb; for a more thorough approach you have to make detailed use of the results in Section 8.2.

Since determining an antilogarithm is simply the reverse process of determining a logarithm, the following rule will come as no surprise to you. *When determining an antilogarithm, there should be as many significant figures in the resulting number as there are decimal places in the mantissa of the logarithm.*

∏   If log x = 2.556, what is the value of x?

We need to determine the antilogarithm of 2.556 or, in other words, $10^{2.556}$. With a pocket calculator the answer is given as 359.74933. Since there are 3 decimal places in the mantissa, there should be 3 significant figures in the antilogarithm. The answer is therefore given as 360 (or, if we wish to be totally clear, as 3.60 × 10^2).

∏   If pH = $-\log[H^+]$ and pH = 3.56, what is the value of $[H^+]$.

We need to determine $10^{-3.56}$. By using the pocket calculator, we get the answer 2.75422 × 10^{-4}. Since there are 2 decimal places in the mantissa, there should be 2 significant figures in the antilogarithm. The answer is therefore 2.8 × $10^{-4}M$. Yes, we know that some would be happier giving the answer as 2.75 × 10^{-4}. However, rules are rules! Furthermore, the more one thinks about evaluating a $[H^+]$ from a pH measurement, the more one becomes convinced that an answer consistent with the significant figure convention is the only one that an honest analytical chemist could tolerate. In normal pH measurements uncertainties of the order of ±0.02 pH units are quite common. When we determine the $[H^+]$ of a solution whose measured pH is 3.56, we should be aware that the pH is likely to be in the range 3.54 to 3.58: this implies that the $[H^+]$ could be in the range 2.9 × 10^{-4} to 2.6 × $10^{-4}M$. Clearly, 2 significant figures is appropriate when evaluating $[H^+]$ from normal pH measurements.

We have been concerned above exclusively with logarithms to base ten and exponentiation of the form 10^x. When using natural logarithms $\ln(x)$ or exponentials e^x, then basically the same rules of thumb apply, except that the last place of the mantissa of $\ln(x)$ is significantly more uncertain than the last figure in x and, conversely, the last figure in e^x is substantially less variable than the last decimal place in x.

8.1.4. The Presentation of Results

Our significant figures convention calls on us to record results so that they contain the figures known with certainty plus the first uncertain figure. A question arises, 'how uncertain are we of the uncertain figure?' Some automatically assume (even after a course in statistics?) that when a result is quoted to the correct number of significant figures, one can safely assume an uncertainty in the uncertain figure of ± 1. Such thinking is mistaken! Let us reflect upon why this is so. Consider an analyst who carries out four replicate titrations and obtains the following results: 23.05 cm^3, 23.09 cm^3, 23.07 cm^3 and 23.06 cm^3. The mean titre is 23.0675 cm^3. This when expressed to the correct number of significant figures is 23.07 cm^3. Can we safely assume that the uncertainty in this result is ± 0.01 cm^3? The answer, clearly, is no. The results are scattered about their mean and two of them differ from the mean by 0.02 cm^3. The origin of this scatter is, of course, the random error of measurement. If we are interested in giving magnitude to the uncertainty associated with a result, we shall have to turn our attention to measures of random error. (If there is one thing we know about after a course in statistics, it must surely be the standard deviation!)

It is customary to give the mean of replicate measurements as an estimate of the quantity measured, and the standard deviation as a measure of the uncertainty associated with the measurement. In statistical terms, the sample mean $\bar{x}$ is an estimate of the population mean μ and the sample standard deviation s is an estimate of the population standard deviation σ. In the example given above, the mean titre is 23.07 cm^3 and the standard deviation 0.017 cm^3. (Check that we have calculated s correctly.) Some would argue that, since we quote the mean to two places of decimals we should give s as 0.02

cm^3. To do that is to give s to only one significant figure. We prefer to keep the extra figure in a case like this. One way to report the result is simply to quote the values obtained for the mean and for s. It is informative also to report the number of replicate readings (4 in this case).

Alternatively, one can report the results by specifying a confidence interval, in the form below.

$$x \pm t_{\alpha/2}s/\sqrt{n}$$

For a 95% confidence interval in the above example, using $t_{0.025,3}$ = 3.182, we have

$$23.07 \pm (3.182 \times 0.017)/\sqrt{4} = 23.07 \pm 0.027 \text{ cm}^3$$

In reporting a result in this fashion it is important to make clear what level of confidence is invoked (95% here) and how many replicates were taken ($n = 4$ here).

8.2.   THE PROPAGATION OF RANDOM ERROR

In analytical chemistry, the quantity to be determined is often calculated from several *independently* measured quantities. Now each of these measured quantities has associated with it an uncertainty due to random error. These uncertainties will all contribute to the uncertainty in the calculated quantity. The way in which uncertainties combine, or are *propagated*, depends upon the type of arithmetical relationship between the measured quantities and the calculated result.

In the following sub-sections we shall study how errors are propagated. This will allow us to give quantitative estimates of the uncertainty in the final calculated quantity. The combination rules we present here are much more thorough-going than the 'rules of thumb' which we outlined above for the propagation of numbers of significant figures.

8.2.1. Propagation of Random Error in Addition and Subtraction

Addition and subtraction are present in what is known as a linear combination. We can write a generalised linear combination in the form $y = k + k_a a + k_b b + k_c c + \ldots$ where k, k_a, k_b, k_c, etc are constants (whose values we presume are known exactly). Let us assume that y is the quantity to be determined and that it is calculated from independent measurements of $\bar{a}$, $\bar{b}$, $\bar{c}$, etc. Measurements would yield us, as estimates of a, b, c, etc, the means $\bar{a}$, $\bar{b}$, $\bar{c}$, etc. The estimate of y would be calculated from $k + k_a \bar{a} + k_b \bar{b} + k_c \bar{c} + \ldots$ We should also obtain from our measurements, s_a, s_b and s_c etc. These are estimates of σ_a, σ_b and σ_c etc. Now, when we are dealing with means we should always remember the relationship $\sigma_{\bar{x}} = \sigma/\sqrt{n}$. From this we see that $s_a/\sqrt{n}$, $s_b/\sqrt{n}$ and $s_c/\sqrt{n}$ are estimates of $\sigma_{\bar{a}}$, $\sigma_{\bar{b}}$ and $\sigma_{\bar{c}}$ respectively. For simplicity, we shall call these estimates $s_{\bar{a}}$, $s_{\bar{b}}$ and $s_{\bar{c}}$.

When combining independent measurements the variance σ^2 has an interesting property: it is additive. It can be shown that the variance in y is given by the expression below.

$$\sigma_y^2 = k_a^2 \sigma_a^2 + k_b^2 \sigma_b^2 + k_c^2 \sigma_c^2 + \ldots$$

Since the sample variances are estimates of the population variances, we can write $s_y^2 = k_a^2 s_{\bar{a}}^2 + k_b^2 s_{\bar{b}}^2 + k_c^2 s_{\bar{c}}^2 + \ldots$. If we take the square root of both sides of this equation we obtain

$$s_y = \sqrt{k_a^2 s_{\bar{a}}^2 + k_b^2 s_{\bar{b}}^2 + k_c^2 s_{\bar{c}}^2 + \ldots} \tag{8.1}$$

This equation shows how we can estimate the uncertainty in a calculated quantity from the uncertainties (due to random error) present in the independently measured quantities. *When a result is obtained by the addition and/or subtraction of independently measured quantities, the variance in the result is the sum of the individual variances.*

∏ In a titration, the initial burette reading is 0.52 cm³ and the final reading 23.75 cm³. It is known from past experience that the reading uncertainty (measured by the standard deviation) associated with use of the burette is ±0.016 cm³. Report the titre value with its associated uncertainty.

The titre is given by $V_{final} - V_{initial}$. This is $23.75 - 0.52 = 23.23$ cm^3. (We are permitted 4 significant figures.) We are told from past experience that the standard deviation (σ) associated with a burette reading is $\pm$ 0.016 cm^3.

$$\sigma_{titre} = \sqrt{\sigma^2_{final\ reading} + \sigma^2_{initial\ reading}}$$

$$= \sqrt{(0.016)^2 + (0.016)^2}$$

$$= \sqrt{5.12 \times 10^{-4}}$$

$$= 0.023$$

(Note that the k's of Eq. 8.1 are here equal to 1.0.)

The uncertainty, as measured by the standard deviation, in the titre, is ± 0.023 cm^3.

∏   A result is obtained by carrying out the calculation $r = a - b + c$. The independently measured values for a, b and c and their associated standard deviations are given below.

$\bar{a} = 6.75, \qquad s_a = 0.05$

$\bar{b} = 2.41, \qquad s_b = 0.03$

$\bar{c} = 2.15, \qquad s_c = 0.04$

Calculate r and the uncertainty associated with its determined value.

$r = \bar{a} - \bar{b} + \bar{c} = 6.75 - 2.41 + 2.15 = 6.49$

(We are permitted 3 significant figures.)

The variance in r the result is the sum of the individual variances.

$$s_r^2 = s_{\bar{a}}^2 + s_{\bar{b}}^2 + s_{\bar{c}}^2$$

$$= (0.05)^2 + (0.03)^2 + (0.04)^2$$

$$= 50 \times 10^{-4}$$

$$s_r = 0.071$$

Since the contributing standard deviations were all given to only one significant figure we should quote $s_r = 0.07$.

That was easy enough! Let us calculate another and its associated uncertainty, one where a new and important issue arises.

∏   The relationship between, A, x, y and z is: $A = x + y - z$. In order to determine A, independent measurements of x, y and z were carried out. The results were:

    $\bar{x} = 35.21$ $s_{\bar{x}} = 0.063$

    $\bar{y} = 23.17$ $s_{\bar{y}} = 0.045$

    $\bar{z} = 11.1$ $s_{\bar{z}} = 0.16$

    Calculate A and the uncertainty associated with its determined value.

The value of A is given by $35.21 + 23.17 - 11.1 = 47.28$. Since there is only one decimal place in the value for z, this result should be recorded with 3 significant figures, ie as 47.3. The standard deviation of A is given by:

$$s_A = \sqrt{s_{\bar{x}}^2 + s_{\bar{y}}^2 + s_{\bar{z}}^2}$$

$$= \sqrt{(0.063)^2 + (0.045)^2 + (0.16)^2}$$

$$= \sqrt{315.94 \times 10^{-4}}$$

$$= 0.1777 = 0.18$$

The standard deviation can be reported as $s_A = 0.18$. There is nothing complicated in the above calculations. However, embedded in them is a most important message for analytical chemists. This message is best discussed in terms of the inherent precision of the methods used to determine x, y and z. If we were content to have a result for A which carries with it an uncertainty of about 0.2, then we have probably wasted both time and money in estimating both x and y to a greater precision than is necessary. It was futile to seek such good precision in the results for x and y when the best that we could obtain in the measurement of z was a standard deviation of $\pm$ 0.16. The uncertainty in A is nearly all due to the uncertainty in the measurement of z. If, however, we would have preferred to quote a result for A that had 4 significant figures, a result which had associated with it much less uncertainty, then we know what we have done wrong. We have used a method for determining z which did not have the necessary precision. If we want to determine A more precisely, then we must seek a more precise way of determining z. Until we have found such a method, any attempt to decrease the uncertainty in A by improving the precision associated with the determination of x and y is, to be charitable, simply silly!

When we apply Eq. 8.1 the largest contributing standard deviation dominates the result.

8.2.2. Propagation of Random Error in Multiplication and Division

We can call both types of operation 'multiplicative'. Let us assume that y is a quantity to be determined and that it is related to a, b, c and d by the expression $y = kab/(cd)$, where k is a constant. Independent measurements would yield, as estimates of a, b, c and d, the means $\bar{a}$, $\bar{b}$, $\bar{c}$ and $\bar{d}$. The estimate of y could be then obtained from $k\bar{a}\bar{b}/(\bar{c}\bar{d})$. We would also obtain from the independent measurements, s_a, s_b, s_c and s_d. These are estimates of σ_a, σ_b, σ_c and σ_d and are measures of the uncertainties present in the respective estimates of a, b, c and d. It can be shown that the variances in the calculated quantity and the independently measured quantities are related as follows.

$$\left(\frac{\sigma_y}{y}\right)^2 = \left(\frac{\sigma_a}{a}\right)^2 + \left(\frac{\sigma_b}{b}\right)^2 + \left(\frac{\sigma_c}{c}\right)^2 + \left(\frac{\sigma_d}{d}\right)^2$$

ie

$$\left(\frac{\sigma_y}{y}\right) = \sqrt{\left(\frac{\sigma_a}{a}\right)^2 + \left(\frac{\sigma_b}{b}\right)^2 + \left(\frac{\sigma_c}{c}\right)^2 + \left(\frac{\sigma_d}{d}\right)^2} \quad (8.2)$$

Since the sample variances are estimates of the population variances and the sample means estimates of the population means, we can write:

$$\frac{s_y}{y} = \sqrt{\left(\frac{s_{\bar{a}}}{\bar{a}}\right)^2 + \left(\frac{s_{\bar{b}}}{\bar{b}}\right)^2 + \left(\frac{s_{\bar{c}}}{\bar{c}}\right)^2 + \left(\frac{s_{\bar{d}}}{\bar{d}}\right)^2} \quad (8.3)$$

What is the significance of this equation? Well first let us answer the question, 'What is the significance of say $s_{\bar{a}}/\bar{a}$?' We have met such a term in the first part of the unit. It is the *relative standard deviation*, RSD. What we see inside the square-root sign is the sum of the squares of the relative standard deviations of the independently measured quantities. The square root of this sum gives the relative standard deviation of the calculated quantity, y. This is all that we need notice. Let us familiarise ourselves with the use of the above relationships with a few exercises.

Π   The quantity y is related to a, b and c by $\dfrac{ab}{c}$.

Independent measurements of a, b and c yield the following estimates.

$\bar{a} = 13.87 \qquad s_{\bar{a}} = 0.035$

$\bar{b} = 26.26 \qquad s_{\bar{b}} = 0.013$

$\bar{c} = 2.245 \qquad s_{\bar{c}} = 0.002$

Calculate y and estimate its *relative* uncertainty, as measured by its relative standard deviation and its uncertainty, as measured by its standard deviation.

The quantity y is obtained from $\bar{a}\bar{b}/\bar{c}$

This is $\dfrac{13.87 \times 26.26}{2.245} = 162.239 = 162.2$

The relative standard deviation in y is given by:

$$\frac{\sigma_y}{y} = \sqrt{\left(\frac{\sigma_a}{a}\right)^2 + \left(\frac{\sigma_b}{b}\right)^2 + \left(\frac{\sigma_c}{c}\right)^2}$$

This can be estimated from

$$\frac{s_y}{y} = \sqrt{\left(\frac{s_{\bar{a}}}{\bar{a}}\right)^2 + \left(\frac{s_{\bar{b}}}{\bar{b}}\right)^2 + \left(\frac{s_{\bar{c}}}{\bar{c}}\right)^2}$$

The relative standard deviations squared (the relative variances) are:

$$\left(\frac{s_{\bar{a}}}{\bar{a}}\right)^2 = \left(\frac{0.035}{13.87}\right)^2 = 6.37 \times 10^{-6}$$

$$\left(\frac{s_{\bar{b}}}{\bar{b}}\right)^2 = \left(\frac{0.013}{26.26}\right)^2 = 0.245 \times 10^{-6}$$

$$\left(\frac{s_{\bar{c}}}{\bar{c}}\right)^2 = \left(\frac{0.002}{2.245}\right)^2 = 0.794 \times 10^{-6}$$

The sum of these relative variances is 7.41×10^{-6}. This is the value of $\left(\dfrac{s_y}{y}\right)^2$. Taking the square root we obtain

$$s_y/y = \sqrt{7.42 \times 10^{-6}} = 2.72 \times 10^{-3} = 0.0027$$

This is the relative standard deviation in y. The absolute standard deviation, s_y, is obtained by multiplying by the calculated value of y, ie

$$s_y = 162.2(0.00272) = 0.44$$

Note that the RSD in the measurement of a is the major contributor to the uncertainty in y. If the uncertainties in b and in c were to vanish altogether (ie if $s_b = s_c = 0$) then s_y/y would still be 2.5×10^{-3} (or 0.25%) and s_y would be 0.41. Again the point can be made that there is little point in pursuing high precision in the measurement of b or of c. To improve the precision in the estimate of y effort would have to be concentrated on reducing s_a. *It is worth collecting together the different standard deviations and relative standard deviations in this example.*

	$\bar{a}$	$\bar{b}$	$\bar{c}$	$y = \bar{a}\bar{b}\bar{c}$
st.dev.	0.035	0.013	0.002	0.44
% RSD	0.25	0.050	0.09	0.27

Remember that when an additive operation was involved the standard deviation of the final result closely reflected that of the measurement with the greatest individual standard deviation. With multiplicative operations this is no longer true; it is a feature of the RSD instead.

∏ A 20.00 cm^3 aliquot of a bromide solution was titrated with $0.1000M$ silver nitrate: the titre was 23.26 cm^3. What is the molarity of the bromide solution? It is thought that the following standard deviation values are applicable:

to the molarity of the standard Ag^+ solution: $\sigma(M_{Ag^+}) = 0.00010M$

to the volume delivered by a 20 cm^3 pipette: $\sigma(V_{Br-}) = 0.020$ cm^3

to the volume read using a burette: $\sigma(V_{Ag^+}) = 0.020$ cm^3

Find the relative standard deviation, and the absolute standard deviation in the calculated molar concentration of the bromide solution.

The molarity of the bromide solution is obtained from the expression below.

$$M_{Br^-} = \frac{(M_{Ag^+})(V_{Ag^+})}{V_{Br^-}}$$

ie

$$M_{Br^-} = \frac{0.1000 \times 23.26}{20.00} = 0.1163 \text{ mol dm}^{-3}$$

The relative uncertainty (relative standard deviation) in the molarity of the bromide solution is given below.

$$\frac{\sigma(M_{Br^-})}{M_{Br^-}} = \sqrt{\left(\frac{\sigma(M_{Ag^+})}{M_{Ag^+}}\right)^2 + \left(\frac{\sigma(V_{Ag^+})}{V_{Ag^+}}\right)^2 + \left(\frac{\sigma(V_{Br^-})}{V_{Br^-}}\right)^2}$$

The relative variances are;

$$\left(\frac{\sigma(M_{Ag^+})}{M_{Ag^+}}\right)^2 = \left(\frac{0.00010}{0.1000}\right)^2 = 1.0 \times 10^{-6}$$

$$\left(\frac{\sigma(V_{Ag^+})}{V_{Ag^+}}\right)^2 = \left(\frac{0.020}{23.26}\right)^2 = 0.74 \times 10^{-6}$$

$$\left(\frac{\sigma(V_{Br^-})}{V_{Br^-}}\right)^2 = \left(\frac{0.020}{20.00}\right)^2 = 1.0 \times 10^{-6}$$

The sum of the relative variances is 2.74×10^{-6}. Thus relative standard deviation in the molarity of the bromide solution is.

$$\frac{\sigma(M_{Br^-})}{M_{Br^-}} = \sqrt{2.74 \times 10^{-6}} = 1.7 \times 10^{-3} = 0.0017$$

If we express the result as a percentage we have $\%RSD(M_{Br^-}) = 0.17\%$. We can see from this value why it is that titrimetric methods are considered so precise. Notice in this case how similar in magnitude were the three contributing relative variances.

The absolute uncertainty in the molarity of the bromide solution is:

$$\sigma(M_{Br-}) = 0.1163(0.0017) = 0.00020 \text{ mol dm}^{-3}$$

If you have had little trouble in following this exercise, you should quite easily manage the SAQ which follows.

SAQ 8.2a

A chemists is asked to prepare a standard solution of sodium chloride. He weighed out, by difference, 0.5850 g of the salt, dissolved it in water and made the solution up to 100.0 cm³ in a graduated flask.

(*i*) What is the concentration of the solution in g dm⁻³?

(*ii*) What is the molar concentration of the solution? (The relative molar mass of NaCl is 58.443.)

(*iii*) If the standard deviation of weighing is known from past experience to be 0.00012 g, what is the standard deviation in the weight of NaCl taken? What is the relative standard deviation?

(*iv*) If the standard deviation associated with the use of the graduated flask is known from past experience to be 0.050 cm³, what is the relative standard deviation associated with the volume of reagent prepared?

(*v*) What is the relative standard deviation in the concentration (g dm⁻³) of the solution? What is the relative standard deviation of the molar concentration of the solution?

$\longrightarrow$

SAQ 8.2a
(cont.)

(*vi*) What is the standard deviation in the molarity of the solution? Give the 99% confidence interval for the molarity.

8.2.3. Propagation of Random Error in Logarithmic Relationships

Logarithmic relationships are quite common in analytical chemistry. Two commonly met examples are the definition of pH and the relationship of spectrophotometric absorbance to transmittance. The propagation of random error in such relationships therefore requires our attention. Consider two quantities, x and y, related by $y = k \ln(x)$, where k is an accurately known constant. We have chosen first to express the relationship in a natural log form because this leads to a simpler formula. In fact it can be shown* that:

if $\quad y = k \, ln(x), \qquad$ then $\quad \sigma_y = k \, \sigma_x/x.$ $\qquad$ (8.4)

Notice *the absolute standard deviation in* y *is proportional to the relative standard deviation in* x.

What about common logarithms (ie logs to base 10)? We simply recall that common logs can be converted into natural logs by using the relationship $\log(x) = (1/2.303)\ln(x)$. We can then apply Eq. 8.4. The upshot is:

if $\quad y = k \log(x), \qquad$ then $\quad \sigma_y = (k/2.303) \, \sigma_x/x$ $\qquad$ (8.5)

We are ready to try some examples.

$\Pi \qquad$ The hydrogen-ion concentration $[H^+]$ of a solution is quoted as 0.0010 M, with $\sigma = 0.00010$ M. Give the pH of the solution and its standard deviation.

Since $pH = -\log[H^+]$, we obtain $pH = -\log(0.0010) = 3.00$.

Do you agree with the number of significant figures we quote (two after the point)? Well let's put it to the test. The standard deviation is given by Eq. 8.5.

---

* We have not derived this or any of the other rules governing error propagation. They can in fact be obtained by using quite elementary calculus. For multiplicative and logarithmic operations the rules are strictly valid only in the limit where the RSD's are small.

$$\sigma_{pH} = (1/2.303)\sigma_{|H^+|}/[H^+]$$

$$= (1.0 \times 10^{-4})/(2.303 \times 1.0 \times 10^{-3}) = 0.04$$

Yes, the second place of decimals is uncertain, though we have done much better than simply decide upon a number of significant figures. We have a value for the standard deviation, so we could quote quantitative confidence intervals for the pH. Notice that here k from Eq. 8.5 was -1 yet we dropped the sign; by convention we quote the standard deviation as a positive value.

∏   The pH of a solution is known by measurement to be 5.55 within a standard deviation of 0.030. Find the hydrogen-ion concentration and its standard deviation.

We have to reverse the calculation of the previous example.

Since $-\log[H^+] = 5.55$

$$[H^+] = 10^{-5.55} = 2.82 \times 10^{-6}$$

We quote this as 2.8×10^{-6} mol dm^{-3}. Do you agree?

Now for the standard deviation.

$$\sigma_{pH} = (1/2.303)\sigma_{|H^+|}/[H^+]$$

$$\therefore \sigma_{|H^+|} = 2.303\sigma_{pH}[H^+]$$

$$= (2.303) \times (0.030) \times (2.8 \times 10^{-6}) = 1.9 \times 10^{-7}$$

So we can handle random error propagation in converting pH into $[H^+]$ and vice versa. We imagine that you are 'straining at the leash' to get at random error in spectrophotometric measurements! In order to satisfy you on this score, let us quickly review the physical relationship involved.

Suppose that we are attempting to relate transmittance readings, T, perhaps from an ir spectrometer, with the concentration, c, of an absorbing species. We further suppose that the Beer-Lambert law applies straightforwardly to the situation in the form

$$A = \epsilon cl$$

where A is the spectrophotometric absorbance, ϵ is the molar absorptivity at the wavelength of interest and l is the cell path-length. We also suppose that ϵ and l are known exactly (not a situation we can always rely upon in practice). The absorbance, A, is related to the transmittance T, which, we suppose, is what is measured, through $A = -\log T$. We thus have the equation below.

$$c = A/\epsilon l = -(1/\epsilon l)\log(T)$$

Hence from Eq 8.5 $\sigma_c = \left(\dfrac{-1}{2.303\epsilon l}\right) \cdot \left(\dfrac{\sigma_T}{T}\right)$

If we divide the second equation by the first we get a tidy expression for the relative standard deviation in c.

$$\frac{\sigma_c}{c} = \frac{\sigma T}{2.303\ T \log(T)} = \frac{\sigma_T}{T \ln(T)} \tag{8.6}$$

In the last line we have taken advantage of the fact that $2.303 \log(T) = \ln(T)$. Let us put this relationship to use.

SAQ 8.2b

A chemist claims to have invented a simple 'breathometer' based on a spectrophotometric method. It has a linear scale for transmittance readings and is claimed to have an uncertainty of 0.010 transmittance units for any transmittance reading. $\longrightarrow$

SAQ 8.2b
(cont.)

The legal limit for ethanol in the breath is found to give a transmittance reading of 0.80. What is the relative error in the concentration at the legal limit?

SAQ 8.2c

The chemist mentioned in the preceeding SAQ sells his instrument to an overseas government. The country in question has a legal limit of 35 μg ethanol/100 cm^3 of breath. (This is the same legal limit as in the UK.) The country also has a large number of lawyers, who first qualified in chemistry (we have some too!). Consequently, the powers-that-be decided that prosecutions would take place only if the breathometer gave a reading equivalent to 37 μg/100 cm^3 of breath. Is their ruling likely to be challenged in the courts?

SAQ 8.2c

8.3. THE PROPAGATION OF SYSTEMATIC ERROR

It is about time that we turned our attention to the topic of systematic error. In particular, we shall examine the ways in which systematic errors, present in independently measured quantities, combine or are propagated to give an overall systematic error in a calculated quantity. Before we discuss such matters, it is useful to revise some of the things we have learned about the structure of error. You will recall that for any measurement, the absolute error (E_{abs}) is related to the different types of error present by the equation below:

$$E_{abs} = x - T = e_r + e_s + e_b$$

where

x = the value of the measurement

T = the true value

e_r = the random error

e_s = the systematic error

and e_b = the error due to blunders.

For replicate measurements we have

$$E_{abs} = \bar{x} - T = e_r + e_s + e_b$$

where $\bar{x}$ is the mean of replicate measurements. For n replicate measurements e_r is expected, an average, to decrease in inverse proportion to $\sqrt{n}$.

If we assume that blunders are absent, we can write $E_{abs} = \bar{x} - T = e_r + e_s$. The implications of these equations are, by now, clear enough. If systematic error is absent, then measurements are normally distributed about the true value. If a very large number of measurements are made, the mean tends towards coincidence with the true value. For a small number of measurements, any difference between the mean and the true value is due to random error. If systematic error is present, then measurements are normally distributed about a value which is defined by the true value plus (or minus) the magnitude of the systematic error. If the systematic error caused a positive bias, then the measurements will be normally distributed about a value greater than the true value: if it causes a negative bias, the distribution will be about a value less than the true value. The difference between the mean of a set of measurements and the true value will incorporate the effects of both systematic and random error.

When a quantity to be determined is calculated from several independently measured quantities, the uncertainties (due to random error) in the measured quantities are propagated into the final result. In an analogous way, any bias (due to systematic error) in the results of the measured quantities will combine to give an overall systematic error in the calculated quantity. The way in which these systematic errors combine can be summarised as follows:

(a) First consider linear combinations (addition and subtraction). Here, the systematic error associated with each measured quantity is transmitted directly into the overall systematic error. Consider a linear combination $y = k + k_a a + k_b b + k_c c + ...$, where the k's are known constants.

If the systematic errors associated with the measurement of a, b and c etc are Δa, Δb and Δc, etc, then the overall systematic error is given by:

$$\Delta y = k_a(\Delta a) + k_b(\Delta b) + k_c(\Delta c) + \ldots \quad (8.7)$$

Unlike random errors, systematic errors are constant in size, though unfortunately, generally of unknown size! They are not randomly variable but affect every replicate measurement equally. Not only does a systematic error have a constant magnitude, it also has a constant *sign*. These signs must be included in applying Eq. 8.7, as must the signs of the k's. In this respect systematic errors combine quite differently from random errors.

(*b*) For multiplicative relationships it is the *relative* systematic error which is transmitted additively into the overall *relative* systematic error. Consider, for example, the relationship $y = kab/(cd)$, where k is a known constant. The overall relative systematic error can be written.

$$\Delta y/y = \Delta a/a + \Delta b/b - \Delta c/c - \Delta d/d \quad (8.8)$$

The expression relies on the individual relative systematic errors being small (eg a few percent). In contrast to random errors (Eq. 8.2), no squares or square roots are involved. Care must be taken with signs. Notice that for divisors the relative systematic error is *subtracted*. The signs of the errors Δa, etc must not be neglected when substituting into the above equation.

(*c*) For the logarithmic relationship $y = k \log x$ we have

$$\Delta y = (k/2.303) \times (\Delta x/x) \quad (8.9)$$

We can also repeat the trick we applied when we considered the effect of random errors in transmittance readings (Eq. 8.6). If we divide Eq. 8.9 by the defining equation $y = k \log x$, we obtain a quite neat expression for the transmitted *relative systematic error*.

$$\Delta y/y = 1/(2.303 \log x) \times (\Delta x/x)$$

$$\Delta y/y = (1/\ln x) \times (\Delta x/x) = \Delta x/(x \ln x) \quad (8.10)$$

The relative systematic error in y equals the relative systematic error in x, divided by the natural log of x. Again, the accuracy of Equations 8.9 and 8.10 depends on $\Delta x/x$ being small. Again the sign of Δx must not be neglected.

Now, if we discover the actual values of systematic errors we tend to lose interest in directly applying Equations 8.7–8.10. We rush instead to correct the value for each measurement and then to recalculate the final result, which should thus hopefully be free of bias due to systematic error.

The rub is that we do not generally know the magnitude of individual systematic errors, sometimes not even their signs. This is where Equations 8.7–8.10 come into their own. If we discover that a systematic error of a particular size is present in the final result of an analysis, we can hypothesise about the presence and magnitude of systematic error in each of the independently measured quantitites, and we can then determine whether these, when propagated into the final result, would explain the bias obtained. Alternatively, before making measurements, we can estimate how a given supposed bias, in one or more of the independently measured quantities, would be propagated into the final result (see SAQ 8.3a). Another important thing which the equations do is to point up the somewhat different way in which systematic errors are propagated compared to random errors. These differences in behaviour result mainly from the inclusion of a definite sign for each given systematic error.

Most of the attention of our Unit has been directed towards random errors, for, once we understood something of the behaviour of random errors, there was much we could achieve in defining and in improving the value of results subject to them. We must take the opportunity afforded by this section to stress once more how important it is to be fully alert to the likely presence of systematic errors and to take steps to minimise them, to circumvent them, or to quantify them. We discussed these issues at greater length in Part 1. The relatively scant attention which systematic error has had since then does not mean that it is in any sense unimportant. Random error is always with us if we are working anywhere near the limits of performance in an analytical method. Statistics greatly helps us to quantify its effects. Treatment of systematic errors depends very much more on the particular analytical process. In essence they must be the subject of detective work. Once they are identified they can be corrected for.

In the next section we shall return to the third general source of error we identified in Part 1, blunders. They happen to the best of us! There the aim will not be to estimate the size of the error, it will be to recognise when a blunder has occurred. Once that is done our action is limited to burying the evidence!

SAQ 8.3a

A standard solution is made by dissolving a weighed quantity of solid in a solvent and making the solution up to the mark in a 250 cm^3 graduated flask. A 25 cm^3 aliquot is then taken for titration against a test solution. Find the effect on the calculated molarity of the test solution if:

(*i*)   the solid used contains 2.0% by weight of a non-reactive impurity;

(*ii*)   The way in which the end-point is detected leads to a consistent 'overshoot' of 0.25 cm^3 (a typical titration requires 20 cm^3 to reach the end-point);

(*iii*)   both (*i*) and (*ii*) operate.

SAQ 8.3b

A mixture contains three compounds A, B and C. Two analytical studies determine the content of A as 80.0% by weight and that of B as 15.0%. It is therefore concluded that the material contains 5.0% by weight of C. Suppose that the determinations of A and B may both have been affected by relative systematic errors of +1.0%. What is the size of the relative systematic error in the concentration of C determined?

SAQ 8.3c

A pH meter operates with a positive bias of 0.10 pH units. What is the systematic error in the hydrogen-ion concentration obtained when the meter reads 4.10?

SAQ 8.3c

8.4. THE DILEMMA OF DISCORDANT RESULTS

We shall start this topic by asking you to recall the first time you ever encountered a quantitative analytical procedure. Yes, it was probably when you were introduced to titrimetry. The time and place of this occurrence obviously varies with the individual. The programme followed, however, was probably the same for all of us. We were assembled in a laboratory, reminded of the principles underlying the method, given instruction in the use of pipettes, burettes and indicators etc. and were told to get on with it! Oh yes, there was one other thing that happened. We were told to do one rough titration, in order to get the approximate position of the endpoint and to follow this with replicate titrations until we got two consecutive titres which agreed with one another. In retrospect, have you any comments to make? Yes, there is nothing wrong with doing a rough titration. Yes, there is nothing wrong with encouraging precise work habits. Yes, 'two consecutive titrations which agree with one another' is, in the light of random error, an incredible demand to place on anyone, especially novices. However, there are differing interpretations of the word 'agree'. Is there any other comment which is worth making? Yes, there is: it is that the instructions given to the students actually encourage bad practice. An explanation is obviously in order.

Let us assume that a student is given a titrand and a titrant which should result in a titre value of 23.80 cm^3. In statistical terms, $\mu =$ 23.80 cm^3. Let us further assume that the random error present in the method results in a standard deviation of 0.02 cm^3, ie that $\sigma =$ 0.02 cm^3. From our knowledge of statistics, we are aware that for such a measurement system, 95% of all titres will fall in the range $\mu \pm 1.96 \ \sigma = 23.80 \pm 1.96 \ (0.02) = 23.80 \pm 0.04$ cm^3. Let us now follow the student as he carries out his instructions (we shall assume that systematic error is absent). The first titre value is 24.1 cm^3. There is no need to dignify a rough titre with 4 significant figures. The student has obviously overshot, but this happens with rough titres. The second titre value is 23.76 cm^3. We are aware that this is the sort of result that could be expected from the measurement system in question. The student is, of course, oblivious of this fact. The third titre value is 23.83 cm^3. Again, we are not surprised by this result; it is in the range 23.80 $\pm$ 0.04 cm^3. The student, however, is a little upset. Remember, he is seeking results which are 'in agreement'. He carries out a fourth titration. As chance would have it, this gives a titre value of 23.84 cm^3. The student is now happy; the last two results have been in close agreement. After consultation with his teacher, he decides to report his result as (23.83 + 23.84)/2 = 23.835 $\approx$ 23.84 cm^3. What has happened? Yes, he has a result which is +0.04 cm^3 in error. That is not at all bad for a novice. However, it is also true that the error would have been even smaller had the student not rejected the second titre value.

The mean of the last three titre values is (23.76 + 23.83 + 23.84)/3 = 23.81 cm^3. This represents an error of only +0.01 cm^3. By following the instructions given, the student ended up with a greater error in his result. The instructions were wrong. However, before anyone starts writing letters to the newspapers, try to think up some better instructions to give young chemists. Yes, the mean of the last three titres values would have served our student better. However, what about the student who obtained the results 24.2 cm^3, 23.90 cm^3, 23.83 cm^3, 23.82 cm^3? Clearly, we need to give the matter further thought.

Often, when carrying out replicate determinations, one of the results will differ considerably from the others. Should this result be rejected or retained? Another way of asking this question is: 'Will the mean result after rejection be a better estimate of the truth than the mean result after retention?' There are occasions when an analyst can answer this question in the light of his experience. For example, if he suspects the 'odd one out' was a result of a blunder, then rejection is undoubtedly the correct course of action. However, blunders may occur without our knowing it. What can be done about this? A few more determinations could be carried out and the results compared with those previously obtained. This should give us a better idea of the validity of the suspect result. However, such a procedure is time consuming, Furthermore, we haven't always got the time. Think of an analysis associated with a hospital emergency or with a continuous process in industry. Moreover, there are situations where the sample analysed is no longer available, eg the adrenelin in an athlete's urine immediately after winning the London marathon. In these circumstances, we are in need of aids to help us decide whether or not a discordant result or outlier should be rejected or retained. It will not surprise you that such aids exist and that they have a statistical basis. Let us discuss a few of them.

8.4.1. Dealing with Outliers: Use of the Median

Some chemists, doubtful about the wisdom of rejecting an outlier, use the median of a set of results rather than the mean. This is because a discordant result has less influence on the median that it does on the mean. The median is the middle result in an odd number of results or the average of the middle two results in an even number. In order to determine the median, we simply arrange the results in an ascending order of magnitude and use our arithmetical skills!

The following results were obtained for the % Fe_2O_3 in an ore: 32.10, 31.96, 32.13, 32.18, 32.15. Note that the result 31.96 appears to be the odd one, ie an outlier. If these results are arranged in order, we have 31.96, 32.10, 32.13, 32.15, 32.18. The median is 32.13%. The mean of these results is 32.10%. If we use the median as the estimate rather than the mean, we are essentially saying that there is a discordant result, that we do not know whether or not it is the result of a mistake, that, consequently, we have decided to retain it, but that we will not let it distort our estimate too much!

8.4.2.   Rejecting an Outlier by Using the Method Standard-deviation

This is an approach which has probably already suggested itself to you. If it has not, then the unit on statistics has failed you! A chemist, who has considerable experience in using a given method to analyse materials similar to the one under test, will probably have a good idea of the precision of his determinations, ie of σ. If σ is known and if it is assumed that $\bar{x}$ is a reasonable estimate of μ, then the range $\bar{x} \pm 2\sigma$ could serve as the basis for deciding whether or not to reject an outlier. Any result which fell outside $\bar{x} \pm 2\sigma$ would be rejected. In the example given above, if $\sigma = 0.04\%$, then the range of acceptable results would be $32.10 \pm 0.08\%$. On this basis, the result 31.96% would be rejected. Once a result has been rejected, the value of the sample mean obviously changes. In the example given, the new mean is 32.14%. This is a result similar to that obtained when the median was used as the estimate of μ. Before leaving this approach to dealing with outliers, it is worthwhile mentioning that when we use $\bar{x} \pm 2\sigma$ as a criterion for rejection, we actually err a little on the side of retention. The initial mean $\bar{x}$ is calculated by using the outlier and hence is distorted in its direction. The consequences of this are apparent.

8.4.3.   Rejecting an Outlier by Using the Sample Range

The range of a set of results is the difference between the largest and the smallest. As has been discussed previously (in Section 7.4.2) the range is, like the standard deviation, a measure of the spread of results. Further, it is actually possible to make estimates of a sample standard deviation and therefore of σ in the light of knowledge of the sample range. The relationship used is $s = R/d_n$, where R is the range and d_n is a parameter whose magnitude depends on the sample size. We first met d_n in our discussion of the use of sample ranges in quality control in Section 7.4.2.

Several tests and 'rules of thumb' have been developed which utilise the sample range in a rejection criterion for outliers. For example, in the same way as it is possible to set up criteria for rejection based on the confidence interval $\bar{x} \pm z\sigma$ or, if we wish to, on $\bar{x} \pm ts$, it is possible to set up confidence intervals based on the range, R. These have the form $\bar{x} \pm t_R R$, where t_R plays the same role as the z or the t in the more familiar confidence intervals. We shall not elaborate on this other than to say that tables of t_R values for various levels of confidence exist and that they are used. Why should anyone wish to use them? For the simple reason that R is so easy to calculate. Imagine that you have lost your pocket calculator and needed a quantitative measure of the spread of a set of results. Which would be easiest – the calculation of s or of R? It should be noted, however, that the value of R exhibits more variability than that of s. Even more!

By far the most widely used of the tests for outliers based on the range is the Dixon Q test. This test is carried out in the following fashion.

(*i*)   The results are arranged in order and the range noted.

(*ii*)   The difference between the suspect value and its nearest neighbour is also noted.

(*iii*) The value obtained in (*ii*) is divided by the value obtained in (*i*), ie by the range. This ratio is the calculated Q-value.

(*iv*)   Reference is made to a table of Q-values. These values have been calculated for different sample sizes and at different levels of confidence. If the calculated Q exceeds the tabulated Q-value, then the suspect result may be rejected with the appropriate level of confidence.

The Q-values for rejection at the 90% confidence level appear in Fig. 8.4a.

Number of Measurements	$Q_{0.90}$
3	0.94
4	0.76
5	0.64
6	0.56
7	0.51
8	0.47
9	0.44
10	0.41

Fig. 8.4a. *Rejection Quotient, Q at 90% Confidence Level, Adapted from R. B. Dean and W. J. Dixon, Anal Chem 23, 636 (1951)*

Let us make use of this table before attempting to get a feel for the rationale behind the test.

∏ Use a Q-test to investigate the suspect result in the set of measurements 31.96, 32.10, 32.13, 32.15 and 32.18.

This is the same set of values which we first introduced as replicate analyses for % Fe_2O_3 in an ore. The range is $32.18 - 31.96 = 0.22$. The low value, 31.96, is the suspect one. The difference between the suspect result and its nearest neighbour is $32.10 - 31.96 = 0.14$. The Q-ratio is $0.14/0.22 = 0.64$ (to two significant figures).

The tabulated Q value for $n = 5$ at the 90% confidence level is 0.64. Yes, life is often like that! Should we reject or accept the suspect value? Let's reject it! The calculated Q-value is right on the rejection limit and from our vast experience, we know the suspect result was probably caused by a blunder! Then again, perhaps we should retain the outlier and report the median!

We are left dithering! But this is not unreasonable. Our outlier lay right on the rejection limit for 90% confidence. That means there was a 10% probability that in the absence of any blunder a set of five measurements would, quite by chance, give as large a Q-value as this. What made us choose a 90% confidence level anyway? Some workers use the table of Q for 95% confidence; they would have ruled unequivocably in favour of retaining the suspect value. (The critical Q for 5 measurements at 95% confidence is 0.72.) Who is right and who is wrong? The answer is the usual one – we do not know. In the absence of further information all we can say is that there is only a 10% probability that so large a Q-value would have arisen by chance. We are in control in the sense that we must decide what level of confidence we shall require before deciding to reject. If the stakes are high then the best option, where feasible, would be to make further measurements.

We have already noticed that rejecting an outlier in a small sample can have a significant effect on the calculated mean. It has an even more dramatic effect on the sample standard deviation. For our ion ore example, if we retain all 5 values we calculate $\bar{x} = 32.10\%$ and $s = 0.086$, whereas if we reject the suspect result of 31.96%, the remaining four values give $\bar{x} = 32.14\%$ and $s = 0.034$, ie s falls by a factor of almost three! Whichever way we jump (deciding to reject, or to retain) we shall be left with some doubt about the validity of our s-value. The only way to resolve things, as always, would be further measurements, or pre-knowledge of σ.

∏　In Section 8.4.2 it was supposed that $\sigma = 0.04\%$ for the iron ore determination. Is this value compatible with $s = 0.086$, the figure obtained when all five results are retained?

How do we test standard deviation? Yes, this is an item for an F-test.

Now　$F = (s/\sigma)^2 = (0.086/0.04)^2 = 4.6.$

What about the degrees of freedom? $\nu_1 = 4$, $\nu_2 = \infty$. Do you agree?

Consult, for example Fig. 4.5d: you should find that $F_{0.01(4,\infty)} = 3.32$.

Thus at a 1% significance level the calculated F is larger than the critical F-value. There is thus less than a 1% probability that so large a s-value could have arisen by chance. Program "STAT" tells us that this probability is indeed a good deal less than 1%. For $F = 4.6$, the probability is given as 0.12%. There is essentially no doubt on the basis of this evidence that the outlier should be rejected.

If we know σ we are in a much stronger position to make decisions about outliers in small samples. Often, however, we shall not have such knowledge. Then a Q-test is about as good a test as any other we can apply.

Some recommend that one should never reject an outlier from a set of only three readings unless one knows the value of σ. This view is based on the quite high probability that two values in a set of three might, quite fortuitously, lie very close together. If one accepts this advice then, if one strongly suspects the presence of an outlier in a set of three replicates, the only reasonable course of action is to make a further measurement.

What is the rationale behind the Q-test? It helps to present the matter visually.

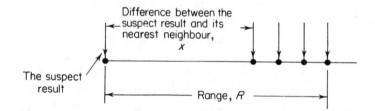

Fig. 8.4b. *Determining Q for a set of measurements*

In Fig. 8.4b. the results of 5 measurements are presented as points on an axis. Four of the results are close together, and the other appears to be an outlier. If all the results were in close proximity then the ratio $Q = x/R$ would be a small fraction. The further an outside result is from its nearest neighbour, the larger the value of the fraction defined by x/R. There must be some value of the fraction x/R which can be used in deciding whether or not the outside result can be retained as a member of the set of replicate measurements. The tabulated values of Q, worked out on a statistical basis, give the value of this fraction, ie give a rejection criterion for x/R and hence for the suspect value.

SAQ 8.4a

Use Q-test(s) to investigate the following data: 1.26, 1.31, 1.23, 1.10, 1.23, 1.24.

There is a rule of thumb, based on the Q-test, to reject an outlier in a sample of size n if the calculated Q-value exceeds $\sqrt{2/(n-1)}$. In fact the formula $\sqrt{2/(n-1)}$ gives a reasonable approximation to the tabulated values for Q for 95% confidence. This is therefore a way of approximating a Q-test at 95% confidence, without the need to look up tables. This, like other tests, will sometimes result in a wrong decision. However it is, for those in a hurry, a nice quick test!

Clearly the existence of a variety of tests for dealing with outliers is evidence that outliers are not at all uncommon. A chemists who confronts them armed with some knowledge of statistics is less likely to make mistakes than those who come armed only with prejudice!

8.5.   THE LIMIT OF DETECTION

We have dealt with significant figures and with how to estimate the
level of error in a result dependent on a number of different measurements. We have considered how to deal with discordant results.
All these things help us to present, as reasonably as possible, our
conclusion as to the result of our analysis and our assessment of the
error in that result.

We now turn our attention towards obtaining the very first significant figure in a determination! More and more often chemists are
being asked to identify the presence of substances at trace concentrations. Sophisticated measurement techniques are being pushed to
their limits. We ask now, for a given instrumental method, just what
is the lowest concentration of a given analyte which is detectable?
If we have an answer to this question we shall know at what level of
concentration we can make a first stab at analysis. We shall find the
lowest concentration at which it is just possible to say: 'Yes, there is
something there, its concentration is of the order of . . .' If we obtain
only one significant figure then our first significant figure is also our
last significant figure, its value is uncertain! But its order of magnitude (ie its power of ten) is not. We are finding a level at which an
analytical technique can first begin to yield results of value.

Let us state our problem in some detail. Our instrument gives a signal whose strength varies with the concentration of analyte present.
Let us suppose that the signal increases with increasing analyte. We
shall, in any instance, have a calibration curve relating instrument
reading, y, to the analyte concentration, x. This calibration may be
a straight-line regression relationship, as discussed at length in Part
5, or it may be a curve. The part of the curve for small x-values is
of interest to us. Fig. 8.5a. gives a sketch of the situation. So, what
is our problem, you may ask? For any measured value of y we can
read off a value of the concentration x. The value y_o in the figure is
the instrument reading corresponding to zero concentration,. Hence
for any measured y-value greater than y_o we shall derive a x-value
bigger than zero. It would be nice if that was all there was to it,
but unfortunately, we have not yet taken account of the existence of
errors of measurement.

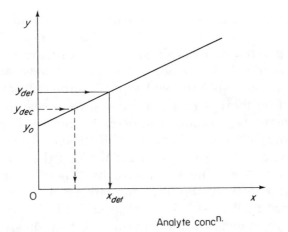

Analyte conc^{n}.

Fig. 8.5a. *An instrument calibration-line at low concentrations and close to the limit of detection*

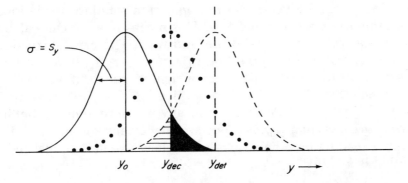

Fig. 8.5b. *Distribution of readings for a blank and for solutions at the limits of decision and of detection*

Let us suppose that systematic errors are absent (this can often be achieved in the use of calibration curves). We shall still be left, however, with random error. If we take a solution with zero analyte concentration and measure y, we are not likely to obtain exactly the value y_0. Instead, replicate readings would be distributed about y_0, sometimes larger sometimes smaller: y_0 should be the mean of the population and it will be characterised by a standard deviation s_y. In the context of our discussions in Part 5, s_y is what we called *the standard error of determination*. The distribution is sketched as the solid curve in Fig. 8.5b. Notice that on 50% of occasions, a reading on a blank solution will give a value greater than y_o. How can we be sure, therefore, if a solution gives a reading in excess of y_o, that it actually contains a non-zero concentration of analyte? Do you recognise the problem? We need to set up a hypothesis test. We take as the null hypothesis the claim that there is zero analyte present. H_o: $x = 0$. The alternative hypothesis is that analyte is present, H_1: $x > 0$. We apply a one-tail test at some chosen level of significance α. We ask what measured critical y-value, will lead us at significance level α, to reject the null hypothesis and to assert that there is indeed analyte present. The critical y-value required will be that for which there is a tail-area of α in the normal curve centred at y_o. This critical value is labelled y_{dec} in Fig. 8.5b. and the tail-area α is heavily shaded.

If the calibration curve has been based on many observations then the value s_y can be taken as the population standard deviation, σ, applicable to measurements of y. This means that we can calculate our critical value in terms of the statistic z. The critical z-value is z_α. The critical y-value would be $y_{dec} = y_o + z_\alpha s_y$. If, for instance, we set our criterion at a 5% level of significance, ie if we take $\alpha = 0.05$, then $z_\alpha = 1.645$. That would give $y_{dec} = y_o + 1.645 s_y$. We call y_{dec} *the limit of decision*. If in any measurement we obtain a reading greater than y_{dec} we declare that analyte is present. If we obtain a reading less than y_{dec} then as far as we are concerned the solution is a blank.

We are still not quite home and dry. We now know what threshold y-value will lead us to decide: 'Analyte is present.' But, *what level of analyte concentration can we be confident of detecting?* Look back at Fig. 8.5a. We see that y_{dec} has been indicated on the y-axis. You might be tempted to answer our latest question by suggesting that we simply use our calibration line to read off the x-value corresponding to y_{dec}, ie follow the dashed arrows on the figure. But look again at Fig. 8.5b. If we had a solution with such an analyte concentration, what measurements would we obtain from it? Again there would be a normal distribution with the same standard deviation s_y as before, but centred now about $y = y_{dec}$. (See the dotted curve in the figure.) Do you see what would happen? Half of the measurements would lie *below* y_{dec} (and half above). There is a 50% probability that any measurement on such a solution will give a value below our chosen limit of decision – we would then declare that no analyte was present! We have a 50% chance of failing to detect analyte at this concentration.

Let us therefore consider measurements on a solution with a somewhat higher level of concentration. These would again have a distribution of the same shape as before, but centred about a correspondingly higher y-value. We have drawn such a distribution in Fig. 8.5b. centred about a value which we have labelled y_{det}. Now there is a reduced probability that a measurement will give a value below the limit of decision, a probability represented by the lightly shaded tail-area in the figure. If we choose y_{det} so that this area is also of size α, then we can have confidence $(1-\alpha)$ that a measurement on such a solution will give a value above the limit of decision, ie that the presence of analyte will indeed be detected. For this to happen we require that $y_{det} = y_{dec} + z_\alpha s_y$. Notice that y_{det} is twice as far from y_o as was y_{dec}. In fact

$$y_{det} = y_o + 2z_\alpha s_y \tag{8.11}$$

If we refer back, yet again, to Fig. 8.5a. we see that we can now read off *the detection limit*, x_{det} by using the calibration line. We have concluded that x_{det} is the lowest concentration which we can detect at the chosen significance level α, by using the limit of decision y_{dec}. (Recall that y_{dec} itself has been fixed so that a blank solution will be correctly identified as blank at this same level of significance.)

Many workers define the limit of decision as that concentration y_{det} corresponding, on the calibration curve, to an instrument reading of $y_{det} = y_o + 3s_y$. In other words it is the concentration of a solution which should give a mean reading three standard deviations above that for a blank solution. The limit of decision is then $y_{dec} = y_o + 1.5s_y$. Any reading of y_{dec} or greater is taken as proof of the presence of analyte. The choice corresponds in our discussion to taking $z_\alpha = 1.5$ (see Eq. 8.11). This in turn corresponds to an α-value of 0.067. The level of significance is 6.7%.

SAQ 8.5a   The calibration curve for an instrument, at low concentration, conforms with the regression line

$$y = 2.70 + 5.23 x$$

where x is the concentration (in mg dm^{-3}) of an analyte. The standard error of determination s_y is 0.043. Find the limit of detection.

The limit of detection for a given analytical procedure is an important piece of information to have available when attempting to decide what method to use to tackle a particular analytical problem. It is, of course, not the only piece of important information; we would want to know about the precision of the method over the whole range of concentration we might be concerned with. We would also want to know about likely systematic errors, about the expense and convenience of the method, etc. We should certainly need to be reassured, however, that the limit of detection is low enough for us to be able to analyse at the lowest concentration of likely interest.

Throughout this section we have implicitly assumed that we have been taking only a single measurement on a solution. It is on this basis that the limit of detection has been defined. It is in fact possible to detect analyte concentrations somewhat below the so-called detection limit, by making replicate determinations. The mean of n replicate measurements of y will be distributed with a standard deviation of $s_y/\sqrt{n}$. Thus the normal curves in Fig. 8.5b. would be correspondingly narrowed and we could with acceptable confidence assert the presence of analyte in solutions holding concentrations somewhat less than the 'official' x_{det} limit. Repeat determinations always improve our precision, but it is not always practicable to make them.

8.6. STATISTICS IN THE LIFE OF AN ANALYTICAL CHEMIST

Well, at long last, we are there. We promised to introduce you to statistical ideas in the context of analytical chemistry. We believe that we have fulfilled our promise. We have just seen statistical ideas used in order to help to decide whether or not to reject a suspect result. Statistical ideas are also germane to the way in which we present our final results – recall $\bar{x} \pm t_\alpha s/\sqrt{n}$ etc. Similar ideas are present in the thinking behind control charts and the thinking used in getting the best-fit line through a set of data-points (name a modern quantitative method which does not require a calibration curve). Statistical concepts permeate practically everything we do in the laboratory. They even make themselves useful in deciding whether or not the requirements of an analytical programme are possible, in other words, at the very beginning of our thinking. We thought that this would be a good place to end – at the beginning! What follows are not exercises or SAQs; rather they are the scribblings of a numerate analyst responding to the imperatives of clients. If you can follow them, congratulations! If you cannot, return to Page 1!

Client.

I am going to bring you some important samples, whose determined concentration is of the order of 10 ppm. I want the results that you report to me to be accurate to 0.2ppm. I'll accept a 95% level of confidence. Can you handle it?

Chemist. (To himself)

Help! What I think he wants (!) is to have a 95% level of confidence that μ is within the range $\bar{x} \pm 1.96\sigma/\sqrt{n}$, where $\pm 1.96\,\sigma/\sqrt{n} = \pm0.2$ ppm. Unfortunately the precision of my method is, at the 10 ppm level, ±0.4 ppm. What can I do? Oh yes – now I see it!

$$1.96(0.4)/\sqrt{n} = 0.2,$$

$$\sqrt{n} = 3.92, \quad n = 15.4. \text{ Take } n = 16$$

Now, if we can minimise systematic error, we can do the job.

Chemist. (aloud)

Yes!

Second Client.

I make money if an ore contains 3.00% of copper. I lose money if I process an ore which has only 2.80% of copper. I don't like losing money! I want to be 99% confident that I reject ore that has 2.80% of copper and an equal probability that I accept ore that has 3.00% of copper. Can you help?

Chemist.

I'll get back to you!

Later that night.

Chemist. (to himself)

It is something to do with that 'Power of the test' stuff. Probability of correct rejection $= 1 - \beta = 0.01$. This is the probability of false rejection. I need to make this equal to probability of false acceptance.

$\beta = 0.01$. Now, if $\alpha = 0.99$.

0.99 is the probability of correct acceptance.

Standard deviation of his method for determining copper is 0.10%.

Draws rough diagram.

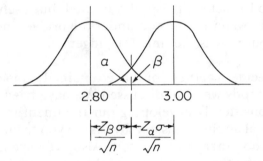

For $\alpha = \beta = 0.01$, $z_\alpha = z_\beta = 2.326$ (tables).

$$3.00 - 2.80 = 0.20 = z_\alpha\sigma/\sqrt{n} + z_\beta\sigma/\sqrt{n}$$

$$= (z_\alpha + z_\beta)\sigma/\sqrt{n}$$

$$= \frac{2(2.326)(0.10)}{\sqrt{n}}$$

Therefore $\quad \sqrt{n} = \dfrac{2(2.326)(0.10)}{0.20} = 2.326$

$$n = 5.4 \approx 6$$

Next day. Chemist to Client.

> Do six replicate determinations. If the result is less than
> 2.90% reject the shipment, if it is greater than 2.90% you
> can accept the shipment, You can be 99% sure that your
> decision will be the correct one.

Chemist. (to himself)

> Isn't statistics great!

Well, isn't it?

8.7. OVERVIEW AND BIBLIOGRAPHY

As we put it in the Study Guide: 'This Unit concentrates on helping
you obtain a good practical grasp of a limited, but useful, range of
methods.' A lot of water has flowed under the bridge since you read
that statement, but now is the time for judgement.

Only you can decide whether you have obtained 'a good practical
grasp' of the methods we have discussed.We have tried our best in
this regard by consistently developing our treatment in the context
of specific practical applications and by asking you to check our calculations and to do other ones yourself. Also, we have made rather
a point of discussing the ins and outs of the results and the niggling
doubts which often persist regarding interpretation.

As far as the 'usefulness' of the methods is concerned, we would
vigorously insist that the nature and variety of the examples we have
used ought really to have convinced you that our studies are of vital
relevance to the practising analyst. In our quotation from the Study
Guide it is the word 'limited' which may cause you to wince just
a little. Have we really spent all this effort obtaining a grasp of a
'limited' range of methods? Yes, we are afraid that is true. But the
good news is that we believe things are not nearly so bad as that
comment makes them sound.

This Unit aimed to achieve more than simply to introduce a few methods. The Study Guide also gave prominence to helping you to 'understand clearly the issues addressed' and to 'develop, through practice, a feeling for the significance of conclusions reached'. We have devoted much time to probing the subtlety of the issues which arise when we try to draw conclusions from the results of our measurements and we have frequently agonised over cases where matters were not clear cut. Going through all that is, in our view, the most important step in mastering statistics, at least from the point of view of a practising analyst. We have sacrificed comprehensive coverage in favour of careful discussion and we do not apologise for so doing. We hope we have inculcated a kind of state of mind which will stand you in good stead whether you are applying the methods we have discussed or delving into other ones.

Although there are very many applications of statistics to chemical analysis which we have not dealt with, in practice you are very unlikely to meet more than a small subset of these. It all depends on the kind of work you find yourself doing. We hope you will find yourself in a reasonably strong position to pick up a book and learn about a new method, as and when you need to. The question of which books to consult we shall come to below.

Just what then have we omitted? Too much to allow us to go into detail, but we should attempt a brief survey. Our whole treatment has strongly emphasised methods based on the mean and standard deviation and examples where the normal distribution is expected, at least approximately, to be applicable. This approach is the natural one when we are dealing with continuous variables – instrument readings, titration values and the like. There is a whole class of techniques known as 'non-parametric methods', which pay more attention, for instance, to the median, or to properties of the results arranged in rank order. These methods often give rise to tests which are arithmetically rather simple; our Q-test for outliers was one example. They are also applicable to situations where other than continuous variables are involved.

We might be concerned with relating the chemical composition of complex materials to their success rate when subjected to some mechanical test. Or we might be dealing with particle-size distribution, measured by noting the proportion of a material passing through various sieves. There is a diversity of possible examples.

A number of the topics we dealt with were unashamedly described as introductions. We have only scratched the surface of a vast field concerning the analysis of variance (known in the trade as 'ANOVA'). In fitting calibration lines we concentrated on linear relationships and on a single independent variable. We merely pointed the way towards fitting curves and to multiple regression analysis. In quality control we concentrated on inspection by variables to the neglect of the field involving inspection by attributes. In each of these topics we feel we have paved the way for further study and that we have left you in a good position to judge when such further work would be needed.

There are a number of areas which we have simply not touched upon at all. *Strategies in obtaining samples*, for instance, are very important and require a statistical viewpoint to be taken. The issues involved are addressed in the Unit entitled *Samples and Standards* in this Analytical Chemistry series, as well as in several of the texts which we mention below. There is also the large area of *experimental design*, covering such things as how to avoid bias in the processing of samples and how to plan one's attack in situations where a large number of interacting factors are at play and it is impractical to study every possible different combination of variables.

Another major developing field which we have not covered is associated with the growing power and sophistication of instrumentation. The processing of electronic signals requires an informed statistical approach. The 'signal-to-noise' ratio can be improved, signals can be smoothed or re-scanned and averaged, instrumental parameters may be varied to seek to optimise the strength of the response to an analyte and to minimise interference effects. These areas are touched upon in the ACOL Unit on *Microprocessor Applications*. More advanced study in this field requires a stronger mathematical background than is probably needed in any of the other areas we have indicated.

Much of what we have just mentioned, and more, is grouped under the general title of *Chemometrics*. Chemometrics encompasses the range of techniques which is used to process large quantities of analytical data with the aim of extracting as much useful chemical information as possible. The processing almost always in practical terms needs to be done by computer.

There we go, we have mentioned the computer again! In this Part, as in the preceding one, computer practice seems to have been fading into the background. Well, that has much more to do with the way we have chosen to present our material than with anything else. Indeed, the further we go in statistics, the more data we need to process, the more advanced our intrumentation becomes, so the computer looms larger and larger as an indispensable tool. You started this Unit with a detailed introduction to programming in BASIC and then we gradually set to work building a fairly complex package "STAT" which we completed by the end of Part 4. In Parts 5 and 6 we developed a shorter and separate program "LINE". We hope that you have made use of these programs and that you have experimented with making modifications of your own. We believe that you should feel ready to write small programs of your own to help you do specific jobs with measurements which you are confronted with in your work. You should feel that the ice has been broken as far as programming is concerned. With any luck we should also have convinced you of the usefulness of large statistical program-packages. Our "STAT"is a fairly make-shift affair, with its structure designed to enable it to be typed in progressively in reasonably manageable chunks. If you find you need to use the computer for a wide range of statistical operations, you should aim soon to gain experience of using a professionally written package available for your machine. If on the other hand you find yourself involved in routine and repetitive manipulations particular to a given work project, then you may well find it advantageous to construct a purpose-written program of your own.

And so, to end, we give a brief bibliography. There are very many books we could mention, but we have limited ourselves to a fairly short but varied list. When you begin browsing through library shelves to find them you will come across many others. Do not let our list inhibit you from scanning those. When you are looking for a book you will probably have a particular purpose in mind. Different books vary enormously in their topic coverage, in depth of treatment, whether they deal in mathematical rigour or whether there is an emphasis on description. It is worth browsing to find the text best suited to your need. But to start, here are brief comments on the books listed in the Bibliography at the beginning of our Unit.

1. J. C. Miller and J. N. Miller, *Statistics for Analytical Chemistry*.

   This is a very nice book written specifically for analytical chemists and at an introductory level. In many ways it is complementary to our Unit, making introductory incursions to a wider range of topics, including chemometrics.

2. C. Chatfield, *Statistics for Technology*.

   This is a more general introductory text, written from the rather wider perspective suggested by its title. It is a very useful text.

3. D. Cooke, A. H. Craven and G. M. Clarke, *Basic Statistical Computing*.

   This gives an excellent account of the application of computing to statistics at an introductory level. The text includes well-contructed listings of program segments which can be taken as they stand and incorporated into purpose-designed programs of your own.

4. I. M. Kolthoff and P. J. Elving, *Treatise on Analytical Chemistry*. Part 1. Volume 1.

   This gives a rigorous treatment of errors, accuracy and precision.

5. O. L. Davies and P. L. Goldsmith, *Statistical Methods in Research and Production*.

   Written as a Company book by staff members of ICI, this is an authoritative and generally readable work and it includes a thorough treatment of quality control.

6. R. E. Walpole and R. H. Myers, *Probability and Statistics for Engineers and Scientists*.

A feature of this book is a good treatment of analysis of variance (ANOVA).

7. E. L. Grant and R. S. Leavenworth, *Statistical Quality Control*.

If you wish to extend your knowledge of quality control this is a good text to choose.

8. C. Liteanu and I. Rica, *Statistical Theory and Methodology of Trace Analysis*.

This is a rather sophisticated book, strong on information and signal processing.

9. M. R. Spiegel and R. W. Boxer, *Schaum's Outline Series: Theory and Problems of Statistics*.

With its concise and business-like presentation, this is always useful for review purposes.

There is much you can already usefully do, but there is much more still to learn. Statistics has much to offer to the working analyst. We hope it serves you well.

Objectives

As a result of completing Part 8 you should feel able to:

- deal competently with the number of significant figures retained during calculations and when reporting results;

- follow the propagation of random errors in addition, subtraction, multiplication, division and in taking logarithms (for cases where different contributions to the random errors are independent);

- similarly follow the propagation of systematic errors, noting the parallels, but also the differences relative to the behaviour of random errors;

- decide whether a particular measurement should be disregarded, as an *outlier*;

- distinguish the concepts of *limit of decision* and *limit of detection* and be able to calculate a limit of detection with the aid of Eq. 8.11;

- claim that you have 'a good practical grasp' of what is involved in the statistical assessment of analytical results, that you have built a reasonable level of experience in confronting typical difficulties which arise, that you could with reasonable confidence begin to delve more deeply into any more specialist application which might become relevant to your work and that you would not shirk from using the computer as your work horse!

Self Assessment
Questions and Responses

<table>
<tr>
<td>SAQ 1.3a</td>
<td>In the determination of lead in a sample of water by atomic absorption spectrophotometry, four absorbance readings were taken: their values were 0.207, 0.210, 0.208, and 0.211.

Calculate:

(i)   the mean absorbance value,

(ii) the mean deviation and the percentage relative mean deviation,

(iii) the standard deviation and the percentage relative standard deviation.</td>
</tr>
</table>

Response

(i)   The mean absorbance value is 0.209.

If you did not obtain this result, there is a strong likelihood that you pressed the wrong button on your pocket calculator. The advent of the pocket calculator has been a godsend to analytical chemists. Whenever possible you should make use of one. However, be careful; like computers, one of their major limitations is us!

(*ii*)   The mean deviation is 0.0015.

If by chance you divided the sum of the absolute deviations (0.006) by 3 rather than by 4, you will have the incorrect answer of 0.002. Any other answer is again probably the result of clumsy fingers!

The percentage relative mean deviation is

$$\frac{100 \times 0.0015}{0.209} = 0.72\%$$

If you used 0.002 for the mean deviation, then you will have obtained the incorrect answer for the % RMD of 0.96%.

(*iii*)   The correct result for the standard deviation is 0.0018

If you got this, well done! Nonetheless, you are strongly advised to read on.

You may have obtained the result 0.0016. The reason for this is that rather than dividing the sum of the squared deviations (10×10^{-6}) by 3 you divided by 4, ie you divided by n rather than ($n - 1$). Now it may be that you determined the standard deviation with the aid of the statistics functions on your pocket calculator. There is a problem which arises here.

The calculator might be programmed to carry out this calculation in a fashion which parallels our method and which therefore involves a division by ($n - 1$). Some calculators however are programmed to divide by n. If you have such a calculator, you will have obtained the result 0.0016. You might say, if dividing by n is good enough for the calculator surely it is good enough for me. Well if n is a large number, there is hardly any difference between the standard deviation obtained by a step involving a division by n and one which involves a division by ($n - 1$). If, however, n is a small number, then there is a difference and the *result obtained by dividing by* (n − 1) *is preferred*. You will have to wait until later in the Unit for an explanation of this.

You may have obtained the result 0.0011 for the standard deviation. If you did, you have made one of the common mistakes of newcomers to the standard deviation. You have divided the square root of the sum of the squared deviations by $(n − 1)$, ie you have carried out the calculation

$$\frac{\sqrt{\sum d_i^2}.}{n − 1}$$

Keep in mind that the $(n − 1)$ is inside the square root sign, ie the calculation is

$$\sqrt{\frac{\sum d_i^2.}{n − 1}}.$$

If you obtained the result 0.0008, then you have calculated,

$$\frac{\sqrt{\sum d_i^2}}{n}.$$

Here there are two mistakes in one calculation. It is fortunate that no one is looking!

The % RSD is obtained from $100s/\bar{x}$ and is 0.18/0.209 = 0.86%.

If you had the incorrect answer for the standard deviation, then it will have been carried forward into this calculation. Just for practice, recalculate the % RSD.

****

SAQ 1.3b	Use a computer program to repeat the calculations of SAQ 1.3a above. You should already have created a suitable program when studying Section 2.4 of the Appendix. It requires only very minor alteration to produce all required quantities.

Response

You can get a large part of the way simply by substituting our lead analysis data into the program you developed in SAQ 2.4a of the Appendix. To complete the job you simply need to add lines to compute the mean deviation and the %RMD and %RSD.

We took as our starting point the program listed in our response to SAQ 2.4a of the Appendix. Our modified version is given below.

```
 98  REM ** STANDARD DEVIATION
 99  REM ** SUM IS S
100  S=0
110  READ N
120  FOR I=1 TO N
130      READ X
140      S=S+X
150      NEXT I
160  M=S/N
165  PRINT "NO. OF ITEMS     = ";N
170  PRINT "MEAN VALUE       = ";M
198  REM ** POINTER BACK
200  RESTORE 410
210  S=0
215  T=0
220  FOR I=1 TO N
230      READ X
240      S=S+(X-M)↑2
```

```
245    T=T+ABS(X-M)
250    NEXT I
260  V=S/(N-1)
270  S=SQR(V)
273  D=T/N
276  PRINT "MEAN DEVIATION      = ";D
280  PRINT "STANDARD DEVIATION      = ";S
283  D1=100*D/M
286  S1=100*S/M
290  PRINT "% REL. MEAN DEV.      = ";D1
295  PRINT "% REL. ST. DEV.      = ";S1
300  END
399  REM ** DATA LIST
400  DATA 4
410  DATA 0.207,0.210,0.208,0.211
```

The changes we have made are:

(*i*)   new lines 215 and 245 arrange for the deviations to be summed (as well as their squares);

(*ii*)   we replaced the original line 270 by a line explicitly calculating *s* (S in the program);

(*iii*) lines 273 and 276 compute and print the mean deviation (D);

(*iv*)   lines 283–295 obtain and print the %RMD and %RSD values;

(*v*)   the lead analysis data have been substituted in lines 400 and 410.

When we ran our program we got the results below.

NO. OF ITEMS	= 4
MEAN VALUE	= 0.209
MEAN DEVIATION	= 1.50000001E-3
STANDARD DEVIATION	= 1.82574187E-3
% REL. MEAN DEV.	= 0.717703356
% REL. ST. DEV.	= 0.873560704

What did your program produce? If you have problems check carefully. Remember that all individual deviations are added as positive numbers in calculating the mean deviation (line 245). Are you sure you have nowhere typed the number 0 or 1 where you meant to type the letter O or I, respectively? If you have not made that mistake yet, you probably will sometime! If you ever have difficulty tracing an error, add lines to print out intermediate results. You'll be surprised how rapidly you can become quite good at 'bug-hunting'. Your computer might well differ from ours in the number of figures it gives for the answers. The number of figures produced is an annoyance which we shall deal with in later programs.

An interesting detail is that the %RSD is calculated as 0.87% (to two significant figures), whereas we quoted a value of 0.86% as our answer in SAQ 1.3a, when we did the arithmetic by hand. The computer has caught us out! The value 0.86% was obtained because we rounded the value of s to two significant figures before we calculated %RSD. 0.87% is the correct result, though the difference is of little significance to us. This does, however, reveal a point which can be more important in other contexts, namely that rounding off an intermediate value in a calculation to a given number of significant figures will not ensure that the final result is correct to that number of figures.

****

| SAQ 1.5a | It is ordinarily a good thing to have as many replicate measurements as possible when attempting a determination. Four chemists have between them provided a total of 16 measurements for a titration. Their values are listed in Fig. 1.5a, and the 'true' result is known to be 23.04 cm^3. Use your program "STAT1" to calculate the mean, the standard deviation and the %RSD for the 16 values taken as a single set. Comment on the value of doing this. |

Response

All that is needed is to recast lines 4000 and 4010 of "STAT1" to include all 16 values as set out below.

```
4000 DATA 16
4010 DATA 23.09,23.04,23.07,23.08,22.81,22.96,23.18,23.13
4011 DATA 22.82,22.84,22.87,22.83,22.74,22.61,22.95,22.82
```

Notice that we have split the data between two lines. Most computers will allow you to include all sixteen values on an extended line 4010. That is the easiest thing to do if it works.

We then obtained the following results.

```
No of values    = 16
Sample mean     = 22.928
Est. st. dev.   = 0.16
% Rel.st.dev.   = 0.696
```

If we compare these results with those in Fig. 1.5b, we can see that in every respect they are substantially worse than those of Chemist 1 taken alone. It may well have been clear to you from the start that this would be so, but we do not apologise for taking you through the exercise because it gives an opportunity to stress an important point. The four chemists clearly differ markedly in competence; when we pool their values we are mixing wheat with chaff. As we progress through this Unit we shall often extol the benefits of having several replicate readings. In doing so, however, we shall be taking it as understood that comparable levels of skill and care are exercised in obtaining every individual reading.

Here, as we know the 'true' result, it is easy to identify the results of the first chemist as being by far the most reliable. More often we work in the dark, unaware of what the 'true' value is. Had we been in that position here we should still give little credence to the work of chemists 2 and 4. They condemn themselves by their own gross imprecision and so their results can only be regarded as unreliable. Chemists 1 and 3, however, present more of a problem. Each has worked with very similar consistency, but their results are in clear disagreement with one another. If we were not in a position to investigate this discrepancy further, we could only average all 8 of their results. We should be aware in so doing, however, that the estimate we were obtaining was almost certainly inferior to the individual results of one or other of these chemists. The rub would be that we would not know which one. To quote an overall standard deviation for the 8 values would be less meaningful than to give details for each chemist separately and to highlight the discrepancy between them.

With the computer it was very easy actually to do the calculation in this SAQ; it would have been quite a lot of work to do it manually. Getting good at doing calculations, and becoming adept at applying increasingly sophisticated statistical tests, will however never absolve you from the need to consider whether it is in fact sensible to treat particular results in a given way. Computer scientists have a jargon word for it – GIGO. That stands for 'garbage in – garbage out'. You should bear the point in mind whenever you work with statistics, with or without a computer.

****

SAQ 1.5b	The following titration values were all obtained by a single analyst, applying a standard technique to separate aliquot samples of a homogeneous starting material. The procedure involves a number of pretreatment stages before the final titration, and it is these which are the main source of the spread in the results.
	Titre/cm^3: 27.86, 27.68, 28.02, 28.05, 27.60, 27.55, 27.03, 27.76, 27.73, 27.75, 27.77, 27.53.
	$\longrightarrow$

SAQ 1.5b (cont.)	Twelve values are reported in all. Use program "STAT1" to compute the mean, $\bar{x}$, the standard deviation, s, and the %RSD for the whole set. Compare the results with what would have been obtained if the investigation had been stopped (*i*) after the first *two* measurements, (*ii*) after the first *four*, (*iii*) after the first *six*.

Response

The way we tackled the calculation was to type all 12 values into DATA line 4010, then we ran the program with the datum on line 4000 set at 12, then at 2, then 4, and finally 6. In the three latter runs only the first 2 (or 4, or 6) titre values will be read by the program.

These are realistic data and none of the reservations we had in the last SAQ apply here. The results generated by "STAT1" are given below.

Values taken:	First 2	First 4	First 6	All 12
Mean, $\bar{x}$/cm^3	22.770	27.902	27.793	27.694
St. dev., s/cm^3	0.127	0.170	0.215	0.265
% RSD	0.458	0.610	0.774	0.957

We instinctively feel that our estimates are more likely to be reliable the more readings we include. We are right to feel so, as we shall confirm later. Now the values we have been using are in fact representative of a series where the 'true' experimental result is 27.66 cm^3 and the technique as applied has a standard deviation of 0.32 cm^3. You can see that when all twelve readings are considered we get quite close to this. Notice, however, that our calculated mean after four readings is significantly more in error than it was after only two. This is because, as it turned our, readings 3 and 4 were the two highest in the whole series.

This kind of behaviour is relatively unusual, but it will happen to us from time to time. In fact, if we habitually perform an analysis such as the above, after four readings we can expect to get a mean result as bad as this one about one time in eight. (We shall soon be in a position to show that that is so.) Large errors are however more likely the smaller the number of replicates we use. You might, for instance, have done only two determinations, but have been landed with results 3 and 4!

****

SAQ 1.7a   Although the gravimetric method of analysis was crucial to the development of chemistry, it is, even in the hands of experienced analysts, often subject to large errors.

Gravimetric analysis often involves steps similar to the following.

(*i*)   A weighed amount of a sample containing the determinand is dissolved in a suitable solvent and the conditions of the solution (eg pH) are appropriately adjusted.

(*ii*) A weighed amount of the precipitant is dissolved in a suitable solvent to give a solution of known concentration.

(*iii*) The solution containing the precipitant is added slowly and with stirring to the determinand solution, which is often at an elevated temperature. This is continued until the precipitant is in excess. The solution is allowed to cool.   ⟶

SAQ 1.7a (cont.)

(*iv*) The precipitate formed in (*iii*) is filtered off and washed with a suitable solvent.

(*v*) The washed precipitate is heated or dried to constant weight and by using the appropriate stoichiometric relationships the percentage of determinand in the sample is calculated.

Try to identify which of the steps are likely to introduce significant errors (you can assume that blunders are absent) and suggest the category of error which is likely to be present.

Response

In this response we shall be a little diffident about saying what is correct. We shall rather discuss briefly each of the steps. We hope that you have met the gravimetric method. We also hope, for the purpose of this question, that you, like us, have suffered a little with the method!

Step (*i*). The random error in weighing is likely to be relatively small. Systematic errors might be large; samples may interact with the atmosphere, and weighing errors are possible from temperature and buoyancy effects. (When you have a moment, you could look these up in a standard text book.) Such errors are minimised by using stoppered weighing vessels which are at the same temperature as the balance, and by weighing by difference. Dissolving the sample in a solvent is not likely to introduce errors, unless, of course, we work carelessly. Failure to adjust the solution conditions could introduce, at a subsequent stage of the method, a significant methodological systematic error. The solution conditions for precipitation are often crucial – see step (*iii*).

Step (*ii*). The weighing and dissolving of the precipitant should not, if we work carefully, introduce much error. The precipitant concentration need not be exact; the precipitant is, in step (*iii*) to be added in excess.

Step (*iii*). If you recognised this step as being associated with potentially large errors, it is likely that you have studied and practised gravimetric analysis at some time. If the precipitant is not added in excess then obviously not all the determinand is precipitated. This causes a methodological systematic error.

The conditions for obtaining pure precipitates are often very stringent. Many precipitates, particularly gelatinous ones, contain impurities (Keep in mind that the sample comprises other species besides the determinand.) If these impurities are not removed than we shall have a serious methodological systematic error.

The elevated temperature is to prevent too much supersaturation which is a cause of the growth of gel-like precipitates and hence contamination with impurities. Cooling should bring all the precipitate out of the solution; however, sometimes this does not happen and hence there is a systematic error.

Step (*iv*). Filtering and washing are processes which require skill, and two important questions arise. Has all the precipitate been transferred to the crucible? Have we in washing the precipitate, actually redissolved some of it? Obviously, the opportunities to introduce systematic errors here are many. We have not mentioned random errors for some time. Well, uncertainties are obviously present, but we have concentrated on systematic errors because they seem to be the major cause of our problems with gravimetric analysis.

Step (*v*). Drying/heating to constant weight seems, in principle to be relatively easy. Provided the precautions mentioned in step (*i*) are taken and that we are patient in our search for a constant weight, then this step should introduce little error.

From the above it is seen that steps (*iii*) and (*iv*) are likely to have many potential sources of systematic error and that these are associated with the method employed. It is often useful to identify possible sources of error before carrying out an analysis. If we know what could go wrong, we shall be in a better position to cope with the situation when things do go wrong.

∗∗∗∗∗∗∗∗∗∗∗∗∗∗∗∗∗∗∗∗∗∗∗∗∗∗∗∗∗∗∗∗∗∗∗∗

SAQ 1.8a

> Do you consider the following statement to be (*i*) true, (*ii*) sometimes true (*iii*) false.
>
> If systematic error is removed, then the results will be accurate.

Response

'Sometimes true' is the best response. We are not hedging our bets here! Accuracy, which can be measured by the absolute error, depends upon both random and systematic error. We are happy if systematic error is absent; it means that our results are subject only to random errors. If this is so, and the precision of our measurements is good, for example in the titrimetric method where % RSD is often of the order of 0.1%, then the random error is small, and we can expect results which are very accurate. However, precision is not always so good. Analysts working in areas such as atomic absorption spectrophotometry, X-ray fluorescence spectroscopy, quantitative chromatography, routinely report % RSD of 1% or more. They have nothing to be ashamed of here. They are probably getting the most precise results possible from their instruments. If they pushed such instruments to their limits it is likely that they would obtain precision of the order of ±10% or more. Recall that even the analytical balance, when pushed to its limits gives poor precision. In the light of this, it is simply wrong to believe that random error always makes a small contribution to the total error and that if systematic error is absent then results will be accurate.

****

SAQ 2.1a

(*i*) What is the probability of throwing a 4 with a fair die?

(*ii*) What is the probability of throwing either a 4 or a 5 with a fair die?

(*iii*) A die is rolled and a 4 is obtained. What is the probability of throwing a 5 on the next roll of the die?

(*iv*) What is the probability of throwing a 4 with one die and a 5 with a second die?

(*v*) A coin is tossed three times in a row. What is the probability of obtaining

(*a*) 3 heads,
(*b*) 2 heads and a tail,
(*c*) 2 of a kind?

Response

(*i*) The correct response here is obviously 1/6. It is hard to imagine that anyone could have got this wrong!

(*ii*) Here we are presented with an application of the addition law. The different outcomes – a 4 or a 5 – are mutually exclusive. We can therefore add the probabilities of each occurring ie 1/6 + 1/6 = 1/3. Hands up those who read this question as the probability of throwing a 4 and a 5! Too embarrassed to admit it? Yes, it is a little bit daft to imagine that the two numbers could be thrown simultaneously.

(*iii*) We have been very careful as to how we phrased this question. The first sentence is there as a distraction. The question appears in the second sentence and the answer to it is obviously 1/6. Each time a die is rolled the probability of obtaining a given number is 1/6, it does not matter what was obtained on the last throw of the die. If you got this one wrong, you have probably interpreted the question as being the same as the one which follows. Let us turn to that question and then have a general discussion.

(*iv*) The correct response to this question is 1/6 $\times$ 1/6 = 1/36.

We are dealing with independent events and are asked the probability of *both* events occurring. The events are *not* mutually exclusive; they can occur simultaneously. Therefore, if by chance you used the addition law you can see the error of your ways!

(*v*) This, as you have no doubt discovered, is an interesting question. Let us first list the possible outcomes.

Outcome	Type of Event
HHH	3 heads
HHT HTH THH	2 heads and a tail
TTH THT HTT	2 tails and a head
TTT	3 tails

Now we can answer the questions.

(*a*) There are eight possible outcomes. One of these results is three heads. Therefore the answer to is 1/8. Equally, you could arrive at this conclusion by arguing that the probability is (1/2) × (1/2) × (1/2).

(*b*) There are three outcomes which result in the event two heads and a tail. The probability of this event is therefore 3/8.

(*c*) Two of a kind means either the event two heads and one tail or the event one head and two tails. There are three ways of the first event occurring and three ways of the second event occurring. The probability of either one of these events occurring is 3/8 + 3/8 = 6/8.

In the above questions, there is a strong liklihood that you somehow miscounted. Do not let this upset you! The question was originally set with four coins and the questioner changed it when he found that he had miscounted. Some methods to avoid this sort of mistake will be given to you shortly.

If you got all the above correct, you are undoubtedly grasping the fundamental ideas about probability. Let us stay with this topic for a time: there is much benefit in doing so.

****

SAQ 2.3a	What percentage of the total area under a standardised normal curve lies between $z = -2$ and $z = +2$?

Response

The area in the tail for $z = 2.00$ is obtained from the table and is 0.02275. Therefore the area under the curve bounded by $z = 0$ and $z = 2$ is $0.5000 - 0.02275 = 0.47725$.

Keeping in mind the symmetry of the normal curve, the area under the curve from $z = -2$ to $z = +2$ is $2 \times 0.47725 = 0.9545$, ie 95.45% The areas are shown graphically in Fig. 2.3j.

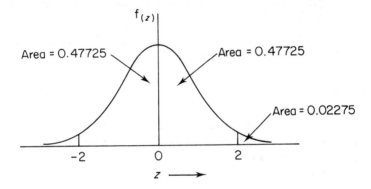

Fig. 2.3j. *Areas of regions under the standardised normal curve involving $z = \pm 2$*

Did you get the right answer? If not, let us think of possible reasons. You might have read the table incorrectly. Many students of introductory statistics do! If you did, go back to the table and make sure you can read it correctly. Now, assuming that you did get the correct area in the tail, did you remember to subtract the tail from 0.5000 – ie, subtract the tail from half the area under the normal curve? This is a not uncommon mistake. Always keep in mind the question being asked. We wish to know the area bounded by the curve and $z = 0$ and $z = +2$. Once we have this we can double the number to get the area bounded by $z = -2$ and $z = +2$. In order to help yourself to remember the question, always sketch a curve and indicate on it the important points – see Fig. 2.3j. Other than forgetting to double the area bounded by $z = 0$ and $z = +2$ in order to obtain the area bounded by $z = -2$ and $z = +2$, it is hard to imagine what further mistakes you could have made.

****

SAQ 2.3b

> A normal curve has $\mu = 800$ and $\sigma = 10$. What is the area under the curve bounded by $x = 805$ and $x = 830$?

Response

As usual it is helpful to represent the problem in the form of a sketch. We wish to know the area of the shaded part of the normal curve in Fig. 2.3p (*i*).

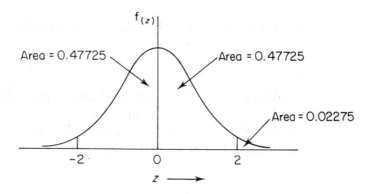

Fig. 2.3p. *(i) The untransformed problem*

We now transform the normal curve into a standardised normal curve and show the required area by shading – see Fig. 2.3p (*ii*).

For $x = 805$, $z = 0.50$ and for $x = 830$, $z = 3.00$.

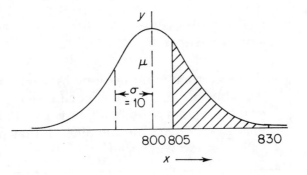

Fig. 2.3p. *(ii) The transformed problem*

Now the area in the tail when $z = 0.50$ is 0.3085 and the area in the tail when $z = 3.00$ is 0.00135. The required area can be obtained by subtracting the tail when $z = 3.00$ from the tail when $z = 0.50$, ie $0.3085 - 0.0013 = 0.3072$. Therefore, the area under the normal curve with $\mu = 800$ and $\sigma = 10$, bounded by $x = 805$ and $x = 830$ is 0.3072, ie 30.72%.

If you obtained this answer, then well done! If you did not, there are several places where you might have made a mistake. The first is in the transformation process, ie using the operation $z = (x - \mu)/\sigma$. What can one say? All of us can make mistakes. Assuming that you obtained the correct z-values, the next possible mistake could have occurred in looking up the areas in the tails associated with the two z-values. Be careful in using the table. Sometimes it seems that the better we become at statistics the more careless we are at using tables! The most likely reason for not obtaining the correct answer is the failure to realise what had to be subtracted from what!

****

SAQ 3.1a

This is a revision exercise on deriving probabilities from a normal distribution curve. The means, $\bar{x}$, for samples of 12 values for RND(1) are expected to fit a normal distribution with mean $\mu = 0.5$ and with standard deviation $\sigma_{\bar{x}} = 1/12$. Find the probability that a given value of $\bar{x}$ lies in each of the following ranges: 0.00–0.05, 0.05–0.15, 0.15–0.25, 0.25–0.35, 0.35–0.45, 0.45–0.55, 0.55–0.65, 0.65–0.75, 0.75–0.85, 0.85–0.95 and 0.95–1.00.

This question could be done by hand, by using the table in Fig. 2.3g. It is much easier, however, to use program STAT2 which you developed in Part 2. (The data on line 3900 of that program will need to be adjusted.)

Response

The probabilities should turn out as follows:

Band	Probability
0.00–0.05	0.00000
0.05–0.15	0.00001
0.15–0.25	0.00134
0.25–0.35	0.03458
0.35–0.45	0.23832
0.45–0.55	0.45150
0.55–0.65	0.23832
0.65–0.75	0.03458
0.75–0.85	0.00134
0.85–0.95	0.00001
0.95–1.00	0.00000

The results quoted in the text, for the expected totals in each band for means from samples of 12 values of RND(1), are in each instance 1000 times these probabilities, rounded to the nearest integer (the predictions were based on 1000 samples).

We got these answers by loading program STAT2, and changing line 3900 to

3900 DATA 0.5,0.08333333

so that it gives the mean and standard deviation for our distribution (1/12 = 0.083333). We then ran the program selecting option 1 and asking successively for the areas of the tail beyond each of the band boundaries below.

Boundary	Probability of exceeding this value (%)
0.95	0.000
0.85	0.001
0.75	0.135
0.65	3.593
0.55	27.425
0.45	72.575
0.35	96.407
0.25	99.865
0.15	99.999
0.05	100.000

The required probabilities are then obtained by subtraction of successive values. For the band 0.55–0.65, for instance, we have a probability of (27.425 − 3.593) = 23.832% or, as a decimal 0.23832.

****

SAQ 3.1b   It is found from long experience that a machine
produces components whose lengths are nor-
mally distributed with mean 20.00 cm and standard deviation 0.15 cm. A random sample of 9
components is taken from the production line.

(*i*)   What is the probability that the mean of
the sample will be greater than 20.05 cm?

(*ii*)   What percentage of samples of size 9 will
have a mean length less than 20.05 cm?

(*iii*) What percentage of samples of size 9 will
have a mean length in the range 20.00 $\pm$
0.10 cm?

Response

(*i*)   The population mean is 20.00 cm; the population standard
deviation is 0.15 cm. (It is important to write down a statement
such as this right at the start of an attempt to solve a problem.
You can, of course, simplify matters by writing $\mu = 20.00$ cm
and $\sigma = 0.15$ cm.) The distribution of means for samples of
size 9 is normal with $\mu = 20.00$ cm and $\sigma_{\bar{x}} = \sigma/\sqrt{n} = 0.15/\sqrt{9} = 0.05$ cm.

The distribution of sample means is represented graphically in
Fig. 3.1f.

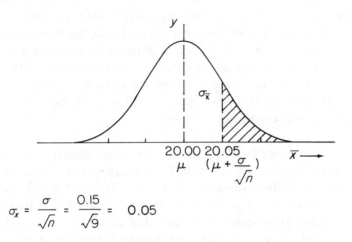

Fig. 3.1f. *SAQ 3.1b(i) depicted graphically*

The first question to be answered is 'What is the probability that the mean of the sample will be greater than 20.05?' With this in mind we have shaded the area under the curve beyond $\bar{x} = 20.05$. Now, when $\bar{x} = 20.05$,

$$z = \frac{\bar{x} - \mu}{\sigma/\sqrt{n}} = \frac{20.05 - 20.00}{0.15/\sqrt{9}} = \frac{0.05}{0.05} = 1.0$$

When $z = 1.0$ the area in the tail of a standardised normal distribution is 0.1587.

Therefore in answer to the question, the probability that a sample of size 9 will have a mean greater than 20.05 cm is 15.87%. Did you get this answer? If not, where did you go wrong? Did you realise that the population mean was 20.00 cm and the population standard deviation was 0.15 cm? Always read a question carefully and write down the information which has been given. Did you remember that when dealing with a distribution of sample means that the standard deviation is $\sigma_{\bar{x}} = \sigma/\sqrt{n}$. Many beginners forget that σ is not the same as $\sigma_{\bar{x}}$. Was your mistake associated with calculating the z value from the appropriate $\bar{x}$ value? Once we have a z value it is simply a matter of looking up or computing the area of the corresponding tail of the standardised normal curve.

(*ii*) This is an easy one! It calls on us to determine the unshaded area in Fig. 3.1f above. Since we know that the shaded area is 0.1587, the unshaded are is $1 - 0.1587 = 0.8413$. Therefore the answer to the question is that 84.13% of samples of size 9 will have mean length less than 20.05 cm. If you found part (*i*) of the question easy, part (*ii*) should have been commonsense.

(*iii*) Let us represent this question graphically (Fig. 3.1g).

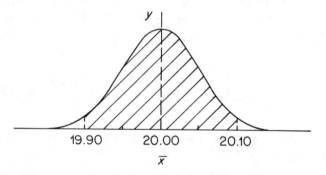

Fig. 3.1g. *Graphical representation of SAQ 3.1b(iii)*

The area of interest has been shaded. Now when $\bar{x} = 20.10$,

$$z = \frac{\bar{x} - \mu}{\sigma/\sqrt{n}} = \frac{20.10 - 20.00}{0.15/\sqrt{9}} = 2.00$$

When z is 2.0 the area in the tail is 0.02275. From the symmetry of the curve we know that the total area in the two tails is $2 \times (0.02275) = 0.04550$. The shaded area is therefore $1 - 0.0455 = 0.9545$. The percentage of samples of size 9 which have a mean length in the range 20.00 $\pm$ 0.10 cm is therefore 95.45%. Did you get this one correct? If so, congratulations! If not you should by now be aware of where you went wrong. The problem is similar to some that we met in Part 2; there we worked with $z = (x - \mu)/\sigma$; here we are working with $z = (\bar{x} - \mu)/(\sigma/\sqrt{n})$. You will probably have noted that in this SAQ we have not sketched the standardised normal distribution, but rather have simply calculated the required z values. If you still need to sketch a z-distribution, do so. We believe that in this question it was not necessary; in other questions, where we feel it would help we shall go back to doing so.

You might just be a little concerned by the fact that we used a non-chemical example in the above SAQ. Don't worry; there will be many examples drawn from analytical chemistry before we have finished the Unit. We used the example of the length of machine components because we believed that it would be an easy one to visualise. The method that we have applied is, of course, generally applicable. Once again, be patient!

****

SAQ 3.2a

A manufacturer has developed a modification of the familiar combustion-analysis apparatus. It permits replicate determinations to be carried out quickly and it is claimed that it will determine the percentage of carbon in an organic compound with a standard deviation of 0.20%. A chemist decided to evaluate the apparatus. He asked a colleague to provide him with a 'pure' organic compound but not to inform him of its identity. The chemist subsequently carried out four determinations on the compound which yielded the following results: 68.83% C, 68.85% C, 68.83% and 68.82% C. Give an estimate, at the 95% confidence level, of the percentage of carbon in the compound.

Response

We have been given the following information:

$$\sigma = 0.20\%$$

$$n = 4$$

and

$$\bar{x} = \frac{68.83 + 68.85 + 68.83 + 68.82}{4}$$

$$= 68.83(3) \% \text{ C}$$

For a 95% confidence interval, $z = 1.96$. Therefore, when we substitute into the Eq. 3.1

$$\mu = \bar{x} + z\sigma/\sqrt{n} \qquad\qquad (3.1)$$

we obtain

$$\mu = 68.83 \pm 1.96(0.20/\sqrt{4}) = 68.83 \pm 0.20 \% \text{ C}$$

This is the 95% confidence interval for the population mean: we are 95% confident that the percentage of carbon in the compound lies in the range 68.83 ± 0.20.

This is a relatively easy SAQ. The equation used is straightforward enough and the arithmetic almost trivial. Nevertheless you might have got the wrong answer. What could have gone wrong? Yes, once again because it may not always be clear what to call what! If, for instance, you found yourself confused by the statement 'that it will determine the percentage of carbon in an organic compound with a standard deviation of 0.20%', do not be embarrassed. Many chemists have the same difficulty when first confronted with statistics. To someone experienced with the problem in the SAQ, the 0.20% could be nothing other than the population standard deviation. Novices, however, can be paralysed into inaction by such questions. If this comment applies to you, be optimistic; things will get better! Much better!

$$*********************************$$

SAQ 3.4a	A barrel of wine was analysed for its ethanol content by gas chromatography. Four replicate measurements were made on a random sample of the wine. The results were: 14.16, 14.23, 14.20 and 14.29% of ethanol (v/v).

(*i*) Calculate the sample mean and the sample standard deviation.

(*ii*) Give a 90% and a 95% confidence interval for the ethanol content of the wine.

(*iii*) If it was known from past experience that the precision of the chromatographic method, as measured by the standard deviation, is 0.065% ethanol (v/v), how would this affect your reporting of the 95% confidence interval for the ethanol content?

Response

(*i*) The mean is obtained from

$$\bar{x} = \frac{\sum_i x_i}{n}$$

This, in our example, gives

$$\bar{x} = \frac{56.88}{4} = 14.22\%$$

ethanol by volume. It is unlikely that you made a mistake here!

The sample standard deviation is obtained from the formula

$$s = \frac{\sum_i (x_i - \bar{x})^2}{n - 1} = \frac{\sum_i d_i^2}{n - 1}$$

The standard deviation is, to 3 places of decimals, 0.055%. A mistake here is always a possibility. Did you use a calculator with statistics functions to obtain the standard deviation? If so, did you by chance obtain a value for s of 0.047%? If you did, it is likely that you used the 's_n' button rather than the 's_{n-1}' button. You might recall that we discussed this matter in the context of SAQ 1.3a.

If you calculated s by using the formula above, you should have found that the term $\sum_i d_i^2$ worked out to be 90×10^{-4}. This when divided by $(4 - 1)$ gives 30×10^{-4}. The square root of 30×10^{-4}. The square root of 30×10^{-4} is 5.477×10^{-2} or 0.055 to 3 decimal places. Of course, you could use program "STAT1" to obtain both $\bar{x}$ and s. If you did, and if you typed the DATA lines correctly, you should not have got the wrong answer!

(*ii*) In order to write out a confidence interval we need $\bar{x}$, s and the appropriate value of t. We obtain the latter value from knowledge of the confidence interval required and the sample size.

For a 90% confidence interval the area in a tail is 0.05. For $n = 4$, $\nu = 3$. From the table we see that in the column headed 0.05, when $\nu = 3$, $t = 2.353$.

The 90% confidence interval is thus:

$$\bar{x} \pm 2.353 \frac{s}{\sqrt{n}} = 14.22 \pm \frac{2.353(0.055)}{\sqrt{4}}$$

$$= 14.22 \pm 0.06\% \text{ of ethanol v/v.}$$

The estimate of the population mean based on a 95% confidence interval is worked out in a similar fashion. The value of t which gives an area in the tail of 0.025 for 3 degrees of freedom is 3.182. The 95% confidence interval is therefore:

$$14.22 \pm \frac{3.182(0.055)}{\sqrt{n}} = 14.22 \pm 0.09\% \text{ ethanol v/v}$$

Did you get both of these answers?

If not, did you by chance divide by $\sqrt{3}$ instead of $\sqrt{4}$. If you did, watch it! Assuming that you used the correct values for $\bar{x}$, s and n, any mistakes which you might have made here have their origin in the use of the t-distribution table. Our advice to you is the familiar 'practice makes perfect'. Get back to the t-table and practise, even social scientists ultimately master it!.

(*iii*) If we know σ it is easy to write out a 95% confidence interval: it is given by

$$\bar{x} \pm 1.96\sigma/\sqrt{n}.$$

In our case this would give

$$14.22 \pm \frac{1.96(0.065)}{\sqrt{4}} = 14.22 \pm 0.06\% \text{ of ethanol v/v}$$

Note that this 95% confidence interval is smaller than the 95% confidence interval which was obtained by using the expression $\bar{x} \pm ts/\sqrt{n}$.

We shall leave it to you to ponder why this is so. Hint, the answer lies in a comparison of $z\sigma$ with ts!

****

SAQ 3.4b	Use program "STAT3" to obtain the 95% confidence range for the mean derived from the first three replicate measurements noted in DATA line 4010. Compare the results obtained by using the sample standard deviation with those based on the population standard deviation. Repeat the process considering in turn the first 4, 6, 8, and finally, 10 measurements.

Response

The only change that is required in the program is to redefine n through line 4000. Changing this to read:

4000 DATA 3

will ensure that the first three values on line 4010 will be considered (*viz* 27.86, 27.68, and 28.02). The program can then be RUN for a confidence level of 0.95, selecting either option 1 or 2 to use the sample or the population standard deviation. The process can be repeated for $n = 4, 6, 8$ and 10. The results obtained are listed below.

95% confidence range for mean

n	by using s	by using σ
2	26.65 – 28.89	27.33 – 28.21
3	27.43 – 28.28	27.49 – 28.22
4	27.63 – 28.17	27.59 – 28.22
6	27.57 – 28.02	27.54 – 28.05
8	27.42 – 27.97	27.47 – 27.92
10	27.50 – 27.91	27.50 – 27.90
12	27.53 – 27.86	27.51 – 27.89

The results for $n = 2$ and $n = 12$ have also been listed. They were quoted earlier in the text. Notice how the width of the 95% confidence interval, 'using σ', steadily narrows as more measurements are included. The predictions 'using s', however, behave somewhat more erratically. This is largely because the calculated s value changes significantly as n increases. The true population mean $\mu = 27.66$ does in fact lie within the predicted range for every one of the above calculations.

****

SAQ 3.4c	The level of manganese in a sample of steel was investigated by atomic absorption spectroscopy. Ten replicate determinations gave the following results.

% Mn: 0.957, 0.922, 0.839, 0.803, 0.724,
        0.857, 0.816, 0.918, 0.767, 0.747

Use program "STAT3" to derive the 99% confidence interval for the mean Mn level in the steel. Examine also the 90%, 95% and the 99.9% confidence intervals.

Response

All that is needed is to type the Mn data into lines 4000 and 4010.

4000 DATA 10
4010 DATA 0.957,0.922,0.839,0.803,0.724,0.857,0.816,0.918,0.767,0.747

It is then simply a matter of running the program, giving 0.99 for the required confidence interval. The population standard deviation is not available so option 1 must be selected. The program will do a t-calculation based on the calculated sample standard deviation. The results obtained should be as follows:

> No. of values = 10
> Sample mean = 0.835
> Est. st. dev.   = 7.9E-2
> % Rel.st.dev. = 9.44
> It is predicted with confidence 0.99
> that the mean lies in the range:
>   0.75 to 0.92

To obtain the alternative confidence ranges requested it is simply matter of running the program a further three times giving 0.90, 0.95 and 0.999, respectively, for the required level of confidence. The results are below.

Specified confidence level	Range for mean% Mn
0.90	0.79 – 0.88
0.95	0.78 – 0.89
0.99	0.75 – 0.92
0.999	0.72 – 0.95

The interval widens as we demand an increased level of confidence. This is as we should expect but in this example, with 10 replicate determinations available (ie with $\nu = 9$) the widening is not dramatic. You might like to investigate what would happen with a smaller number of replicates. If, for instance, you set $n = 3$ (ie 4000 DATA 3) so that only the first three determinations are considered, then the 90% confidence range is 0.80% to 1.01%, whereas the 99.9% range is given as from -0.28% to 2.09%. In this latter case the 99.9% confidence range includes the possibility that there is no Mn present in the steel at all!

****

SAQ 3.6a

> A manufacturer produces dry cells which under a standardised test are claimed to have a life-time of 28.0 hours with a standard deviation of 1.50 hours.
>
> (*i*) Test the manufacturer's claim in the light of the knowledge that a random sample of 25 dry cells was found to have a mean life-time of 27.45 hours. Use a 5% level of significance for the test.
>
> (*ii*) The manufacturers's research team believe that a slight change in the production procedure will result in dry cells with an improved life-time. The changes in the production procedure are effected and a random sample of 25 dry cells was found to have a mean life-time of 28.60 hours. Carry out an appropriate test at both the 5% and 1% level of significance.

Response

(*i*) First we need to state the null hypothesis; it is H_0: $\mu = 28.0$ hours. Next, there is a need to agree upon an alternative hypothesis. There are three possibilities: they are H_1: $\neq 28.0$ hours, H_1: < 28.0 hours, and H_1: > 28.0 hours. The first of these is a reasonable alternative hypothesis. In essence it states that evidence which suggests that the population mean might be either larger or smaller than the asserted value will be accepted in drawing the conclusion that the population mean is not 28.0 hours. The second of the possible alternative hypotheses is probably a better choice. One could ask the question: 'Is the manufacturer likely to have underestimated the life-time of his dry cells?' An equally pointed question would be: 'Why are we testing the manufacturer's claim?' Surely, it must be in order to draw the conclusion, on rejection of the null hypothesis, that the mean life-time of the dry cells is less than 28.0 hours. The third possibility for an alternative hypothesis is not reasonable. There is nothing in the statement of the problem to suggest that the conclusion to be drawn, on rejection of the null hypothesis, is that the mean life-time is greater than 28.0 hours. It seems therefore that there are two reasonable alternative hypotheses. We shall carry out test for both of them.

Taking H_1: $\mu \neq 28.0$ hours requires a two-tailed test. For a 5% level of significance we must allow for an area of 0.025 in each tail. The appropriate z-value is 1.96 and the acceptance region is defined by

$$28.0 \pm 1.96(1.5)/\sqrt{25} = 28.0 \pm 0.59 \text{ hours.}$$

The critical values are 27.41 and 28.59 hours. The experimental sample mean is 27.45 hours. Consequently, we accept the null hypothesis, ie at the 5% significance level the sample mean is not incompatible with the population mean asserted.

Taking H_1: $\mu < 28.0$ hours requires a test of the one tail type viz a lower-tail test. For an area of 5% in a tail, the z-value is 1.645. Consequently the critical value is $28.0 - 1.645(1.5)/\sqrt{25} = 27.51$ hours.

A mean of a sample of size 25 which is greater than this critical value will result in the acceptance of the null hypothesis: a sample mean which is less than the critical value will result in the rejection of the null hypothesis and the acceptance of the alternative hypothesis, H_1: $\mu < 28.0$ hours. Since the experimental sample mean was 27.45 hours, we must conclude at the 5% significance level that $\mu < 28.0$ hours. Yet again we are dealing with a borderline case. Whether we accept or reject the manufacturer's claim depends on how exactly we frame our test.

Does the above argument coincide with your thinking on the matter? If you answered the question correctly by assuming either H_1: $\mu \neq 28.0$ hours or H_1: $\mu < 28.0$ hours, well done! If you found the choice of an alternative hypothesis a little difficult, you should, by now, be convinced that you were right to be a little hesitant. The problem was stated in a slightly ambiguous fashion. This is not so in part (*ii*) of the question.

(*ii*) Here, again, the null hypothesis is H_0: $\mu = 28.0$ hours. When it comes to the alternative hypothesis, we should be guided by the fact that we are interested in the possibility that the new production procedure results in an enhanced life-time. This suggests that evidence which results in the rejection of the null hypothesis should lead to the conclusion that $\mu > 28.0$ hours. What we are dealing with here is an upper-tail test, an alternative hypothesis stated as H_1: $\mu > 28.0$ hours. For significance level of 5%, the area in the upper tail is 5% and the associated z-value is 1.645. The critical value is thus $28.0 + 1.645(1.5)/\sqrt{25}$ $= 28.49$ hours. Samples of size 25 which have mean life-time of less than this value will result in the acceptance of the null hypothesis; if they are greater than this value the null hypothesis will be rejected and the conclusion will be drawn that $\mu > 28.0$ hours. Since our sample mean (28.60 hours) is greater than the critical value, we conclude that there is evidence that the new procedure does result in dry cells with a mean life-time of over 28.0 hours. Did you follow this? We suspect that you did. If by chance you missed the clue which led to the stated alternative hypothesis, read the question again.

For a significance level of 1%, the area in the upper tail is 1% and the appropriate z-value is 2.326. The critical value is thus $28.0 +$ $2.326(1.5)/\sqrt{25} = 28.70$ hours. Since the experimental mean is 28.60, this time we accept the null hypothesis. This means that at the 1% significance level the sample mean is compatible with a population whose mean is 28.0 hours. This implies that we lack evidence that the new procedure enhances the life-times of dry cells. Yet again the two tests we have suggested lead us to contradictory conclusions. Which should we accept? One is encouraged to suspect that the mean lifetime of the dry cells has been enhanced by the new procedure. One always tends to give the benefit of the doubt to hard-working fellow scientists or technologists! Let us display the wisdom of Solomon and suggest that more tests be carried out!

****

SAQ 3.7a   Associate each of the terms (i) to (v) below with the most appropriate item from the list (a) to (f).

(i) null hypothesis,
(ii) power of a test,
(iii) significance level,
(iv) type II error,
(v) type I error.

(a) accepting a batch which is off specification,
(b) rejecting a batch which is off specification,
(c) accepting a batch which conforms to specification,
(d) rejecting a batch which conforms to specification,
(e) the assumption that a batch conforms to specification,
(f) the assumption that a batch is off specification.

Response

The correct matching is as follows.

Term:	(i)	(ii)	(iii)	(iv)	(v)
Item:	(e)	(b)	(d)	(a)	(d)

If you have difficulty agreeing with the match given in any example you should re-read the section in which the term concerned was first introduced. Term (*i*) was introduced in Section 3.5.2, term (*iii*) and term (*v*) in Section 3.5.3, term (*iv*) in Section 3.5.4 and term (*ii*) in Section 3.5.5.

****

SAQ 3.7b

> A new manufacturer claims that his caustic solution is 51.0 (w/v) NaOH. There is reason to doubt this assertion! Nine replicate determinations are carried out on a random sample: the results are $\bar{x} = 50.75$ (w/v) and $s = 0.30$ (w/v). Test the manufacturer's assertion at the 5% significance level.

Response

The null hypothesis is H_0: $\mu = 51.0$. The way the question was phrased suggests that a lower one-tail test is in order. Why? Because it is highly likely that the doubts about the manufacturer's claim relate to the fact that everyone else seems to produce caustic solutions with a specification of 50.0 NaOH(w/v)! If you disagree with this, then it is likely that you have carried out a two-tail test with the alternative hypothesis H_1: $\mu \neq 51.0\%$. If this were an examination, a generous marker would hardly penalise you at all! We are not generous; we planned a one-tail test and a one-tail test it shall be! At the 5% significance level, a one-tail test means that the area in the tail is to be 5%. For an area in a tail of 5% and for 8 degrees of freedom, the t value is 1.860. Since this is a lower one-tail test the critical value is therefore $51.0 - 1.860(0.30)/\sqrt{9} = 50.81\%$(w/v).

Since the mean percentage for the sample is 50.75%(w/v), we reject the null hypothesis and accept the alternative hypothesis that the mean NaOH content of the solution is *less* than 51.0%. Sorry new manufacturer! The graphical representation summarises the situation. We have been skipping graphical representations for a while. If they help you, always use them.

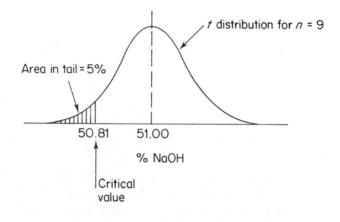

****

SAQ 4.1a

> A new method for the determination of the aspirin content of analgesic tablets is being developed. The method was applied to tablets which were known to contain 300 mg of aspirin: the results obtained on four tablets were 308 mg, 307 mg, 304 mg and 301 mg. Is there any evidence of a systematic error?

Response

We obviously require a *t*-test here.

The null hypothesis is H_0: $\mu = 300$ mg.

What about the alternative hypothesis? Some would argue that a two-tail test is in order: this leads to an alternative hypothesis of H_1: $\mu \neq 300$ mg. Others would disagree. They would argue that systematic error leads to results being either high or low, ie the results would show either a positive or a negative bias. Therefore, a more appropriate alternative hypothesis would be H_1: $\mu > 300$ mg (this is an after the event decision – the results were all greater than 300 mg).

Just to keep in practice, we shall carry out both tests and do so at the 5% level of significance.

(*i*) *Two-Tail Test*

$$H_o : \mu = 300 \text{ mg}$$
$$H_1 : \mu \neq 300 \text{ mg}$$

Significance level = 5%, therefore we need the t-value associated with 0.025 in a tail and 3 degrees of freedom, ie $t_{0.025,3}$ which is 3.182. The critical values of t are therefore ± 3.18.

$\bar{x} = 305$ mg   (We shall no longer comment on the calculation of means!)

$s = 3.16$ mg   (If you got the answer 2.74 mg, then either you or your calculator is dividing by n rather than by $(n-1)$ in the equation for the standard deviation.)

$t_{calc} = \dfrac{\bar{x} - \mu}{s/\sqrt{n}}$ (Based on acceptance that H_o is true.)

$$\alpha = (305 - 300)/(3.16/\sqrt{4}) = 3.16.$$

Since t_{calc} is within the critical range for t (just), we accept the null hypothesis. Therefore at the 5% significance level we reject the evidence of systematic error.

(*ii*) *One-Tail Test*

$$H_o : \mu = 300 \text{ mg.}$$
$$H_1 : \mu > 300 \text{ mg.}$$

Significance level is 5%, therefore we need the t associated with a tail area of 5% and 3 degrees of freedom, ie $t_{0.05,3}$, which is 2.35.

$$t_{calc} = 3.16.$$

Since this is greater than the critical t-value, we reject the null hypothesis and accept the alternative hypothesis: at the 5% significance level there is evidence that $\mu > 300$ mg. This implies that the method, when applied to these tablets, exhibits a systematic error: the results are higher than they should be.

Yet again we have taken an example in which the answer we get depends on our decision as to which test to apply. If we work out p-values, it can be shown that the significance level for the one-tail test is 2.55%.* If there is no systematic error then there was just over a 2.5% chance that our measurements would result in a mean value as high as 305 mg. For the two-tailed test the p-value must be doubled. If the null hypothesis is true, then there was about a 5.1% probability of obtaining a mean differing from 300 mg by as much as occurred (ie either as high as 305.0 mg or as low as 295.0 mg).

We hope that by now we have convinced you that when designing a statistical test it is important to give careful thought as to exactly what your criterion for decision should be. In practice you will not meet borderline cases as persistently as we have been presenting them, but they will arise.

---

* To verify this run program STAT3. Select option 4 under 'tables' and put $\nu = 3$, $t = 3.16$. The program reports a 'probability of EXCEEDING this value' of 2.547%.

****

SAQ 4.2a   In each of the following examples information is given for two samples. For each derive a 95% and a 99% confidence interval for $(\mu_1 - \mu_2)$, the difference between the means of the populations sampled.

(*i*) $\bar{x}_1 = 550, \bar{x}_2 = 480, n_1 = 8, n_2 = 6$, and $\sigma_1 = \sigma_2 = 10$.

(*ii*) $\bar{x}_1 = 155$ ppm, $\bar{x}_2 = 148$ ppm, $n_1 = 32, n_2 = 36, s_1 = 8$ ppm, and $s_2 = 6$ ppm.

(*iii*) $\bar{x}_1 = 13.50\%, \bar{x}_2 = 13.00\%, n_1 = 10, n_2 = 10, s_1 = 0.10\%$ and $s_2 = 0.08\%$

(*iv*) $\bar{x}_1 = 0.1015M, \bar{x}_2 = 0.01010M, n_1 = 15, n_2 = 10, s_1 = 0.0002M$, and $s_2 = 0.0003M$.

Response

(*i*)   Since both σ_1 and σ_2 are known and are equal we can use Eq. 4.6.

$$\mu_1 - \mu_2 = (\bar{x}_1 - \bar{x}_2) \pm z_{\alpha/2}\sigma\sqrt{(1/n_1) + (1/n_2)}$$

For a 95% confidence interval $z = 1.96$.

$$\therefore \quad \mu_1 - \mu_2 = (550 - 480) \pm 1.96(10)\sqrt{(1/8) + (1/6)}$$

$$= 70 \pm 10.6.$$

For 99% confidence interval $z = 2.576$

$$\mu_1 - \mu_2 = 70 \pm 2.576(10)\sqrt{(1/8) + (1/6)} = 70 \pm 13.9$$

(*ii*) Here σ_1 and σ_2 are not known. However, since the sample sizes are large we can use s_1 and s_2 as reasonable estimates of the σ's.

$$\mu_1 - \mu_2 = (\bar{x}_1 - \bar{x}_2) \pm z_{\alpha/2}\sqrt{(\sigma_1^2/n_1) + (\sigma_2^2/n_2)}$$

For a 95% confidence interval $z = 1.96$.

$$\mu_1 - \mu_2 = (155 - 148) \pm 1.96\sqrt{(8^2/32) + (6^2/36)}$$

$$= 7.0 \pm 1.96\sqrt{3}$$

$$= 7.0 \pm 3.4 \text{ ppm.}$$

For a 99% confidence interval $z = 2.576$.

$$\mu_1 - \mu_2 = 7.0 \pm 2.576\sqrt{3}$$

$$= 7.0 \pm 4.5 \text{ ppm.}$$

(*iii*) In this example σ_1 and σ_2 are not known and the sample sizes are small. If σ_1 and σ_2 are equal we can use Eq. 4.8.

$$\mu_1 - \mu_2 = (\bar{x}_1 - \bar{x}_2) \pm t_{\alpha/2}s\sqrt{(1/n_1) + (1/n_2)},$$

Let us now assume that $\sigma_1 = \sigma_2 = \sigma$. It follows that s_1 and s_2 are both estimates of σ and that an even better estimator is s_p. In the example given, the sample sizes are the same. Hence, Eq. 4.9 reduces to $s_p^2 = (s_1^2 + s_2^2)/2$

$$s_p = \sqrt{\frac{(0.10)^2 + (0.08)^2}{2}} = 0.0906$$

The number of degrees of freedom in this estimate is $n_1 + n_2 - 2 = 10 + 10 - 2 = 18$. For a 95% confidence interval, the t value sought is thus $t_{0.025,18}$; this is 2.101.

The confidence interval is thus:

$$\mu_1 - \mu_2 = (13.5 - 13.0) \pm 2.101(0.0906)\sqrt{1/10 + 1/10}$$

$$= 0.5 \pm 0.085\%.$$

For a 99% confidence interval, the t value sought is $t_{0.005,18}$; this is 2.878, leads to a confidence interval of $0.5 \pm 0.12\%$.

(*iv*) In this question we have, once again, σ_1 and σ_2 unknown and small sample sizes. Can we assume that σ_1 and σ_2 are equal? Methods appropriate to the determination of molarities of approximately 0.1, eg titrations, are likely to have similar precisions. Indeed, it is likely that both $\bar{x}_1$ and $\bar{x}_2$ were determined by a titrimetric method. We can assume therefore that both s_1 and s_2 are estimates of $\sigma = \sigma_1 = \sigma_2$ and that s_p is an even better estimator.

The appropriate relationship is

$$\mu_1 - \mu_2 = (\bar{x}_1 - \bar{x}_2) \pm t_{\alpha/2}s_p\sqrt{(1/n_1) + (1/n_2)}$$

First it is necessary to calculate s_p

$$s_p^2 = \frac{(n_1 - 1)s_1^2 + (n_2 - 1)s_2^2}{n_1 + n_2 - 2}$$

$$= \frac{14(2 \times 10^{-4})^2 + 9(3 \times 10^{-4})^2}{15 + 10 - 2}$$

$$= 5.96 \times 10^{-8}$$

$$s_p = 2.44 \times 10^{-4}.$$

For a 95% confidence interval and 23 degrees of freedom, the t value required is $t_{0.025,23}$; this is 2.069.

$$\therefore \quad \mu_1 - \mu_2 = (0.1015 - 0.1010) \pm$$

$$2.069(2.44 \times 10^{-4})\sqrt{1/15 + 1/10}$$

$$= 0.0005 \pm 0.00021M.$$

For a 99% confidence interval and 23 degrees of freedom, $t_{0.005,23}$; $= 2.807$.

$$\therefore \quad \mu_1 - \mu_2 = 0.0005 \pm 2.807(2.44 \times 10^{-4})\sqrt{1/15 + 1/10}$$

$$= 0.0005 \pm 0.00028M.$$

****

SAQ 4.3a

An agricultural research laboratory carries out a large number of protein-nitrogen determinations by the Kjeldahl method. This method involves digesting a substance with concentrated H_2SO_4, which converts the nitrogen into $(NH_4)_2SO_4$, and then adding an excess of NaOH to liberate NH_3. The NH_3 is distilled into standard acid and is determined by means of a back titration of the excess of acid. Knowing the amount of NH_3 produced enables the nitrogen content to be calculated (and the protein content of the sample to be estimated). The laboratory has heard reports that the NH_3 in Kjeldahl digests can be determined by an 'ammonia electrode'. It decides to compare this procedure with the distillation/back-titration procedure. $\longrightarrow$

**SAQ 4.3a
(cont.)**

> A batch of barley is used for the test. Ten Kjel-
> dahl digests are prepared; five have their NH_3
> content determined by the ammonia-electrode
> procedure and five by the distillation/back-
> titration procedure. The results, reported as %
> N, are below.
>
> | Ammonia Electrode | Distillation-Back-Titration |
> |:---:|:---:|
> | 1.20 | 1.25 |
> | 1.30 | 1.24 |
> | 1.25 | 1.26 |
> | 1.25 | 1.22 |
> | 1.26 | 1.27 |
>
> Can we conclude at the 5% level of significance
> that the two procedures for nitrogen determina-
> tion give the same result?

Response

Here both σ_1 and σ_2 are unknown and the sample sizes are small.
We shall assume that $\sigma_1 = \sigma_2$ (whether or not we are justified in
doing so will be discussed later). If $\sigma_1 = \sigma_2 = \sigma$, then s_1 and s_2,
which we can calculate, are estimates of the same thing, ie of σ, and
the pooled standard deviation s_p, which can be calculated from s_1
and s_2, is considered an even better estimate of σ.

For the ammonia electrode procedure:

$$\bar{x} = 1.252 \text{ and } s_1 = 0.0356$$

For the back-titration procedure:

$$\bar{x} = 1.248 \text{ and } s_2 = 0.0192$$

There is quite a bit of arithmetic in getting these values. The formulae you must apply are given below.

$$\bar{x} = \sum_{i=1}^{5} x_i/5 \quad \text{and}$$

$$s = \sqrt{\sum_{i=1}^{5}(x_i - \bar{x})^2/4}$$

The null hypothesis is H_o: $\mu_1 = \mu_2$.

The alternative hypothesis is H_1: $\mu_1 \neq \mu_2$. We need a two-tail test at the 5% level of significance and we have 8 degrees of freedom, $(5 + 5 - 2)$. The critical t-values for the test are thus $\pm t_{0.025,8} = \pm 2.31$.

Now to calculate the t-value resulting from the test. First we need s_p and, since $n_1 = n_2$ here, we have

$$s_p = \sqrt{\frac{s_1^2 + s_2^2}{2}} = \sqrt{\frac{(0.0356)^2 + (0.0192)^2}{2}} = 0.0286$$

Now we can obtain t_{calc}:

$$t_{calc} = \frac{\bar{x}_1 - \bar{x}_2}{s_p\sqrt{1/n1 + 1/n_2}} = \frac{1.252 - 1.248}{0.0286\sqrt{(1/5) + (1/5)}} = 0.22$$

t_{calc} lies within the critical range (easily) so we accept the null hypothesis: we have found that there is no significant difference between the two procedures for determining the NH_3 content of the digest.

****

SAQ 4.5a	Measurements were made on two samples. On sample A we made 8 replicate measurements and this yielded a sample standard deviation s_A = 0.12. On sample B we made 4 replicate measurements and this gave s_B = 0.24. Is this result consistent, at a 5% significance level, with both samples conforming to the same population standard deviation (ie is $\sigma_A = \sigma_B$)?

Response

The results *are* consistent with $\sigma_A = \sigma_B$ at the 5% level of significance. Did you reach this conclusion? We wonder if you agree with the detailed argument we give below? Don't worry if you found this rather difficult as your first SAQ using an F-test; it is designed to make you refer back to the argument in Section 4.5.2 if need be. Do this when necessary as you read our solution. Once you have cracked one problem with an F-test you should have little difficulty in applying it elsewhere.

We think that, from the way the question was framed, you should have attempted a two-tail test, ie that your alternative hypothesis should be $\sigma_A \neq \sigma_B$. This means that we must check upper and lower tails, each at a level of significance of 2.5%.

In the event s_B is larger than s_A so, to obtain a value >1, we define $F = s_B^2/s_A^2$. Hence you should find that the test results give $F = 4.0$ (remember to square the s values).

What about the critical value for F? First let's be sure we look up the right table. The level of significance for each tail is 2.5% so our table is that for a tail area of 0.025, ie Fig. 4.5c. Next we must sort out the degrees of freedom ν_1 and ν_2; we must be sure to get these the right way round. We have defined F as s_B^2/s_A^2, so we must take $\nu_1 = \nu_B$ and $\nu_2 = \nu_A$. Now $n_A = 8$ and $n_B = 4$, so $\nu_1 = 3$ and $\nu_2 = 7$. That is tricky! Remember the number of degrees of one less than the sample size, and sample B has been adopted as sample '1'. So the critical F-value is $F_{0.025(3,7)}$. Look up Fig. 4.5c; the value is $F = 5.89$.

Now our conclusion: the calculated F for our sample is smaller than the critical value of F for the test we are applying. Therefore we accept the null hypothesis: the results are consistent with the assumption $\sigma_A = \sigma_B$.

Did you perhaps try a one-tail test? We think this was the wrong choice in the context of the question, but it should still have led you to the same conclusion. The calculated F for the sample is unchanged at 4.0, but the critical F corresponding to $\sigma_B > \sigma_A$ becomes $F_{0.05(3,7)}$ (all the 5% significance is now associated with the single tail). For a tail area of 0.05 we must this time consult Fig. 4.5b. The resulting critical value for F is 4.35. This is still greater than 4.0 so the null hypothesis would still stand.

****

SAQ 4.5b

In each of the following examples, test the null hypothesis H_o: $\sigma_1 = \sigma_2$ against the stated alternative hypothesis at the given level of significance.

(*i*)   Sample A: $n_A = 6, s_A = 0.0715$.
Sample B: $n_B = 4, s_B = 0.0220$.
Level of significance = 5%: one-tail test.

(*ii*) Sample A: $n_A = 6, s_A = 0.0715$.
Sample B: $n_A = 4, s_B = 0.0220$.
Level of significance = 5%: two-tail test.

$\longrightarrow$

SAQ 4.5b (cont.)	(iii) Sample A: $n_A = 11, s_A = 0.0220$. Sample B: $n_B = 16, s_B = 0.0715$. Level of significance = 5%: two-tail test.
	(iv) Sample A: $n_A = 11, s_A = 0.0220$. Sample B: $n_B = 16, s_B = 0.0715$. Level of significance = 2%: two-tail test.

Response

(i)   Null hypothesis, $H_O : \sigma_A = \sigma_B$.

Alternative hypothesis, $H_1 : \sigma_A > \sigma_B$.

Sample A has the greater sample variance, therefore we shall calculate F from s_A^2/s_B^2, ie

$$F_{calc} = \left(\frac{0.0715}{0.0220}\right)^2 = 10.56$$

Level of significance = 5%. Since this is one-tail test, there will be 5% in the upper tail. The critical value for F is obtained by looking up $F_{0.05(5,3)}$. It is 9.01 (see Fig. 4.5b).

Since the calculated value for F exceeds the critical value, we reject the null hypothesis and accept the alternative hypothesis, ie $\sigma_A > \sigma_B$. This means that at the 5% level of significance, we have evidence that the samples come from populations with unequal variances.

If you got this one right, well done! There are several places where you could have gone wrong. Defining the critical F-value often throws novices. Remember; the greatest sample variance 'goes on top'; a one-tail test defines the rejection area as being the level of significance; when looking up a critical value for F in the table always ensure that the first degree of freedom mentioned refers to the ν_1 heading. For example, for $F_{(5,3)}$ we take $\nu_1 = 5$; for $F_{(3,5)}$ we take $\nu_1 = 3$.

If you did get it wrong, try again!

(ii)   Null hypothesis, $H_o : \sigma_A = \sigma_B$

   Alternative hypothesis, $H_1 : \sigma_A \neq \sigma_B$

From (i) we know that $F_{calc} = s_A^2 / s_B^2 = 10.56$.

Level of significance $= 5\%$. Since this is a two-tail test, there will be 2.5% in each tail. The critical value of F is obtained by looking up $F_{0.025(5,3)}$. By consulting Fig. 4.5c, you should find that this corresponds to $F = 14.88$.

Since the calculated value is smaller than the critical value of F, we accept the null hypothesis: $H_o : \sigma_A = \sigma_B$. At the 5% level of significance, the sample variances are such that they are compatible with belonging to populations whose variances are equal. The one-tail test of part (i) above was a more severe test than its two-tail counterpart.

(iii)   Here we have tried to fool you by presenting first the sample with the smallest variance. We hope that this did not upset you!

   Null hypothesis, $H_o : \sigma_A = \sigma_B$.

   Alternative hypothesis, $H_1 : \sigma_A \neq \sigma_B$.

Here $F_{calc} = s_B^2 / s_A^2 = 10.56$.

Level of significance $= 5\%$. This is a two-tail test, hence the critical value of F sought is $F_{0.025(15,10)}$. Convince yourself that this is the right description of the critical value, and that Fig. 4.5c gives $F_{0.025(15,10)} = 3.52$.

This time the calculated value greatly exceeds the critical value, ie $10.56 > 3.52$. We reject the null hypothesis, and accept the alternative hypothesis, $H_1 : \sigma_A \neq \sigma_B$. The evidence at the 5% level of significance is that the samples belong to populations with unequal variances.

It has probably not escaped your notice that we used the same values of sample standard deviation in this item as were used in (*ii*). We did this to make a point. With small sample sizes, the large difference in sample standard deviations was tolerated; we accepted the null hypothesis. The same standard deviations led to a rejection of the null hypothesis when the sample sizes were larger. You might wish to recall here, that as sample sizes get larger there is a tendency for s to settle down and to become a better estimate of σ. A large difference in the sample standard deviations is more tolerable for small sample sizes when we test $\sigma_A = \sigma_B$ than it is when we carry out the same test with large sample sizes.

(*iv*)   The null hypothesis is $H_O : \sigma_A = \sigma_B$.

The alternative hypothesis is $H_1 : \sigma_A \neq \sigma_B$.

As in (*iii*), $F_{calc} = s_B^2 / s_A^2 = 10.56$.

The level of significance is 2%, allowing 1% in each tail. The critical F-value is $F_{0.01(15,10)} = 4.56$ (see Fig. 4.5d). The critical value for F is greater than it was in item (*iii*), but it is still much smaller than the calculated F-value. We must, therefore, still reject the null hypothesis; at the 2% level of significance we conclude that the samples belong to populations with unequal variances. In fact it can be shown that, even if the level of significance is lowered to 0.2%, the null hypothesis should still be rejected.

We have been looking at the effect of two factors here. In item (*iii*) above, when the level of significance was set at 5%, the critical F-value was 3.52. In item (*iv*) a 2% significance level applied and the critical F grew to 4.56. The lower we make the level of significance for a test, the more tolerant it becomes from the point of view of accepting the null hypothesis. In spite of this, however, we still had to reject the null hypothesis in (*iv*), whilst in (*ii*), with the *same* values for the sample variances and a *larger value* for the level of significance, we had actually accepted the null hypothesis. A more potent factor changed when we progressed beyond case (*ii*); the *sample sizes* had increased. As sample sizes become larger, the range of acceptable F-values closes in towards the value one should expect for $\sigma_A = \sigma_B$. What value would one expect? Well, for very large samples s becomes a very reliable estimate of σ, so if indeed $\sigma_A = \sigma_B$, then we should find $s_A \approx s_B$. In other words the value of F will have to turn out very close to 1.0.

****

SAQ 4.5c

A new technician started a month ago in the quality-control lab. Since he arrived things seem to have gone swimmingly. His results indicate consistently that the plant's production is almost exactly on specification. He seems to be very efficient too, having plenty of time between samples to pursue his interest in reading the fine print on page 3 of his newspaper. Suddenly a rash of complaints start coming in from customers. Their quality checks have led them to refuse delivery of certain batches. Is something going wrong during the transport of the product to the customers? Or could it be that the technician is 'cooking the books'?

The analytical method being used has, from long experience, a standard deviation of 0.05%. The last ten results obtained by the technician, from this morning's production, are found to have a sample standard deviation of 0.015%. Is there evidence of 'cookery'? Test the data at a 1% level of significance.

Response

The Production Manager is not amused! The evidence is not favourable to the technician.

The data are below.

Sample A: $n_A = 10$, $s_A = 0.015\%$

Sample B: $n_B = \infty$, $s_B = \sigma_B = 0.05\%$,

Our null hypothesis is $H_0 : \sigma_A = \sigma_B$.

Our alternative hypothesis is $H_1 : \sigma_A < \sigma_B$.

Now $s_B > s_A$, so we define F as $s_B^2/s_A^2 = (0.05/0.015)^2 = 11.1$. Sample B is 'sample 1' and sample A is 'sample 2' so for a one-tail test at a 1% level of significance, our critical F-value is $F_{0.01(\infty,9)}$. Notice the order, it is ν_1 which is infinite this time. The F-value read from Fig. 4.5d is 4.31.

The calculated F-value exceeds the critical F-value, so the technician is condemned at the 1% level of significance (by a wide margin).

****

SAQ 4.6a

(*i*) Use your computer program to calculate the F-value, at a significance level $\alpha = 0.025$, when $\nu_1 = 8$ and $\nu_2 = 1$. Repeat this calculation for the same values of α and ν_1, but for successively larger ν_2 values, viz for $\nu_2 = 2, 3, 5, 10, 20, 30$ and 10000. Compare your values with those tabulated in Fig. 4.5c. Comment. $\longrightarrow$

SAQ 4.6a (cont.)	(*ii*) For the same series of values of ν_1 and ν_2 ie for $\nu_1 = 8$ and $\nu_2 = 1, 2, 3, 5, 10, 20, 30,$ and 10000, use the F-values tabulated in Fig. 4.5c (viz $F = 956.66, 39.37, ...$etc) to obtain corresponding one-tail probability values from the program. Comment on the accuracy of these values.

Response

(*i*) You must RUN your program, selecting the 'tables' option. Within that take sub-option 5 ('prob $- > F$'). You should obtain the results below.

ν_1	Data ν_2	prob	Program response: calculated F	From Fig. 4.5c: tabulated F
8	1	0.025	1114	956.7
8	2	0.025	41.98	39.37
8	3	0.025	15.43	14.54
8	5	0.025	7.00	6.76
8	10	0.025	3.87	3.85
8	20	0.025	2.91	2.91
8	30	0.025	2.65	2.65
8	10000	0.025	2.19	2.19

The results illustrate the general comment made in the text that for small values of ν_2 the program's computed F-values may be somewhat in error. For $\nu_2 = 1$ the discrepancy in the calculated F-value is 16.5%. This error decreases rapidly as ν_2 increases. It has already fallen to 6.6% by $\nu_2 = 2$, and it more or less disappears above $\nu_2 = 5$. Note that the larger discrepancies occur for large values of F, ie far out on the tail of distributions which have very long tails. The reason why we claim that these errors are of little practical importance will become clearer from the answer to (*ii*) below.

(*ii*) This time you must select 'tables' option 6: '$F->$ prob'. You should obtain the results below.

ν_1	ν_2	Data F (Fig. 4.5c)	Program response: calculated probability (%)
8	1	956.66	2.653
8	2	39.37	2.420
8	3	14.54	2.560
8	5	6.76	2.671
8	10	3.85	2.546
8	20	2.91	2.514
8	30	2.65	2.504
8	10000	2.19	2.518

We think that here at least you will have little difficulty in agreeing that the calculated probabilities are always adequately close to the expected value of 2.5%. The agreement here looks more reassuring than does that (for small ν_2) in (*i*) above where F was being calculated. But what, in fact, are we really interested in in our study of statistics? Our concern is to judge the statistical significance of the conclusions drawn from a given set of measurements. In other words it is *probabilities* which are of direct interest to us. F-values themselves are merely tools in our work. It will not trouble us unduly if our program estimates some F-values a little inaccurately, so long as we obtain satisfactory accuracy in predicting probabilities. Our program achieves that reasonably well.

****

SAQ 4.6b

A manufacturer produces dry cells which under a standardised test have an average life-time of 28.0 hours with a standard deviation of 1.50 hours. It is suggested that a slight change in the production process will lead both to an increase in mean life-time and to reduced variability in performance. The production process is duly changed and a random sample of 16 dry cells on testing gives the following life-times (in hours).

29.8   28.3   27.4   29.2   29.5   29.2   27.5   29.2

27.9   30.3   27.5   30.3   28.7   28.4   29.2   29.7

Use your computer program to assess the claims made for the modified process.

Response

The modified process seems to score on both counts. The mean life-time of the 16 cells tested is 28.88 hours and the sample standard deviation is 0.96 hour. There is evidence (to a significance level of 2.4%) that the true (population) standard deviation is now lower than the original value of 1.50 hours. By applying a t-test using the sample standard deviation we gain evidence to a significance level of 0.11% that the new true (population) life-time is greater than the previous 28.0 hours.

In more detail, to tackle this problem we must first modify the DATA lines in our program to give these below.

3900 DATA 28.0,1.50
4000 DATA 16
4010 DATA 29.8,28.3,27.4,29.2,29.5,29.2,27.5,29.2,27.9,30.3,27.5,
                                              30.3,28.7,28.4,29.2,29.7

It is preferable first to test the evidence that the standard deviation of the life-time of the dry cells has decreased. RUN the program, and take 'sample' option 4. The output should report that the sample standard-deviation from the test is 0.963 hour, compared with a population value of 1.50 hour previously. Comparison of these values gives an F-value of 2.43, which is reported to give a one-tail significance level of 2.385%. If, therefore, our null hypothesis is that the new process gives no improvement in standard deviation, ie $H_O : \sigma = 1.50$ hour, and the alternative hypothesis is that the new process yields an improvement, $H_1 : \sigma < 1.50$ hour, then we reject the null hypothesis at any significance level greater than 2.385%. This is pretty convincing evidence.

Turning to a test of the effect on the mean ('sample' option 3), the above conclusion means that we should select the test based on the sample standard deviation. When we do this we are told that the sample mean is 28.88 hours, compared with the original value of 28.0 hours, and that with $s = 0.963$ hour (for our sample of size 16) this difference has a one-tail significance level of 0.114%. Thus we have a null hypothesis, $H_O : \mu = 28.0$ hour, and an alternative hypothesis $H_1 : \mu > 28.0$ hour, and we reject the null hypothesis right down to a significance level of 0.114%. The evidence is extremely strong; if the null hypothesis had been true there would have been only just over one chance in a thousand that our sample would yield such a high mean life-time.

Had you chosen to test the change in mean life-time on the assumption that the population standard deviation was still 1.50 hours (ie selecting the population standard deviation in running 'sample' option 3), then the program would still suggest that there was an improvement. The evidence would hold down to a significance of 0.939%. This test does not seem an appropriate one, however, since there is rather strong evidence that the population standard deviation has been reduced to less than 1.50 hours.

Notice the strength of the conclusions which we are able to draw in this example in spite of the fact that four of the sample of 16 cells from the new process still had individual life-times below the original population mean of 28.0 hours. You might like to experiment with the effect of sample size in this investigation. This can be very easily done. For instance, retyping line 4000 to read

    4000 DATA 4

would mean that when the program was RUN it would act as though the test sample consisted of only the first four dry cells. Could you have drawn any clear evidence of improvement with such a small sample?

****

SAQ 4.6c

The cadmium content of various tissue samples was investigated by two procedures using atomic absorption spectrometry. Comparative results (for Cd in ppm) are given below.

Procedure 1	Procedure 2
1.6	1.3
4.1	4.4
2.8	3.3
5.7	6.5
1.9	1.4
2.4	2.5
4.2	3.5
10.3	10.2
11.3	12.8
10.4	10.7
3.5	4.5

Procedure 2 gives a larger result for 7 of the 11 samples reported. Is there evidence that this due to a systematic discrepancy between the two methods?

Response

A paired-difference test is in order. You should find that the measurements by procedure 1 are on average 0.264 ppm lower than the corresponding values from procedure 2. The standard deviation of the differences is, however, somewhat larger than this, viz 0.664 ppm. For a test of the null hypothesis $H_0 : \mu_d = 0$, the t-value is 1.32, and there are 10 degrees of freedom. The program reports a one-tail significance-level of 10.85%.

We were asked to test the proposition that procedure 2 systematically gave the higher result. That implies a one-tail test. It therefore follows that we accept the null hypothesis at any level of significance below 10.85%. In the absence of any systematic discrepancy between the methods there was better than a one in ten chance that the results for procedure 2 would have turned out this much larger on average.

Had the question been phrased as a test of whether there was evidence of *any* systematic discrepancy between the two procedures, ie regardless of direction, then a two-tail test would have been indicated. The null hypothesis would then be retained for any level of significance below 21.7% (ie 2 × 10.85%).

Using the program to obtain the above answers simply required substituting the two sets of measurements, and the size of each set (11) through DATA lines 4000-4030. All that is then necessary is to RUN the program selecting 'two samples' and 'paired-difference of means'.

****

SAQ 4.6d

Two laboratories were asked to apply a standard method to determine the total phosphate concentration in a sample of river water. The following replicate results were reported, on a basis of $\mu g \ l^{-1}$ of phosphate.

Lab A: 20.7   27.5   30.4   23.9   18.1   24.1
        24.8   28.9
Lab B: 20.9   21.4   24.9   20.5   19.7

(*i*) Are the results from the two labs consistent with one another?

(*ii*) How does it affect your conclusions if it is known from long experience that the analytical method, properly applied to such samples, has a standard deviation of 2.0 μg l^{-1}?

Response

(*i*) We do not know the true result, nor at this stage can we assume that the population standard deviation is known. We have, therefore, simply to compare the means and standard deviations derived from the measurements of each laboratory.

Lab A: Mean = 24.80 $\mu g \ l^{-1}$ st. dev. = 4.11 $\mu g \ l^{-1}$

Lab B: Mean = 21.48 $\mu g \ l^{-1}$ st. dev. = 2.01 $\mu g \ l^{-1}$

Inspection of these values suggests that Lab B has worked to substantially greater precision than Lab A; the sample standard deviations differ by more than a factor of two. There is also a substantial difference in the means. Lab A, with 8 replicate measurements, reports a mean value which is higher than Lab B's by over 1.5 times Lab B's standard deviation. But such argument is simple hand-waving! We need to make our judgements quantitative. It then turns out that neither the difference in the standard deviations, nor that in the means, is very significant.

The first step is to enter the values into our program.

>   4000 DATA 8
>   4010 DATA 20.7,27.5,30.4,23.9,18.1,24.1,24.8,28.9
>   4020 DATA 5
>   4030 DATA 20.9,21.4,24.9,20.5,19.7

It is then a matter of running 'two samples' option 3 to test the difference in the standard deviations. The F-value is revealed as 4.18 and, for the sample sizes involved, this has a reported one-tail significance of 9.339%. We could not damn Lab A on this evidence, the labs could be working equally precisely and this sort of thing would be expected to happen by chance almost one time in ten. In terms of a two-tail test, we would have accepted this result as consistent with the labs being equally precise at any significance level below 19% (strictly 2 × 9.339%).

Now the difference in means can be tested (option 1). The pooled standard deviation can be used, and its value is reported as 3.498 μg l^{-1}. The corresponding t-value is 1.66, and there is a total of 11 degrees of freedom. The one-tail significance level is 6.208%. If we test the laboratories against one another, on a two-tail basis, at any significance level below 12.4%, we shall accept that the results are consistent with there being no systematic bias between their respective measurements.

(*ii*) Now we are allowed access to a very useful piece of further information. We are told that the population standard deviation for the standard method applied to waters containing Zn at these levels is 2.0 μg l^{-1}. This is inserted in the population DATA.

>   3900 DATA 0,2.0
>   3910 DATA 0,2.0

We do not know the population mean, nor are we going to ask the program to use it, so we have simply given it as zero in the DATA lines. We would not need to employ two labs on this analysis if we knew what the answer was! Knowing σ immediately renews our suspicion about the precision of Lab A's work. So this time we RUN the 'one-sample' menu and test the sample standard deviation against the population value. Fortunately, the 'one sample' selected under this menu is always sample A. If you do this you should find that there is quite a dramatic change in the news. The F-value (s_A^2/σ^2) is 4.23, not very different from before. The real change is in the relevant numbers of degrees of freedom (still 7 for sample A but infinite for the population). The one-tail significance is quoted as 0.014%. If the population standard deviation of 2.0 μg l^{-1} had been applicable to Lab A's procedures, there would be only a little more than one chance in 10,000 that they would have found so large a value for their sample standard deviation. Lab A has not been working to the precision expected of the analytical method!

So what about the difference in means between the labs?

Easily tested. RUN the 'two-samples' menu option 1 again, but now we can ask for the population standard deviations to be used. This allows a z-test. The z-value is reported as 2.91 and this leads to a one-tail significance of 0.18%. In terms of the claimed population standard deviation the difference in means is judged highly significant. There is a systematic bias leading to Lab A obtaining a higher result than Lab B.

Be careful not to jump to conclusions! The temptation is to send a snooty letter to Lab A and to congratulate Lab B. That would probably be judging the situation correctly. But this is a tricky business and we cannot be absolutely sure. Do you remember our discussion, long ago at the start of the Unit, of the chemist who was accurate but imprecise, and of the chemist who was precise but inaccurate? It is possible that it is Lab B which is making the systematic error, even though it is working with good precision! Perhaps Lab A's work is accurate, even though it is imprecise. In this game niggling doubts always remain. If they really worry us, we might think of employing a third lab ... !

****

SAQ 5.0a	An instrumental analysis of an unknown involves a number of stages, including measurements on standard samples of known concentration, construction of a calibration curve, and use of the curve in conjunction with measurements made on the unknown.
	(*i*) What different sources of random errors can be identified?
	(*ii*) Can you think of any reasons why systematic errors may be avoided?

Response

Your answer is likely to be considerably different from that below. It could possibly be better! The point of asking the question was, however, simply to encourage you to realise that, from the point of view of assessing errors, using a calibration curve is a moderately complex exercise. So something has been achieved even if you gave up because you were at a loss for an adequate answer!

(*i*) In an analysis making use of a calibration curve, random errors may be associated (*a*) with inaccuracies in the curve itself, (*b*) with the actual measurement made on the unknown. The former will be minimised if the calibration curve is constructed on the basis of a large number of measurements made on standards. The latter will be reduced by making several repeat determinations on the unknown.

(*ii*) Systematic errors might in many circumstances be expected to *cancel*. Provided the same procedure is meticulously followed, both with the standard samples used in constructing the calibration curve and with the unknown, then any constant systematic error should affect both equally and be automatically eliminated in reading the calibration curve. The same would apply to any systematic error whose size varies only with analyte concentration.

We still need to beware, however, of any possible systematic error affecting only the calibration curve or only the unknown. If, for example, all the standards used in a calibration were obtained by dilution of a single stock solution, and the concentration of that stock was in error, then determinations of unknowns would be subject to a systematic error whose presence would not be at all obvious. We also need to take careful note of the guarded phrase: 'provided the same procedure is meticulously followed.' Are we dealing with the same operator, was the calibration graph constructed a week ago, has the instrument's response wandered ... ? Remember our recent discussions of repeatability and reproducibility! We shouldn't be too complacent. It is assumed in the above that the analytical procedure handles possible interferences in the sample.

****

SAQ 5.1a	With the aid of a ruler, draw a straight line to conform with the data in Fig. 5.1a. Read off the gradient and intercept for your line. Check to see whether your values lie within the ranges specified in the SAQ responses section.

Response

For your line you should have found values in the following ranges:

$$0.026 < m < 0.031 \quad \text{and} \quad 0.14 < c < 0.20$$

If your values do not lie within these ranges check again whether your line looks a reasonable fit to the four points. Redraw it if necessary. If you still cannot agree with the above values then read on.

If you extrapolate your line through the points you should find that it cuts the vertical axis somewhere within the main scale division below $y = 0.20$. It is the value read off here (ie at $x = 0$) which is referred to as the 'intercept'.

If you are not used to measuring *gradients* this may have given you more trouble. This term is defined in the text immediately before the question - did you try to evaluate that? Be sure that you read off the coordinates of the two chosen points on your line in terms of the x- and the y-scales as labelled on the graph. For the sake of argument suppose you drew your line through the first and last of the plotted points. The coordinates of these are (8,0.39) and (20,0.73), so that the gradient formula gives:

$$m = (0.73 - 0.39)/(20 - 8) = 0.34/12 = 0.028.$$

****

SAQ 5.1b   Derive the least-squares straight line to fit the data below.

x	8.00	12.00	16.00	20.00
y	0.388	0.523	0.624	0.734

Obtain the values of m and of c using Eqs. 5.1 and 5.2. Plot the resulting line on Fig. 5.1a.

Response

There are four points given, so we have $n = 4$. For the various sums you should find:

$$S_x = 56, \quad S_y = 2.269, \quad S_{xx} = 864 \text{ and } S_{xy} = 34.044.$$

In more detail, for example:

$$S_{xy} = 8(0.388) + 12(0.523) + 16(0.624) + 20(0.734) = 34.044$$

Note the distinction between S_{xx} and S_x^2

$$S_x^2 = 56 \times 56 = 3136, \text{ whereas } S_{xx} = 864.$$

Substitution for these sums in Eqs. 5.1 and 5.2 gives

$$m = 0.02848 \text{ and } c = 0.1686$$

To draw the line we simply need to identify any two points on it. It is best to choose these far apart. Thus, at $x = 0$, $y = 0.1686$ and, at $x = 25$, $y = 0.02848 \times 25 + 0.1686 = 0.881$. Draw the line joining $(0,0.169)$ and $(25,0.881)$

****

SAQ 5.1c

Determine the x-on-y regression line for the chloride analysis data given in SAQ 5.1b. Use Eqs. 5.3 and 5.4 to obtain the value of m' and c', and then add this further line to Fig. 5.1a.

Response

The method is precisely similar to that in the previous SAQ. The only new sum needed is

$$S_{yy} = \sum_{i=1}^{n} y_i^2 = 1.3522.$$

Substituting into the formulae given yields $m' = 0.02858$ and $c' = 0.1671$.

When you draw this line on to Fig. 5.1a, you should find that it is almost indistinguishable from the *y*-on-*x* regression line drawn after SAQ 5.1b.

****

SAQ 5.1d

> The *y-on-x* and the *x-on-y* regression equations obtained in the previous two SAQs for the chloride analysis data are given below.
>
> $$y = 0.028475\ x + 0.16860$$
>
> $$\text{and } y = 0.028584\ x + 0.16707$$
>
> Solve these as two simultaneous equations to find the *x*- and the *y*-value at their crossing-point. Hence confirm that the crossing point is indeed the mean point $(\bar{x}, \bar{y})$ of the data. (You may notice that a fifth significant figure has been added in quoting the *m* and *c* values. These two equations are so similar that this is needed in order to obtain an answer with sufficient precision.)

Response

Subtract the second equation for the first to eliminate *y*, yielding:

$$0.0000 = -0.000109\ x + 0.00153$$

from which we find $\bar{x} = 14.0$. Substituting back into the first equation gives

$$\bar{y} = 0.028475(14.0) + 0.16860 = 0.56725$$

Now $S_x = 56$ and $S_y = 2.269$ (see the response for SAQ 5.1b).

Hence $\bar{x} = 56/4 = 14.0$ and $\bar{y} = 2.269/4 = 0.56725$

****

SAQ 5.2a

> Can you think of the reason why we use $(n-2)$ in the denominator of the expression defining s_y, the standard error of the estimate of y?
>
> (Hint. Recall that we divide by $(n-1)$ when we estimate the standard deviation from a set of repeated measurements of a single quantity when the population mean is unknown.)

Response

When finding the standard deviation from a simple set of n readings, we divide by $(n-1)$ whenever the mean was also determined from the data. The act of computing the mean uses up one item of information from the n independent measurements, leaving only $(n-1)$ further degrees of freedom. If we had only *one* reading we should have no measure at all of dispersion.

When we fit a line to n data points, we determine *two* parameters, m and c, from the data. Hence there are only $(n-2)$ degrees of freedom remaining to measure the dispersion about the regression line. If we had only two data points, the only rational choice of line would be that line passing through the two points, leaving no measure of dispersion.

****

SAQ 5.2b	Calculate the value of s_y for the chloride-ion example. (The necessary data are tabulated in Fig. 5.2a.)

Response

By using Eq. 5.5 we have:

$$s_y = \sqrt{\frac{(-0.0084)^2 + (0.0126)^2 + (-0.0003)^2 + (-0.0042)^2}{(4 - 2)}}$$

$$= \sqrt{0.000124} = 0.0111$$

Notice that, with such a small set of points, the calculated standard error of estimate is almost as large as the largest residual.

****

SAQ 5.2c	A 'regression' line is drawn when there are only two calibration points. (*i*) What will the values of the residuals be? (*ii*) What result would you obtain for the standard error of the estimate?

Response

Since the line chosen will be that joining the two calibration points, both points will lie exactly on the line. Hence both residuals are zero.

(See Fig. 5.1b if you need reminding of what a residual is). Since the number of points, n, equals two, then there are zero degrees of freedom for determining s_y. Eq. 5.5 gives

$$s_y = \sqrt{\left(\sum_{i=1}^{n} d_i^2\right)/(n-2)} = \sqrt{(0+0)/(2-2)} = 0/0 = ?$$

ie we have an undetermined result. It is then impossible to give any estimate at all for the precision of the line drawn.

****

SAQ 5.2d	Substitute the value $x_o = 0$ into the formula given in Fig. 5.2c for the standard error of the regression line. Show that for this value of x the formula reduces to that given for the standard deviation of the intercept c.

Response

The result has to follow, since the intercept c simply *is* the value of y at $x = 0$. Substituting $x_o = 0$ into the formula in row (d) of Fig. 5.2c gives the standard deviation as:

$$s_y \sqrt{\frac{1}{n} + \frac{\bar{x}^2}{S_{xx} - S_x^2/n}}$$

$$= s_y \sqrt{\frac{S_{xx} - S_x^2/n + n\bar{x}^2}{n(S_{xx} - S_x^2/n)}}$$

Now $\bar{x} = S_x/n$, so the second and third terms in the numerator cancel, giving

$$s_y\sqrt{S_{xx}/(nS_{xx} - S_x^2)},$$

which is indeed the formula given for the standard deviation of the intercept c.

****

SAQ 5.2e

> Consider once more the calibration line for chloride determination. (The data are tabulated in Fig. 5.2a.) An unknown sample, when treated in the standard way, gives an absorbance, $y = 0.660$. Estimate the concentration, x, and evaluate the standard error of this prediction supposing:
>
> (*i*) the value of y is the result of a single determination,
>
> (*ii*) $y = 0.660$ is the mean of four replicate determinations.

Response

Substitute $y = 0.660$ in the regression equation:

$$y = 0.02848x + 0.1686$$

Solve to obtain: $x = (0.660 - 0.1686)/0.02848 = 17.25$

The estimated concentration is thus 17.25 mg dm^{-3}

To find the standard error of the prediction of x, we need to substitute

$$s_y = 0.0111, \quad m = 0.02849, \quad n = 4,$$
$$\bar{y} = S_y/n = 2.269/4, \quad S_{xx} = 864 \text{ and } S_x = 56$$

into the formula in Fig. 5.2c. (To confirm these values check back to SAQs 5.1b and 5.2b.)

(*i*) For a single determination we have $N = 1$. Hence you should find:

$$\text{standard error} = 0.390\sqrt{(1.0 + 0.25 + 0.133)}$$

$$= 0.46 \text{ mg dm}^{-3}$$

(*ii*) Taking $N = 4$ the first term under the square root becomes 0.25, and the standard error becomes 0.31 mg dm^{-3}. This is a significant improvement.

****

SAQ 5.2f

(*i*) For the example discussed in SAQ 5.2e, in which four determinations of an unknown chloride solution gave a mean absorbance of 0.660, give the 95% confidence interval for the chloride concentration.

(*ii*) Give the 95% confidence interval for the intercept c for our chloride regression line.

(*iii*) Can you, at a 1% level of significance, assert that the true value of the intercept is greater than zero?

Response

(*i*)   The standard error for the prediction of x at $y = 0.660$ was calculated in SAQ 5.2e(*ii*) to be 0.31 mg dm^{-3}. We have four points and so this estimate has only two degrees of freedom. The regression line gives the chloride concentration as 17.25 mg dm^{-3}.

For a 95% confidence interval we want 2.5% probability in each tail. so you should have looked up $t_{0.025,2}$ for which Fig. 3.4b gives the value 4.30. Hence the 95% confidence interval is (17.25 $\pm$ 4.30 $\times$ 0.31) mg dm^{-3}. In other words the chloride concentration lies in the range (15.9–18.6) mg dm^{-3}.

We are quite severely limited by the smallness of the number of points on which the regression has been based.

If the analysis figure for the unknown was based on a single measurement instead of four, as was suggested in SAQ 5.2e(*i*), then the standard error for the prediction would be 0.46 mg dm^{-3}. This would significantly widen the confidence interval to (17.25 $\pm$ 4.30 $\times$ 0.46) mg dm^{-3}, ie to the range (15.3–19.2) mg dm^{-3}

(*ii*)   The calculated intercept of the regression line is 0.1686 (SAQ 5.1b). The expression for the standard deviation given in Fig. 5.2c is:

$$s_c = s_y\sqrt{S_{xx}/(nS_{xx} - S_x^2)}$$

Now $s_y = 0.0111$ (SAQ 5.2b), $S_{xx} = 864$, $S_x = 56$, $n = 4$ (SAQ 5.1b)

Substituting these values in the formula gives $s_c = 0.0182$. The required value of t is 4.30 (as in part (*i*) above). Thus the 95% confidence range is

$$c = 0.1686 \pm 4.30 \times 0.0182$$

$$viz 0.090 < c < 0.247$$

(*iii*) The null hypothesis we should test is $H_0: c = 0$. The alternative hypothesis is $H_1: c > 0$. So our test is to be one-tailed.

Our estimated c-value is 0.1686 and its estimated standard deviation is 0.0182 (see part (*ii*) above). Hence the calculated t for $c = 0$ is

$$t_{calc} = 0.1686/0.0182 = 9.26.$$

The regression is based on four points, so we have two degrees of freedom. The critical value of t for a one-tail probability of 1% is thus $t_{0.01,2}$. By looking this up you should find $t_{crit} = 6.97$.

Since the calculated t-value exceeds the critical value we reject the null hypothesis at the 1% level of significance and conclude that the true c-value is indeed positive. We have more to say about this conclusion in Section 5.3!

****

SAQ 5.2g Evaluate the 95% confidence limits of the chloride regression line at $x = 14$ mg dm^{-3} (ie at $x = \bar{x}$). How much smaller is this range than that at $x = 0$?

Response

The relevant formula in Fig. 5.2c is:

$$\text{standard error of regression} = s_y\sqrt{\frac{1}{n} + \frac{(x_0 - \bar{x})^2}{S_{xx} - S_x^2/n}}$$

At $x_0 = \bar{x}$ the second term under the square root gives zero. Now $s_y = 0.0111$ (SAQ 5.2b) and $n = 4$.

Substitution gives:

standard error of regression = 0.00555

For 95%. confidence with two degrees of freedom $t = 4.30$ (as in SAQ 5.2f).

Now we are at the mid-point of the calibration. The point on the line is $(\bar{x}, \bar{y})$. So the y-value given by the line is

$$\bar{y} = S_y/n = 2.269/4 = 0.5673 \text{ (see SAQ 5.1d)}$$

The resulting range for y is:

$$y = 0.5673 \pm 4.30 \times 0.00555$$

$$viz\, 0.543 < y < 0.591.$$

Hence the width of the 95% interval at the mean-point is 0.591 − 0.543 = 0.048.

Compare this with the width of the 95% interval for the intercept, ie at $x = 0$, which is $(0.247 - 0.090) = 0.157$ (see SAQ 5.2f(*ii*)).

The uncertainty at the intercept is over three times larger than that at the mean point of the regression.

****

SAQ 5.3a   The regression line in Fig. 5.2d has been calculated on the basis of four calibration points
for chloride concentrations in the range 8–20
mg dm^{-3}. It has been argued that, has many
more calibration points been included, all in the
range 8–20 mg dm^{-2}, then a very similar regression line would probably have resulted, but that
its 95% confidence limits would have been considerably narrowed. Give arguments for and/or
against these claims.

Response

There are various ways of tackling this question and there is no short
obviously 'correct' answer. The claims made could only in the end
be supported either by making further measurements or by some
evidence of the size of the random error inherent in making any
given reading. Our own defence of our claims starts by arguing that
random measurement errors appear to be small.

The fuller evidence of Fig. 5.2d, with the additional points at 0, 2
and 4 mg dm^{-3}, gives a fairly clear indication that the true relationship is curved. More than that, it looks as though a smooth curve
could be drawn which would pass very close to all the points. Random scatter seems to be small. It therefore appears likely, that if
many addition calibration points were obtained between 8 and 20
mg dm^{-3} they would roughly follow the existing points, all of which
lie close to the existing regression line. A newly calculated regression line would therefore not be expected to differ very much from
the existing one; it would surely lie well within the 95% confidence
region of Fig. 5.2d. Equally, the scatter of points (if all lie between
8 and 20 mg dm^{-3}) about the revised line would be expected to
be little different, ie the residuals would be of similar typical sizes.
thus we should expect little change in the value of s_y, the standard
error of estimate. So why do we suggest that the confidence intervals for the new line would be 'considerably narrowed'? Look at

the formula for the standard error of regression in Fig. 5.2c. It is the old story again. We have more points, and as n increases so the standard deviation in an estimate decreases. Here the formula for the standard deviation involves the square root of the sum of two terms. Both of the terms decrease as n increases. Check it, convince yourself!

****

SAQ 5.4a

(i) Use the piezoelectric crystal data for SO_2 analysis to construct a linear regression graph, including 95% confidence lines (ie construct the analogue of Fig. 5.2d for this system). The data given were as follows.

SO_2/ppm	f/Hz
5	45
10	55
15	67
20	75
25	90
30	101
35	113
40	122
45	143
50	150
55	159
60	177
65	192
70	195
75	215

$\longrightarrow$

SAQ 5.4a
(cont.)

> (*ii*) An emission is required to conform to a standard which imposes a upper limit of 50 ppm of SO_2. Four measurements are made, giving readings of 154, 159, 162 and 155 Hz. Is the limit being breached? Test at a 5% level of significance.

Response

(*i*) For 95% confidence ranges we require 2.5% probability in each tail. There are 15 calibration points, therefore 13 degrees of freedom. You should find $t_{0.025,13} = 2.160$. (2.161 according to "STAT"). Running program "LINE" with the given data should produce first the results listed in section 5.4, followed by an invitation to enter an x-value to determine the corresponding standard error of regression. By using $x = 20$ ppm for instance, you should find that the predicted y is 78.06 Hz, with a standard error of regression of 1.156. The 95% confidence range is thus:

$$78.06 \pm 2.160 \times 1.156$$

ie   from 75.56 to 80.56 Hz.

Repeating the calculation at various x-values should give you the following figure. Notice how well-defined the position of the line is compared to that in Fig. 5.2d.

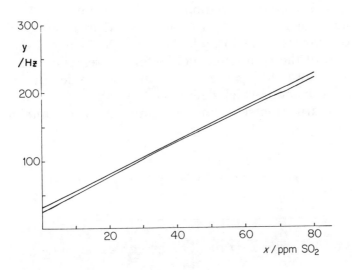

(*ii*) The mean of the four readings given is 157.5 Hz. Run "LINE" with the SO_2 data. The final stage of the program calculates standard errors of prediction. Give $y = 157.5$, and specify 4 for the number of repeats. You should find that the predicted x is 52.73 ppm, with a standard error of 0.789 ppm.

Since the critical value for the emission standard is 50 ppm the t-value corresponding to this finding is $(52.73 - 50.00)/0.789 = 3.46$.

The null hypothesis to test is that the emission just meets the 50 ppm limit; the alternative hypothesis is that the limit is exceeded. The test is thus one-tailed and so, for 5% significance, the critical t-value is $t_{0.05, 13}$ which has a value of 1.77. The calculated t-value exceeds the critical t-value, so the null hypothesis is rejected. The limit for SO_2 emission has been breached. Program "STAT" will tell you that a t-value of 3.46 with 13 degrees of freedom has a one-tail significance level of 0.21%, so the conclusion of the test is very clear cut.

You might wonder why we take 13 degrees of freedom. Perhaps you are thinking that, since the reading of y involved 4 repeats, we have only 3 degrees of freedom? Not so, remember that the number of degrees of freedom is to do with the reliability of a (sample) estimate of a standard deviation. Now we have not used the spread of y-values for the unknown in any way. Our estimated standard deviation, through the formula in Fig. 5.2c, is based entirely on the spread of the calibration points for the regression line. There were 15 of those, though the points were used to determine the two parameters m and c which define the line. This left 13 independent items of information reflecting the scatter about the line, hence 13 degrees of freedom.

****

SAQ 5.4b	*This SAQ should be regarded as optional!*
	Incorporate the regression line calculation as a facility within program "STAT". Note that all variables in program "LINE" have been named so as to avoid clashes with others in "STAT". Also note that some gaps in line numbers have been left in "STAT", which might prove convenient for this task.

Response

We made this SAQ optional, but we hope you find time to give it a try. We have suggested the task in the spirit of our suggestion at the end of Part 4 of the Unit that you should experiment with various possible extensions to "STAT". We intend to leave you, to some extent, to your own devices in tackling this, but we are prepared to make one or two suggestions which might help you on your way!

Notice that the arrays X, Y and D are just the same as were specified in "STAT" (line 80). But note that "STAT" reads in the x's and y's separately, through DATA lines 4000–4030.

The regression calculation would most naturally come as a new option under the 'two samples' menu. Line 930 is left blank for a statement such as:

930 PRINT"4＝regression of y on x"

Lines 950 and 960 would have to be amended to allow for the new option being selected (ie I＝4 after the INPUT line 940). For the regression calculation the GOSUB instruction could cause a transfer to, say, line 2900. From that line number, the body of program "LINE" could be inserted (omitting the contents of lines 10, 20 and 80 whose effects will already have been catered for within "STAT").

Once you get the regression option to work, then you might like to see how you can extend it so that confidence limits can be calculated! Remember that the subroutine starting on line 1300 can generate the t-value for any desired single tail area α. (The program symbols are T for t, and P for the probability α.)

If you suceed with this SAQ, then we feel you are truly beginning to find your feet as a programmer.

****

SAQ 5.5a Look carefully at the different lines (a), (b) and (c) in Fig. 5.5a. Note some brief critical comments that occur to you, then compare your observations with those we give in our response.

Response

This was an open-ended question which was principally intended to encourage you to look closely at Fig. 5.5a. Several comments can be made. The following seem most pertinent to us.

(*i*) The straight line (*a*) gives a *better* fit than either of the curves (*b*) or (*c*) over the interval 8-20 mg dm^{-3}. It is, however, hopeless below about 7 mg dm^{-3}.

(*ii*) Curve (*b*) gives a significantly better fit than (*c*). This might at first sight seem surprising as these two were both derived to give optimum fits of the form $y = ax + bx^2$. The explanation is that they are 'best fits' in different senses. Curve (*b*) gives the best possible fit to *y* in the graph of *y versus x* (ie in Fig. 5.5a). Curve (*c*) gives the best possible fit to *y/x* in the graph of *y/x versus x* (ie in Fig. 5.5b). The fact that these two lines differ significantly is a fair indication that the 'true' relationship between *y* and *x* is not ideally well represented by an equation of this form. The form $x = ay + by^2$ looks more satisfactory (see curve (d)).

(*iii*) Notice the widely different predictions of (*a*), (*b*) and (*c*) when they are extrapolated beyond 20 mg dm^{-3}. It gives yet another opportunity to warn against the dangers of extrapolation!

****

SAQ 5.5b The expanded set of calibration data for the chloride determination case study is given below.

x mg dm^{-3}	*y*
2.0	0.113
4.0	0.218
8.0	0.388
12.0	0.523
16.0	0.624
20.0	0.734

The best fit to these data in Fig. 5.5a is curve (*d*), which has an equation of the form.

$$x = ay + by^2 \qquad \longrightarrow$$

SAQ 5.5b **(cont.)**	Obtain values for parameters *a* and *b*. (Hint: put the equation into linear form: $x/y = a + by$. Then use program "LINE").

Response

By using the approach suggested we get $a = 15.14$ and $b = 16.03$.

The equation to fit is $(x/y) = a + by$. The values are:

y	0.113	0.218	0.388	0.523	0.624	0.734
x/y	17.70	18.35	20.62	22.94	25.64	27.25

When you enter these data into "LINE" you have to be careful.

y needs to be put into array *X* and (x/y) into array *Y*! (Compare the above equation with the standard $y = mx + c$).

The DATA lines are

1000 DATA 6
1010 DATA 0.113,17.70,0.218,18.35,0.388,20.62,0.523,22.94,0.624,
25.64,0.734,27.25

When RUN with these data the program reports a gradient of 16.03 and an intercept of 15.14. Again care is needed correctly to identify $a = 15.14$ and $b = 16.03$. (Look at the equation again: *a* is the intercept and *b* the gradient).

Multiplying by *y* we get:

$$x = 15.14y + 16.03y^2$$

You might like to check that this is in very good agreement with curve (d) in Fig. 5.5a. (The agreement is not quite exact because the equation for curve (d) was produced by a least-squares fit to the y *versus* x graph, not to a graph of (x/y) *versus* y. See comment (ii) in the response to SAQ 5.5a)

****

SAQ 5.5c

> Our original linear regression [line (a) in Fig. 5.5a], appeared to be a reasonable fit to the four calibration points for the analysis of chloride in the range 8–20 mg dm^{-3}. The standard error of the estimate, s_y for the line was 0.0111 (SAQ 5.2b). If we take the value 0.005 as the accepted population standard deviation for a simple measurement of y, is there evidence, even over this concentration range, that the calibration curve is non-linear?

Response

We have $s_y = 0.0111$, with 2 degrees of freedom. (There were 4 calibration points, so $n - 2 = 2$.) Also, we are told that $\sigma = 0.005$ for a simple measurement of y.

Hence $F = s_y^2/\sigma^2 = (0.0111/0.005)^2 = 4.9$

Suppose we test at a 1% significance level. This means that the critical F-value is $F_{0.01(2,\infty)}$. You should find that this value is 4.6. Hence the F-test shows that, even for the four points between 8 and 20 mg dm^{-3}, there is highly significant evidence for non-linearity. There is just less than a 1% chance that results as errant as these would have arisen had the true relationship (between 8 and 20 mg dm^{-3}) been linear.

****

SAQ 6.0a   We have not yet introduced any method of measuring correlation. Nonetheless you are invited to stick your neck out! Which of the following would be expected to show the stronger evidence for correlation, and for which would a correlation study be more profitable?

(*i*) Data for absorbance *versus* concentration in the standard spectroscopic method for determination of chloride.

(*ii*) Data relating the extent of tooth decay to the concentration of fluoride in drinking water.

Explain your conclusions.

Response

The data from (*i*) would be expected to show the best correlation (we shall confirm this later), but (*ii*) would be the study in which correlation analysis would be more appropriate.

In (*i*), if our method is well designed, we believe there to be a unique true relationship between concentration and absorbance. Correlation between these properties should therefore be exceedingly good. However, confirming the existence of a 'good' correlation in such a study is of relatively little value to us. What we need (and what we obtained in Part 5) is a best estimate of the calibration line, enabling us to make determinations of chloride concentration levels with known precision, from measurements of absorbance.

In (*ii*), on the other hand, it is expected that tooth decay may depend on may factors other than fluoride in drinking water. Sugary diets and poor tooth care, for instance, might be expected to be significant influences. There is no question of being able to construct a calibration curve predicting, say, the number of fillings which will be required as a function of fluoride level. On the other hand it will certainly be of interest if it can be shown that fluoride level does, on average, have some effect on the matter.

****

SAQ 6.1a   Study Fig. 6.1a.

(*i*)   From the regression line estimate, on average, the decrease in SMR associated with a 1 mmol dm^{-3} increase in total water hardness.

(*ii*) Comment critically on the validity of your conclusion.

Response

(*i*)   From the regression line the average SMR drops approximately 15 points per 1 mmol dm^{-3} increase in water hardness. One way to obtain this figure is to read off the intercepts of the line on the *y* and the *x*-axis. Thus the SMR shows a drop of just under 60 units (from around 125 to a little over 65) whilst hardness increases by just under four units (from zero to just under 4 mmol dm^{-3}. Thus the rate of fall of SMR is approximately 60/4 = 15 units per mmol dm^{-3} increase in hardness.

(*ii*) This was another open-ended question and there are various points you could quite validly have made. In setting the question we hoped that you would raise at least some question mark over the reliability of the answer to (*i*) in view of the spread of the data reported in Fig. 6.1a. Our own comments are below.

A 15% decrease in cardiovascular mortality as a result of a relatively modest increase in water hardness seems a very important and useful finding. But how valid is it? There is enormous scatter in the data. There is an overall net tendency for there to be more deaths from heart disease where drinking water is softer. Is water softness thus identified as a *cause* of heart failure, or could this finding have simply arisen by chance? Many other factors might be expected to have a bearing on heart disease. Do these explain the wide deviations of many data points from the regression line?

We address questions such as these in later sections.

****

SAQ 6.2a

Evaluate the correlation coefficient between tooth decay and the fluoride concentration in drinking water, by using the data quoted from Dean's study. Use Eq. 6.6.

(i) It takes a little labour to work out the different sums. Check at least one of the following results:

$$S_x = 14.55,$$
$$S_{xx} = 21.27,$$
$$S_y = 111.7,$$
$$S_{yy} = 719.1,$$
$$S_{xy} = 45.31.$$

(ii) Use the values to evaluate r. What conclusion do you draw from your result?

Response

(i) If you have any doubts what the summation symbols mean consult SAQ 5.1b and its answer. You will realise that the task becomes a fairly laborious one if there is a large number of data points, so you will appreciate why we have not asked you to evalute the corresponding sums for the heart disease data!

(ii) All that is involved is straight substitution into the formula below.

$$r = (nS_{xy} - S_x S_y)/\sqrt{(nS_{xx} - S_x^2)(nS_{yy} - S_y^2)}$$

$$= (951.5 - 1625.2)/\sqrt{(235.0)(2624)} = -673.7/785.2$$

$$= -0.86$$

The fact that r is negative means that on average the incidence of tooth decay is lower in communities with higher fluoride levels in their water. A value of $r = 1.0$ would represent the most extreme negative correlation possible; all of the data points would have had to lie exactly on one straight line to achieve that. A value of $r = 0$ would represent no correlation at all. The fact that the result here is much closer to -1.0 than 0.0 indicates that the correlation is a pretty strong one.

$$**********************************$$

SAQ 6.3a	From the results obtained in SAQ 6.2a, test whether there is truly a negative correlation between fluoride level and tooth decay.

Response

The calculated correlation coefficient for the fluoride data was -0.86. The sample size was 21, so there are 19 degrees of freedom (ie n-2). The calculated t-value is given by:

$$t = r\sqrt{(n - 2)/(1 - r^2)} = -0.86\sqrt{19/(1 - 0.86^2)} = -7.3$$

The smallest α-value quoted in our t-table (Fig. 3.4b.) is 0.001 (ie 0.1%). The corresponding critical value $t_{0.001,19}$ is 3.60. The magnitude of the calculated t handsomely exceeds this. At a 0.1% level of significance, therefore, we reject the null hypothesis (that there is no correlation) and accept that fluoride and tooth decay are indeed negatively correlated. The evidence is overwhelming; program "STAT" gives the one-tail significance as 3×10^{-7}, ie the null hypothesis would be rejected at any significance level above 0.00003%.

****

SAQ 6.3b

> The data in Fig. 6.1a relating heart disease mortality to water hardness give a correlation coefficient of -0.67. The data are much more widely scattered than for our fluoride example. Show, however, that the evidence for the existence of true negative correlation is very highly significant indeed. (The data cover 243 towns.)

Response

We can compute the appropriate t-value for the data below.

$$t = -0.67\sqrt{(243 - 2)/(1 - 0.67^2)} = -14.0$$

By consulting the t-table (Fig. 3.4b.) you will find that for the number of degrees of freedom available (241) the t-value for 0.1% significance is about 3.1. This is so much smaller than 14.0 that we can assert with essentially total conviction that a negative correlation does truly exist. (Program "STAT" finds the tail area so small that it cannot distinguish it from zero.) Other tests (not explained in the Unit lead to the conclusion, with about 96% confidence, that $r < -0.60$.

****

SAQ 6.4a   Consider the full set of data for the spectroscopic determination of chloride, discussed at length in Part 5.

$x([Cl-]/\text{mg dm-3})$	y (absorbance)
0.0	0.000
2.0	0.113
4.0	0.218
8.0	0.388
12.0	0.523
16.0	0.624
20.0	0.734

Derive the value of the correlation coefficient, r.

Response

When we inserted the chloride data into program "LINE" we got the following output:

x	y	residual
0	0	−5.1E-2
2	0.113	−1.04E-2
4	0.218	2.23E-2
8	0.388	4.76E-2
12	0.523	3.79E-2
16	0.624	−5.8E-3
20	0.734	−4.05E-2

Gradient = 3.61749147E-2
Intercept = 5.10221842E-2
St err of estimate = 4.14281771E-2
St dev of gradient = 2.26394588E-3
St dev of intercept = 2.54415258E-2
Correlation coefft = 0.990349832
t for correlation = 15.9787015

The correlation coefficient is just greater than 0.99!

****

SAQ 6.4b

Draw a scatter diagram for Dean's study relating fluoride levels to tooth decay. The data are listed in Section 6.2. Add the y-on-x regression line to your graph by making use of the fact that our program found that this line has gradient −2.87 and intercept 7.31.

(Hint: look back at Fig. 6.1a; you are being asked to construct the equivalent diagram for Dean's study.)

Response

Look ahead to Fig. 6.6a – you should find that you have just drawn
your own version of it! The (x,y) points are plotted directly from
the list in Section 6.2. As for the regression line, you can draw that
once you have found any two points on it. When $x = 0$, $y = 0 + c$,
so one point to choose could be (0,7.31). When $y = 0$, we have $0 =
mx + c$, so $x = -c/m = 7.31/2.87 = 2.55$. Hence (2.55,0) could
be our second point.

****

SAQ 6.6a

Three possible prescriptions might be suggested
with the aim of decreasing the incidence of tooth
decay. They involve increasing the fluoride concentration in fluoride-deficient water supplies:

(*i*)   to as high a level as is economically feasible,

(*ii*) to about 2.5 mg dm^{-3},

(*iii*) to about 1 mg dm^{-3}.

Comment critically on all of these suggestions.

Response

Suggestion (*i*) rests simply on the evidence of the existence of negative correlation. This suggests that the greater the fluoride level
the less prevalent is the problem of tooth decay. Fig. 6.6a. shows,
however, a distinct tendency towards diminishing returns. At higher

fluoride levels the indications are that further improvement of dental health becomes minimal, if it exists at all. In fact the earliest interest in the effect of fluoride in drinking water arose because of evidence that people from localities with *very* high levels of fluoride suffered from a distinctive tooth disfigurement!

Suggestion (*ii*) might be arrived at if we gave overdue attention to the regression line. If it represented a 'true' relationship between dental caries and fluoride concentration, then at a concentration of just over 2.5 mg dm^{-3} tooth decay would be totally eliminated. Of course this is nonsense, but it is the sort of prescription we might have been tempted to make had we simply computed the regression line and not bothered to draw or examine Fig. 6.6a.

Suggestion (*iii*) seems eminently sensible. The data suggest that up to about this level, adding fluoride should have a quite dramatic effect on dental health, but beyond it any benefit appears rather marginal. We should still need to satisfy ourselves, however, that such action would actually *cause* the expected effect and that it would not produce other undesirable effects.

****

SAQ 6.6b

Scan Fig. 6.6b and, from within the main subsections, select examples of factors which you might expect to be quite strongly correlated with one another. State whether you would expect such correlations to be positive or negative.

(*i*) Water quality factors.

(*ii*) Climatic factors.

(*iii*) Socio-economic factors.

Response

Below are listed some of the strongest inter-correlation effects.

(i)   There are substantial correlations between many of the water quality factors. There were particularly strong correlations between total hardness and

(a) calcium level (the survey reports $r = +0.97$),
(b) carbonate hardness ($r = +0.94$),
(c) conductivity ($r = +0.93$).

A strong negative correlation ($r = -0.80$) was found between water hardness and the percentage of water from upland sources.

(ii)   The three rainfall measures would, for instance, be expected to be quite strongly positively correlated with one another.

(iii)   The first three socio-economic factors listed would be expected to be strongly positively correlated with one another and each would be expected to be negatively correlated with the number of cars per household.

Some strong correlations might be expected also between factors in different categories. Upland water is positively correlated with rainfall; mean temperature and percentage unemployed are both correlated with latitude within the UK.

All-in-all it is clearly not a trivial problem to pick out a limited list of factors which best explain the SMR data, nor is it easy to identify the relative importance of different factors.

****

SAQ 6.6c	(i)	Examine Fig. 6.6c. and note the optimum characteristics of the community you would like to live in if you cared only about trying to minimise your risk of suffering an untimely death due to heart disease.
	(ii)	Is exercise good or bad for your health? Comment.

Response

The purpose of this question and of the previous one, has been principally to encourage you to scrutinise tables of information carefully and critically. Making full use of information from tables and figures is a valuable skill in its own right. And it is a quite vital skill if you are to reach sensible conclusions from studies involving multiple correlation analysis!

(i) It seems that the most healthy community to live in from the point of view of your heart is in a warm and dry suburb with at least average water hardness and populated by white-collar commuters. Perhaps it would be safest actually to be one of the white-collar commuters living there.

(ii) Manual workers who do not own a car seem to be at some disadvantage, compared to car-owning professionals. The former would be expected to be involved in greater physical exertion than the latter. However, it would be completely unwarranted to reach the conclusion that the evidence implied that exercise is bad for the heart. Many other differences in typical lifestyle exist between the two groups. This question leads us back to questions about causal connections and spurious correlations.

****

SAQ 6.6d	How were the figures in the last sentence arrived at? Criticise the argument used.

Response

From the continuous line in Fig. 6.6d. it appears that the average SMR for towns with water hardness in the 0.1–0.5 mmol dm^{-3} range is ca 121 whereas that in the 2.1–2.5 mmol dm^{-3} range is ca 87. Taking these mean figures as representative of towns in the middle of each range (0.3 and 2.3 mmol dm^{-3} respectively) there appears typically to be a SMR reduction of 34 between them (34 is a little under 30% of 121.

The validity of the argument can be attacked on a number of grounds. The most critical of these is that the argument *assumes* that there is indeed a causal connection. Even accepting that, the argument neglects the possibility that part of the reduction may be due to other factors which are in turn correlated with water hardness.

****

SAQ 7.2a   Under statistical control a process produces a powder with a mean ferrous-iron content of 6.70% w/w with a standard deviation of 0.18%. Samples of size 4 are routinely analysed by the quality-control lab. The following sequence of results was obtained for the mean percentage of ferrous iron.

6.68, 6.71, 6.68, 6.72, 6.82, 6.54, 6.59, 6.58, 6.71, 6.48, 6.57, 6.61, 6.64, 6.56, 6.69, 6.64, 6.69, 6.57, 6.54, 6.56, 6.55, 6.50, 6.55, 6.51, 6.42.

Construct a Shewhart control chart for these results.

Comment.

Response

The sequence of analyses has been plotted in the figure. The first occasion on which action is signalled is after the very last sample, which lies below the lower action line. This is taken as convincing evidence that the mean ferrous-iron content has fallen below specification. It is noticeable, however, that even though the earlier results individually gave no cause for concern, they show an overall pattern which seems to suggest that the process mean may be falling steadily below its target value.

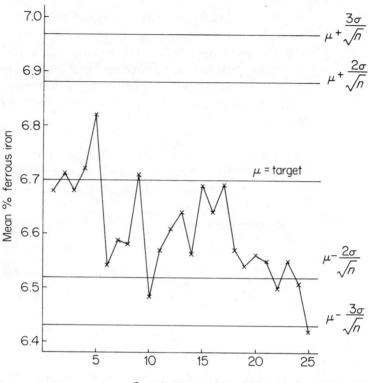

Sample sequence number

Did you manage to draw the diagram easily enough? The central horizontal line is drawn at 6.70%, the value given for the target mean, μ. The process standard-deviation was given as 0.18%, so the standard deviation for the results, which were obtained from samples of size 4, should be $0.18/\sqrt{4} = 0.09\%$. The warning lines are drawn two standard deviations above and below the target line, at $6.70 \pm 2 \times 0.09\%$, ie at 6.52% and at 6.88%. the action lines are drawn one further standard deviation outside these, at 6.43% and at 6.97%. Plotting the readings on to the diagram should present no difficulties.

The 10th value is the first to lie outside a warning line. A number of other points lie quite close to the lower warning line but none lies below it again till point 22. Result 24 also lies below this line but action is still not signalled because no two consecutive values have crossed the line. Result 25, however, falls below the lower action line and is taken as establishing the presence of an assignable cause.

Looking at the pattern of results as a whole suggests further evidence for a drift in the process. Only once after sample 5 is the target mean exceeded. If the process was in a state of statistical control such a sequence of low values would be extremely unlikely. Yet our decision-making criteria did not give us grounds for action till the very end of the series. It looks as though corrective action might have been justified a good deal earlier! This is a common problem with Shewhart charts, it often takes some time before a steady drift in performance leads to a clear conclusion that remedial action is required. The question arises as to whether there is an alternative method of presentation of results which might be more efficient in this regard. There is! We shall introduce such a method in Section 7.3.

****

SAQ 7.2b
> Suppose the operating mean for a process suddenly increases by two standard deviations, ie from μ to $\mu + 2\sigma/\sqrt{n}$. What is the average run-length before a sample mean falls outside an action line?

Response

The answer is 6 (rounded from 6.3).

The main purpose of this question was to test your understanding of the effect of a change in the process mean in shifting the normal distribution curve for sample means. Fig. 7.2d shows the situation when the process mean shifts by one standard deviation. Sketch for yourself a diagram showing a shift of two standard deviations. The new operating mean will be at $\mu + 2\sigma/\sqrt{n}$, and so the upper action line lies just one standard deviation beyond this. The probability, p, of breaching it is given by the tail area of the normal distribution for $z = 1$. This value is 0.1587 (Fig. 2.3g). The average run length is $1/p$; thus ARL = 6.3.

If you found this question difficult you should re-read the discussion in the text of the example illustrated in Fig. 7.2d.

****

SAQ 7.3a	Draw a cusum chart for the example discussed in SAQ 7.2a where 25 consecutive results are reported for the mean ferrous-iron content in a product. Comment on the chart obtained. Assess quantitatively what is happening to the process mean over the period in which the samples were taken.

Response

The cusum chart is shown in the figure. After running roughly horizontally for the first five or six readings the cusum line 'nose-dives'. It falls progressively more steeply, on average, as the measurements proceed. Towards the end of the series the operating mean appears to be around 6.5% of ferrous iron w/w.

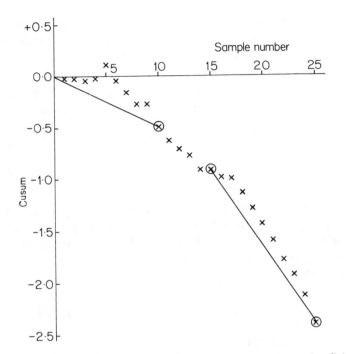

The values obtained for the sample means are given in SAQ 7.2a. The target process mean, T is 6.70%. The calculation giving the cusum points is as follows.

Sample	mean	$\bar{x} - T$	cusum	Sample	mean	$\bar{x} - T$	cusum
1	6.68	−0.02	−0.02	14	6.56	−0.14	−0.91
2	6.71	+0.01	−0.01	15	6.69	−0.01	−0.92
3	6.68	−0.02	−0.03	16	6.64	−0.06	−0.98
4	6.72	+0.02	−0.01	17	6.69	−0.01	−0.99
5	6.82	+0.12	+0.11	18	6.57	−0.13	−1.12
6	6.54	−0.16	−0.05	19	6.54	−0.16	−1.28
7	6.59	−0.11	−0.16	20	6.56	−0.14	−1.42
8	6.58	−0.12	−0.28	21	6.55	−0.15	−1.57
9	6.71	+0.01	−0.27	22	6.50	−0.20	−1.77
10	6.48	−0.22	−0.49	23	6.55	−0.15	−1.92
11	6.57	−0.13	−0.62	24	6.51	−0.19	−2.11
12	6.61	−0.09	−0.71	25	6.42	−0.28	−2.39
13	6.64	−0.06	−0.77				

The behaviour of the chart is very convincing. it becomes abundantly clear, from sample 5 onwards, that the curve is following a downhill path. The extent of the fall from point to point varies widely, but the direction is relentlessly downwards. As we know about the connection between cusum gradient and process mean, we can have little doubt that the operating mean has fallen below the target value. Compare the clear-cut picture revealed by this chart with the Shewhart chart for the same data (shown in the response to SAQ 7.2a). The Shewhart chart gave an out-of-control signal only after the very last sample was taken.

There are various ways in which you could 'assess quantitatively what is happening to the process mean', but they would all involve measuring the slope of the cusum line. For the purposes of discussion we have measured the slope over the first ten samples, and over the last ten. Over the first ten samples we get:

$$\text{Gradient} = (S_{10} - S_0)/10 = (-0.49 - 0.00)/10 = -0.049,$$

giving an estimated process mean as follows:

$$\text{Mean} = T - 0.049 = 6.70 - 0.049 = 6.651\%.$$

For the last ten samples we get:

$$\text{Gradient} = (S_{25} - S_{15})/10 = (-2.39 + 0.92)/10 = -0.147$$

giving the estimated process mean as

$$\text{Mean} = T - 0.147 = 6.553\%.$$

It appears that not only is the process mean low, it is steadily decreasing. This trend is further confirmed by looking at the lines drawn in the figure which correspond to these two periods. For each the intermediate points follow a rough curve above the straight lines, and the gradient of each cusum line appears to be steeper towards the end of the interval than at the beginning. So we have a consistent picture of a steadily decreasing process mean.

The one doubt we are left with at this stage arises from the lack of any statistical ground rules. Certainly the cusum is falling, but how clear must the trend be before we can rule out the possibility that it is a mere matter of chance? How soon in the series would it have been reasonable to conclude that the process mean really was moving downwards? How reliable an estimate of the current process mean can we obtain by measuring the slope of the cusum line? These questions are pursued in later sections.

****

SAQ 7.3b

> For the example introduced in SAQ 7.2a, involving measurements of the ferrous-iron content of a product, obtain 99% confidence intervals for the operative process mean over the periods covered by the first ten, and by the last ten samples. Comment. (Note that estimates of the process means for these periods were obtained in SAQ 7.3a, but without any specified confidence intervals.)

Response

Estimated means were obtained in SAQ 7.3a, from the slope of the cusum plot. For the first ten samples the value obtained was 6.651% w/w of ferrous iron. For the last ten samples the figure was 6.553%.

Now the process standard deviation was given (SAQ 7.2a) as 0.18% and the sample size used was 4. For 99% confidence limits we need $z_{0.005}$, to allow for 0.5% probability in each tail. This value is 2.576 (Fig. 3.4b). For the first ten samples we therefore get:

$$\text{process mean} = 6.651 \pm 2.576 \times 0.18/\sqrt{4 \times 10}$$

$$= 6.651 \pm 0.073\%,$$

and for the last ten samples:

$$\text{process mean} = 6.553 \pm 0.073\%$$

The range for the period of the first ten samples is 6.58% − 6.72%. The target mean of 6.70% lies within this interval, so this result as it stands does not signal that anything is amiss. The confidence range for the last ten samples, however, is 6.48%–6.63%, very clearly below target.

Recall that the cusum plot for this example (shown in the response for SAQ 7.3a) gives a clear indication that the process mean is steadily drifting downwards, from a value initially not far off target. These findings are consistent with that impression, but we have now given it a more quantitative justification. How soon could we be convinced that the mean is drifting downwards? Note that the cusum begins to fall more or less consistently from sample 5. It is instructive to estimate the 99% confidence interval between samples 5 and 10.

$$\text{Process mean} = (S_{10} - S_5)/5 + T \pm z/\sqrt{mn}$$

$$= (-0.49 - 0.11)/5 + 6.70 \pm$$

$$2.576 \times 0.18/\sqrt{4 \times 5}$$

$$= 6.580 \pm 0.104\%$$

The range is from 6.48% to 6.68%, already excluding the target mean! So with this kind of analysis it would have been possible to identify the drift in the process mean at an early stage.

****

SAQ 7.4a

(*i*) Set up Shewhart Control Charts for sample means and sample ranges for a process whose target value is 100 and which, when under statistical control, has a standard deviation of 6. Assume a sample size of 9 and allow for up to 20 data-points on the *x*-axis.

(*ii*) Enter the information tabulated below on the appropriate chart and comment on any interesting occurrences. ⟶

SAQ 7.4a
(cont.)

Sample Number	$\bar{x}$	Range
1	105	19
2	97	13
3	100	25
4	110	40
5	104	36
6	102	10
7	102	27
8	103	20
9	97	16
10	98	10
11	103	12
12	95	20
13	98	20
14	94	23
15	97	12
16	97	19

Response

(*i*) When $\sigma = 6$ and $n = 9$, the value of $\sigma/\sqrt{n} = 2$. The upper action line, which is defined by $\mu + 3\sigma/\sqrt{n}$ is therefore drawn at $100 + 6 = 106$. The upper warning line ($\mu + 2\sigma/\sqrt{n}$) is drawn at $100 + 4 = 104$. The lower warning line ($\mu - 2\sigma/\sqrt{n}$) is drawn at $100 - 4 = 96$. The lower action line ($\mu - 3\sigma/\sqrt{n}$) is drawn at $100 - 6 = 94$.

The control chart for sample means is shown below. The data-points have been plotted on it and it can be seen that the mean for sample 4 falls outside the action line.

Shewhart chart for sample means

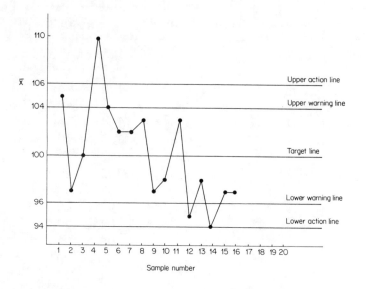

(*ii*) For samples of size 9, the factors from Fig. 7.4c, which, when multiplied by the standard deviation give the values for the control lines, are: $d_2' = 0.96$, $d_1' = 1.55$. $d_1 = 4.70$, $d_2 = 5.90$ and $d_n = 2.970$. The lower action line is therefore drawn at 5.76; the lower warning line at 9.30; the upper warning line at 28.2; the upper action line at 35.4; and the central line at 17.82. The control chart for sample ranges is shown below. The data-points have been plotted and indicate an 'out-of-control' signal at sample 4 and at sample 5.

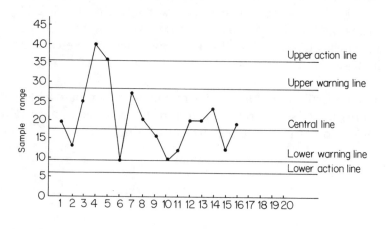

With these samples the range is larger than would be expected from a process with standard deviation, $\sigma = 6$. Another way of saying this is that the standard deviation for the process appears to have increased (become worse) for samples 4 and 5. This is probably the reason why we received an 'out-of-control' signal at sample number 4 on the chart for sample means. The sample ranges appear to be well behaved after sample 5. This suggests that the warning signals for sample 12 and 14 on the sample means control chart are probably due to a downward change in the process mean rather than a change in the process standard deviation.

****

SAQ 8.1a

Each of the following measurements is given to the correct number of significant figures.

$a = 23.70$
$b = 64.82$
$c = 250.0$
$d = 70.77$
$e = 363$

Give the answer to the following calculation to the correct number of significant figures.

$$\frac{(a/b) \times c + d}{e}$$

Response

The calculation is $\dfrac{(23.70/64.82) \times 250.0 + 70.77}{363}$

Let us carry it out on a pocket calculator. The first operation is $(23.70/64.82) \times 250.0$

The calculator gives the answer 91.406973. We know that there are only 4 significant figures permitted for such a calculation. However, let us continue to push the buttons on our calculator! We next add 70.77. The calculator gives the answer 162.17697. Again, we know that all these figures are not significant. There are only 4 significant figures permitted. However, let us continue to push buttons! The final step, ie the division by 363, gives an answer of 0.4467685. Many novices leave the matter here. However, we know that there are only 3 significant figures in the divisor. Hence, the answer should have 3 significant figures and is therefore given as 0.447.

It is likely that you got this answer. You might have approached the question by rounding off to the correct number of significant figures at each step of the calculation. In the example given, this would still have yielded the answer 0.447. However, it is not always thus. If you are forced to transcribe intermediate results in a calculation you should always be sure to carry at least one additional significant figure. The time for rounding is when you have the final result.

****

SAQ 8.2a

A chemists is asked to prepare a standard solution of sodium chloride. He weighed out, by difference, 0.5850 g of the salt, dissolved it in water and made the solution up to 100.0 cm³ in a graduated flask.

(*i*) What is the concentration of the solution in g dm⁻³?

(*ii*) What is the molar concentration of the solution? (The relative molar mass of NaCl is 58.443.)   ⟶

SAQ 8.2a
(cont.)

> (*iii*) If the standard deviation of weighing is known from past experience to be 0.00012 g, what is the standard deviation in the weight of NaCl taken? What is the relative standard deviation?
>
> (*iv*) If the standard deviation associated with the use of the graduated flask is known from past experience to be 0.050 cm^3, what is the relative standard deviation associated with the volume of reagent prepared?
>
> (*v*) What is the relative standard deviation in the concentration (g dm^{-3}) of the solution? What is the relative standard deviation of the molar concentration of the solution?
>
> (*vi*) What is the standard deviation in the molarity of the solution? Give the 99% confidence interval for the molarity.

Response

(*i*) Concentration $= \dfrac{\text{weight}}{\text{volume}} = \dfrac{0.5850 \text{ g}}{0.1000 \text{ dm}^3} = 5.850$ g dm^{-3}

(*ii*) The molar concentration is therefore

$$\frac{5.850 \text{ g dm}^{-3}}{58.443 \text{ g mol}^{-1}} = 0.1001 \text{ mol dm}^{-3}$$

Note that it is in order to quote the molarity to 4 significant figures.

(*iii*) The weight of NaCl used was ascertained by subtracting the balance reading for the final weight from the reading for the initial weight. There is therefore a standard deviation of 0.00012 g affecting each reading. (This is something which we often forget. Did you?) The combined standard deviation in the weight σ_w, is obtained below.

$$\sigma_w = \sqrt{\sigma^2_{init} + \sigma^2_{final}}$$

$$= \sqrt{(1.2 \times 10^{-4})^2 + (1.2 \times 10^{-4})^2}$$

$$= \sqrt{2.88 \times 10^{-8}}$$

$$= 1.7 \times 10^{-4}$$

This is the standard deviation in the weight taken. The relative standard deviation is

$$\frac{\sigma_w}{w} = \frac{0.00017}{0.5850} = 2.9 \times 10^{-4}$$

(*iv*) The relative standard deviation in the volume prepared.

This is obtained from $\frac{\sigma_v}{V}$. It is $\frac{0.050}{100.0} = 5.0 \times 10^{-4}$

(*v*) As you have undoubtedly discovered, this part of the question is meant to assess your ability to handle the propagation of random error in multiplication/division operations. The concentration is obtained by dividing the weight by the volume, ie $c = \frac{w}{v}$. For this type of relationship we have:

$$\sigma_c = \sqrt{\left(\frac{\sigma_w}{w}\right)^2 + \left(\frac{\sigma_v}{v}\right)^2}$$

Now, we have already determined values for σ_w/w and σ_v/v.

They are 2.9×10^{-4} and 5.0×10^{-4} respectively.

$$\therefore \quad \frac{\sigma_c}{c} = \sqrt{(2.9 \times 10^{-4})^2 + (5.0 \times 10^{-4})^2} = 5.8 \times 10^{-4}$$

$$= 0.00058$$

The absolute standard deviation in the concentration, σ_c, is obtained by multiplying the RSD by the concentration.

$$\sigma_c = 0.00058 \times 5.850 = 3.4 \times 10^{-3} \text{ g dm}^{-3}.$$

We were also asked to give the RSD for the molar concentration. Since the molarity, m, equals the concentration, c, in g dm^{-3}, divided by the relative mass, M, we have:

$$\frac{\sigma_m}{m} = \sqrt{\left(\frac{\sigma_c}{c}\right)^2 + \left(\frac{\sigma_M}{M}\right)^2}$$

Now $M = 58.443$ g mol^{-1}, but we were not given σ_M. However, σ_M is certainly no larger than 0.001, so $\sigma_M/M < 0.001/58.443$, ie 1.7×10^{-5}.

This is thus smaller than σ_c/c by more than a factor of ten. When these terms are squared the difference becomes more than a factor of a hundred. In short, the effect of uncertainty in the molar mass is negligible, so the RSD for the molarity becomes identical to that for the concentration in weight terms, ie

$$\frac{\sigma_m}{m} = 5.8 \times 10^{-4}$$

(*vi*) The absolute standard deviation in the molarity is

$$\sigma_m = (5.8 \times 10^{-4}) \times (0.1001) = 5.8 \times 10^{-5} \text{ mol dm}^{-3}$$

Notice that we have been using population standard deviations (σ's) throughout this question. This means that we can obtain confidence intervals by using the relationship

$$\mu = \bar{x} \pm z_{\alpha/2}\sigma/\sqrt{n}$$

For a 99% confidence range we need 0.5% probability in each tail, so we need $z_{0.005}$, which is 2.576. (You might look this up as $t_{0.005,\infty}$) We have made up a single solution, so $n = 1$. Our 99% confidence interval for the molarity is thus:

$$\text{molarity} = 0.1001 \pm (2.576) \times (5.8 \times 10^{-5})$$

$$= 0.1001 \pm 0.00015 \text{ mol dm}^{-3}$$

From the point of view of precision, this is an excellent result! It demonstrates that the random error associated with the preparation of a typical standard solution can be very small indeed. If we can prepare such solutions with excellent precision and then use them in high precision titrations, it is not surprising that volumetric procedures are so popular. That is not to say that volumetric work is always accurate. We must not forget the possibility of systematic errors. We shall turn our attention to systematic errors in Section 8.3.

****

SAQ 8.2b

A chemist claims to have invented a simple 'breathometer' based on a spectrophotometric method. It has a linear scale for transmittance readings and is claimed to have an uncertainty of 0.010 transmittance units for any transmittance reading.

The legal limit for ethanol in the breath is found to give a transmittance reading of 0.80. What is the relative error in the concentration at the legal limit?

Response

$$\frac{\sigma_c}{c} = \frac{\sigma_T}{T \ln T} = \frac{0.010}{0.80 \ln(0.80)} = -0.056$$

This expressed as a percentage relative standard deviation is 5.6%. Would you be happy to be on the wrong side of this instrument? No? Why not? It is actually quite a good instrument for its purpose.

****

SAQ 8.2c

> The chemist mentioned in the preceeding SAQ sells his instrument to an overseas government. The country in question has a legal limit of 35 μg ethanol/100 cm^3 of breath. (This is the same legal limit as in the UK.) The country also has a large number of lawyers, who first qualified in chemistry (we have some too!). Consequently, the powers-that-be decided that prosecutions would take place only if the breathometer gave a reading equivalent to 37 μg/100 cm^3 of breath. Is their ruling likely to be challenged in the courts?

Response

The instrument has a relative standard deviation in concentration determinations of 0.056, ie

$$\frac{\sigma_c}{c} = 0.056$$

There are several ways in which we could make use of this value in order to test the fairness of the government's policy. A simple way is to ask whether or not a suspect could be found guilty even though his ethanol-in-breath level was below 35 μg/100 cm^3. Let us test the possibility of this happening for a suspect whose true ethanol-in-breath level is 34 μg/100 cm^3. The absolute uncertainty of the instrument at this concentration is given by:

$$\sigma_c = 0.056 \times (34) = 1.9 \ \mu g/100 \ cm^3.$$

The 95% confidence interval for a reading of 34 μg/100 cm^3 is 34 $\pm$ 1.96(1.9) = 34 $\pm$ 3.7 μg/100 cm^3. There is therefore a distinct possibility that an 'innocent' man could be convicted. The probability of this happening is easy to calculate:

$$z = \frac{x - \mu}{\sigma} = \frac{37 - 34}{1.9} = 1.58$$

For $z = 1.58$, the area in the tail is 0.0571 or 5.7%. Obviously the lawyers will have a 'field-day'.

You might not have argued in the same way as we did, but if you came to the same conclusion, then well done. Breathometers, normal curves and the law - we have come a long way!

****

SAQ 8.3a

A standard solution is made by dissolving a weighed quantity of solid in a solvent and making the solution up to the mark in a 250 cm^3 graduated flask. A 25 cm^3 aliquot is then taken for titration against a test solution. Find the effect on the calculated molarity of the test solution if:

(*i*)   the solid used contains 2.0% by weight of a non-reactive impurity; $\longrightarrow$

SAQ 8.3a (cont.)	(*ii*) The way in which the end-point is detected leads to a consistent 'overshoot' of 0.25 cm^3 (a typical titration requires 20 cm^3 to reach the end-point);
	(*iii*) both (*i*) and (*ii*) operate.

Response

The bald answers are that the relative systematic error in the calculated molarity will be (*i*) +2.0%, (*ii*) −1.25% and (*iii*) +0.75%.

Perhaps you felt that you were left a little short of information. It need not upset us that we do not know the names of the substances involved: Let us simply call the standard substance A and the test substance B. You might have been alarmed, however, not to know the stoichiometry of the titration reaction. Suppose it is

$$n_A A \ + \ n_B B \ \longrightarrow \ \text{products}$$

At the end-point the numbers of moles of A and B present should accord with this equation:

$$\frac{\text{moles of B}}{\text{moles of A}} \ = \ \frac{V_B[B]}{V_A[A]} \ = \ \frac{n_B}{n_A}$$

Thus
$$[B] \ = \ n_B V_A[A]/(n_A V_B)$$

The calculation required is entirely multiplicative. Therefore relative systematic errors add (or subtract in the case of division). Possible systematic errors arise in the values of V_A (25 cm^3), V_B (the end-point titre) and in the molarity [A] of the standard solution.

$$\Delta[B]/[B] \ = \ \Delta V_A/V_A \ + \ \Delta[A]/[A] \ - \ \Delta V_B/V_B$$

Now we can see why it was not necessary for the question to specify the stoichiometry, it has no bearing on the relative systematic error in [B].

The question suggests possible systematic errors affecting [A] and V_B

(*i*)   If the solid weighed out contains 2.0% by weight of impurity then the standard solution contains only 98.0% as much A as we would have calculated on the assumption that the solid was pure. In other words there is a relative systematic error of $+2.0\%$ in [A]. (Do you agree with the sign? We mistakenly believe we have 2.0% more A than is actually present.) If this is the only source of systematic error then the above equation makes clear that the relative systematic error in [B] will also be $+2.0\%$.

(*ii*)   Here it is supposed that there is an overshoot of 0.25 cm^3 in a 20 cm^3 titration. In other words V_B is overestimated by 0.25/20 = 0.0125 or 1.25%, ie $\Delta V_B / V_B = +1.25\%$. If this is the sole source of systematic error then, from the equation above, it follows that the molarity of B is underestimated by 1.25%. Note the sign, V_B is a divisor; $\Delta[B]/[B] = -1.25\%$.

(*iii*) If both effects occur together then the relative systematic errors add, so in total:

$$\Delta[B]/[B] = +2.0 - 1.25 = +0.75\%$$

Where calculations are multiplicative, then individual relative systematic errors combine additively. This question reinforces that simple point. The example also illustrates that systematic errors *may* partially cancel. Of course this does not always happen, but sometimes we may be lucky. With random error, on the other hand, every new source of random variability increases the variability of the final result.

****

SAQ 8.3b	A mixture contains three compounds A, B and C. Two analytical studies determine the content of A as 80.0% by weight and that of B as 15.0%. It is therefore concluded that the material contains 5.0% by weight of C. Suppose that the determinations of A and B may both have been affected by relative systematic errors of +1.0%. What is the size of the relative systematic error in the concentration of C determined?

Response

We are asked to suppose that the determined values of 80.0% by weight for A and 15.0% by weight for B might each be subject to a relative systematic error of +1.0%. The true values would therefore be *smaller* by 0.80% and by 0.15%, respectively. (Do not allow yourself to be confused by taking percentages of percentages!) This would give a reassessed C content as follows.

$$\% \text{ by weight of C} = 100.0 - 79.20 - 14.85 = 5.95\%$$

Whilst unaware of any such systematic error we had calculated that the C content was 5.0%. The relative systematic error in this C value is $(5.0 - 5.95)/5 = -0.19$, or -19%.

This example brings out another point of general importance. If a result depends on a relatively small difference between larger quantities, then apparently small relative errors in the larger values can be greatly magnified in the relative error in the difference. This point applies to random errors also.

****

SAQ 8.3c | A pH meter operates with a positive bias of 0.10 pH units. What is the systematic error in the hydrogen-ion concentration obtained when the meter reads 4.10?

Response

The easy way to do this is as follows.

The meter reading is pH = 4.10, which implies $[H^+] = 10^{-4.10} = 7.94 \times 10^{-5}$ mol dm^{-3}.

If the bias is $+0.10$ pH unit, the correct reading should be 4.00, corresponding to $[H^+] = 1.00 \times 10^{-4}$ mol dm^{-3}.

The systematic error in $[H^+]$ is

$$(7.94 - 10.00) \times 10^{-5} = -2.06 \times 10^{-5} \text{ mol dm}^{-3}.$$

The relative systematic error is $-2.06/7.94 = -0.26$, or -26%. Notice how large this is.

You might well have used Eq. 8.10 to tackle this problem. Applying that to the relationship pH $= -\log[H^+]$ gives the expression below.

$$\Delta(\text{pH})/\text{pH} = (1/\ln[H^+])(\Delta[H^+]/[H^+])$$

You should find that, from pH $= 4.10$, $\ln[H^+] = -9.44$. Also $\Delta(\text{pH})/\text{pH} = 0.10/4.10 = 0.024$. The equation therefore gives the relative systematic error in $[H^+]$ as $-9.44 \times 0.024 = -0.23$, or -23%. Compare this answer with the -26% we calculated more straightforwardly above!

As we said in the text, the equations for the propagation of systematic error in multiplicative and logarithmic operations are approximations valid for small errors. The main point of Eqs. 8.7–8.10 was to give us a general picture of the way systematic errors are propagated. If we know the actual values for the systematic errors it is more straightforward simply to correct the earlier calculation. Not only more straightforward, but more accurate also!

****

SAQ 8.4a	Use Q-test(s) to investigate the following data: 1.26, 1.31, 1.23, 1.10, 1.23, 1.24.

Response

The results, in ascending order, are 1.10, 1.23, 1.23, 1.24, 1.26 and 1.31.

The result 1.10 appears to be suspect.

The range $= 1.31 - 1.10 = 0.21$.

The difference between the suspect value and its nearest neighbour is $1.23 - 1.10 = 0.13$.

$$Q(\text{calculated}) = \frac{0.13}{0.21} = 0.62.$$

The tabulated value $Q_{0.90}$ for $n = 6$ is 0.56.

We shall therefore reject the suspect result. The consequences of this are that the mean value changes from 1.228 (1.23) to 1.254 (1.25) and the median hardly changes at all, ie 1.235 (1.24) to 1.24. If we examine the 5 results which remain, we see that the result 1.31 appears to be an outlier. Let us apply a second Q-test.

The range of the 5 results is $1.31 - 1.23 = 0.08$.

The difference between the suspect result and its nearest neighbour is $1.31 - 1.26 = 0.05$.

The Q(calculated) value is $\dfrac{0.05}{0.08} = 0.625 = 0.62$

The tabulated value $Q_{0.90}$ for $n = 5$ is 0.64. We shall therefore accept the result 1.31 and report the mean of 5 measurements as 1.25. Having done this, we are still left with a slight unease. We rejected one result because of a calculated Q which was just a little higher than the appropriate tabulated value and then we have retained a second result because of a calculated Q which was just smaller than the tabulated value. That's life! It is also in the nature of statistical tests. Remember the decision is ours. Statistics is only an aid to decision making; we remain in charge!

****

SAQ 8.5a

> The calibration curve for an instrument, at low concentration, conforms with the regression line
>
> $$y = 2.70 + 5.23\,x$$
>
> where x is the concentration (in mg dm^{-3}) of an analyte. The standard error of determination s_y is 0.043. Find the limit of detection.

Response

The level of significance to be used has not been defined in the question. This gives you a measure of freedom, though it might also have worried you. If it did we are delighted! It means you are clearly aware that various definitions of the limit of detection are possible. It is an unfortunate fact that different workers have adopted different definitions. The important point is that you should always make clear what definition you are adopting.

If you did not worry about the level of significance, it is likely that you simply adopted the convention that $y_{det} = y_o + 3s_y$, which we described in the text as the basis of the definition used by 'many workers'. Basically, if you did that we approve, we merely stress that you should make a point of stating clearly what your choice is.

Taking the, $y_{det} = y_o + 3s_y$, we obtain

$$y_{det} = 2.70 + 3(0.043) = 2.829$$

It is then simply a matter of substituting in the regression equation to find x_{det}.

$$2.829 = 2.70 + 5.23 \, x_{det}$$

$$x_{det} = (2.829 - 2.70)/5.23 = 0.025 \text{ mg dm}^{-3}$$

Had you chosen, for instance, to work at a 5% level of significance for the decision limit, you would have $z_\alpha = z_{0.05} = 1.645$, so

$$y_{det} = y_o + 2z_\alpha s_y = 2.70 + 2(1.645)(0.043)$$

$$= 2.841.$$

This leads to $x_{det} = 0.027 \text{ mg dm}^{-3}$

Units of Measurement

For historic reasons a number of different units of measurement have evolved to express quantity of the same thing. In the 1960s, many international scientific bodies recommended the standardisation of names and symbols and the adoption universally of a coherent set of units—the SI units (Système Internationale d'Unités)—based on the definition of five basic units: metre (m); kilogram (kg); second (s); ampere (A); mole (mol); and candela (cd).

The earlier literature references and some of the older text books, naturally use the older units. Even now many practicing scientists have not adopted the SI unit as their working unit. It is therefore necessary to know of the older units and be able to interconvert with SI units.

In this series of texts SI units are used as standard practice. However in areas of activity where their use has not become general practice, eg biologically based laboratories, the earlier defined units are used. This is explained in the study guide to each unit.

Table 1 shows some symbols and abbreviations commonly used in analytical chemistry; Table 5 is a glossary of abbreviations used in this particular text. Table 2 shows some of the alternative methods for expressing the values of physical quantities and the relationship to the value in SI units.

More details and definition of other units may be found in the *Manual of Symbols and Terminology for Physicochemical Quantities and Units*, Whiffen, 1979, Pergamon Press.

Table 1 *Symbols and Abbreviations Commonly used in Analytical Chemistry*

Å	Angstrom
$A_r(X)$	relative atomic mass of X
A	ampere
E or U	energy
G	Gibbs free energy (function)
H	enthalpy
J	joule
K	kelvin (273.15 + t °C)
K	equilibrium constant (with subscripts p, c, therm etc.)
K_a, K_b	acid and base ionisation constants
$M_r(X)$	relative molecular mass of X
N	newton (SI unit of force)
P	total pressure
s	standard deviation
T	temperature/K
V	volume
V	volt ($J\ A^{-1}\ s^{-1}$)
$a, a(A)$	activity, activity of A
c	concentration/ mol dm^{-3}
e	electron
g	gramme
i	current
s	second
t	temperature / °C
bp	boiling point
fp	freezing point
mp	melting point
$\approx$	approximately equal to
$<$	less than
$>$	greater than
e, $\exp(x)$	exponential of x
ln x	natural logarithm of x; ln x = 2.303 log x
log x	common logarithm of x to base 10

Table 2   *Alternative Methods of Expressing Various Physical Quantities*

1.   **Mass (SI unit : kg)**

$$g = 10^{-3} \text{ kg}$$
$$mg = 10^{-3} \text{ g} = 10^{-6} \text{ kg}$$
$$\mu g = 10^{-6} \text{ g} = 10^{-9} \text{ kg}$$

2.   **Length (SI unit : m)**

$$cm = 10^{-2} \text{ m}$$
$$\text{Å} = 10^{-10} \text{ m}$$
$$nm = 10^{-9} \text{ m} = 10\text{Å}$$
$$pm = 10^{-12} \text{ m} = 10^{-2} \text{ Å}$$

3.   **Volume (SI unit : m³)**

$$l = dm^3 = 10^{-3} \text{ m}^3$$
$$ml = cm^3 = 10^{-6} \text{ m}^3$$
$$\mu l = 10^{-3} \text{ cm}^3$$

4.   **Concentration (SI units : mol m^{-3})**

$$M = mol \ l^{-1} = mol \ dm^{-3} = 10^3 \ mol \ m^{-3}$$
$$mg \ l^{-1} = \mu g \ cm^{-3} = ppm = 10^{-3} \ g \ dm^{-3}$$
$$\mu g \ g^{-1} = ppm = 10^{-6} \ g \ g^{-1}$$
$$ng \ cm^{-3} = 10^{-6} \ g \ dm^{-3}$$
$$ng \ dm^{-3} = pg \ cm^{-3}$$
$$pg \ g^{-1} = ppb = 10^{-12} \ g \ g^{-1}$$
$$mg\% = 10^{-2} \ g \ dm^{-3}$$
$$\mu g\% = 10^{-5} \ g \ dm^{-3}$$

5.   **Pressure (SI unit : N m^{-2} = kg m^{-1} s^{-2})**

$$Pa = Nm^{-2}$$
$$atmos = 101 \ 325 \ N \ m^{-2}$$
$$bar = 10^5 \ N \ m^{-2}$$
$$torr = mmHg = 133.322 \ N \ m^{-2}$$

6.   **Energy (SI unit : J = kg m² s^{-2})**

$$cal = 4.184 \ J$$
$$erg = 10^{-7} \ J$$
$$eV = 1.602 \times 10^{-19} \ J$$

Table 3 *Prefixes for SI Units*

Fraction	Prefix	Symbol
10^{-1}	deci	d
10^{-2}	centi	c
10^{-3}	milli	m
10^{-6}	micro	μ
10^{-9}	nano	n
10^{-12}	pico	p
10^{-15}	femto	f
10^{-18}	atto	a

Multiple	Prefix	Symbol
10	deka	da
10^{2}	hecto	h
10^{3}	kilo	k
10^{6}	mega	M
10^{9}	giga	G
10^{12}	tera	T
10^{15}	peta	P
10^{18}	exa	E

Table 4   *Recommended Values of Physical Constants*

Physical constant	Symbol	Value
acceleration due to gravity	g	9.81 m s^{-2}
Avogadro constant	N_A	$6.022\ 05 \times 10^{23}$ mol^{-1}
Boltzmann constant	k	$1.380\ 66 \times 10^{-23}$ J K^{-1}
charge to mass ratio	e/m	$1.758\ 796 \times 10^{11}$ C kg^{-1}
electronic charge	e	$1.602\ 19 \times 10^{-19}$ C
Faraday constant	F	$9.648\ 46 \times 10^4$ C mol^{-1}
gas constant	R	8.314 J K^{-1} mol^{-1}
'ice-point' temperature	T_{ice}	273.150 K exactly
molar volume of ideal gas (stp)	V_m	$2.241\ 38 \times 10^{-2}$ m^3 mol^{-1}
permittivity of a vacuum	ϵ_0	$8.854\ 188 \times 10^{-12}$ kg^{-1} m^{-3} s^4 A^2 (F m^{-1})
Planck constant	h	$6.626\ 2 \times 10^{-34}$ J s
standard atmosphere pressure	p	$101\ 325$ N m^{-2} exactly
atomic mass unit	m_u	$1.660\ 566 \times 10^{-27}$ kg
speed of light in a vacuum	c	$2.997\ 925 \times 10^8$ m s^{-1}

Table 5 *Glossary of Symbols used in this Unit*

c	– intercept on y axis (ie on axis for response variable)
d_i	– the ith deviation from the mean, *or* the ith residual
$d_1\ d_2\ d_n$	– control chart limits for sample range
e_b	– an error due to blunder
e_r	– a random error
e_s	– a systematic error
E_{abs}	– the absolute error
E_{rel}	– the relative error
F	– parameter for the F-distribution
H_o	– the null hypothesis
H_1	– the alternative hypothesis
k	– a constant coefficient
m	– gradient of linear regression fit
n	– sample size *or* number of data points
P_i	– probability of event i
Q	– Dixon's Q parameter (for testing an outlier)
r	– coefficient of linear correlation
R	– sample range
s	– sample standard deviation
s_p	– the pooled sample standard deviation
s_y	– standard error of estimate
S_{xx}	– sum of x_i^2 values (similarly S_x, S_y, S_{xy}, S_{yy})
t	– parameter for Student's t-distribution
T	– the 'true' value for a result
U	– an 'uncertainty'
x_i	– the result of the ith measurement
$\bar{x}$	– the sample mean
x_{det}	– the limit of detection
y_i	– the ith instrument reading (value of response)
y_{dec}	– the decision limit for a measured response
z	– parameter for the standardised normal distribution
α	– tail area under a distribution function *or* the level of significance
β	– the probability of type II error
Δa	– the uncertainty in a (similarly Δb, etc)
μ	– the population mean
ν	– the number of degrees of freedom
σ	– the population standard deviation

Appendix

This appendix consists of Sections 2.1–2.5 of the *Microprocessor Applications* Unit of the Analytical Chemistry by Open Learning series. For the current Unit you are specifically asked to complete the Sections represented here.

You may find that this Appendix will break the ice for you in computing to the extent that it may induce you later to decide to study *Microprocessor Applications*, and to introduce yourself to the techniques of interfacing computers to instruments, with all the benefits which that can give to experimental work. The later Sections 2.6–2.9 contained within that Unit would allow you to pursue your introduction to computing to a slightly deeper level. These later sections would help you, if you so wish, to build substantially upon the statistics programs which you will meet in this present Unit. For instance, you might be interested in developing some of the programs to incorporate graphics.

That, however, will be entirely up to you – it rather depends on whether you fall victim to the well-known computing 'bug'. To achieve everything we ask of you in this Unit, however, the Sections 2.1–2.5 which are reprinted here will be adequate.

Different microcomputers have different characteristics and consequently it is not possible to write instructional material which is applicable to every machine. This text has therefore been written as generally and as simply as possible to concentrate on those parts of BASIC that are indeed basic and that are applicable to virtually all available machines. Since it is difficult to guess how much you know already we make few assumptions about knowledge and experience. If certain passages seem rather elementary you will apreciate that someone else may welcome an opportunity for a little revision.

2.1. USING A COMPUTER

2.1.1. Introduction

Computers are instructed or programmed by means of a sequence of numbers known as *machine code*. The computer is able to understand and execute instructions given in this way but the preparation of a machine code program is not an easy task. Because of this a number of more readable *languages* have been developed; these range from *assembly* languages which consist of a number of mnemonics for machine code instructions to *high level* languages such as FORTRAN, PASCAL and BASIC. Different approaches to programming a microprocessor or computer were outlined in Part 1 of this Unit and Part 2 is devoted to a more comprehensive account of the language known as BASIC.

We assume that you have available a microcomputer with a visual display unit (VDU) and either a magnetic disc or cassette system for storing programs. Because of differences in language some of the examples may not be in exactly the right form for your microcomputer but you should find little difficulty in adapting them to suit your particular machine. From time to time material is given for the sake of completeness rather than because it is absolutely essential. At these points some indication will be given so that you can pass to a section of more immediate importance. You should, however, study the omitted material at a later reading.

One of the books on programming listed in the Unit bibliography or any other book which you find readable and suitable to your circumstances should be adequate to support the course. Apart from a book on programming, the manual or guide for the computer must be available. The manual will be required when you have to find out something which is peculiar to your own situation, eg how to switch on the machine or when to use a semicolon (;) rather than a colon (:). We prompt you to find this information at suitable points throughout the course, indicated by the symbol Π. You must yourself check answers to these exercises since the answers will depend on the machine you are using. You should write down the answers in a systematic manner so that you can refer to them at any time.

The symbol Π is also used when certain examples are being discussed to indicate where it is important for you to follow the examples on your machine and to note any points of difference between our procedures and yours.

One of the first things you must do is read enough of the manual to permit you to:

— connect up and switch on your machine;

— enter characters from the keyboard;

— delete a character or characters typed;

— clear the screen.

Very soon you will also have to find out how to:

— start a new program;

— list a program (ie make it appear on the screen);

— stop a program which is being run;

— save a program on tape or disc.

While a program is running it often happens that it has to be stopped and this is usually done by pressing a key marked ESCAPE or BREAK. You should find out how this is done on your machine as soon as possible. At the same time find out whether or not a program is lost when you ESCAPE or BREAK. Of course switching off is always a last resort if there is trouble but the program is usually lost when the power is disconnected.

It may be that you have to find out how to make your machine operate using BASIC. This will be no problem with most micro-computers but some may require a special instruction to be typed in or some special program to be loaded.

Since this is a course in programming many questions ask you to write programs. *Every time* you write a program you should type it into your computer and run it. If your will is strong enough you might even do this before looking at the answer but do not worry if you steal a look after writing and before typing. The answers to the questions often contain useful information. You should also type and run the example programs and program segments in the text.

Our prime objective is to guide you in your efforts to achieve some skill in programming applications of interest to an analyst. You should look upon this part of the Unit as a practical course. *Use your computer as much as possible* and when trying something new keep it simple. Remember that complicated computer programs are usually just a collection of simple procedures.

Throughout the course DOING MUST DOMINATE READING. If you read these notes for more than about ten minutes without either writing or running a program you have stopped following the course.

TYPE AND RUN EVERY EXAMPLE

WRITE, TYPE AND RUN EVERY QUESTION

2.1.2. Initial Practice

If you are not familiar with a computer it is a good idea to use it first as a calculator, provided of course that you machine can operate in this way. When the machine is used in this way it is usually said to be operating in the *immediate mode or direct mode*.

The following example is a good start. Type it into your machine by typing each line in turn and then pressing the RETURN key.

LET X = 12

LET Y = 10

LET Z = X*Y

PRINT X, Y, Z

These four lines are instructions which tell the machine to do the following:

Assign the value 12 to the *variable* called X.

Assign the value 10 to the *variable* called Y.

Multiply the value of X by the value of Y and assign the result to the *variable* called Z.

Show on the screen the values of the *variables* X, Y and Z.

You can learn three important points from these four lines:

> The equals sign (=) has a somewhat different meaning from usual. Here it means *become equal to*. When this sign is used the variable on its left is *assigned* a value.

> The asterisk or star (∗) means *multiplied by*.

> The word PRINT in capital letters means *show on the screen*.

Each of the four lines contains an *instruction* or a *statement*.

The *syntax* or structure of a statement is very important. For example the words LET and PRINT must usually be in capital letters though some machines may allow them to be typed in lower case.

Most computers allow the word LET to be omitted so that the first line above could be simply X = 12.

The word PRINT is used above as a statement of the BASIC language. This word is unfortunate, having been introduced before VDU screens became common and the only way of seeing output from a computer was to have it printed on paper. For real printing on paper the word LPRINT or WRITE is usually used.

While you must be quite rigorous as to spelling, punctuation and structure when typing statements you should feel free to use your imagination in any way that will help you form a picture of what is going on. For example, you may find it useful to think of a variable as a box or location in memory which is labelled with the name of the variable. Then the statement LET X = 12 can be taken as meaning *place the number 12 into the box labelled X*. Similarly, LET Z = X∗Y means *multiply X by Y and place the result in box Z*.

Many microcomputers have keys that reduce the amount of typing required; thus a single key press might result in the word PRINT appearing on the screen, while another key might produce the word INPUT. This kind of facility depends on the particular machine and you must therefore discover any special aids for yourself. The same may be said about any editing facilities. There is usually some means of correcting wrong words or lines typed on the screen but the systems employed are so machine dependent that it is not possible to give general instructions.

If you have little experience of using a computer it is a good idea to do a few calculations like the one above, perhaps finding the sum ($+$) or the difference ($-$) of two numbers, or dividing one by the other ($/$). This will familiarise you with the keyboard and with any special keys.

2.1.3. Scientific Notation

You probably make use of a simple hand calculator and so are already familiar with what is usually called *scientific notation*. For example, the number 6.02E23 is the same as 6.02 multiplied by 10 to the power 23.

While we usually use this system when dealing with very large or very small numbers most computers use it quite often and so we must become quite confident with the notation. In case you feel that you need some revision here is a question. Try it mentally at first!

Appendix SAQ 2.1a

In the tables below, the entry in column X, row A is 2.0 and that in column Y, row B is 4.0. The product of these is 8.0. That is, AX times BY gives 8.0. What other combinations produce the same result, viz 8.0? $\longrightarrow$

**Appendix
SAQ 2.1a
(cont.)**

	X	Y
A	2.0	0.4E-1
B	2.0E1	4.0
C	200E-2	4.0E3
D	0.2E3	40.0E-1
E	0.002E0	4.0E-1

2.1.4. Printing Words

While we shall be mainly concerned with numbers and computation it is also necessary to make the computer store and print characters. This usually requires the use of quotation marks which will be single or double depending on your computer.

Type in the following line and press RETURN:

PRINT "JOHN SMITH"

If an error is signalled try again using single quotation marks. The characters contained inside the quotation marks form a *string*.

Just as we used X and Y as variables with numeric values, so we can have variables with 'values' which are strings of characters. Type the following lines into your machine, pressing the RETURN key at the end of each line. (We assume double quotes though you may need single):

LET A$ = "JOHN"

LET B$ = "SMITH"

LET C$ = A$ + B$

PRINT C$

The variables here all end with the dollar sign ($). This indicates that they are *string* variables. That is, they represent strings of characters rather than numbers.

2.1.5. Variable Types and Names

You can probably read this discussion fairly lightly at first since its utility will depend on the microcomputer you use and on your applications. Do not omit it completely, however, as you will need some of the concepts later.

The two main types of variable are the numeric and the string. However, many computers permit a distinction to be made between *real variables* and *integer variables*. Both of these are numeric but the latter can only have values which are whole numbers. When this distinction is allowed in BASIC it is usual to indicate an integer variable by means of the percent sign (%) as the last character of the variable name, eg:

Names for Real variables: X   A

Names for Integer variables: X% A%

Names for String variables: X$ A$

Valid assignments of values to two of the examples given above would be:

LET A = 5.62

LET A% = 65

An attempt to assign a value 32.9 to the variable A% would be unsuccessful. The fractional part of the number would be discarded and A% would be assigned the value 32.

There are certain advantages in using integer variables. These include:

When integers are used in calculations there are no rounding-off errors because the result of the calculation must also be an integer.

With integer variables you can rely on the last figure given whereas with real variables the number of significant figures is usually less than the number actually displayed. You will have to refer to your manual for details.

Many programs run faster if integer variables are used.

There is some saving in space because integers are usually stored in a smaller number of memory locations. For example with the BBC machine an integer uses four locations or bytes whereas a real variable uses five.

Against these advantages it must be remembered that integer variables cannot be used when fractional values occur. Also, the range of integer variables is normally much smaller than the range of real variables. (For the BBC machine the highest possible real and integer values are respectively 1.7×10^{38} and 2 147 483 647).

| **Appendix** | What type of variable might be employed for the |
| **SAQ 2.1b** | following: |

(*i*)   the date of a month;

(*ii*) the name of a day;

(*iii*) the weight of a sample.

In the example programs above simple capital letters like X and Y were used as numeric variables. The use of single letters is always safe but most computers allow a wide choice of variable name. Some examples of possible names are:

Real variables: A3   SUMM   pay   my_age

Integer variables: YEARS%   QUANTA%

String variables: Names$ TITLE$

Since practice depends on the particular computer only two general rules can be given. These are:

(*a*) The name of a variable must always start with an alphabetical character, never with a number or a punctuation mark.

(*b*) The name must not include a space. (And probably certain other characters).

While it is often useful to be able to use words as variable names care must always be taken not to include some reserved word in the name. For example, the word PLOT usually has a special meaning associated with graphics and therefore must not be used as part of a variable name. This excludes names like XPLOT or OLDPLOT for variables though the corresponding words in lower-case letters may be acceptable. It may be that your machine only objects if a reserved word forms the first part of the variable name (eg it allows XPLOT but not PLOTX); this is another point you can only discover for yourself.

One pitfall is that some computers allow variable names to be of any length but only pay attention to the first two or the first four characters. When this is the case it is advisable to use names of no more than two or four characters.

∏   Find out how variables may be named for your computer.

Does your computer permit a distinction between real and integer variables?

What are the lowest and highest numbers that can be stored (*a*) as real variables and (*b*) as integer variables?

Appendix
SAQ 2.1c
Some of the following statements are unacceptable and some will lead to errors. Comment on each statement.   ⟶

Appendix SAQ 2.1c (cont.)

(*i*)   LET X% = 12

(*ii*)   LET X + Y% = 12

(*iii*)   LET residue = 0.456

(*iv*)   LET PPT = 1.347g

(*v*)   LET B% = 3.456

(*vi*)   LET 3rd% = 4.67

(*vii*)   LET sample% = 3.92

(*viii*) LET Name$ = "Tom Brown"

(*ix*)   LET PRINTER$ = "Caxton"

(*x*)   LET FIVE = 6

2.2. SIMPLE PROGRAMMING

2.2.1. Computer Programs

A *program* is a series of instructions which the computer obeys one after the other. In a BASIC program each line of instruction has a number. When the program runs it takes the line numbers in order and carries out the instruction on the line. The program always goes from one line number to the next higher number unless there is an instruction directing the program to go to a certain line.

The line numbers do not have to go up by one each time. It is usual to number the lines 10, 20, 30 ... so that additional lines can be inserted if desired. This is an important point. You should always assume that more lines will be necessary after you have tested a program. Never use line numbers 1, 2, 3 ... if it can be avoided.

Most computers have a facility for renumbering lines automatically. It is usually also possible to type a line at any point of a program, the computer itself placing the line into its proper position within the program.

Many computers allow several instructions to be placed on one line, the instructions being separated from each other by a colon (eg LET X = 5: LET Y = 7). However, since not all computers permit more than one instruction per line we shall always place only one on each line. This is always a safe procedure. Various short cuts suitable for your own computer are probably available and you will learn these as you gain confidence in simple tasks.

The big advantage of a computer program is that it can be used time after time with different values of the variables.

The following is a simple program:

```
 9 REM ** A 3-LINE PROGRAM
10 LET X = 12
20 PRINT X
30 END
```

Line 9 is simply a reminder. Any line that starts with REM is ignored by the computer. The two stars help to make the reminder stand out when the program listing is printed.

At line 10 a variable X is assigned the value 12.

At line 20 the value of the variable X is printed on the computer screen.

At line 30 the program stops.

∏   Type this program into your computer and run it.

    Even though it may seem extremely simple it will get you familiar with the keyboard and with entering program lines. After the program has been typed in you should LIST it; that is, make the complete program appear on the screen. You can then make any necessary alterations before running the program. The usual method of running a program is to clear the screen and then enter the word RUN.

If a program is stored on a magnetic tape or disc it must first be loaded into the computer before being run. Your instruction manual will describe how a program is loaded from or saved to tape or disc. The keywords LOAD and RUN are normally used but the word CHAIN may mean *load and run*. The name of the program will probably have to be within quotation marks. The keyword SAVE is usually used when transferring a program to tape or disc so that it can be recalled at a later time.

∏   Find out how to LIST, SAVE, LOAD and RUN a program.

    Experiment with a simple program like the one above before it is necessary to save an important program.

2.2.2. Output to Screen

The value of a variable is displayed on the screen when the program meets a statement like:

20 PRINT X

A single statement can output more than one value. Alter the program above by adding line 12 and changing line 20:

12 LET Y = 25
20 PRINT X, Y

In line 20 there is a *list* of two variables, X and Y, which are to be printed or displayed on the screen. The variables are separated by a comma. Most computer obey this instruction by printing the two numbers in different *columns* or *zones* or *fields*, eg:

12   25

Other separators besides the comma may be used with different results. As practice differs considerably you should find the punctuation used by your own machine, paying particular attention to the use of the comma, the semicolon and the colon.

∏   Find out how the numbers 12 and 20 are output by your computer.

What instructions are needed to print the numbers:

(*a*) with 1 space between,

(*b*) in columns or fields,

(*c*) with no spaces (to give 1220)?

Make your computer actually print these numbers in the forms suggested and note carefully the answers to the questions. The three types of format mentioned, *viz* columns or fields, one space, no spaces, are the formats most commonly used. With the column format the number of columns across the screen depends on the particular computer but it may be possible to alter this number.

It is a good idea to experiment with various statements using simple numbers. You could try this program:

```
10 LET X = 12
20 LET Y = 10
30 PRINT X, Y
50 END
```

and run the program with various statements at lines 30 and 40, eg:

```
30 PRINT X, Y          30 PRINT X ;Y

30 PRINT; X;Y          30 PRINT; X,Y

30 PRINT X;            30 PRINT X,
40 PRINT Y             40 PRINT Y
```

and indeed any other combination of PRINT statements that occurs to you.

Later on we shall have to consider how to tabulate data and how to round off numbers. If a column can display no more than 7 characters a number like 121.7346 will encroach on the next column. Unfortunately several computers print out numbers with a large number of figures and as analysts who understand about significant figures we must make any necessary adjustments. We shall deal with this later.

2.2.3. Arithmetical Operations

The computer was developed to do calculations and this is still one of its principal functions. Many program instructions or statements are concerned with arithmetical operations.

Consider this program:

```
 9 REM ** MULTIPLY
10 LET X = 0.5
20 LET Y = 25
30 LET Z = X*Y
40 PRINT X, Y, Z
50 END
```

At line 30 the numbers represented by X and Y are multiplied together and the result assigned to the variable Z. At line 40 the values of X, Y and Z are printed in three different fields.

In algebraic expressions the symbols '+' and '−' have their usual meanings but the following must be noted:

> * means *multiply*
>
> / means *divide*
>
> ^ means *raise to the power of*

Be careful to use the correct sign (/) for divide. The reverse sign (\\) called the *backslash* must not be used.

Brackets are often necessary to ensure that operations are performed in the correct order. eg Suppose X = 4 and Y = 8. Then

> Z = (3*X + Y)/2   gives Z = 10
>
> Z = 3*X + Y/2   gives Z = 16
>
> Z = 3*X^2   gives Z = 48
>
> Z = (2*X)^2   gives Z = 64

Any expression within brackets is evaluated first so that the brackets can be removed. Otherwise the order of operations is:

> Raise to power
>
> Multiply and divide
>
> Add and subtract.

Sometimes the sign '**' can be used in place of the upward arrow for raising to a power, eg:

A*X**2

A*X**(A + B)

This use of '**' is not general. You will have to find out if your machine allows it.

It is often necessary to use more than one pair of brackets as in

$$Z = ((A + B)*(C + D))/(X + Y)$$

If an expression looks complicated it is always a good idea to count the total number of left and the total number of right brackets to ensure that a pair has been used.

Appendix SAQ 2.2a

Problems (*i*) to (*iii*) below should be done by hand (or mentally) and the results checked by computer.

(*i*)

$$Z1 = 100/(X + Y) \qquad Z2 = 100/X + Y$$

If X = 5 and Y = 15, what are the values of Z1 and Z2?

(*ii*)

$$Z3 = 100/X/Y \qquad Z4 = 100/X + Y$$
$$Z5 = 100/(X + Y)$$

If X = 5 and Y = 2, what are the values of Z3, Z4 and Z5?

$\longrightarrow$

**Appendix
SAQ 2.2a
(cont.)**

(*iii*)

$$Z6 = 100/X*Y \qquad Z7 = 100/(X*Y)$$

If $X = 2$ and $Y = 10$, what are the values of Z6 and Z7?

(*iv*)

Write computer-type expressions corresponding to:

$$Z8 = Ax^2 + Bx + C$$

$$Z9 = Ax^{(a + b)} + Bx^2 + 1/C$$

2.2.4. Input from Keyboard

Values of variables are often assigned by being typed in at the keyboard. A program will stop at a statement like:

    10 INPUT X

and wait for a number to be entered. The user does this by typing the number and then pressing the RETURN key. While it is waiting for this to happen the computer usually gives a prompt, that is, it displays a character like ! or > or ? or it displays a flashing line or rectangle.

Π   Try this on your machine:

    10 INPUT X
    20 PRINT 2*X
    30 END

Run this program and when a *prompt* character appears type in a value for X. The number only must be typed, no words like "input" or other characters like "X=".

Note particularly how the computer shows that it is waiting for you to type in the value for the variable X. The prompt that it gives here is likely to be different from the prompt given when it expects a line of program to be entered.

When the computer expects input it is important that the correct kind of information is given. For example, if a number is expected an alphabetical character will not be acceptable. It is a good idea to make some deliberate errors early in your use of a computer to find out how it reacts. If you type in a non-numerical character when input is expected at line 10 above some computers will simple ask you to re-enter, others will stop the program completely, and others will assign a value of zero to the variable. This last result is probably the most dangerous as you are not warned of your error.

It is usually advisable to print a line to remind the user what input is expected and what the output means. We can do this in the program above by including two more lines as *prompts*:

8 PRINT "ENTER A NUMBER"

18 PRINT "DOUBLE THE NUMBER IS ";

Not only do these lines make the program more comprehensible to the user, they help the programmer in the development of the program.

The program can be modified to accept two numbers and print out the product. The simplest modification would be to include another line 12 INPUT Y and alter line 20 to PRINT X*Y. Of course the prompts would have to be changed. It is usually possible, however, to enter or input more than one value in a single statement as in the following program:

```
10 PRINT "ENTER X AND Y"
20 INPUT X, Y
30 PRINT "THE SUM OF X AND Y IS ";
40 PRINT X + Y
50 END
```

X and Y form a *variable list* at line 20. The two values entered are separated by a comma (eg you would type 12,10).

At line 40 the value of the expression X + Y is computed before being printed. The two PRINT statements could be combined in a print list:

40 PRINT "THE SUM OF X AND Y IS ";X + Y

You should modify the program above to make it operate on the numbers entered in different ways, eg make it print the product or the difference of the numbers or the square of one plus the other.

Let us take an example involving a little chemistry.

Suppose we want a program to calculate the molarity (ie mol dm^{-3} concentration) of a sulphuric acid solution after titrating a sample with standard sodium hydroxide.

The *first* thing that must be done is to sort out the chemistry. From the chemical equation for the reaction we arrive at the conclusion that the unknown molarity is given by the expression:

$$\frac{\text{molarity of base} \times \text{volume of base}}{2 \times \text{volume of acid}}$$

(We leave you to justify this result by your own method).

It cannot be emphasised too strongly that the chemistry must be done before the programming. We can now use symbols. For clarity we shall represent variables by two characters rather than one as we have been doing: volume 'VA' of acid requires volume, 'VB' of standard base. If the molarity of the base is 'MB' then the unknown molarity, 'MA' is given by:

$$MA = MB*VB/2/VA$$

If your computer permits it you might prefer to use names like MBASE, MACID, VBASE, VACID for the variables. We are now in a position to write a program which does three things, *viz*:

accepts the necessary data;

calculates MA from the equation;

prints out the result.

A suitable program is:

```
 99 REM ** SULPHURIC ACID - SODIUM HYDROXIDE
100 PRINT "ENTER MOLARITY & VOLUME OF
BASE";
110 INPUT MB, VB
120 PRINT "ENTER VOLUME OF ACID";
130 INPUT VA
140 LET MA = MB*VB/2/VA
150 PRINT "MOLARITY OF ACID IS ";MA
160 END
```

Run this program and perhaps modify it to deal with any acid–base titration by introducing a variable to represent the ratio base/acid in the chemical equation; the ratio is 2 in the example taken since this is the mole ratio $NaOH/H_2SO_4$ in the neutralisation reaction.

Appendix SAQ 2.2b

In the standardisation of acid solutions a known weight of anhydrous sodium carbonate is titrated with an acid. One mole of monobasic acid neutralises 53.00 g of Na_2CO_3.

Write and run a program to calculate the molarity of a hydrochloric acid solution from the weight of sodium carbonate (W g) and the volume of hydrochloric acid (V cm^3) required for neutralisation.

Reminder:

molarity = number of moles per litre of solution.

The inclusion of prompts when input is required is so useful that many computers allow a prompt as part of the INPUT statement. An example is:

    10 INPUT "ENTER VALUE OF X", X

The words in quotation marks form a *prompt-string*. When the statement is reached these words are printed on the screen and the computer waits for the value of X. In this example a comma is placed between the prompt-string and the variable name but practice differs considerably.

∏   Does your computer support prompt-strings?

       If it does support prompt-strings find out the punctuation. In particular note the answers to the following:

       Are single or double quotation marks used?

       What punctuation mark, if any, comes between the prompt-string and the variable list?

       If a question mark appears, does it always appear or can it be suppressed?

       It is a good idea to repeat your answer to SAQ 2.2b, but this time employ prompt-strings if your computer supports them.

When the input required is a single character it may be possible to make use of the string function GET$. This is available in the BASIC of most microcomputers.

GET$ produces a string variable of a single character which is assigned merely by pressing a key. While we shall make little use of the function a brief mention may be of interest.

You will probably have used GET$ or an equivalent function when running a commercial package or game which invites you to select from a 'menu' by pressing a key; as soon as the correct key is pressed the appropriate program runs.

The use of GET$ is machine dependent. With the BBC machine the statement:

    30 A$ = GET$

has the effect of assigning the *character* of the next key pressed to the string variable A$. If the key 'K' is pressed then A$ becomes 'K'. If key '3' is pressed A$ becomes *character* 3. The 'character' assigned may be a space.

When GET$ is used it is not necessary to press RETURN but only one character can be accepted and the character does not appear on the screen. The following simple program will show how GET$ works:

    10 A$ = GET$
    20 PRINT "YOU PRESSED KEY "; A$
    30 GOTO 10

This will allow you to try various keys though you will have to press ESCAPE or BREAK to stop the program.

2.2.5. Mathematical Functions

All computer languages have facilities for working out certain standard mathematical *functions*. Your computer manual will list those available in your particular version of BASIC.

The most frequently used functions are:

V = LOG(X)   V becomes the logarithm of X. (See 2.2.7 for discussion on common and natural logs.)

V = EXP(X)   V becomes the exponential of X.

V = SQR(X)   V becomes the *square root* of X.

V = SIN(X)   V becomes the sine of angle X. COS(X) and TAN(X) similarly. (NB Angles must usually be in *radians*.)

V = ASN(X)   V is the angle whose sine is X. ACS(X) and ATN(X) similarly.

V = RAD(X)   V becomes the value in *radians* of X degrees. (Possibly not always available)

V = ABS(X)   V becomes the *absolute value* of X. (ie V is the same as X but always *positive*.)

V = INT(X)   V becomes the integer less than X.

Be careful when using INT when X is negative as some computers respond to it by returning the integer nearer to zero.

Π   Type PRINT INT(5.4), INT(-5.4) into your machine to find how it operates.

The functions, above give a selection of the more important ones usually available. You computer language may not have all of them but will probably have a number of others. There is often a function for giving a random number and one for the number of characters in a string. Many versions of BASIC also permit the programmer to define *user* functions within a program. This is a useful facility but is unlikely to be used in elementary programming.

You will observe that the *argument* of the function is always enclosed by brackets. In the examples above X represents the argument. A program segment might be:

```
100 INPUT "ENTER A NUMBER ",X
110 LET Y = SQR(X)
120 PRINT "SQUARE ROOT OF ";X;" IS ";Y
```

The argument of a function may itself be a function. A simple example occurs when an angle is known in degrees but a trigonometric function requires radians. If your computer language has the RAD function then the cosine of an angle D degrees is given by:

```
30 V = COS(RAD(D))
```

Here the argument of the COS function is RAD(D). Normally angles have to be in radians when using the SIN, COS and TAN functions and angles are returned in radians by the inverse trigonometric functions. You should check this for yourself.

The argument of a function may be an expression which must be evaluated before the function operates. Here is a simple example:

We require a program to calculate the length of the long side (C) of a right-angled triangle from the lengths (A and B) of the other two sides.

Simple geometry gives:

$$C = \sqrt{(A^2 + B^2)}$$

and this leads to the program:

```
 99 REM ** PYTHAGORAS
100 PRINT "LENGTHS OF SHORT SIDES? ";
110 INPUT A, B
120 LET C = SQR(A^2+B^2)
130 PRINT "LONG SIDE = "; C
140 END
```

The expression in brackets at line 120 is evaluated and then the square root of the result is returned in variable C. Try this program with a few numbers. (A triangle with sides in the ratio $3:4:5$ is right-angled).

The program can be extended to return the two other angles since the sines of the angles must be A/C and B/C:

```
132 X = DEG(ASN(A/C))
134 Y = DEG(ASN(B/C))
136 PRINT "ANGLES ARE "; X ;" AND ";Y;
138 PRINT " DEGREES"
```

Here DEG is a function which converts radians to degrees. We have not used the word LET at lines 132 and 134. This is permissible because the use of the word is usually not essential when assigning values to variables.

A more chemical example is the calculation of the hydrogen ion concentration of a weak acid solution.

If the dissociation constant of a weak acid is K_a and the solution concentration is C mol dm^{-3} then the approximate H$^+$ concentration is given by:

$$[H^+] = \sqrt{(K_a C)}$$

Written as a BASIC statement this becomes:

110 H = SQR(K*C)

The significance of the variables should be obvious. You should use your computer to check that a 0.001 mol dm^{-3} solution of an acid of $K_a = 10^{-5}$ has a hydrogen ion concentration of 10^{-4} mol dm^{-3}.

2.2.6. Rounding and Truncation

One of the less attractive features of many computers and electronic calculators is a tendency to display too many figures. When we know that the result of a calculation is 4.52 it is disconcerting to be presented with 4.519998 or, worse, .4519998E01.

The reason for this kind of behaviour is to be found in the way the computer stores numbers. We need not go into this in any detail but we must know how to get the number of significant figures we want. Some languages have a built-in facility for formatting output but this is not general and so we shall now address ourselves to this problem. Even if your machine can produce output as you want it you will still find this discussion instructive.

The key to the control of numerical output is the function INT. This function discards the fractional part of a decimal number. The number 54.67 is changed to 54. The syntax is

30 Y = INT(X)

The value of Y is equal to the integer which is less than X. That is, X is *rounded down* to give Y.

As mentioned above, there may be a problem with negative numbers and so it is safer to assume that this discussion applies only to positive numbers.

It is most important to remember that the number is rounded *down* or *truncated* to the next lower integer and not to the nearest integer. Thus, following the use of INT:

2.001 becomes 2

2.999 becomes 2

Clearly this can cause problems unless something is done.

To get the *nearest* integer we simply add 0.5 to the number before we use INT:

INT(2.001 + 0.5) = INT(2.501) = 2

INT(2.999 + 0.5) = INT(3.499) = 3

It will be found that this rule works generally, viz., to obtain the nearest whole number add 0.5 and then truncate. The procedure gives the next higher whole number if the fraction is 0.5 or greater and the next lower if the fraction is less than 0.5.

A very important use of the INT function is in the adjustment of the number of decimal places in a number before printing. Suppose we only want two figures after the point. For example, we might require:

12.764 to become 12.76 (round down)

12.757 to become 12.76 (round up)

Let us take A = 12.764. The procedure involves 4 steps:

multiply by 100	(A = 1276.4)
add 0.5	(A = 1276.9)
use INT to truncate	(A = 1276)
divide by 100	(A = 12.76)

If only one figure is required after the point you would multiply and divide by 10 instead of by 100.

**Appendix
SAQ 2.2c**

> Write a program which accepts numbers from the keyboard and prints out the reciprocal of the number correct to 3 decimal places.

2.2.7. Using Logarithms

Of the available functions the one most frequently employed in analytical chemistry is probably LOG or LN. It is particularly important to note whether LOG means the *common* (base 10) logarithm or the *natural* (base e) logarithm. Practice differs: if your computer has both functions LOG probably means *common* log while LN refers to the *natural* log. You will have to find this out for yourself either by reading the manual or by trial. If only one of the log functions is available the following conversions will be found useful:

$$\log_{10}(X) = 0.4343*\log_e(X)$$

$$\log_e(X) = 2.303*\log_{10}(X)$$

A simple application of the LOG function is to the conversion of transmittance to absorbance:

```
 9 REM ** T to A
10 PRINT "ENTER TRANSMITTANCE (FRACTION)"
20 INPUT T
29 REM ** ASSUME LOG TO BASE 10
30 LET A = LOG(1/T)
40 PRINT "ABSORBANCE = ";A
50 END
```

The value of $1/T$ is calculated from the known T value. This result becomes the argument of the LOG function. If $T = 0.5$ then $1/T = 2$ and the log (base 10) of 2 is 0.301. The value of A is therefore 0.301.

The recovery of a number from its logarithm involves exponentiation. If X is the common log of a number then the number is 10^X. A computer statement would be:

$$110 \ Y = 10^{\wedge}X$$

A pH of 6.5 means a hydrogen ion concentration of $10^{-6.5}$ mol dm^{-3} and so the statement:

110 H = 10^(-P)

gives variable H the value of the H^+ concentration of a solution having a pH equal to variable P. The transmittance of a solution can be calculated from its absorbance in a similar manner.

While the sign '^' (or perhaps '**') must be used to recover the 'antilog' from a common or base 10 logarithm, the function EXP is usually used to recover a number from its natural logarithm. If X is the natural log of a number and E the base of natural logs (E = 2.7183 ...) then the number is given by either of the two expressions:

Y = EXP(X)   Y = E^X

Though these two should be equivalent, the second is never used.

Appendix SAQ 2.2d

Write programs to calculate:

(*i*)   absorbance from percent transmittance;

(*ii*) percent transmittance from absorbance;

(*iii*) pH from hydrogen ion concentration;

(*iv*) hydrogen ion concentration from pH.

Reminder:

absorbance = $\log_{10}(100/\%\text{transmittance})$

pH = $-\log_{10}[H^+]$

SAQ 2.2d

2.2.8. String Functions

Several functions use a string as argument. Many computer languages include a function which produces a new string which is part of the old (argument) string. Thus, if one string is "CHEMISTRY" a function may produce a new string, "CHEM", by selecting the first four characters of the original string. You can find what your own computer does if you are interested in these things; we shall only mention three functions that involve strings.

We sometimes want to regard a sequence of numerical characters as a string rather than a number or *vice versa*. The conversion from one variable type to the other is accomplished by means of the functions STR$ and VAL.

To illustrate, suppose we have the number 22.4 and that this is the value currently assigned to variable V. We can assign a value to a string variable, A$, by means of either of the statements:

$$A\$ = \text{"22.4"} \quad \text{or} \quad A\$ = STR\$(V)$$

It must be remembered that A$ represents the *characters* of 22.4 and not the actual number.

Provided the characters of a string represent a number it is possible to convert the string into a numeric variable. The function VAL does this:

$$X = VAL(A\$)$$

If A$ = '22.4' then the value of numeric variable X would become 22.4.

Appendix SAQ 2.2e

We wish to write a program which asks a child his/her name and age and then prints the message

"(NAME) WILL BE (Y) NEXT YEAR"

with the correct name and number.

(*i*) Write the program on the assumption that the child will always enter the age as a number.

(*ii*) Write the program to allow the age to be entered as a number followed by a word like 'years'. For this case assume that the VAL function can be used to return a number which appears before any non-numeric character.

SAQ 2.2e

The string function GET$ has already been mentioned in connection with the input of data at the keyboard. This function produces a string variable of a single character which is assigned merely by pressing a key.

The GET$ function can be useful if you are having trouble with a program and you cannot establish where the difficulty is located. At certain points of the program you arrange that variable values are printed and place the statement Z$ = GET$. When the program reaches this point it waits for a key to be pressed before proceeding. Z$ should not have any particular significance. In this way you introduce a pause which allows you to check variable values as you step through a program. This is part of the process known as "debugging".

2.3.   PROGRAM CONTROL (1)

2.3.1.   Simple Loop

It is often necessary to repeat a section of a program. Consider a program to calculate the concentration of a substance from the absorbance of a solution.

We know by experiment that a certain solution containing an iron complex has an absorbance of 0.6 for a 3 mg dm^{-3} concentration of iron when measured using a 1 cm cell. Beer's Law holds and so absorbance (A) is proportional to concentration (C) and to path length (b):

$$A = E*C*b$$

Putting $A = 0.6$, $C = 3$ mg dm^{-3} and $b = 1$ cm we find the proportionality constant $E = 0.2$ dm^3 mg^{-1} cm^{-1}. (Note that the concentration is expressed in *milligrams* per dm^3.)

We can now use this E value to determine the percentage iron in different samples.

Suppose that a weight W g of sample dissolved in 1 dm^3 of solvent produces a solution which has an iron concentration of C mg dm^{-3}. The percentage of iron in the sample is:

$$\% \text{ iron} = \frac{10^{-3} C}{W} 100$$

or:

$$\% \text{ iron} = \frac{0.1*C}{W}$$

Replacing C by the ratio A/(E*b) from the relationship between concentration and absorbance we get:

$$\% \text{ iron} = \frac{0.1*A}{W*E*b} = K*A/W$$

where $K = 0.1/E/b$ is a constant for the particular system. The value of the constant must be found by experiment.

At this point the problem passes from chemistry to computing. We represent percent iron by a variable P and then write a skeleton program:

INPUT values for K, W and A

Calculate

PRINT value of P

From this a simple program is produced:

```
 9 REM**PERCENT IRON
10 PRINT "ENTER CONSTANT K";
20 INPUT K
30 PRINT "ENTER WEIGHT, ABSORBANCE";
40 INPUT W, A
50 LET P = K*A/W
60 PRINT "PERCENT IRON = ";P
70 END
```

Π   Type this into your computer and run it to see that it gives correct results.

This program suffers from a disadvantage that it must be run every time a calculation is required and the value of the constant, K, must be entered each time. If several calculations are to be done with different values for W and A each time the simplest way of repeating the program is to include a line like:

65 GOTO 30

After printing the value of P the program returns to line 30 where it receives the next pair of W and A values. It continues round the loop again and again. Note that the value of K is entered before the loop starts because its value remains the same for every pass through the loop. Put the modified program into your computer and see that it runs as described.

While the program keeps repeating lines 30 to 60 there is no way of stopping it other than by using the ESCAPE or BREAK keys or by switching off. One way of avoiding this is to include a *conditional* statement like one of the following:

        35 IF W>90 GOTO 70
        35 IF W>90 THEN 70
        35 IF W>90 THEN GOTO 70

It is then possible to stop the program under control by entering 99,99 when input is requested at line 30. When the conditional statement is reached the computer applies the appropriate test (in this case it tests to see if W is greater than 90) and then acts as instructed.

Of course the actual number is selected to suit the situation. For example, one could use IF W<O ..., the loop being stopped by entering a negative value for W.

It is sometimes necessary to avoid conditions which test for equality (eg IF W = 2) because it may be difficult to ensure that two variables have *exactly* the same value. The reason for this lies in the manner in which a computer stores numbers.

When a variable has a value 10 the computer may hold this in its memory as a number 9.99998 or 10.00002. If integer variables can be employed it is possible to ensure that the value is exactly 10 but meantime suppose that the value held is not an exact integer. When a test is applied to find if the value of some other variable is *equal* to this value the computer may take the result quite literally and register a failure even if the difference is only 0.00002. It is as well to remember this any time a test is included in a program and make sure that the test allows a little latitude. For example, IF W > 2.2 is a safer test that IF W = 2 when W is being assigned values which are whole numbers.

Many computer scientists and programmers do not like the statement GOTO as it appears in line 65 above. They do not mind too much if it is accompanied by a conditional statement as in line 35 but otherwise they recommend that GOTO should not be used. As far as we are concerned we shall not worry too much as all our programs are relatively short and are unlikely to contain many such statements.

2.3.2. FOR NEXT Loops

The FOR-NEXT loop is a common method of repeating a section of program. When appropriate, it is a much more elegant method than using the IF ... GOTO structure as in the last program.

In its simplest form a FOR-NEXT loop causes a variable to increase by a certain amount each time the loop is executed:

```
 9 REM ** FOR-NEXT
10 FOR I=1 TO 10 STEP 1
20    PRINT I
30    NEXT I
40 END
```

∏   This program will print the numbers 1, 2, 3 ... 10 (one per line) and then stop. Run it on your computer.

In the program listing the lines of the loop following FOR ... are indented to make the program more readable.

The variable I is the *index* or *control variable*.

The *end value* or *limit* of the *index* is 10.

The *step* or *increment* is 1 because the index increases by 1 each time the loop is traversed.

When you have it working correctly, alter the program to make it print out both I and the square of I in column format. Changing line 20 to 20 PRINT I,I^2 should achieve this.

The last line of the loop is line 30 NEXT I. Many microcomputers allow the variable name to be omitted so that the line could be simply 30 NEXT. Since not every computer allows this to be done however, and since the omission of the name can often lead to confusion, we shall make it our practice to include the variable name.

The step size may be a whole number or a fractional number, and it may be positive or negative. With most computers it is not necessary to state the step size if it is +1 but any other size of step must be stated.

There are sometimes restrictions on the name and type of the control variable. One popular microcomputer requires that the control variable be represented by a single letter like I or X or P. Another insists that the control be a *real* variable and not an integer despite the fact that this is the reverse of the requirement of some high level languages. You will soon discover any restrictions imposed by your machine.

Sometimes it is convenient to replace the start value or the end value or the increment by an expression which yields the proper value when evaluated. Though this is not recommended for the beginner an example is:

110 FOR I = 1 TO 3*P STEP 2

A loop starting with this statement would require to have a value assigned to variable P before the statement is reached. The computer calculates the value of 3*P before the loop is entered and the control variable, I, is compared with this value between passes round the loop.

While it is usually possible to use expressions instead of the index or the limits or the step size, you should keep things as simple as possible until you have considerable experience in programming. Use either proper constants or variables with assigned values.

Finally, remember that a test is applied before the start of each cycle round a loop and remember how a computer stores its numbers. On occasion it may be necessary to write a statement like:

120 FOR I = 2 TO 20.1 STEP 2

Here the end value is made slightly higher than 20 in case the value of I is held as, say, 20.00002 when the test before the last cycle is made. The problem can be avoided by using integer variables since there would then be no doubt about exact values:

```
10 A% = 2
20 B% = 20
30 S% = 2
40 FOR I% = A% TO B% STEP S%
    . . .
    . . .
```

Of course your machine must allow the control variable to be an integer for this to be practicable.

This is a good point at which to pause and assess progress. Try SAQ 2.3a and study the answer carefully. Then review the ground we have covered until you feel quite at home with all the parts of programming we have looked at. When you can answer SAQ 2.3a without help you are ready to continue.

**Appendix
SAQ 2.3a**

Write programs to display on the screen:

(*i*)   the squares of the numbers 1 to 10 one per line;

(*ii*)   the squares of the numbers 1 to 10 in a row with a space between each number shown;

(*iii*)   the *even* numbers from 0 to 10 and their squares in two columns (take 0 as an even number.)

(*iv*)   as (*iii*) but starting with 10 and finishing with 0.

SAQ 2.3a

2.3.3. Loops and Calculations

The FOR-NEXT loop is particularly useful when it is necessary to operate on a series of numbers, as for example in doing a simple addition. As always you should remember that the sign '=' is best interpreted as meaning *become equal to*.

Here is a simple adding program:

```
 9 REM ** ADD 10 NUMBERS
10 PRINT "PROGRAM TO ADD 10 NUMBERS"
20 PRINT
30 LET S=0
40 FOR I=1 TO 10 STEP 1
50    INPUT "ENTER NUMBER ",X
60    LET S=S+X
70    PRINT S
80    NEXT I
90 END
```

A REM statement is simply a reminder. The program ignores this.

The variable X may take a different value each time the loop is traversed depending on the value input at line 50.

The variable S represents the sum of all the X values. S starts with the value zero at line 30.

At line 50 a value for X is entered and at line 60 this value is added to S to give a new value for S. The form of line 60 often causes problems because we have been conditioned to expect the sign '=' to link two equal values. In the present context it is best to translate the effect of line 60 as: *Add the value found in box X to the value found in box S and place the result in box S.*

At line 70 the value of S is printed. Lines 50, 60 and 70 are repeated ten times so that S finally becomes the sum of ten numbers.

∏   Run the program, entering your own numbers when prompted. Keep the numbers simple so that you can check the result.

The program as written can only deal with 10 numbers because it goes round the loop 10 times. To make it more general the number of items to be added could be entered before the loop commences. The following changes will achieve this:

```
10 PRINT "HOW MANY NUMBERS?"
15 INPUT N
40 FOR I=1 TO N STEP 1
```

∏   Alter your program in this way and run it again.

Another method of exerting some control is to use a *conditional statement* so that when a certain value of X is entered the program stops. Look back to Section 2.3.1 (program line 35) and then answer the next question.

**Appendix
SAQ 2.3b**

(*i*)   Alter the adding program of Section 2.3.3 so that the program stops and prints the sum when a negative number is entered. The negative number should not be included in the summation.

(*ii*) Alter the program so that the number of items to be added is entered at the keyboard before the loop commences. The program must add this number of items and finish by printing not only the value of the sum, S, but also the value of the control variable, I.

Is the value of I the same as the number of items added?

If I is different from the number of items added why is this?

Sometimes it is necessary to check that an operation within a loop does not lead to disaster. The following is an example worth studying.

You will recall that the factorial of an integer N is the continued product:

$$N! = N(N-1)(N-2)(N-3) \ldots (3)(2)(1)$$

The program below will calculate a factorial for any value of N greater than zero. It has been written in such a way that it keeps asking for an integer and printing the factorial until you type in 0 (zero). This saves having to type RUN for every factorial you want.

```
 99  REM ** FACTORIAL PROGRAM
100  PRINT "FACTORIAL CALCULATION"
110  PRINT
120  PRINT "ENTER INTEGER (0 TO STOP)";
130  INPUT N
140  IF N<1 GOTO 210
150  F=1
160  FOR I=1 TO N
170     F=F*I
180     NEXT I
190  PRINT "FACTORIAL ";N;"="F
200  GOTO 110
210  END
```

The step size has been omitted at line 160 since the increment is +1.

∏ Run the program on your computer and use it to evaluate factorials. Enter larger and larger numbers each time input is requested until the computer gives up because it is being asked to compute a number that is outside its range. (This will give you a very rough idea of the range of your machine). You can then alter the program so that if such a number is entered it simply repeats a request for input.

You will notice that the program as written continues to request input until it receives a value of zero. It then moves to line 210 and stops. An alternative stopping method would be omit line 210 and re-write line 140:

    140 IF N<1 THEN END

To make the program refuse a number that is too large we first find the number which makes the program "crash" by entering larger and larger numbers. If this number is M we can make the program refuse a number larger than M−1 by inserting the lines:

    145 IF N>(M-1) GOTO 220

    220 PRINT M;" IS TOO LARGE"

    230 GOTO 100

∏   The calculation part of this program could be done in reverse. Instead of starting with F = 1 and then multiplying by 1, 2, 3, ... N we could start with F = N and multiply by N−1, N−2, ... 1. The principal change would be to line 160. Why not rearrange the program to do this and see if you get the same answers?

Here is an SAQ which you can regard as a key assessment exercise. That is, you can judge your progress so far by the ease with which you can write and run the program.

Appendix SAQ 2.3c

Write a program which:

(*i*)   accepts a maximum of 20 numbers;

(*ii*) computes the sum of the numbers;

(*iii*) computes the sum of the squares of the numbers;   ⟶

Appendix SAQ 2.3c (cont.)

> (*iv*) stops when a negative number is entered;
>
> (*v*) calculates the total number of items;
>
> (*vi*) prints out the sum, the sum of squares, the total number of items and the mean of the numbers entered.
>
> Run the program using the following numbers:
>
> 89 82 67 72 86 80 86 91 80 80

2.3.4. Limiting Loop Size

In some of the examples above the programs jump out of loops. In general this procedure is frowned upon though it does no great harm provided the program stops almost immediately.

While jumping out of a loop may be simply bad practice 'jumping in' is a major error. A GOTO statement must *never* direct a program to go into a loop from outside the loop.

When the size of a FOR-NEXT loop cannot be determined before the loop is entered it may be possible to use a different type of loop. Some computers support loops that are governed by the word UNTIL. One possibility is:

REPEAT

...

(lines to be repeated)

...

UNTIL $N < 1$

...

As with FOR-NEXT loops some condition is checked each time the loop is traversed but the REPEAT-UNTIL loop has the merit of adjusting the size of the loop automatically to the size required. Since loops of this type are not so universally used as are FOR-NEXT loops, we shall not employ them in this course except for illustrative purposes. It is not difficult to see when they can be employed to advantage and if your computer language supports this kind of structure the manual will give the proper syntax. Some languages also support loops which are similar to the REPEAT-UNTIL but in which the test UNTIL is replaced by the test WHILE. Again, these can be very useful but we shall not employ them in the course.

When only the conventional FOR-NEXT loop can be used it is good practice to avoid jumping out in case problems result through an accumulation of incomplete loops. This is best done by setting the proper end value of the index before the loop is entered. It may be that this is not possible however, and then another strategy must be used. We give below one method of making sure that all passes round a loop are completed and, for comparison, an example of a REPEAT-UNTIL loop. After studying these methods you might care to modify your answer to SAQ 2.3c; proper operation of the program will provide assessment.

Let us suppose that we are going to add some numbers and are pretty confident that we shall not need to add more than 100 items. Also, no number will be greater than 998.

(*a*) All passes are made.

In the following program a decision is made at line 190 in every pass round the loop:

```
 99  REM ** COMPLETING LOOPS
100  PRINT "ADD UP TO 100 NUMBERS"
110  PRINT " LESS THAN 999"
120  PRINT
130  PRINT "TO STOP ENTER 999"
140  PRINT
150  S=0
160  N=0
170  INPUT "ENTER FIRST NUMBER ",X
180  FOR I=1 TO 100
189      REM ** CHECK FOR TERMINATION
190      IF X>998 THEN 240
200      S=S+X
210      N=N+1
220      PRINT "NEXT NUMBER ";
230      INPUT "(999 TO STOP) ",X
240      NEXT I
250  PRINT
260  PRINT "SUM OF ";N;" ITEMS = ";S
270  END
```

Variable S represents the sum, variable N the number of items added. Input is requested at line 170 initially, and then at line 230 inside the loop. Provided the number input (X) is 998 or less all operations in the loop are completed. If, however, the number entered is greater than 998 the conditional statement at line 190 ensures that lines 200–230 are missed in the remainder of the 100 loop cycles.

This kind of program can be wasteful of computer time if the number of steps set is much greater than the number actually required. It could be improved by using integer variables if this is possible, particularly at line 190 where a test is made in every pass round the loop.

(*b*) Loop size not specified.

The REPEAT-UNTIL or REPEAT-WHILE structure is better than (*a*) if your computer language supports one of these. Here is the same program written for a language which allows REPEAT-UNTIL:

```
 99  REM ** UNTIL LOOP
100  PRINT "ADD UP TO 100 NUMBERS"
110  PRINT "LESS THAN 999."
120  PRINT
130  PRINT "TO STOP ENTER 999"
140  PRINT
150  S=0
160  N=0
170  INPUT "ENTER FIRST NUMBER ",X
180  REPEAT
190     S=S+X
200     N=N+1
210     PRINT "NEXT NUMBER ";
220     INPUT "(999 TO STOP) ",X
240     UNTIL X>998
250  PRINT
260  PRINT "SUM OF ";N;" ITEMS = ";S
270  END
```

In the REPEAT program a test is applied automatically at the end of each cycle (line 240). This is clearly a much tidier program. It does not employ a GOTO and its construction requires much less thought than the previous version.

You should note the feature which is common to both these methods of control:

Input is taken before entering the loop and then again at the end of each loop cycle.

A third method of controlling loop size which is sometimes used involves changing the index or control variable inside a loop. This method cannot be recommended generally as some languages do not allow any of the loop parameters to be changed after they have been specified. If it is permissible line 190 (and 189) of program (*a*) would be omitted and a new line 235 inserted:

235 IF X>998 then I = 101

After X is assigned a value greater than 998 this line changes the control variable so that when it is tested against the limit or and value all passes appear to have been completed and the program passes to line 250 and out of the loop.

We must repeat, however, that changing the control variable cannot be recommended as a general method for controlling a loop.

2.4. DATA CONTROL

2.4.1. Read, Data, Restore

The DATA statement is a convenient way of supplying data to a program. It is particularly useful when a program requires a lot of data or uses the same set of data many times.

The idea is that instead of the program prompting you to input a number or string at the keyboard the program reads the item itself from a list that you have prepared. One of the great advantages of this method of supplying data is that you can check the data at your leisure before starting the program. In this way, if you make a mistake when typing the data, you have an opportunity of correcting the mistake before the program runs. Also, the data supplied stays with the program and can be checked for accuracy if there are any doubts about what the program actually read. Here is a simple example:

∏   We want to add several numbers and find the mean. Type
      and run this program. Do not destroy it as we shall develop
      it later.

```
 99  REM ** READ DATA
100  S=0
110  READ N
120  FOR I=1 TO N
130      READ X
140      S=S+X
150      NEXT I
160  M=S/N
170  PRINT "MEAN VALUE="; M
300  END
399  REM ** DATA LIST
400  DATA 9
410  DATA 10,11,12,13,14
420  DATA 15,16,17,18
```

The data items are on lines 400 to 410. The comma is used to separate the items. It does not matter how many lines are used for the
data so long as each line starts with the word DATA.

DATA statements may be located at any point in the program; it is
usual to place them at the end of the program.

In the above program the first item of data (line 400) gives the number of items to be added. This number (9) is read into variable N at
line 110 by a READ statement.

The READ statement at line 130 causes each succeeding number in
turn to be read into the program as though it had been typed in at
the keyboard.

Instead of reading the value of N one could use a large loop (eg
FOR I=1 TO 1000) and a conditional statement after line 130 to
test for, say, a negative item being read. The last item of the list must
then be negative and care would have to be taken not to count this
item as one of the numbers to be added.

When the computer reads from a DATA statement a 'pointer' moves from one item to the next. The pointer starts at the first item when the program begins. If the same data list is to be used more than once it is necessary to restore the pointer to point to the first item again. This is done by the simple statement RESTORE.

RESTORE by itself returns the pointer to the very first DATA item. However, many versions of BASIC permit the pointer to be moved to the first item of a particular line by a statement like:

174 RESTORE 410

This has the effect of making the next READ statement take in the first item on line 410 of the program and continue to read the following items as before. RESTORE without a following line number would move the pointer to the very first item on line 400 of the program.

∏   This facility to re-use items of data can be quite useful. To check how it works let us add the following lines to the program:

```
172  PRINT
174  RESTORE 410
176  S=0
178  FOR I=1 TO N
180     READ X
182     D=X-M
184     PRINT X;" DEVIATES FROM MEAN BY ";D
186     S=S+D
190     NEXT I
200  PRINT "SUM OF DEVIATIONS= ";S
```

The effect of these additional lines is to read again the data items (but not N) and subtract the mean value from each one. The deviation of each item from the mean is calculated and printed. At the same time variable S is used to sum the deviations. As a final check the value of S is printed.

Following this example it only needs a few more steps to calculate the *standard deviation* of the data items. Before tackling the next question recall that the standard deviation of a number of measurements is defined as the square root of the *variance* and that the variance is obtained by dividing the sum of the squares of the deviations by one less than the number of items:

$$\text{variance} = V = (\text{sum of } D^2)/(N - 1)$$

$$\text{standard deviation} = \text{square root of } V$$

Appendix SAQ 2.4a

Write a program which includes DATA statements and which:

(*i*) reads the number of items from the data list;

(*ii*) calculates and prints out the mean of several numbers;

(*iii*) restores the data pointer;

(*iv*) calculates and prints out the standard deviation of the numbers.

Test the program using the following numbers which were obtained by 11 analysts for the concentration (g dm^{-3}) of copper in a solution:

49.89	49.82	49.67	49.72	49.86	49.80
49.86	49.96	49.80	49.80	49.83	

SAQ 2.4a

2.4.2. Tabulation of Output

Sometimes the columns or fields or zones which are obtained by using the comma to separate items of the print list are not the right length for the material we have to print. It is then necessary to use a TAB statement. This takes various forms depending on the computer, eg TAB(5) or TAB5, to place the first character of the item in position 5 from the left (starting at zero).

After TAB(5) or its equivalent one usually places the punctuation mark that means 'leave no space' but practice differs and it may be that no mark is required. In this example the semicolon is used though it may not be necessary:

     110 PRINT "SAMPLE ";N$;TAB(20);"WEIGHT= ";W

This statement would cause the following to be printed:

     SAMPLE ....WEIGHT= ...

with the string N$ in the first space and the value of variable W in the second.

Since the syntax is critical and as there are many variations you must find details for your own machine.

The TAB statement can also be used to start printing data at any character position on the screen. A statement like

    110 PRINT TAB(10,15); X

would have the effect of placing the value of X on the screen starting at character position 10 horizontally and 15 vertically. Text character positions normally start at top left. That is, the first character at the top left of the screen is normally in position 0,0. It may be worth checking this for your machine in case its first screen position is 1,1.

String variables may be used like any other in tabulation as, for example, in the headings of a table:

    110 R$ = "READING"
    120 P$ = "CONC IN PPM"
    130 PRINT TAB(15); R$;TAB(25); P$

A last word on tabulation concerns the use of the word *column*. A 40 column screen means a screen which displays 40 characters horizontally. The *columns* are usually numbered from 0 to 39. Similarly, an 80 column display means that 80 characters are shown horizontally. This kind of display is often used by word processors.

Appendix SAQ 2.4b

In a book on quantitative analysis there are tables of data on common acids. Typical entries are:

	% w/w	kg/litre	mol/litre
Hydrochloric acid	35	1.18	11.3
Sulphuric acid	96	1.84	8.0

Write a program which employs TAB statements to produce a table like this from data in DATA statements.

SAQ 2.4b

2.4.3. One-dimensional Array

It frequently happens that you want to take a series of readings and store them for later computation. For example, you might take a series of pH or absorbance readings over a period of time and then investigate the pattern of behaviour by plotting against time.

For the moment, let us not worry about how the data are collected but simply consider how we might store and manipulate the data. This is where an *array* is useful.

You will be familiar with labelling a series as:

$$X_1 , X_2 , \text{etc.}$$

An array does the same thing but the subscripts are in parentheses, eg

$$X(1), X(2), X(3), \ldots$$

Each of these elements represents a variable. We can write a program statement:

30 LET X(9) = A

The value of the variable X(9) would then be the same as variable A. Again, a variable can be used to assign a value to an array element as in:

40 LET X(I) = B

The value of element X(I) becomes equal to the value of variable B. Of course the computer would require to know the values of subscript I and variable B when the program reaches this statement.

An array is very convenient when a set of numbers is to be stored or is to be operated on in a repetitive manner. The same segment of a program can be used repeatedly on different numbers simply by changing the subscript within a FOR-NEXT loop.

But let us start at the beginning. We must first decide the maximum number of elements the array can have. The computer will then reserve the correct amount of memory space. Since it will not recognise more then the number of elements we decide upon we must make sure that the size of the array is big enough before we start to use it. Suppose we know that we shall need no more than 100 elements. We then *dimension* the array by the statement

10 DIM X(100)

This makes the computer set aside memory space (boxes!) for, usually, 101 variables called X(0), X(1), X(2), etc. We say *usually* because some computers will set aside only 100, the variable X(0) not being allowed. This is a point that you should check from your computer manual.

If the numbers to be stored are integers and your computer can make a distinction between real and integer arrays it is often a good idea to use an integer array. This save time and space. The normal convention for naming integer variables is followed, ie the name of the array has a % sign as the last character, eg DIM X%(100).

Your computer probably supports *string arrays*. That is, a series of strings can be held as variables with names like A$(1), A$(2), etc. However, we shall not make use of this kind of array.

With some languages the dimension statement must be made at the start of the program while with others it may be made at any point before the array is used. You should check this point but in any case it is always good practice to dimension arrays early in a program to ensure that enough memory space is made available. The same array must not be dimensioned more than once in a program.

∏ Suppose we want to collect a set of numbers. Let us start with X(1) rather than X(0). Here is a segment of program:

```
 99 REM ** SIMPLE ARRAY
100 DIM X(100)
110 PRINT "HOW MANY VALUES?";
120 PRINT " (UP TO 100) ";
130 INPUT N
140 FOR I=1 TO N
150    PRINT "ENTER NUMBER ";I
160    INPUT X(I)
170    NEXT I
180 END
```

Run this program and check it using a few simple numbers. The loop should collect N values and place them in locations labelled X(1), X(2), ... X(N). To see if this happens add the next segment (line 180 being replaced):

```
179 REM ** CHECK ENTRIES
180 FOR I=1 TO N
190    PRINT X(I)
200    NEXT I
210 END
```

The numbers entered should now be printed out in the same order as they were entered. Make sure this happens and then alter line 180 to

180 FOR I=N TO 1 STEP -1

Run the program again. The numbers should now be printed in the reverse order.

It is necessary to include the step size in the altered line because it is not +1.

A FOR-NEXT loop and an array are often used together as in the program developed above. You should become quite familiar with the technique as it will prove to be extremely useful.

**Appendix
SAQ 2.4c**

Write a program to accept up to 10 real numbers into an array. After testing the program with simple numbers (eg 2,4,6, ...) alter the latter part of the program to make it print out:

(*i*) the numbers and their cubes in column format;

(*ii*) the numbers, and their reciprocals correct to 3 decimal places

2.4.4. Arrays and Data

It is often useful to use DATA statements when assigning data items to the elements of an array. When an array, a FOR-NEXT loop, and DATA statements are used together in a program you have a very powerful combination. Items from the data list are read into the array and then computations are done as required. The next question illustrates a typical application.

Appendix SAQ 2.4d

In a gas chromatography experiment the solvent peak was recorded at a retention time of 0.85 minutes and successive peaks were recorded at retention times of 3.96, 4.33, 5.20, 6.72, 7.58, 9.28 and 10.75 minutes. Write a program which will:

(*i*) read the experimental data from DATA statements, placing the peak times in an array;

(*ii*) compute retention times relative to peak number 4 (6.72 minutes); that is, for each peak compute the ratio

$$\frac{\text{retention time} - 0.85}{\text{peak 4 time} - 0.85}$$

(*iii*) display the retention times and the relative retention times in a table.

The array is particularly useful when several calculations require the same data at different points in a program. To follow up this point you might like to look again at SAQ 2.4a which required the use of the RESTORE statement in the calculation of a standard deviation. The program could be written to read the data into an array and then calculate the sum of squared deviations without restoring the data pointer.

2.5.   PROGRAM CONTROL (2)

2.5.1.   IF ... THEN Control

The material of Section 2.5 is not essential to the study of elementary programming. You should read the section over lightly to see what it is about and return for more detailed study later.

In several programs we have employed statements that start with the word IF. As this word implies these are *conditional* statements.

Every conditional statement involves at least one *test* by one of the *relational operators* '=' or '>' or '<'. Following the test a certain course of action is taken by the program. This action is usually easily seen:

        IF X<0 THEN GOTO 250

        IF X>0 THEN END

        IF X<22 THEN X = 22

        IF X% = Y% THEN PRINT "EQUALITY"

We have already used statements like the first two. The third statement is a means of ensuring that the value of X does not fall below a certain value (in this case 22). This kind of limitation is very useful when plotting points on the screen to guard against attempting to plot a point outside the correct range.

Let us discuss the last example in more detail. The test can be stated in words '*Is the value of variable X% the same as that of variable Y%?*'. If the answer to the question is 'yes' the test is successful and the program prints the word "EQUALITY". If the answer to the question is 'no' the test fails. What is the program to do in this case? In most cases the program simply moves to the next *numbered* line.

It is very important to note that control passes to the *next numbered* line.

The BASIC of most microcomputers allows more than one statement to be placed on a numbered line, the statements being separated from each other by colons. When this is the case the IF statement acts as a 'gate' to the rest of the line. For example, consider a program line which contains two statements:

110 IF A%>B% THEN PRINT A%;" IS BIGGER THAN ";B% : M% = A%

Provided A% is greater than B% the message is printed *and also* M% is made equal to A%. (Perhaps the intention is to find the maximum of a series of numbers.)

It is important to note that the second instruction, M% = A%, will not be executed unless A% is greater than B%.

Here is an example which illustrates a common trap:

Suppose a program includes the two lines:

100 IF X>0 THEN Y = SQR(X)

110 A% = B%

In the first of these lines care has been taken to avoid trying to find the square root of a negative number.

If it is possible to write more than one statement on each line there is sometimes a temptation to make the program neater with a line like:

```
100 IF X>0 THEN Y = SQR(X) : A% = B%
```

Unfortunately, the second statement can only be reached if X is greater than zero because of the 'gate' effect of the IF statement. As this example demonstrates it is essential to consider the consequences of placing more than one statement on a line. This is particularly important when a conditional statement is involved.

To illustrate the use of IF statements let us write a program to find and print out the maximum and minimum of a sequence of numbers entered at the keyboard.

```
 99 REM ** HIGH & LOW
100 PRINT "TO FIND HIGHEST & LOWEST OF N
            NUMBERS"
110 PRINT
120 PRINT "HOW MANY NUMBERS ";
130 INPUT N
140 PRINT
150 PRINT "ENTER FIRST NUMBER ";
160 INPUT X
168 REM ** H = HIGHEST L = LOWEST
169 REM ** MAKE FIRST NO. BOTH H AND L
170 H = X
180 L = X
190 FOR I = 2 TO N
200     PRINT
210     PRINT "NEXT NUMBER ";
220     INPUT X
229     REM ** TEST AND CHANGE
230     IF X>H THEN H = X
240     IF X<L THEN L = X
250     NEXT I
260 PRINT
269 REM ** NOW REPORT
270 PRINT "HIGHEST = ";H
280 PRINT "LOWEST = ";L
290 END
```

The first number (N) input at line 130 gives the total number of items to be entered. The first of the numbers is taken to be both the highest (H) and the lowest (L) initially (lines 170 and 180). As each subsequent number is input these are altered as necessary (lines 210 to 240).

2.5.2. Combining Operators

The relational operators can often be combined as in the statement

110 IF X $>=$ A THEN ...

The use of combinations helps to simplify programming. But be careful. Check your manual before you use a combination like this since your machine may be fussy about whether '$=>$' or '$>=$' is used to test for 'equal to or greater than'.

Again be careful. The computer takes things literally. It may consider a number like 1.99999 to be less than 2 even though you would like it to be taken as equal to 2. Problems of this nature are more likely when using real variables rather than integer variables and you always have to be on your guard.

Sometimes the NOT operator is used to reverse the effect of a test. An example is:

```
100  INPUT X
110  IF NOT (X = 6) THEN PRINT;X;" IS NOT SIX"
```

An alternative to this example uses the combination $<>$ which means *not equal to*:

```
100  INPUT X
110  IF X<>6 THEN PRINT;X;" IS NOT SIX"
```

A common use of the conditional statement is to ensure that the computer is not asked to perform an impossible calculation. Common errors that must be guarded against include:

trying to find the square root of a negative number;

trying to divide by zero;

trying to take the logarithm of zero or a negative number.

**Appendix
SAQ 2.5a**

Write a program which prints out the reciprocal and the common logarithm of any number entered at the keyboard and which warns without crashing when the task set is impossible. All output should be correct to two places of decimals.

Apart from using combinations like > = or < =, two or more conditions may be tested in one statement:

    110 IF X>0 AND X<10 THEN ...

This will ensure that the action following THEN is taken only if X has a value which is greater than zero and less than 10. The AND must be in upper-case.

Another example uses OR:

    110 IF X<5 OR X>10 THEN ...

The action which follows THEN will take place if X is less than 5 (but not 5) or if X is greater than 10 (but not 10). If we wanted to include the value 5 and 10 we must use the < = and > = signs.

While AND and OR can be very useful some care must be taken when using them. Frequently the NOT operator comes in handy as an aid to our thinking. The meaning of the following line is quite clear:

    110 IF NOT (X = 0 OR X = 10) THEN ...

It is normally permissible to use more than one AND or OR:

    110 IF A>B AND X>Y AND C<>D THEN ...

We shall meet AND and OR again when we study memory locations.

One application of conditional statements is to test a string before attempting to use the VAL function. Every keyboard character has a code number, generally referred to as its ASCII code number (from American Standard Code for Information Interchange). Thus, the ASCII code for the letter 'A' is 65 and that for 'a' is 97. Within a program the code for the first character of a string X$ can be obtained by means of a statement like:

    102 C = ASC(X$)

The reverse operation of obtaining the character corresponding to a given code number uses a statement like:

    104 X$ = CHR$(C)

You should check the syntax required by your machine for both of these statements.

**Appendix
SAQ 2.5b**

Write a program which accepts words or numbers greater than -1000 from the keyboard and which:

(*i*)   prints the message 'POSITIVE' if the first character of the input string is one of the 'numeric' characters 0 to 9;

(*ii*) prints 'NEGATIVE' if the first character is the minus sign ($-$);

(*iii*) prints 'FRACTION' and the character of ASCII code 7 if the first character is the period (.);

(*iv*) stops accepting input when the character 'Q' is entered;

(*v*)   reports the total number of entries made, the sum of all the numerical entries and the highest negative numerical entry.

SAQ 2.5b

2.5.3. IF ... THEN ... ELSE

It was stated previously that the program proceeds to the next line number when a test fails. This default action can be prevented by the use of ELSE. Consider the program line:

120 IF X<0 THEN END ELSE GOTO 30

The program stops if X is negative, otherwise it jumps to line 30.

This kind of structure can be very useful and leads to some neat programming. If we did not use ELSE in the line above two lines would be required to produce the same effect:

120 IF X<0 THEN END
130 GOTO 30

It is often possible to place several statements after the THEN but before the ELSE. This is one way of overcoming the 'gate' effect of IF mentioned earlier. Also, more that one IF ... THEN may be 'nested' as in:

110 IF X<0 THEN IF X>Q THEN Q = X

Before attempting to nest conditional statements you should consult
the computer manual to see what is allowed. It is also a good idea to
run a short experimental program to check that nested statements
have the effect you want.

You will appreciate that the use of these structures can lead to some
very complicated tests. Properly used, they can impose all sorts of
control on a program. They may, however, lead to programs which
are difficult to understand and you are therefore well advised to
avoid such complications if at all possible. If you find yourself writing a statement involving several tests and branches it is a good idea
to pause and consider if they are all really necessary.

**Appendix
SAQ 2.5c**

Provided your computer allows the structure answer SAQ 2.5a again but using IF ... THEN ...
ELSE.

2.5.4. Nested Loops

If a program contains several loops it is very important to ensure that the loops are self-contained. If all steps of one loop have not been completed when the next loop starts the computer thinks that the new loop is inside or *nested* in the previous one. Provided the second loop is completed before the first this causes no problem. There is, however, a limit to the number of nested loops permitted and if this number is exceeded an error will be signalled. The number of loops which may be nested depends on the particular version of BASIC supported by the computer.

∏ As a simple illustration of nesting here is a program which produces three multiplication tables:

```
 9 REM ** NESTING
10 FOR I = 1 TO 24
20    FOR J = 1 TO 3
30       PRINT I*J;
40       NEXT J
50    PRINT
60    NEXT I
70 END
```

Try this program. It should produce something like:

```
1    2    3
2    4    6
3    6    9
.    .    .
.    .    .
```

Each of the three tables should extend from 1 to 24. You may have to modify the punctuation at line 30 to make the output suit your computer. Integer variables would be better than real for a program like this.

Appendix
SAQ 2.5d

The potential of an ion-selective electrode may depend on the activities of two ions:

$$E = E^{\circ} + 58*LOG(A1 + K*A2) \qquad \text{(mV at 292K)}$$

A1 is the activity which the electrode is designed to measure, A2 the activity of an interfering ion. K is the selectivity ratio or selectivity constant.

Write a program to create a table showing, for any value of K, how the second term on the right depends on A1 and A2. Suitable ranges might be:

A1: 0.001 to 0.005 mol dm^{-3} in 5 steps
A2: 0 to 0.01 mol dm^{-3} in 5 steps

Output should be given to the nearest millivolt.

....

An example of output for K = 10 and A2 = 0.002 mol dm^{-3} is given below:

INTERFERING ION AT 2 MMOL

ION 1 AT 1 MMOL	TERM = −97
ION 1 AT 2 MMOL	TERM = −96
ION 1 AT 3 MMOL	TERM = −95
ION 1 AT 4 MMOL	TERM = −94
ION 1 AT 5 MMOL	TERM = −93

SAQ 2.5d

2.5.5. Double Array

A *double array* or an array of two dimensions is often convenient because it saves some programming. However, as such an array is hardly ever essential and as the associated programming requires careful thought it would not be unreasonable to skip this section until you feel ready for it. The subject is dealt with here simply for completeness.

Suppose that we have a series of ten measurements to make each day for five days. We could dimension an array A(5,10) and place the data in five DATA statements, one for each day. The following program segment could then be used to read the data into a program:

```
199  REM ** DAILY TEN
200  DIM A(5,10)
299  REM ** SELECT DAY
210  FOR D = 1 TO 5
219     REM ** TEN READINGS
220     FOR I = 1 TO 10
230        READ A(D, I)
240        NEXT I
249     REM ** TEN TAKEN
250     NEXT D
259  REM ** FIVE DAYS READ
499  REM ** KEEP DATA AT END
500  DATA M1,M2,M3,... M10
510  DATA Tu1,Tu2,Tu3, ... Tu10
520  DATA W1,W2,W3, ... W10
530  DATA Th1,Th2,Th3, ... Th10
540  DATA F1,F3,F3, ... F10
```

Here M1, M2, etc. represent the measurements made on Monday, Tu1, Tu2, etc. those made on Tuesday and so on.

Once all the data have been read in they may be used in calculations. The alternative to using the two-dimensional array would be to use five different one-dimensional arrays, M(10), T(10), etc. and this would lead to a larger program.

To check if the program segment works all right it might be followed by a segment to print out the data entered in another format. This may seem trivial but it can be quite useful as a check on the program:

```
259 REM ** CHECK READ LOOPS
260 FOR D = 1 TO 5
270    PRINT "DAY ";D
280    FOR I = 1 TO 10
290        PRINT A(D, I),
300        NEXT I
310    PRINT
320    NEXT D
330 END
```

This should print out the data in column or zone format:

```
DAY 1
M1      M2    M3    M4
M5      M6    M7    M8
M9      M10
DAY 2
Tu1     Tu2   Tu3   Tu4
... etc.
```

Note the PRINT statement at line 310. This ensures that the next day starts on a new line.

The most important rule in using arrays of more than one dimension is to make sure that the loops do not intersect. This has already been emphasised in connection with nested loops. In the program above NEXT I must come *before* NEXT D. That is, the I loop must be properly *nested* within the D loop. There will be chaos if proper nesting is not ensured. Those computers which allow the statement NEXT but do not require the index variable to be specified would appear to have certain advantages in this respect. This facility must be used with great caution however because errors due to incorrect nesting may occur without any indication being given. The computer cannot rectify poor thinking.

Appendix SAQ 2.5e	To assess a new analytical method four samples of a product were taken and each sample was divided into two parts. Four different analysts determined the purity using the old method for one part and the new method for the other. The results were listed as they were reported by the analysts, always in the order:

method, sample, % purity. $\longrightarrow$

**Appendix
SAQ 2.5e
(cont.)**

The results so listed were:

```
2  2  98.8      1  4  98.4
1  3  98.1      2  4  98.1
1  2  98.1      2  3  98.9
1  1  98.6      2  1  98.6
```

Write a program to read these results into a two-dimensional array and present them in something like this table:

SAMPLE	1	2	3	4
METHOD 1	98.6	98.1	98.1	98.4
METHOD 2	98.6	98.8	98.9	98.1

Self Assessment
Questions and Responses
to Appendix

Appendix SAQ 2.1a (cont.)

In the tables below the entry in column X, row A is 2.0 and that in column Y, row B is 4.0. The product of these is 8.0. That is, AX times BY gives 8.0. What other combinations produce the same result, viz 8.0?

	X	Y
A	2.0	0.4E-1
B	2.0E1	4.0
C	200E-2	4.0E3
D	0.2E3	40.0E-1
E	0.002E0	4.0E-1

Response

The following products give the result 8.0:

     AX by BY   CX by BY   BX by EY   CX by DY

     AX by DY   DX by AY   EX by CY

You should test yourself and your computer by making it accept and print some of these numbers. A suitable procedure is to type in lines like:

     LET P = 0.2E3

     LET Q = 0.4E-1

     PRINT P*Q

          ****

**Appendix
SAQ 2.1b**

What type of variable might be employed for the following:

(*i*)   the date of a month;

(*ii*) the name of a day;

(*iii*) the weight of a sample.

Response

(*i*)   A date must always be a whole number and therefore either an integer or a real variable would be suitable.

(*ii*) A name must be held by a string variable.

(*iii*) A weight is unlikely to be a whole number and should therefore be represented by a real variable.

****

Appendix SAQ 2.1c

Some of the following statements are unacceptable and some will lead to errors. Comment on
each statement.

(*i*)   LET X% = 12

(*ii*)   LET X + Y% = 12

(*iii*)   LET residue = 0.456

(*iv*)   LET PPT = 1.347g

(*v*)   LET B% = 3.456

(*vi*)   LET 3rd% = 4.67

(*vii*)   LET sample% = 3.92

(*viii*) LET Name$ = "Tom Brown"

(*ix*)   LET PRINTER$ = "Caxton"

(*x*)   LET FIVE = 6

Response

(*i*)   LET X% = 12 is all right since 'X%' is an integer variable.

(*ii*)   LET X + Y% = 12 is not correct because it includes the plus
sign, 'XY%' or 'X_Y%' would probably be acceptable as
integer variables.

(*iii*)   LET residue = 0.456 is all right, 'residue' being the name of a real variable.

(*iv*)   LET PPT = 1.347g will not do. 'PPT' is all right as the name of a real variable but the 'g' after the number will cause an error.

(*v*)   LET B% = 3.456 would be accepted by the computer but the value held by integer variable 'B%' would be 3, the number being truncated to the lower integer.

(*vi*)   LET 3rd% = 4.67 has two errors. The computer would not accept this assignment because a variable name must not start with a number. Also, the % sign indicates an integer variable which is unsuitable for a value which includes a fractional part.

(*vii*)   LET sample% = 3.92 would be accepted but the integer variable 'sample%' would have the value 3.

(*viii*)   LET Name$ = "Tom Brown" is all right. Note that this string includes 3 spaces.

(*ix*)   LET PRINTER$ = "Caxton" would not be accepted because the name starts with the reserved word, PRINT. The corresponding name in lower case, *viz* "printer$", would probably be acceptable.

(*x*)   LET FIVE = 6 is all right as far as the computer is concerned since FIVE can be the name of a real variable. In the interests of clarity and sanity, however, an assignment like this is inadvisable.

****

**Appendix
SAQ 2.2a**

Problems (i) to (iii) below should be done by hand (or mentally) and the results checked by computer.

(i)

$$Z1 = 100/(X + Y) \qquad Z2 = 100/X + Y$$

If X = 5 and Y = 15, what are the values of Z1 and Z2?

(ii)

$$Z3 = 100/X/Y \qquad Z4 = 100/X + Y$$
$$Z5 = 100/(X + Y)$$

If X = 5 and Y = 2, what are the values of Z3, Z4 and Z5?

(iii)

$$Z6 = 100/X*Y \qquad Z7 = 100/(X*Y)$$

If X = 2 and Y = 10, what are the values of Z6 and Z7?

(iv)

Write computer-type expressions corresponding to:

$$Z8 = Ax^2 + Bx + C$$

$$Z9 = Ax^{(a + b)} + Bx^2 + 1/C$$

Response

Each expression is evaluated in parts. Pay particular attention to the order in which each step is done.

(*i*) $Z1 = 100/(5 + 15) = 100/20 = 5$

          $Z2 = 100/5 + 15 = 20 + 15 = 35$

(*ii*) $Z3 = 100/5/2 = 20/2 = 10$

          $Z4 = 100/5 + 2 = 20 + 2 = 22$

          $Z5 = 100/(5 + 2) = 100/7 = 14.3$

(*iii*) $Z6 = 100/2*10 = 50*10 = 500$

          $Z7 = 100/(2*10) = 100/20 = 5$

(*iv*) $Z8 = A*x^2 + B*x + C$

          $Z9 = A*x^{(a+b)} + B*x^2 + 1/C$

Examine Z6 and Z7 closely to see the effect of divide and multiply without brackets.

The lower-case X has been used in (*iv*). Some computers require all variable names to be in upper-case. You will soon discover what your machine allows.

When expressions are evaluated the order of priority is usually a matter of mathematical common sense, *viz*:

Expressions within parenthesis (brackets).

Exponentiation, ie Raise to a power ($^\wedge$).

Multiplication and Division ($*$ and $/$).

Addition and Subtraction ($+$ and $-$).

Watch the division sign (/). Some computers have the sign (\) called the *backslash*. It does *not* mean divide.

****

Appendix SAQ 2.2b

In the standardisation of acid solutions a known weight of anhydrous sodium carbonate is titrated with an acid. One mole of monobasic acid neutralises 53.00 g of Na_2CO_3.

Write and run a program to calculate the molarity of a hydrochloric acid solution from the weight of sodium carbonate (W g) and the volume of hydrochloric acid (V cm^3) required for neutralisation.

Reminder:

molarity = number of moles per litre of solution.

Response

The essential theory is that the number of moles acid is W/53.00, the volume of acid is V/1000 dm^3 and the molarity is the number of moles per dm^3. The molarity of the acid is therefore 1000W/(53.00V) or, using computer notation, the molarity is represented by variable C in:

$$C = 1000*W/53.00/V$$

Since 1000/53.00 always has the same value we may define a constant K by

$$K = 1000/53.00$$

We have now done the chemistry and can concentrate on the computation. It is essential that problems be tackled in this order: a computer does not know any chemistry. The program to be written must:

> accept values of weight and volume;
>
> calculate concentration;
>
> print out concentration.

We can see clearly how to do each of these and so we are in a position to write a program.

There are several possible ways of handling the data. The following is a very short program without prompts:

```
 99 REM ** SHORT MOLARITY
100 LET K = 1000/53.00
110 INPUT W
120 INPUT V
130 LET C = K*W/V
140 PRINT C
150 END
```

The inclusion of some prompts improves the program:

```
 99 REM ** MOLARITY
100 LET K = 1000/53.00
110 PRINT "WEIGHT (GRAMMES)?";
120 INPUT W
130 PRINT "VOLUME (CUBIC CENTIMETRES.)?";
140 INPUT V
150 LET C = K*W/V
160 PRINT "CONCENTRATION = ";C
170 END
```

You should check the program with known data. A weight of 0.1325 g and a titration of 25.00 cm^{-3} will give a 0.1 mol dm^{-3} solution. Try several weights and titrations but do not worry if the output gives a large number of figures. We shall tidy this up later.

In these two programs a variable K was defined to combine values that are constant. The programs could be made more general by allowing K to be defined from the keyboard. Why not work out what value K must have if the standard substance is borax or if the titrant is sulphuric acid?

If you have applications requiring it, an obvious extension is to more complex calculations involving oxidising and reducing agents. You could ask for the molar mass and the number of electrons involved in the reaction before working out the value of K.

****

Appendix SAQ 2.2c	Write a program which accepts numbers from the keyboard and prints out the reciprocal of the number correct to 3 decimal places.

Response

We use the INT function with an expression as argument. For 3 decimal places we multiply and divide by 1000. In general, for n decimal places we multiply and divide by 10 to the power n.

```
 99 REM ** 1/X TO 3 PLACES
100 INPUT X
110 Y = 1/X
119 REM ** NOW ADJUST
120 Z = INT (1000*Y+0.5)/1000
130 PRINT Z
140 END
```

Even with formatting like this some computers may not make it obvious that there has been any control. Thus, if an answer is 0.6 exactly this may be printed 0.6 instead of 0.600 as we might expect and desire. The computer manual must be consulted to sort out such problems.

In the above program three lines, 110 to 130, are used where one would do. We could have used a single statement:

PRINT INT(1000/X + 0.5)/1000.

****

Appendix SAQ 2.2d

Write programs to calculate:

(i)   absorbance from percent transmittance;

(ii) percent transmittance from absorbance;

(iii) pH from hydrogen ion concentration;

(iv) hydrogen ion concentration from pH.

Reminder:

absorbance $= \log_{10}(100/\%\text{transmittance})$

$pH = -\log_{10}[H^+]$

Response

(i)   In this program the value of absorbance first found is adjusted by means of the INT function to give 2 figures after the decimal point.

```
 99 REM ** PERCENT T to A
100 PRINT "ENTER % TRANSMITTANCE"
110 INPUT T
119 REM ** ASSUME LOG( ) IS TO BASE 10
120 A = LOG(100/T)
130 A = INT(100*A+0.5)/100
140 PRINT "ABSORBANCE IS ";A
150 END
```

Notice how the same variable name, *viz* A, can be re-used after adjusting the number of decimal places: the INT function obtains a value which is 'placed in the box labelled A'.

If several conversions are to be made it would be useful to change line 150 to:

150 GOTO 100

To stop the program, however, you would have to use the ESCAPE or BREAK key. Better methods of controlling a program are discussed in the next Section.

(*ii*) In this program a prompt-string is used at the INPUT line and the transmittance is given to one decimal place.

```
 99 REM ** A to PERCENT T
100 INPUT "ENTER ABSORBANCE", A
110 T = 100*10^(-A)
120 T = INT(10*T+0.5)/10
130 PRINT "TRANSMITTANCE IS ";T;"%"
140 END
```

At line 110 the base 10 is raised to the power $-A$ and the result multiplied by 100. Then INT is used at line 120.

An alternative procedure would be to multiply A by -2.303 and then take the exponential.

110 T = 100*EXP(-2.303*A)

The factor 2.303 converts the common log to the natural log.

(*iii*) We revert to 2 decimal places though whether or not this is justified will depend on the precision with which the hydrogen ion concentration is known.

```
 99 REM ** pH CALCULATION
100 INPUT "ENTER HYDROGEN ION CONCN", H
110 P = -LOG(H)
120 P = INT(100*P+0.5)/100
130 PRINT "pH = "; P
140 END
```

(*iv*) This program does not attempt to fix the decimal point.

```
 99 REM ** pH to H ion
100 INPUT "ENTER pH", P
110 H = 10^(-P)
120 PRINT "H+ CONCENTRATION IS "; H
130 END
```

****

**Appendix
SAQ 2.2e**

We wish to write a program which asks a child his/her name and age and then prints the message

"(NAME) WILL BE (Y) NEXT YEAR"

with the correct name and number.

(*i*)   Write the program on the assumption that the child will always enter the age as a number.

(*ii*) Write the program to allow the age to be entered as a number followed by a word like 'years'. For this case assume that the VAL function can be used to return a number which appears before any non-numeric character.

Response

Some parts of the programs listed below may not be suitable for your machine because of the way it controls spacing and prompts for input. Thus, if your machine does not use the question mark as a prompt a '?' should be placed at the end of the string at line 100.

(*i*) Since the age is expected to be a whole number an integer variable, X%, is advisable to hold the number of years in the first program.

```
 99  REM ** NAME & AGE
100  PRINT "WHAT IS YOUR NAME ";
110  INPUT N$
120  PRINT
130  PRINT "HELLO ";N$;" HOW OLD ARE YOU ";
140  INPUT X%
150  PRINT
160  PRINT N$;" WILL BE ";X%+1;" NEXT YEAR"
170  END
```

The PRINT statements at lines 120 and 150 help the layout on the screen.

When running programs like this one of the chief problems is to ensure that input data are of the right type. The program above would probably fail through a child entering '5 years' or 'six and a half'. Provided no arithmetic had to be done a string variable, X$, could be used instead of X% at line 130 but then it would only be possible to repeat exactly what was entered:

```
140 PRINT N$;" IS ";X$
```

The use of the VAL function may get round this difficulty.

(*ii*) The following program will only be successful if the VAL function accepts the numerical part of a string like '12 years old' as a numeric variable.

```
 99  REM ** NAME & AGE
100  PRINT "WHAT IS YOUR NAME ";
110  INPUT N$
120  PRINT
130  PRINT "HELLO "; N$;" HOW OLD ARE YOU ";
140  INPUT X$
142  X% = VAL(X$)
150  PRINT
160  PRINT N$;" WILL BE "; X% + 1;" NEXT YEAR"
170  END
```

****

Appendix
SAQ 2.3a

Write programs to display on the screen:

(i) the squares of the numbers 1 to 10 one per line;

(ii) the squares of the numbers 1 to 10 in a row with a space between each number shown;

(iii) the *even* numbers from 0 to 10 and their squares in two columns (take 0 as an even number.)

(iv) as (iii) but starting with 10 and finishing with 0.

Response

(i) Since the step size in the FOR-NEXT loop is +1 it may be omitted at line 10 below.

```
99 REM ** TABLE OF SQUARES
10 FOR I = 1 TO 10 [STEP 1]
20 LET X = I^2
30 PRINT X
40 NEXT I
50 END
```

It is usually possible to replace lines 20 and 30 by a single line:

30 PRINT I^2.

(*ii*) If your computer uses the semicolon to make next printing follow on directly from the previous then line 30 should end with ;" "; (that is, semicolon, space, semicolon).

(*iii*) To select even numbers only line 10 becomes:

10 FOR I = 0 TO 10 STEP 2

To get column or zone format replace line 30 by:

30 PRINT I , X

(*iv*) To go down in steps of two from 10 replace line 10 by:

10 FOR I = 10 TO 0 STEP -2

Note that the only step that may be omitted is STEP 1. Any other step must be stated.

****

Appendix
SAQ 2.3b

(*i*)   Alter the adding program of Section 2.3.3 so that the program stops and prints the sum when a negative number is entered. The negative number should not be included in the summation.

(*ii*) Alter the program so that the number of items to be added is entered at the keyboard before the loop commences. The program must add this number of items and finish by printing not only the value of the sum, S, but also the value of the control variable, I.

Is the value of I the same as the number of items added?

If I is different from the number of items added why is this?

Response

(*i*)   The following is one possibility:

```
 99 REM ** ADDING PROGRAM
100 LET S = 0
110 FOR I = 1 TO 100
120     INPUT X
130     IF X<0 GOTO 160
140     LET S = S + X
150     NEXT I
160 PRINT S
170 END
```

At line 110 "STEP 1" is understood.

This program limits the number of items to 100 but it could easily be modified to make it more general.

The program is made to 'jump out' of the loop by entering a negative value for X. The negative value is not included in the sum.

In general 'jumping out' of a loop is not recommended. It is acceptable with a short program in which we exit the loop simply to stop the program but with longer programs the language may object if several loops are not completed. This is discussed more fully in 2.3.4.

(*ii*) The necessary changes are:

```
105  INPUT N
110  FOR I = 1 TO N
130  (OMIT)
160  PRINT S, I
```

To cancel line 130 you simply type 130 and press RETURN.

The final value of I will indicate how the loop operates.

After the first pass the value of I is increased by the step size (1 in this case); then I is tested to see if its value exceeds the end value which is N in the program above. If the program continues until N items are entered the final value of I will be one more than N. When I becomes N + 1 the test at line 110 shows that I is greater than the end value and so execution of the loop terminates and the program continues at line 160.

****

Appendix SAQ 2.3c

Write a program which:

(*i*)   accepts a maximum of 20 numbers;

(*ii*) computes the sum of the numbers;

(*iii*) computes the sum of the squares of the numbers;

(*iv*) stops when a negative number is entered;

(*v*)   calculates the total number of items;

(*vi*) prints out the sum, the sum of squares, the total number of items and the mean of the numbers entered.

Run the program using the following numbers:

89 82 67 72 86 80 86 91 80 80

Response

In the program below variable S represents the sum of the numbers, variable P the sum of the squares:

```
 99 REM ** MEAN PROGRAM
100 S = 0
110 P = 0
119 REM ** MAKE LOOP BIG ENOUGH
120 FOR I = 1 TO 20
130    PRINT "ENTER NUMBER (-1 TO STOP)";
140    INPUT X
150    IF X<0 GOTO 190
160    S = S+X
170    P = P+X^2
180    NEXT I
189 REM ** I IS 1 TOO MANY
190 N = I-1
199 REM ** DIVIDE TO GET MEAN M
```

```
200  M = S/N
209  REM ** ROUND OFF M
210  M = INT(10*M + 0.5)/10
220  PRINT "SUM OF NUMBERS = ";S
230  PRINT "SUM OF SQUARES = ";P
240  PRINT "TOTAL ENTRIES = ";N
250  PRINT "MEAN VALUE = ";M
260  END
```

Note that the number of items is one less than the final value of I. The final number input was the negative number to stop the program and we do not count this.

Using the numbers given our answers were:

$$S = 813 \quad P = 66591 \quad N = 10 \quad MEAN = 81.3$$

The mean is the only number which has been rounded though it was not really necessary in this particular calculation because the sum was divided by 10; had there been 11 samples, however, rounding would have been very desirable.

****

Appendix SAQ 2.4a

Write a program which includes DATA statements and which:

(*i*)   reads the number of items from the data list;

(*ii*) calculates and prints out the mean of several numbers;

(*iii*) restores the data pointer;

(*iv*) calculates and prints out the standard deviation of the numbers. $\longrightarrow$

Appendix SAQ 2.4a (cont.)	Test the program using the following numbers which were obtained by 11 analysts for the concentration (g dm^{-3}) of copper in a solution:

49.89 49.82 49.67 49.72 49.86 49.80
49.86 49.96 49.80 49.80 49.83

Response

In the program below the variable S is used first to sum the data and then to sum the squares of the deviations from the mean. Re-use of the same variable is all right and indeed is advisable if a microcomputer has a low memory but you must be sure that the variable is set to the proper value when it is used for the second time.

```
 98  REM ** STANDARD DEVIATION
 99  REM ** SUM IS S
100  S = 0
110  READ N
120  FOR I = 1 TO N
130      READ X
140      S = S + X
150      NEXT I
170  M = S/N
180  PRINT "NO. OF ITEMS = ";N
190  PRINT "MEAN VALUE = ";M
198  REM ** POINTER BACK
200  RESTORE 410
209  REM ** S NOW SUM OF SQUARED DEVS.
210  S = 0
220  FOR I = 1 TO N
230      READ X
240      S = S + (X-M)^2
250      NEXT I
260  V = S/(N-1)
270  PRINT "VARIANCE = ";V
```

```
280  PRINT "STANDARD DEVIATION = ";SQR(V)
300  END
399  REM ** DATA
400  DATA 11
410  DATA 49.89,49.82,49.67,49.72
420  DATA 49.86,49.80,49.86,49.96
430  DATA 49.80,49.80,49.83
```

Results with this program are:

MEAN = 49.82 VARIANCE = 0.00615

STANDARD DEVIATION = 0.0784

(These results have been tidied up to get rid of extra figures. Probably the program should include some rounding but this has been omitted to concentrate on other points).

You will notice that N is not read after the RESTORE statement because the pointer is restored only to line 410.

This program might be worth developing. You could add a section of program to restore the data pointer again and then count the number of items which are more than 2 and 3 standard deviations from the mean. Such calculations can be very useful to check for systematic errors. Thus, following a large number of measurements of the same property it is expected that about 95% of the results will be within two standard deviations of the mean result and that about 68% will be within one standard deviation. If a series of results departs markedly from these percentages some kind of strange error or blunder should be suspected, eg there was an unusual variation of the base-line in a spectroscopic measurement or the wrong pipette was used in a volumetric analysis. Checks of this nature are particularly important in the processing of data obtained from on-line instruments.

An alternative, and possibly better, method of programming the same kind of calculation uses an array. This is discussed in Section 2.4.3.

****

Appendix **SAQ 2.4b**	In a book on quantitative analysis there are tables of data on common acids. Typical entries are:			

	% w/w	kg/litre	mol/litre
Hydrochloric acid	35	1.18	11.3
Sulphuric acid	96	1.84	8.0

Write a program which employs TAB statements to produce a table like this from data in DATA statements.

Response

The program below allows for expansion by placing the names of the acids in DATA statements and reading the names into a string variable. The first data item gives the number of entries in the table. This number would be updated every time data for another acid are added.

```
 98 REM ** ACID DATA
 99 REM ** START WITH HEADINGS
100 PRINT TAB(18);"% W/W";
110 PRINT TAB(25);"KG/LITRE";
120 PRINT TAB(35);"MOL/LITRE"
130 PRINT
139 REM ** HOW MANY?
140 READ N
150 FOR I = 1 TO N
160    READ N$,A$,B$,C$
170    PRINT N$;
180    PRINT TAB(20);A$;
190    PRINT TAB(27);B$;TAB(36);C$
200    NEXT I
210 END
299 REM ** DATA
```

```
300  DATA 2
310  DATA "HYDROCHLORIC ACID"
320  DATA 35,1.18,11.3
330  DATA "SULPHURIC ACID"
340  DATA 96,1.84,18.0
```

Note that the print lists at lines 100, 110, 170 and 180 finish with the punctuation mark which suppresses an 'end of line' message but those at lines 120 and 190 do not have this mark.

****

Appendix SAQ 2.4c

Write a program to accept up to 10 real numbers into an array. After testing the program with simple numbers (eg 2,4,6,...) alter the latter part of the program to make it print out:

(i) the numbers and their cubes in column format;

(ii) the numbers, and their reciprocals correct to 3 decimal places

Response

(i) To print out numbers and cubes in column format.

```
 99  REM ** TABLE OF CUBES
100  DIM X(10)
110  PRINT "HOW MANY VALUES? (UP TO 10) ",N
120  INPUT N
130  FOR I = 1 TO N
140      PRINT "ENTER ITEM ";I
150      INPUT X(I)
160      NEXT I
```

```
169  REM ** ALL IN. CALCULATE & PRINT
170  FOR I = 1 TO N
180       PRINT X(I), X(I)^3
190       NEXT I
200  END
```

(*ii*) To print numbers and reciprocals correct to 3 decimal places
the only change really necessary is at line 180. This is changed
to:

$$180 \text{ PRINT } X(I), \text{INT}(1000/X(I) + 0.5)/1000$$

If the second item in the print list look rather complex you can
work it out in stages. Starting with the reciprocal, $1/X(I)$, multiply
by 1000, add 0.5, truncate to the integer and finally divide by 1000.

```
*************************************
```

**Appendix
SAQ 2.4d**

In a gas chromatography experiment the solvent
peak was recorded at a retention time of 0.85
minutes and successive peaks were recorded at
retention times of 3.96, 4.33, 5.20, 6.72, 7.58, 9.28
and 10.75 minutes. Write a program which will:

(*i*)   read the experimental data from DATA
statements, placing the peak times in an array;

(*ii*) compute retention times relative to peak
number 4 (6.72 minutes); that is, for each
peak compute the ratio

$$\frac{\text{retention time} - 0.85}{\text{peak 4 time} - 0.85}$$

(*iii*) display the retention times and the relative
retention times in a table.

Response

The DATA statements are kept well away from the main program. As these statements will be altered every time the program runs the line numbers should be readily remembered.

```
 99  REM ** RETENTION TIMES
100  DIM T(12)
109  REM ** NO. OF PEAKS, WHICH REF.
110  READ N,P
119  REM ** READ SOLVENT TIME
120  READ S
129  REM ** TIMES INTO ARRAY
130  FOR I = 1 TO N
140      READ T(I)
150      NEXT I
159  REM ** CALCULATE & PRINT
160  FOR I = 1 TO N
170      R = (T(I)-S)/(T(P)-S)
180      R = INT(100*R + 0.5)/100
190      PRINT TAB(4);I;TAB(6);T(I);TAB(12);R
200      NEXT I
210  END
499  REM ** DATA LIST
500  DATA 7,4,0.85
510  DATA 3.96,4.33,5.20,6.72,7.58,9.28,10.75
```

The first READ statement (line 110) takes in the number of peaks (excluding the solvent) and the number of the peak which is to be used as reference. The second READ takes in the solvent time.

The peak retention times are read into an array in a loop from lines 130 to 150. The calculations and printing are done in the loop from line 160 to line 200.

The following output was obtained on running this program:

1	3.96	0.53
2	4.33	0.59
3	5.2	0.74
4	6.72	1
5	7.58	1.15
6	9.28	1.44
7	10.75	1.69

Some additional formatting of output is required to line up the decimal points in the second column. Your computer probably has some means of achieving this but as the method is likely to be specific to the particular machine you will have to consult your manual for details.

It might be a good idea to look again at SAQ 2.4a (on calculating a standard deviation) and perhaps re-write the program using an array. Restoring the data pointer will not then be necessary. With this kind of calculation the use of an array facilitates the examination of individual data items. For example, one might want to count those items which are more than two standard deviations from the mean. Such calculations are easily programmed using arrays.

****

Appendix SAQ 2.5a

Write a program which prints out the reciprocal and the common logarithm of any number entered at the keyboard and which warns without crashing when the task set is impossible. All output should be correct to two places of decimals.

Response

The program below assumes that LOG( ) gives or returns the common logarithm. We guard against input of zero for the reciprocal and against zero and negative values for the log. There is, however, no check for a number which is too large or too small for the computer.

```
 99 REM ** 1/X & LOG(X)
100 INPUT "ENTER NUMBER X ";X
110 PRINT
120 IF X>0 THEN PRINT "LOG(X)=";INT(1000*
    LOG(X)+0.5)/1000
130 IF X< = 0 THEN PRINT "LOG IMPOSSIBLE"
140 PRINT
150 IF X<>0 THEN PRINT "1/X = ";INT(1000/X +
    0.5)/1000
160 IF X = 0 THEN PRINT "RECIPROCAL INFINITE"
170 PRINT
180 GOTO 100
```

This program makes no pretence at elegance. Apart from anything else it can only be stopped by pressing the ESCAPE or BREAK key or by making some input error which the language cannot tolerate.

In the answer to question 2.5c a somewhat better program is suggested.

****

**Appendix
SAQ 2.5b**

Write a program which accepts words or numbers greater than -1000 from the keyboard and which:

(*i*)   prints the message 'POSITIVE' if the first character of the input string is one of the 'numeric' characters 0 to 9; $\longrightarrow$

Appendix
SAQ 2.5b
(cont.)

> (*ii*) prints 'NEGATIVE' if the first character is
>   the minus sign (−);
>
> (*iii*) prints 'FRACTION' and the character of
>   ASCII code 7 if the first character is the
>   period (.);
>
> (*iv*) stops accepting input when the character
>   'Q' is entered;
>
> (*v*)   reports the total number of entries made,
>   the sum of all the numerical entries and
>   the highest negative numerical entry.

Response

It is necessary to note certain ASCII codes:

Character	Code
0	48
9	57
−	45
.	46
/	47

The program below is one possibility:

```
 98 REM ** REM NUMBER TESTING
 99 REM ** SET TOTAL & SUM TO ZERO
100 T = 0
110 S = 0
119 REM ** H = HIGHEST −VE AT START
120 H = −1000
129 REM ** N = CURRENT HIGHEST −VE
```

```
130 N = H
140 PRINT "ENTER A NUMBER >-1000 OR A WORD."
150 PRINT "TO STOP THE PROGRAM ENTER Q"
160 PRINT
170 INPUT "ENTER (Q TO STOP)"X$
180 IF X$ = "Q" THEN 280
190 C = ASC(X$)
200 IF C>47 AND C<58 THEN PRINT "POSITIVE"
210 IF C = 45 THEN PRINT "NEGATIVE"
220 IF C = 46 THEN PRINT "FRACTION";CHR$(7)
229 REM ** USE VAL ONLY IF NUMERIC
230 IF C>44 AND C<58 AND C<>47 THEN S = S +
    VAL(X$)
239 REM ** CHANGE H IF NECESSARY
240 IF C = 45 THEN N =VAL(X$)
250 IF N>H THEN H = N
260 T = T + 1
270 GOTO 160
280 PRINT
290 PRINT "TOTAL ENTRIES ";T
300 PRINT
310 PRINT "SUM OF NUMBERS ";S
320 PRINT
330 IF H<>-1000 THEN PRINT "HIGHEST
    NEGATIVE";H
340 IF H = -1000 THEN PRINT "NO NEGATIVES
    ENTERED"
350 END
```

This program could be made much tidier and more understandable by employing the IF ... THEN//ELSE structure discussed in 2.5.3.

****

**Appendix
SAQ 2.5c**

Provided your computer allows the structure answer SAQ 2.5a again but using IF ... THEN ... ELSE.

Response

It is highly unlikely that your program is exactly the same as the one below as there are many possible ways of using IF ... THEN ... ELSE. The construction is used in lines 160 and 190 to prevent errors. Also, in line 130 the input is tested to see if it is time to stop. The input is taken in as a sequence of characters and assigned to string variable, X$. If the string is not the letter Q it is assumed to be a number and is converted to a value of numeric variable X using VAL. This is one way of testing input for a 'stopper' or 'terminator'.

```
 98 REM ** BETTER 1/X & LOG(X)
 99 REM ** ASSIGN 2 STRINGS
100 A$ = "IMPOSSIBLE"
110 B$ = "INFINITY"
120 INPUT "ENTER NUMBER X "X$
130 IF X$ = "Q" THEN END ELSE X = VAL(X$)
140 PRINT
150 PRINT "LOG(";X;") = ";
160 IF X>0 THEN PRINT; INT(1000*LOG(X)+0.5)/1000
    ELSE PRINT A$
170 PRINT
180 PRINT "1/";X;"= ";
190 IF X<>0 THEN PRINT;INT(1000/X+0.5)/1000 ELSE
    PRINT B$
200 PRINT
210 INPUT "NEXT NUMBER OR Q TO STOP "X$
220 PRINT
230 GOTO 130
```

Note how strings A$ and B$ are given 'values' and then used as required. This technique is particularly valuable when the same string of characters is to be printed several times in a program and space is at a premium. This is not the case in the present program but the use of A$ and B$ keeps lines 160 and 190 reasonable short.

Strictly, one should test a string before using VAL to ensure that conversion to a numeric will be successful. Whether or not this kind of test is worth doing depends on the consequences of an impossible or incorrect conversion.

****

**Appendix
SAQ 2.5d**

The potential of an ion-selective electrode may depend on the activities of two ions:

$$E = E^o + 58*LOG(A1 + K*A2) \qquad \text{(mV at 292K)}$$

A1 is the activity which the electrode is designed to measure, A2 the activity of an interfering ion. K is the selectivity ratio or selectivity constant.

Write a program to create a table showing, for any value of K, how the second term on the right depends on A1 and A2. Suitable ranges might be:

A1: 0.001 to 0.005 mol dm^{-3} in 5 steps
A2: 0 to 0.01 mol dm^{-3} in 5 steps

Output should be given to the nearest millivolt.

············

An example of output for K = 10 and A2 = 0.002 mol dm^{-3} is given below:

INTERFERING ION AT 2 MMOL

ION 1 AT 1 MMOL   TERM = −97
ION 1 AT 2 MMOL   TERM = −96
ION 1 AT 3 MMOL   TERM = −95
ION 1 AT 4 MMOL   TERM = −94
ION 1 AT 5 MMOL   TERM = −93

Response

While the program below answers the question there are many other methods of formatting the output.

```
 99 REM ** ION-SELECTIVE
100 INPUT "ENTER SELECTIVITY CONST. "K
110 FOR A2 = 0 TO 10 STEP 2
120     PRINT "INTERFERING ION AT "; A2;" MMOL"
130     PRINT
140     FOR A1 = 1 TO 5
150         E = 58*LOG(A1/1000 + K*A2/1000)
160         E = INT(E + 0.5)
170         PRINT "ION 1 AT "; A1;" MMOL","TERM
            = dq; E
180         NEXT A1
190     PRINT
200     A$ = GET$
210     NEXT A2
220 END
```

The GET$ at line 200 is a convenient method of controlling output on the VDU screen. A key must be pressed to allow the next A2 loop to be completed and displayed.

****

**Appendix
SAQ 2.5e**

To assess a new analytical method four samples of a product were taken and each sample was divided into two parts. Four different analysts determined the purity using the old method for one part and the new method for the other. The results were listed as they were reported by the analysts, always in the order:

method,   sample,   % purity.

The results so listed were:

2 2 98.8		1 4 98.4	
1 3 98.1		2 4 98.1	
1 2 98.1		2 3 98.9	
1 1 98.6		2 1 98.6	⟶

Appendix SAQ 2.5e (cont.)	Write a program to read these results into a two-dimensional array and present them in something like this table:

SAMPLE	1	2	3	4
METHOD 1	98.6	98.1	98.1	98.4
METHOD 2	98.6	98.8	98.9	98.1

Response

We know that there are eight sets of data to be read. The first two items of each set give the method and the sample (I and J). The third item is the result. When reading the data we must ensure that I and J are known *before* the result is assigned to an array element.

In the following program it has been assumed that TAB can take an argument which includes variables. That is, the tabulation can depend on I as in line 160.

```
 99 REM ** 2-ARRAY
100 DIM R(2,4)
109 REM ** READ IN 8 SETS OF DATA
110 FOR N = 1 TO 8
120     READ I, J, R(I, J)
130     NEXT N
139 REM ** ALL READ IN
140 PRINT "SAMPLE";
150 FOR I = 1 TO 4
160     PRINT TAB(6 + I*6); I;
170     NEXT I
180 PRINT
190 PRINT
200 FOR I = 1 TO 2
210     PRINT "METHOD "; I;
220     FOR J = 1 TO 4
230         PRINT TAB(6 + J*6); R(I, J);
```

```
240        NEXT J
250     PRINT
260     NEXT I
270 END
399 REM ** DATA LIST
400 DATA 1,2,98.1,2,3,98.9,1,1,98.6
410 DATA 2,1,98.6,2,2,98.8,1,4,98.4
420 DATA 1,3,98.1,2,4,98.1
```

On running this program the table reproduced in the question was obtained.

****